Artificial Intelligence for 6G

Haesik Kim

Artificial Intelligence for 6G

 Springer

Haesik Kim
VTT Technical Research Centre of Finland
Oulu, Finland

ISBN 978-3-030-95043-9   ISBN 978-3-030-95041-5   (eBook)
https://doi.org/10.1007/978-3-030-95041-5

This Springer imprint is published by the registered company Springer Nature Switzerland AG
The registered company address is: Gewerbestrasse 11, 6330 Cham, Switzerland

To my wife Hyeeun,
daughter Naul,
son Hanul
and
mother Hyungsuk.

Preface

The cellular systems have been incrementally evolved, and the old and new network equipment co-exists for a certain period. Likewise, the 4G equipment will continuously roll out, adopt a new feature, and evolve to 5G systems. 5G systems are deployed progressively. The transition to 5G may take longer time than 4G because many different features should be included. While 5G systems are now deployed, research groups of cellular communications and networks started investigating beyond 5G systems and conceptualizing 6G systems. 6G will revolutionize the wireless communications and networks more intelligently with higher requirements than 5G systems. In the era of 6G, we need a game-changing approach. 6G systems will redefine the communications and networks depending on the required services. 6G business will not play a game but change a game. In order to support new requirements and services, a new blood technology is required. Artificial intelligence (AI) and machine learning (ML) will be one of key technologies for 6G. They are now matured technologies and improve many other research fields significantly. AI and ML make our day-to-day life easier. They pervade every aspect of our life. For example, we use mobile phone apps such as maps and navigation, facial recognition, autocorrect text, search recommendation, and so on. In addition, Chabot, social media, social media, and Internet banking are widely used. They all are based on AI and ML technologies. Thus, many experts and businessmen expect that they can dramatically improve the efficiencies of our workplaces as well as create new applications and services. AI will play a critical role in wireless communications and networks and change how we design and manage 6G communications and networks. We expect that AI makes communications and networks design and management smarter and safer. Key question is not about whether but when and how to implement AI in 6G communication systems.

This book introduces AI techniques for wireless communications and networks and helps audiences find an optimal, sub-optimal, or trade-off solution for each communications and networks problem using AI techniques. The target audiences are senior undergraduate students, graduate students, and young researchers who have a background about fundamentals of wireless communications and networks and start studying AI and ML techniques. From this book, audiences understand how to obtain a solution under specific conditions and realize the limit of the solution. This book introduces, in a step-by-step manner, AI techniques such as unsupervised learning, supervised learning, reinforcement learning, and deep learning and explains how they are used for wireless communications and networks system. The organization of the book is as follows: In Part I, AI techniques are introduced. It will provide audiences with a mathematical background about AI algorithms. Unsupervised learning includes hierarchical clustering, partitional clustering, association rule mining, and dimensionality reduction. Supervised learning covers decision tree, K-nearest neighbouring, and support vector machine. Linear regression, gradient descent algorithms, and logistic regression are discussed. In reinforcement learning, both model-based approaches and model-free approaches are investigated. Deep learning is discussed from a perceptron to neural networks, convolutional neural networks, and recurrent neural networks. In Part II, 6G communication and network systems are designed and optimized using both wireless communications and networks techniques and AI techniques. In physical layer, data link layer, and network layers, key algorithms are selected and introduced. AI techniques are adopted in wireless communications and networks systems. We look into how AI techniques help them to improve the performance. 6G systems are now under discussion in academy, standardization body, and industry. 6G use cases, requirements, and key enabling techniques are discussed as a preliminary. In physical layer, channel model, signal detection, channel estimation, error control coding and modulation, and MIMO are explained. In data link layer, we focus on resource allocation techniques as one selected research topic. In network layer, cellular system is introduced. We focus on network traffic prediction techniques as one of key AI-enabled network layer techniques.

I am pleased to acknowledge the support of VTT Technical Research Centre of Finland and Springer and also the valuable discussion of my colleagues and experts in EU project 5G-HEART. I am grateful for the support of my family and friends.

Oulu, Finland   Haesik Kim

Contents

Abbreviations

3GPP	The 3rd Generation Partnership Project
5G	The fifth generation
5GC	5G core
6G	The sixth generation
ABC	Artificial bee colony
ADC	Analog-to-digital converter
AGI	Artificial general intelligence
AI	Artificial intelligence
AID	Automatic interaction detection
AMF	Access and mobility management function
ANI	Artificial narrow intelligence
ANN	Artificial neural network
API	Application programming interface
APSK	Amplitude phase shift keying
AQM	Active queue management
ARIMA	Autoregressive integrated moving average
AS	Access stratum
ASI	Artificial super intelligence
ASIC	Application-specific integrated circuit
AuC	Authentication centre
AWGN	Additive white Gaussian noise
BackCom	Backscatter communications
BCC	Blocked calls cleared
BCD	Blocked calls delayed
BCH	Bose–Chaudhuri–Hocquenghem
BER	Bit error rate
BF	Bit flipping
BLER	Block error rate
BP	Belief propagation
BPTT	Backpropagation through time
BS	Base station

BSS	Base station subsystem
BTS	Base transceiver station
BW	Bandwidth
CAPEX	Capital expenditures
CART	Classification and regression tree
CBR	Constant bit rate
CDMA	Code-division multiple access
CEPT	European Conference of Postal and Telecommunications Administrations
CLS	Concept learning system
CN	Core network
CNN	Convolutional neural network
CoMP	Coordinated multipoint
COST	Commercial off-the-shelf
CPCC	Cophenetic correlation coefficient
CPU	Central processing unit
CQP	Convex quadratic programming
C-RAN	Cloud or centralized radio access network
CRC	Cyclic redundancy check
CSI	Channel state information
CSMA/CA	Carrier sense multiple access/collision avoidance
CU	Centralized unit
CUPS	Control plane and user plane separation
CWT	Continuous wavelet transform
DAS	Distributed antenna systems
DCI	Downlink control information
DM-RS	Demodulation reference signal
DNN	Deep neural network
DoS	Denial-of-service
DP	Dynamic programming
DPSK	Differential phase shift keying
DU	Distributed unit
DWT	Discrete wavelet transform
E2E	End-to-end
EC	European commission
EDA	Electronic design automation
eMBB	Enhanced mobile broadband
eNB	Evolved node B
EPC	Evolved packet core
E-UTRAN	Evolved universal terrestrial RAN
eV2X	Enhanced vehicle-to-everything
EXIT	Extrinsic information transfer
FBMC	Filter bank multi-carrier
FCC	Federal communications commission
FCFS	First-come-first-serve

FCM	Fuzzy c-mean
FDD	Frequency division duplex
FDMA	Frequency division multiple access
FEC	Forward error correction
FIFO	First-in-first-out
FN	False negatives
FP	False positives
FPGA	Field-programmable gate array
GDPR	General data protection regulation
GEO	Geostationary satellites
GFMC	Generalized frequency division multiplexing
GGSN	Gateway GPRS support node
GMSC	Gateway mobile switching centre
gNB	Next-generation node B
GOS	Grade of service
GPRS	General packet radio service
GPU	Graphics processing unit
GRAN	GSM radio access network
GRU	Gated recurrent unit
HAPS	High-altitude pseudosatellites
HARQ	Hybrid automatic repeat request
HLR	Home location register
HMM	Hidden Markov model
HSS	Home subscriber server
HW	Hardware
IAB	Integrated access and backhaul
ID3	Interactive dichotomizer 3
IEEE	Institute of electrical and electronics engineers
IMEI	International mobile equipment identity
IoT	Internet of Things
IP	Internet Protocol
IPI	Interpacket interval
IPM	Interior point method
IQ	In-phase and quadrature-phase
ISDN	Integrated services digital network
ITU	International telecommunication union
KNN	K-nearest neighbours
LAA	Licenced assisted access
LDPC	Low-density parity check
LEO	Low earth orbit satellites
LLR	Log likelihood ratio
LSTM	Long short-term memory
LTE	Long-term evolution
LWA	LTE-WLAN aggregation
MAC	Medium access control

MAP	Maximum a posteriori
MC	Monte Carlo
MDP	Markov decision process
MEC	Multi-access edge computing
MF	Matched filter
MIMO	Multiple-input and multiple-output
MIT	Mobility interruption time
ML	Machine learning or maximum likelihood
MLP	Multi-layer perceptron
MME	Mobility management entity
MMSE	Minimum mean square error
mMTC	Massive machine-type communications
MOS	Mean opinion score
MPSK	M-ary phase shift keying
MRP	Markov reward process
MS	Mobile station
MSC	Mobile switching centre
NAS	Non-access stratum
NFV	Network functions virtualization
NMI	Normalized mutual information
NN	Nearest neighbour
NNP	Neural network processors
NOMA	Non-orthogonal multiple access
NP	Non-deterministic polynomial
NR	New radio
OFDMA	Orthogonal frequency division multiple access
OMA	Orthogonal multiple access
OPEX	Operating expenses
O-RAN	Open radio access network
OS	Operating system
OSI	Open systems interconnection
OSTBC	Orthogonal space time block code
PAM	Partitioning around medoid
PCA	Principal components analysis
PCR	Principal component regression
PDCP	Packet data convergence protocol
PGW	Packet data network gateway
PID	Proportional–integral–derivative
PSO	Particle swarm optimization
PSS	Primary synchronization signal
PSTN	Public switched telephone network
QAM	Quadrature amplitude modulation
QoE	Quality of experience
QoS	Quality of service
QPSK	Quadrature phase shift keying

RAN	Radio access network
RB	Resource block
RED	Random early detection
Relu	Rectified linear units
RF	Radio frequency
RFID	Radio frequency identification
RGB	Red, green, and blue
RIC	RAN intelligent controller
RL	Reinforcement learning
RLC	Radio link control
RMS	Root mean square
RMSE	Root-mean-square error
RMSProp	Root-mean-square propagation
RNN	Recurrent neural network
RRC	Radio resource control
RRH	Remote radio head
RSSI	Received signal strength indicator
RTS	Request to send
RU	Radio unit
SARSA	State–action–reward–state–action
SBA	Service-based architecture
SDG	Sustainable Development Goal
SDP	Semi-definite programming
SDU	Service data unit
SEAL	Service enabler architecture layer
SGSN	Serving GPRS support node
SGW	Serving gateway
SIC	Successive interference cancellation
SINR	Signal-to-interference noise ratio
SIoT	Socialized Internet of Things
SISO	Single-input/single-output
SMF	Session management function
SNR	Signal-to-noise ratio
SNS	Smart networks and services
SOCP	Second-order cone programming
SSB	Sum of squares between
SSCP	Sums of squares and cross-products
SSE	Sum of squared errors
SSS	Secondary synchronization signal
STBC	Space time block code
STTC	Space time trellis code
SVM	Support vector machine
SVMs	Support vector machines
SW	Software
TCP	Transmission control protocol

TD	Temporal difference
TDD	Time division duplex
TDMA	Time division multiple access
TN	True negatives
TP	True positive
TPU	Tensor processing unit
TTT	Time-to-trigger
UAV	Unmanned aerial vehicle
UCI	Uplink control information
UDN	Ultra-dense networks
UDP	User datagram protocol
UE	User equipment
UFMC	Universal filtered multi-carrier
UMTS	Universal mobile telecommunications service
URLLC	Ultra-reliable low latency communications
UTRAN	Universal mobile telephone system RAN
V-BLAST	Vertical bell laboratories layered space time
VLR	Visitor location register
VLSI	Very large-scale integration
WCDMA	Wideband code division multiple access
WMMSE	Weighted minimum mean-squared error
XLA	Accelerated linear algebra
ZF	Zero-forcing

Part I
Artificial Intelligence Techniques

Chapter 1
Historical Sketch of Artificial Intelligence

Artificial intelligence (AI) is a popular theme in science fiction books and films. Sometimes, a film deals with singularity risk and AI becomes uncontrollable in a film. People concerns about AI growth. The famous physicist Stephen Hawking said "I fear that AI may replace humans altogether. If people design computer viruses, someone will design AI that improves and replicates itself. This will be a new form of life that outperforms humans." The Tesla's owner Elon Musk said "There is a lot of risk in concentration of power. So, if artificial general intelligence represents an extreme level of power, should that be controlled by a few people at Google with no oversight?" On the other hands, some researchers are negative to come singularity risk or uncontrollable AI soon. In the history of AI research and development, there were multiple AI winters which significantly reduce the interest and research funding of AI topics. Although a big progress of AI research has been made in recent decade and the gap between fact and fiction is getting narrow, they consider that general purpose AI with highly intelligence, conscience, and emotion is still many years away. However, both sides agree that AI is getting an important part of our day-to-day life and industries. For example, some mobile user searches an information using Apple AI Siri. IBM AI Watson helps a doctor to diagnose a variety of diseases. In this chapter, we will introduce to AI topics and discussions and trace back through successive stages of AI developments in technical aspect.

1.1 Introduction to Artificial Intelligence

We have an age-old question about intelligence: What is intelligence? There is no consensus on the definition of intelligence because it can be defined in many different ways. However, most of definitions includes some key words: understanding, self-consciousness, learning, reasoning, creativity, critical thinking, and problem-solving.

© The Author(s), under exclusive license to Springer Nature Switzerland AG 2022
H. Kim, *Artificial Intelligence for 6G*,
https://doi.org/10.1007/978-3-030-95041-5_1

In general, intelligence can be defined as the ability to perceive information and contain it as knowledge to adapt to an environment. More simply, it is the ability to understand, learn from experience, and apply knowledge to adapt an environment. In terms of science and engineering, keywords would be understanding, learning, and making decisions. Intelligence can be defined as understanding an environment and learning from an experience, and making a decision to achieve its goals. The term artificial intelligence was coined by John McCarthy in 1955. He organized the famous Dartmouth conference in 1956 [1]. In this conference, 11 scientists including John McCarthy, Marvin Minsky, and Claude Shannon established AI as an independent research filed. However, they didn't conclude specifically. The main topic was "Every aspect of learning or any other feature of intelligence can in principle be so precisely described that a machine can be made to stimulate it" [1]. The aspects of AI problems are identified as follows [1]: (1) Automatic computers, (2) How can a computer be programmed to use a language? (3) Neuron nets, (4) Theory of the size of a calculation, (5) Self-improvement, (6) Abstractions, (7) Randomness and creativity. These research challenges are still key topics in AI research. John McCarthy defined AI as the science and engineering of making intelligent machines. People attempted to make AI machines and reproduce intelligence using computers. When studying AI researches, we face other age-old questions: "How do humans think?", "How do machines think?", "What is artificial intelligence?", "What are ethical concerns involving AI?", "Can AI have morals?" and so on. These questions are highly related to AI researches and investigated by many different study fields including philosophy, mathematics, economics, linguistics, computer science, and others. Thus, AI is a multidisciplinary research topic with multiple approaches. There is no consensus on artificial intelligence definition as well. In [2], Stuart Russell and Peter Norvig tried to answer the questions. They explored 4 different approaches and defined 4 AI research categories: Think humanly, Think rationally, Act humanly, and Act rationally [2]. Think humanly covers questions: how human think? and what cognitive capabilities are required to produce intelligent performance? In this approach, it is more important to solve problems like a human than to solve problems accurately. Cognitive science and cognitive psychology take this approach and investigate the testable theories of the human mind and human–machine interaction. This topic will not be discussed in this book. Thinking rationally is about laws of thought. The approach covers question: How we should think logically? It deals with problem descriptions using a formal notation and computability of problems. Act humanly is about reproducing human behaviour. It is the Turing test approach. The Turing test is a method of inquiry in AI for determining whether or not a machine's intelligence is equivalent to a human's intelligence. In order to pass the test, a machine should have the following capabilities [2]: (1) natural language processing to communicate with a human, (2) knowledge representation to store information, (3) automated reasoning to answer questions using the information, (4) machine learning to adapt to new environments and explore problems, (5) computer vision to perceive objects, and (6) robotics to manipulate objects. In this book, machine learning will be one of the main topics. Act rationally is about rational agents or actors. In order to achieve the best goal, a rational agent acts under the assumption that its convictions are correct. In this approach, AI is regards as the construction of rational agents. Performing actions in to increase the value of the state of the agents. Agents are expected to operate autonomously, perceive their environment, persist over a prolonged time period, adapt to change, and create and pursue goals [2]. Rational thinking is basically a prerequisite of rational acting. In terms of communication and network system design, AI definition is in line with the AI definition in science and engineering fields. In general, AI is the ability of machines to perform human tasks and imitate human behaviour.

There is another category to research on AI to differentiate performance of AI machines: strong AI and weak AI. The terms were coined by American philosopher John Searle [3]. He said strong AI [3]: "The appropriately programmed computer with the right inputs and outputs would thereby have a mind in exactly the same sense human beings have minds." In this article [3], he sets out Chinese room argument and counters the assertion about strong AI. He asks us fundamental questions about what happens inside a computer? and where is the mind? The Chinese room argument can be summarized as follows: There is one person in a room with two slits, a paper and a book, who does not speak Chinese. Someone outside gives the person some Chinese characters through the first slit. The person inside the room follows the instructions in the book and translates Chinese characters to English characters on the paper. The person gives the resulting sheet to someone outside through the second slit. In terms of people outside world, there is a Chinese speaker inside the room. However, the person inside the room does not understand Chinese because the person just follows the instructions. This suggests that a computer executes a language translation function but there is no computer understanding Chinese language. However, this view of strong AI is criticized by other philosophers. This is a subject still under debate. This is an important argument philosophically. However, in terms of AI machine designers or developers, the argument about the real intelligence or the simulated intelligence is not important. They are more interested in whether or not a AI machine acting intelligently can be created. Generally, strong AI is able to think, act like a human, and learn from experiences. Strong AI machines have their own mind, make independent decisions, and act in different situations without a human intervention. There are no real-life examples about strong AI, but many science fiction books and movies represent strong AI. Weak AI is able to respond to specific situations but cannot think and act like a human for themselves. Weak AI machines are algorithms programmed by a human and simulate human behaviour under meaningful human control. Weak AI machines cannot perform without human intervention. In our real life, Apple Siri, Samsung Bixby, or Amazon Alexa are weak AI machines. They are software programs of voice-activated assistance and respond to user's request limitedly.

Similar to strong AI and weak AI category, there is another AI category: artificial narrow intelligence (ANI), artificial general intelligence (AGI), and artificial super intelligence (ASI). This category is about the level of AI's intelligence. ANI is referred to as weak AI or narrow AI. ANI does not mean narrow intelligence. They are not stupid, but it focuses on a single task of abilities such as image recognition and enables AI machines to outperform humans in some tasks. As of 2020, most of existing AIs can be categorized as ANI. It is the most common form of AI machines.

Machine learning and data mining are considered in ANI. AGI is able to think on the same level of a human mind. It is also known as strong AI, deep AI, or human level AI. There are some issues to debate the human level intelligence. In a way, AGI machines are indistinguishable from a human in a given situation. They have an ability not only to improve efficiency in a single task but also to apply their own knowledge to a wide range of problems. Developing AGI is based on a human brain model. Due to the lack of comprehensive knowledge about the human brain functionality, it might be very difficult to have AGI in near future. ASI implies machine intelligence and behaviour beyond human capabilities. In addition, its intelligence over time would be increased exponentially. Many science fictions and books deal with ASI, but it is controversial whether or not a human develops ASI eventually.

The categories about strong/weak AI and ANI/AGI/ASI are about how much they emulate human capability. Based on functionality, we can have another category of AI: Reactive machines, limited memory, theory of mind, and self-awareness. Reactive machines are the oldest forms and the most basic type of AI machines. They cannot form memories, use experiences to make the current decision, or have the ability to learn. Thus, they have very limited capability and emulate a human ability to respond to currently existing stimuli. Reactive machines act as it is programmed and couldn't grow their ability. IBM's Deep Blue that beats chess champion in 1997 is an example of reactive machines. It is programmed for here and now, and its ability is able to identify a chess board, understand the move of its pieces, and play chess matches. Limited memory machines can store data for a short period, learn from the stored data, make a decision, and update experimental knowledge. Most of AI machines nowadays store a large volume of data in their memory, update a reference model for solving future problems, and are continuously trained by big data. Automated vehicles are an example of limited memory machines. They are capable of storing recent data about distances and speeds of neighbouring cars, using both pre-programmed data and observational data, and driving in the best way. Theory of mind AI machines will be able to understand the needs, emotions, beliefs, and thoughts of other AI machines or humans and interact socially with other AI machines or humans. They are completely different level of AI machines, and many AI researchers and developers are currently engaged in innovations of AI machines. It requires to perceive emotional and behavioural changes of humans and respond a proper emotional and behavioural actions. Developing theory of mind AI machines requires innovation of many different research areas. We are not aware of mechanisms about human emotions and thought process. After understanding these mechanisms, AI machine designers could develop them by emulating human ability. In a science fiction film "Her", AI machine "Samantha" is an example of theory of mind AI machines. Samantha is AI machine personified through a voice. There is very little difference between a human and AI machine. Self-awareness AI machines are a hypothetical concept. It will be able to have their own consciousness, sentiment, and self-awareness. They are smarter than human. This is the ultimate goal of AI developments and beyond the singularity. Self-awareness AI machines may improve civilization significantly or cause catastrophe because AI machines may have their own ideas, judge humans, and spell the end for humanity. In a science fiction film

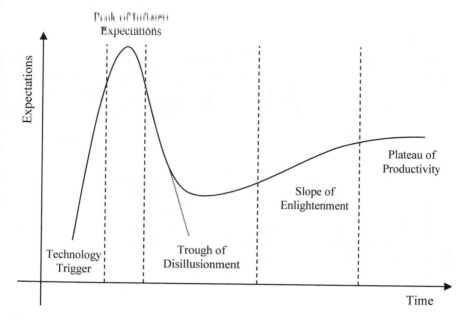

Fig. 1.1 Hype curve of emerging technologies

"Terminator", AI machine "Skynet" is an example of self-awareness AI machines. The film "Terminator" describes that Skynet has their own consciousness and decide to spell the end of the human race.

Gartner's hype curve [4] is well-known graphical presentation about how a technology will evolve over time. Figure 1.1 illustrates the hype curve of emerging technologies.

As we can observe Fig. 1.1, it describes a life cycle of technologies. The emerging technologies are evolving with 5 key stages: Technology trigger, peak of inflated expectations, trough of disillusionment, slope of enlightenment, and plateau of productivity. In the phase of technology trigger, a new technology is conceptualized and early proof-of-concept is complete. Researchers and developers pay attention on a new technology. In the phase of peak of inflated expectations, the technology is implemented and obtains a lot of publicity. In the phase of trough of disillusionment, researchers and developers find problems and disadvantages and they cause a big disappointment in the technology. Some technologies are disappeared or escape the attention due to significant problems. In the phase of slope of enlightenment, some problems are solved and the potential of the technology becomes more broadly understood. In the phase of plateau of productivity, the technology is more widely implemented. Mainstream industry adopts the technology, standardization is considered, and the technology grows continuously. It seems that AI technology developments follow hype curve, but it will be slightly different. In the history of AI, there are two periods of reduced expectations and developments. We call them AI winters. Figure 1.2 illustrates two AI winters in the history.

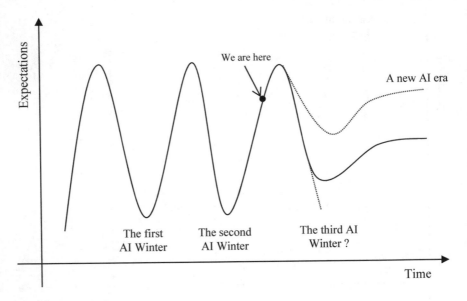

Fig. 1.2 Two AI winters

Two AI winters blocked progress of AI. The first AI winter came in 1973. The first AI boom appeared in 1950s and 1960s. Many researchers investigated and get a funding for AI developments. However, they failed to solve the problems such as limited computing power, exponential computing time, limited common sense knowledge, Moravec's paradox (Complex calculation is easy for AI, but simple task like a human face recognition is difficult for AI.) and others. The Lighthill report [5] in 1973 evaluated the AI research status for the British Science Research Council and made a conclusion that the visions and promises of AI researchers were exaggerated. Defence Advanced Research Projects Agency (DARPA) cuts its funding for AI fundamental research and invested more applied AI projects. It grew quickly and was disappointed deeply. The first AI winter lasted until 1980. The second AI winter came in 1987. The market for Lisp machines that is a specialized hardware for AI application collapsed in 1987. The performance of Lisp machines depends on data. In the 1980s, storages were very expensive. In addition, the complexity and control flow of the system are getting bigger. Expensive complex systems were threatened by affordable simple systems. Desktop computers became more powerful than the expensive Lisp machines even if they couldn't meet all AI requirements. In 1981, Japanese Ministry of International Trade and Industry launched the 5th generation computer project to perform conversation, translate language, interpret images, and achieve humanlike reasoning. However, it failed to deliver the goals. From two AI winters, we can obtain a realistic view of AI capability. Firstly, we now have enough computing power and storage that can handle a big data. In 4G and 5G communications and networks, a huge amount of data are produced and they are stored and transferred in the networks. It allows us to deal with intractable AI problems cost efficiently. In addition, a new AI technology such as deep learning leads to a revolution of AI systems. A human face recognition as one difficulty of the first AI winter is no longer difficult for AI if deep learning technique is used. Secondly, human–machine interaction are now acceptable. In the market, Apple Siri, Amazon Alex, and other are now widely used with limited interaction between human and machine. It is worth for further investments and developments. Thirdly, AI technology contributes to our day-to-day life and are expected to further contributions to 6G communications and networks. AI-empowered systems are now widely adopted such as automated vehicles, emailing filtering, factory automation, marketing, healthcare, food production, and so on. Lastly, do we face the third AI winter? In order to achieve strong AI, AGI, and ASI, there are many problems we should solve. No-free lunch theory in mathematical folklore tells us more lessons. There is no short cut, no general purpose solver, and no universal optimization strategy. This is a big barrier to huddle. Many people assumes that AI will look like T-800 of the film "Terminator" or R2D2 of the film "Starwars". At the moment, it is not possible to achieve them. The possibility of coming the third AI winter is quite high.

1.2 History of Artificial Intelligence

The history of AI may begin in Greek mythology. The concept of AI appeared in literature and actual devices with some degree with intelligence were demonstrated. We can call it AI pre-history. In Greek myths, Hephaestus is the god of the smiths. He manufactured a giant bronze automaton Talos incorporating the concept of AI. Talos became a guardian for the Crete Island. In the fourth century B.C, Aristotle invented syllogistic logic that is a mechanical thought for reasoning. He laid the foundations of modern science including AI research. He believes that the matter of all material substances is distinguished from form of them. This approach allows us to have a basic concept about computation and data abstraction in computer science. It becomes an intellectual heritage of AI. In the thirteenth century, Spanish theologian Ramon Llull invented a combining logical machines for discovering nonmathematical truths. Arab inventor Al-Jazari created the first programmable humanoid robot driven by water flow. In the sixteenth century, many clockmakers created mechanical animals and others. Rabbi Judah Loew ben Bezalel of Prague invented the Golem that is an anthropomorphic thing. In the seventeenth century, a French philosopher, mathematician, and scientist Rene Descartes thought that animals are complex physical machines without experience and cuckoo clocks are less complex machines. He believed that thoughts and minds are properties of immaterial souls and humans with immaterial souls only have subjective experience. A French mathematician Blaise Pascal invented the first mechanical calculator. An English philosopher Thomas Hobbes published "The Leviathan" and presented a mechanical and combinatorial theory of thinking. In the book, he said that reasoning is nothing but reckoning. A German mathematician Gottfried Leibniz improved Pascal's machine and made the Stepped Reckoner to do multiplication and division. In addition, he thought that reasoning could be reduced to mechanical calculation and tried to devise a universal calculus of formal reasoning. In the eighteenth century, plenty of mechanical systems are invented. Automation chess play also known as the Mechanical Turk was constructed in 1770 and astounded the world. However, it was revealed later as a fake machine containing a human chess master hidden inside to operate the machine. In the nineteenth century, Joseph Marie Jacquard invented the Jacquard loom in 1801. It was controlled by a number of punched cards. An English writer Mary Shelley published a science fiction book "Frankenstein or The Modern Prometheus" in 1818. In the book, a young scientist Victor Frankenstein created a hideous monster assembled from old body parts, chemicals, and electricity. English mathematicians Charles Babbage and Ada Lovelace invented a programmable mechanical calculating machines. An English mathematician George Boole published "The Laws of Thought" containing Boolean algebra in 1854. Boolean algebra laid the foundations of digital world. In the twentieth century, a Spanish engineer Leonardo Torres y Quevedo invented the first chess-playing machine El Ajedrecista in 1912 using electromagnets under the board. It was a true automation to play chess without human intervention and is regarded as the first computer game. British mathematicians Bertrand Russell and Alfred North Whitehead published "Principia Mathematica" in 1913. In this book, they presented a formal logic of mathematics. It covers set theory, cardinal numbers, ordinal numbers, and real numbers. A Czech writer Karel Capek introduced the word Robot in his play R.U.R (Rossum's Universal Robots) in 1921. In the play, artificial people called Robot are manufactured in a factory. In 1927, German science fiction film "Metropolis" is released. The film describes a futuristic urban dystopia and became the first robot film. In 1942, American writher Isaac Asimov published a short story "Runaround" and created three laws of robotics in the story [6]. The three rules describe the ways a robot must act [6]: (1) A robot may not injure a human being or, through inaction, allow a human being to come to harm, (2) a robot must obey the orders given it by human beings except where such orders would conflict with the First Law, and (3) a robot must protect its own existence as long as such protection does not conflict with the First or Second Law. In 1943, Warren McCulloch and Walter Pitts published a paper "A logical Calculus of the Ideas Immanent in Nervous Activity" in the Bulletin of Mathematical Biophysics. It described networks of idealized and simplified artificial neurons and inspired to neural networks. In 1950, Alan Turing published a paper "Computing Machinery and Intelligence" describing the Turing test. In the paper, he proposed to consider the question "Can machines think?" and replace the question by another, which is closely related to it and is expressed in relatively unambiguous word [7]. In order to determine if machines can think, he invented the Imitation Game involving three players: an AI respondent, a human respondent, and an interrogator. In the game, an interrogator stays in a room apart front an AI respondent and a human respondent and can communicate with them through written notes. The objective of the game is whether or not an interrogator can determine which of the others is the AI respondent and which is the human respondent by asking questions of them. The role of the AI respondent is to trick the interrogator to make a wrong decision and the role of the human respondent is to assist the interrogator to make a right decision. Figure 1.3 illustrates Turing test.

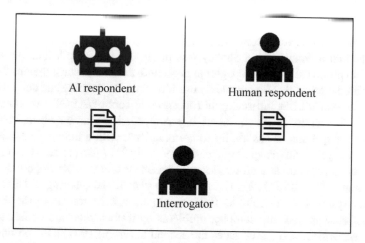

Fig. 1.3 Turing test

In 1951, Marvin Minsky designed Stochastic Neural Analog Reinforcement Calculator (SNARC) that is the first artificial neural network using vacuum tubes and synapse. From 1952 to 1972, AI celebrates golden age. The field of modern AI research was created. The study from various academic areas such as mathematics, psychology, neuroscience, economics, linguistics, and computer science began to create AI machines. In 1952, Arthur Samuel invented the checkers-playing program as the world's first self-learning program. It has enough skill to challenge a world champion. He coined the term machine learning as "The field of study that gives computers the ability to learn without being explicitly programmed". In 1955, the term artificial intelligence was coined. Dartmouth conference was organized in July and August 1956. Many people say that it is the official birthdate of the research field AI. In 1958, Johan McCarthy developed the Lisp language that is the most popular programming language for AI research. In 1961, the first industrial robot Unimate started working on a General Motors assembly line in New Jersey, USA. It was invented by George Devol using his patent [8]. The patent describes the machine as follows [8]: "The automatic operation of machinery, particularly to automatically operable materials handling apparatus, and to automatic control apparatus suitable for such machinery." He founded the first industrial robot company Unimation. The film 2001: Space odyssey featuring a sentient computer HAL 9000 is released in 1961. In 1964, Daniel Bobrow completed his Ph.D. dissertation "Natural language Input for a Computer Problem Solving System" and developed a natural language understanding computer program STUDENT. In his dissertation, he showed that a computer is able to understand natural language well to solve algebra word problems. In 1965, Josep Weizenbaum built an interactive program ELIZA that performs a dialogue in

English language on any topic. He demonstrated communication between human and machine. It was very attractive because people feel a human-like computer. In 1966, the first animated robot Shakey was produced. In 1969, Arthur Bryson and Yu-Chi Ho proposed a back propagation procedure as a multi-stage dynamic system optimization method that is now widely used for machine learning and deep learning. In 1972, the world's first full-scale anthropomorphic robot WABOT-1 was developed at Waseda University in Tokyo, Japan. It is composed of a limb control, vision and conversation systems. It has ability to communicate with a human in Japanese as well as walk, grip, and transport objects. In 1978, the R1 (also called XCON (eXpert CONfigurer)) program as a production rule-based system was developed by John P. McDermott at Carnegie Mellon University. It assists in the ordering of DEC's VAX computer systems. From 1980 to 1987, an expert system emulating the decision-making ability and solving complex problems by if–then rules are widely adopted around the world. AI research faces the second boom. In 1981, IBM produced the first personal computer Model 5150. In 1986, the first automated vehicle Mercedes Benz van equipped with cameras and sensors was built at Bundeswehr University in Munich, Germany. It achieved a speed of 63 km/h. In 1990's, an intelligent agent concept perceiving its environment and acting to maximize its success rates is widely accepted. In 1997, a chess-playing computer Deep Blue of IBM won against a world chess champion. In 2000, Cynthia Breazeal of MIT developed Kismet that can recognize and simulate human emotions. A humanoid robot ASIMO of Honda was developed for practical applications and achieved to walk like a human and deliver a tray to customers in a restaurant. In 2004, the first competition of the DARPA Grand Challenge was held in the Mojave Desert, USA. The goal of competition was to drive 240 km route, but none of the automated vehicles finished the route. From 2010, a new AI research and industry have boomed due to mobile broadband, strong computing power, and massive data. In 2009, Google started developing automated vehicles. In 2013, IBM developed the first commercial question answering computer system Watson for using decision of lung cancer treatment at Memorial Sloan Kettering Cancer Centre, New York, USA. In 2016, Google DeepMind AlphaGo won Go against Go world champion Lee Sedol. As we reviewed the AI definition, classification, and history, AI will impact on our day-to-day life and many research areas and industries. It will be the main driver of emerging ICT technologies such as 6G. In this book, AI techniques are introduced in part I and use cases and applications of AI techniques in 6G are discussed in part II. Table 1.1 summarizes the history of AI. Figure 1.4 illustrates the evolution of key AI techniques.

Table 1.1 AI history

Time	Achievements
AI pre-history	
Greek mythology	– Giant bronze automaton Talos in Greek myths
Fourth century B.C	– Syllogistic logic by Aristotle
Thirteenth century	– Combining logical machines – The first programmable humanoid robot driven by water flow
Sixteenth century	– Golem that is an anthropomorphic thing
Seventeenth century	– The first mechanical calculator by Blaise Pascal – Stepped Reckoner to do multiplication and division
Eighteenth century	– Automation chess play as a fake machine
Nineteenth century	– Science fiction book Frankenstein or The Modern Prometheus by Mary Shelley – Boolean algebra by George Boole
1912	– The first chess-playing machine El Ajedrecista
1913	– Principia Mathematica including a formal logic of mathematics
1921	– The word Robot coined in Rossum's Universal Robots
1927	– Science fiction film Metropolis
1942	– Short story Runaround created three laws of robotics
AI history	
1950	– Computing Machinery and Intelligence describing the Turing test
1951	– The first artificial neural network SNARC using vacuum tubes and synapse
1952	– Checkers-playing program as the world's first self-learning program
1955	– The term artificial intelligence was coined
1956	– The official birthdate of the research field AI in Dartmouth conference
1958	– Lisp language that is the most popular programming language for AI research
1961	– The first industrial robot Unimate started working on a General Motors
1965	– ELIZA that performs a dialogue in English language on any topic
1969	– Back propagation procedure as a multi-stage dynamic system optimization method
1972	– The world's first full-scale anthropomorphic robot WABOT-1
1978	– R1 program as a production rule based system
1981	– The first personal computer Model 5150
1986	– The first automated vehicle with cameras and sensors
1997	– Chess-playing computer Deep Blue of IBM won against a world chess champion

(continued)

Table 1.1 (continued)

Time	Achievements
2004	– The first competition of the DARPA Grand Challenge
2013	– The first commercial question answering computer system Watson for using decision of lung cancer treatment by IBM
2016	– Google DeepMind AlphaGo won Go against Go world champion Lee Sedol

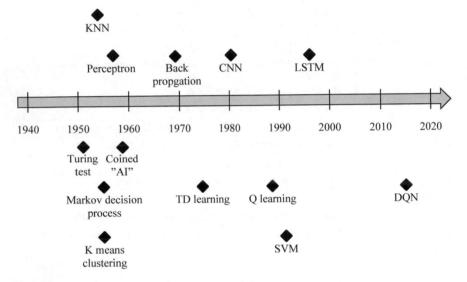

Fig. 1.4 Evolution of key AI techniques

References

1. J. McCarthy, M.L. Minsky, N. Rochester, C.E. Shannon, A proposal for the Dartmouth summer research project on artificial intelligence, August 31, 1955. AI Mag. **27**(4), 12 (2006)
2. S. Russell, P. Norvig, *Artificial Intelligence: A Modern Approach*, 3rd edn (Pearson, 2009)
3. J.R. Searle, Minds, brains, and programs. Behav. Brain Sci. **3**(3), 417–424 (1980)
4. https://www.gartner.com/en/research/methodologies/gartner-hype-cycle
5. J. Lighthill, Artificial intelligence: a general survey, in Artificial intelligence: a paper symposium, Science Research Council, 1973
6. I. Asimov, *Runaround* (Street&Smith, 1942)
7. A.M. Turing, Computing machinery and intelligence. Mind 49, 433–460 (1950)
8. G.C. Devol, Programmed article transfer, US Patent: US2988237A, 13th June 1961

Chapter 2

Artificial Intelligence Ecosystem, Techniques, and Use Cases

AI techniques are evolving rapidly and widely adopted in many different areas because it provides us with many opportunities such as problem solving, decision making, and performance improving. However, some investors are reluctant to jump in AI technique use due to potential risks and uncertainty. In order to adopt a new AI solution, we should require a new hardware (HW) and software (SW) with minimal cost and overcome a lack of effective AI solutions. In this chapter, we review AI ecosystem, AI hardware and software, overall AI techniques, and their workflow and use cases.

2.1 Artificial Intelligence Ecosystem

The world is evolving as a new technology emerges. Personal computers changed our life in many ways. We can store huge amounts of data in our personal computers, calculate complex mathematical problems easily, perform useful applications such as word processors, and communicate with one another through many social media such as Facebook. Smartphones brought a big impact on our life as well. Smartphones are not single-function devices. They are equipped with not only connectivity function but also camera, GPS, accelerometer, gyroscope, and multiple sensors. They enable us to connect with anyone, anytime, and anywhere. In addition, they allow us to stay organized by many applications such as mobile email, map, and office tools. They are helpful for overall health care, security, and others. At the forefront in changing the world, there are AI technologies. It will bring a huge impact on our life and business. Adopting AI in not only communications and networks but also many different areas is sharply on the rise. AI is already all around us. AI is now guiding many decisions about crop harvest and financial investment, improving a product and service, and creating a new business. For example, Apple's Face ID function was included in iPhone X in 2017. Apple developed Bionic neural engine

H. Kim, *Artificial Intelligence for 6G*,
https://doi.org/10.1007/978-3-030-95041-5_2

that is built for a set of machine learning (ML) techniques. The face recognition technology is based on machine learning. Apple's Siri is a voice-controlled AI assistant. In addition, Facebook uses AI for image recognition. A paradigm shift comes from multiple innovations and breakthroughs. Deep learning as one of AI techniques is a buzzword nowadays. However, the basic concept of deep learning was developed in 1970s and the term deep learning was coined in 1986. The recent innovations such as big data, cloud and edge computing, broadband, cryptography, blockchain, IoT, and high computing power have shed new light on deep learning. Most AI techniques are connected to other technology developments. The main difference between AI techniques and other earlier techniques is to interact with environments, learn from previous data, improve a model, and create a new result. The data is key asset and the fuel operating a model. AI is not a plug-and-play technique. In order to adopt AI in a business, companies should invest in AI HW and SW platform, data management, AI system model development, In-house AI engineers, and so on. AI ecosystem is composed of various areas such as AI hardware developments, robotics, computer vision, natural language processing, speech recognition, machine learning, deep learning, and others. There are many AI platforms and components in those areas. They are not isolated but are parts of large scientific workflows and industrial ecosystems. Many AI platforms and components are being developed in many different areas. Similar to electricity, personal computer, and smartphone, AI will bring a huge impact on various industries as well as our day-to-day life. The time seems ripe for investing on AI. Many companies take advantage of the AI techniques due to create better services and products. Large IT companies in industries of electronics, telecommunications and information technologies leads AI technology adoption. Industries of finance, automotive, and factory automation are actively adopting AI technology. Healthcare, food, energy, retail, and education sectors are slowly adopting due to regulatory concerns. We can say that AI ecosystem is composed of data resources acquisition and management, AI SW and HW platform, AI core technologies, AI applications, and AI business. Many start-up companies and large enterprises are working as a part of AI ecosystem. Figure 2.1 illustrates AI ecosystem in layers.

Data resources acquisition and management is a key part of AI implementation. For example, a cellular phone is no longer a tool for a telephone call only. A cellular phone user creates many data including location, information searching, photo, video, and so on. Some healthcare apps on a cellular phone or a smart watch collect individual healthcare data. IT companies such as Apple, Samsung, Google, and Facebook have been able to collect them and use for improving their products and services. These data should be securely stored and shared with people authorized to access data. Depending on type, structure, and complexity of data, they should be managed properly. The performance of AI algorithms relies on sufficient data. In many business areas, sufficient data allows us to have a better product and service. The better product and service brings more users. More users create more data. Data acquisition is the first step of this virtuous circle. Thus, the data acquisition and management is as important as the AI algorithm itself. AI SW and HW platform is a HW architecture and SW framework to run AI algorithms efficiently and perform AI tasks such

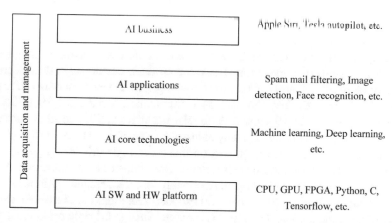

Fig. 2.1 AI ecosystem in layers

as reasoning, learning, perceiving, and solving problems. It will be discussed in the next section. AI core technologies are based on computational models and algorithms such as data science, evolutionary algorithms, machine learning technique, and deep learning technique. Figure 2.2 illustrates AI core technologies.

There are many AI techniques to perform classification, clustering, dimension reduction, regression, and decision making. They continue to evolve their own algorithms and create a new technique to perform a new feature. In this book, AI core techniques are mainly discussed in Part I. Part II covers how they can be adopted in 6G wireless communications and networks. AI has been widely used for character recognition, spam mail filtering, face recognition, image detection, language interpreter, process automation, and others. These AI applications have become key

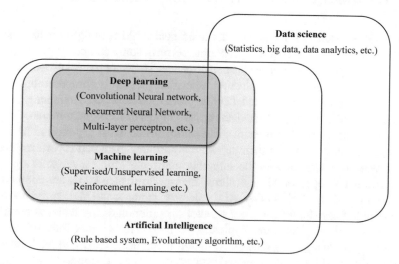

Fig. 2.2 AI core technologies

tools to improve business process and reap higher profits in many business areas. For example, AI application image detection is now widely used for diagnosing cancer. According to one study [1], AI is even more accurate than doctors in diagnosing breast cancer from mammograms. The research team developed a deep learning model for identifying breast cancer from mammograms and evaluated two large data set from USA and UK. They show an absolute reduction of 5.7 and 1.2% (USA and UK) in false positives and 9.4 and 2.7% in false negatives [1]. Those AI applications improve a service and product in many industries and enable us to create a new business model. In [2], AI business impact on industries is described. AI applications will significantly reduce costs, increase revenue, and enhance asset utilization. AI can create values in four areas: (1) Project: Accurately forecast demand, optimize supply, and shape future offerings for success, (2) Produce: Get more out of machines while minimizing maintenance and repairs, (3) Promote: Charge the right price and deliver the right message to the right target, (4) Provide: Give customers a rich, personalized, and convenient experience [2]. Table 2.1 describes examples of a business impact and value creation by AI application in manufacturing and healthcare industries.

As we can observe Table 2.1, a big business impact is expected in each area. When deploying AI applications in their product and service, companies should take into account accessible data, AI HW and SW platform, and capability of AI machine operation. As importance of data acquisition is already explained, data management is a key requirement to drive AI machines. Depending on the target AI application, AI HW and SW platform should be chosen and computing power and connectivity solution should be considered as well. Experienced AI scientists and engineers should be involved in the AI business model development and operation.

2.2 Hardware and Software of Artificial Intelligence

As we briefly discussed in Chap. 1, the main goal of AI techniques is to make AI machines solve the mathematical problems, perform some complex tasks requiring a human intelligence, or improve the performance. There are many key technologies of artificial intelligence. The typical AI algorithm process is composed of 3 steps: (1) The previous data are collected. (2) The previous data become a training data set. A system model is built. (3) New data become inputs of the system model. Outputs are produced as the predicted results. The performance of AI depends on hardware components that incorporate their software algorithms. The better performance of AI hardware provides us with the smoother and more efficient AI software operation. In order to operate AI algorithms, high-performance hardware components such as CPU, GPU, ASIC, FPGA, AI accelerator, storage, and high-speed networks are required. They should carry out parallel operation and simultaneous computation for tasks of AI algorithms. Typically, AI hardware is a mix form of them. In case of deep learning, a big data computation is required. GPU and AI accelerator are primarily used for computing and processing of deep learning. Short-term and long-term storages of huge date should be provided. The network should support a

Table 2.1 Examples of a business impact and value creation by AI application in manufacturing and healthcare industries [2]

		Project	Produce	Promote	Provide
Manufacturing	Value creation	Improve product design yield and efficiency, automate supplier assessment, and anticipate parts requirements	Improve processes by the task, automate assembly lines, reduce errors, limit product rework, and reduce material delivery time	Predict sales of maintenance services, optimize pricing, and refine sales-leads prioritization	Optimize flight planning and route and fleet allocation; enhance maintenance engineer and pilot training
	Business impact	– 10% yield improvement for integrated-circuit products using AI to improve R&D process – 39% IT staff reduction by using AI to fully automate procurement processes	– 30% increase of material delivery time using machine learning to determine timing of goods' transfer – 3–5% production yield improvement	– 13% EBIT (Earnings before interests and taxes) improvement by using machine learning to predict sources of servicing revenues and optimize sales efforts	– 12% fuel savings for manufacturers' customers, airlines, by using machine learning to optimize flight routes
Health care	Value creation	Predict disease, identify high-risk patient groups, and launch prevention therapies	Automate and optimize hospital operations; automate diagnostic tests and make them faster and more accurate	Predict cost more accurately, focus on patients' risk reduction	Adapt therapies and drug formulations to patients, use virtual agents to help patients navigate their hospital journey

(continued)

Table 2.1 (continued)

	Project	Produce	Promote	Provide
Business impact	– $300 billion possible savings in the United States using machine learning tools for population health forecasting – £3.3 billion possible savings in the United Kingdom using AI to provide preventive care and reduce nonelective hospital admissions	– 30 to 50% productivity improvement for nurses supported by AI tools – Up to 2% GDP savings for operational efficiencies in developed countries	– 5 to 9% health expenditure reduction by using machine learning to tailor treatments and keep patients engaged	– $2 trillion–$10 trillion savings globally by tailoring drugs and treatments – 0.2 to 1.3 additional years of average life expectancy

wider and faster channel for large amounts of data. Multicore processors and accelerators in forms of CPU, GPU, ASIC, and FPGA are very useful for AI techniques implementation. The main difference between AI hardware and general hardware is AI accelerators that can compute AI applications faster and easier. The AI accelerators are typically designed as a new dataflow architecture for AI applications, low-precision arithmetic, massive parallel processing, multicore implementation, or in-memory computing. They deliver a better performance and power efficiency and also have new features for AI applications. The key requirements of AI hardware can be summarized as follows: (1) high computational power, (2) cost-efficient solution, (3) cloud computing support, (4) short time-to-market, and most importantly (5) a new hardware architecture for more efficient computation. There are 3 major hardware machines for AI algorithm computation. Firstly, Google developed Tensor Processing Unit (TPU) to process neural networks and accelerate AI tasks such as speech translation in 2016. TPU is highly optimized for large data set and convolutional neural network (CNN) and very helpful for accelerate the performance of CNN while GPU provides us with better flexibility and programmability. Secondly, tensor cores of NVIDIA are designed for computing the tensor or matrix operations which are the key computation function in deep learning. Thus, it is very helpful for accelerating AI tasks such as real time ray tracing. Thirdly, Intel AI solutions are equipped with various hardware packages including FPGA and neural network processors (NNPs). The NNP is firstly designed for the acceleration of AI tasks in 2017 and continuously improved by 2019. As AI techniques are widely adopted in many areas such as health care, factory automation, and others, the demands for AI hardware will grow and those specialized hardware chipsets keep evolving to perform AI tasks and functions more efficiently.

On the top of the AI hardware, we implement AI software such as reinforcement learning and deep learning. AI software platforms include these AI algorithms as a standardized form or support further developments. They help us to create AI systems by implementing AI algorithms, supporting computational graph abstraction, evaluating the models, and deploying in multiple HWs and OSs. AI software is capable of a human intelligent behaviour including reasoning, learning, perceiving, and solving problems. However, implementing AI machines to have a human intelligent behaviour is not a simple task. In order to develop and ensure high-quality AI software, many scientific backgrounds and development skills are required. Fortunately, in the AI software market, there are many AI software development tools and frameworks such as Google TensorFlow, IBM Watson, Scikit-learn, Amazon Web Service machine learning, Apache MXNet, Microsoft AI Azure, Caffe, Theano, Keras, and so on. Using the standardized programming tools and frameworks, they enable AI software designers to develop the creation of AI algorithms and applications easier and faster. The TensorFlow by Google is an open-source AI software library and framework. It focuses on machine learning and deep learning and perform high-performance numerical computations. The library is used for natural language processing, speech recognition, or computer vision. TensorFlow is widely used by major ICT companies such as eBay, AirBnB, Uber, and so on. The time series algorithm of TensorFlow is used for finance or accounting areas. Due to the

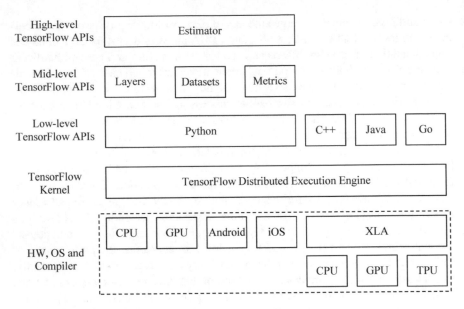

Fig. 2.3 TensorFlow structure

AI frameworks and libraries, it is relatively easier to develop AI machines. The key features of TensorFlow are predictive analytics, speech recognition, easy model building, machine learning production, transfer learning, and workflow automation [3]. However, maintaining AI machines over time is difficult and expensive. Technical debt may be paid down by refactoring code, improving unit tests, deleting dead codes, reducing dependencies, and tightening APIs [4]. Figure 2.3 illustrates TensorFlow structure.

As we can observe Fig. 2.3, TensorFlow Kernel works on the HW and OS machines: CPU, GPU, Android, and iOS. XLA (Accelerated linear algebra) is a compiler for linear algebra accelerating TensorFlow models. It has low-level APIs in multiple languages (Python, C++, Java, and Go) for TensorFlow graph construction and execution. The mid-level APIs are reusable libraries for common model components. TensorFlow estimators are high-level object-oriented APIs. IBM Watson developed in IBM DeepQA project is a question answering computing system. This commercial AI software platform has a massive parallel probabilistic evidence-based architecture to gather data, analyse them, and predict their outcomes or performances. This platform can be used in various business sectors such as health care, energy, finance, education, transportation, and so on. In 2013, this SW platform was commercially used for decision making in lung cancer treatment at Memorial Sloan Kettering Cancer Centre, New York, USA. Figure 2.4 illustrates the high-level architecture of IBM DeepQA.

As we can observe Fig. 2.4, when comparing with documentation search algorithms, the key difference is that they are analysing the questions, generating multiple hypotheses, finding and scoring evidences, making the final decision according to

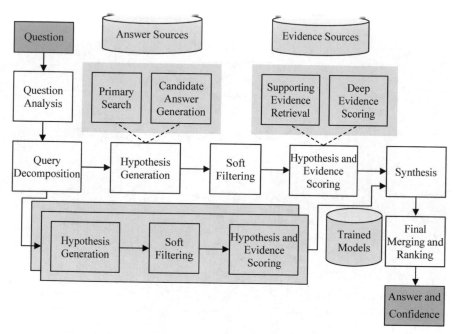

Fig. 2.4 IBM DeepQA architecture

ranked hypotheses, and returning accurate ranked answers. The IBM Watson is continuously evolving and including new technologies: natural language processing, cloud models, machine learning and deep learning algorithms, and customized hardware. The key features of IBM Watson are native language classifier, adaptive customer experiences, visual recognition, targeted recommendations, interaction enrichment, and so on. IBM Watson Studio is an integrated environment to develop AI applications [5]. It supports the end-to-end AI workflows [5]: (1) connect and access data, (2) search and find relevant data, (3) prepare data for analysis, (4) build and train machine learning and deep learning models, (5) deploy models, and (6) monitor, analyse, and manage. Scikit-learn is a machine learning software library and tool for Python programmers [6]. It is widely used in both commercial domains and research domains because it contains many machine learning tools and models such as support vector machines, k-mean clustering, gradient boosting, and so on. The Scikit-learn algorithm cheat sheet diagram [6] as shown in Fig. 2.5 enables us to understand what machine learning algorithms are suitable according to problem types.

The proper algorithm selection is a key part of AI implementation. Figure 2.5 provides us with good guideline. We will briefly discuss those algorithms and algorithm selection in the next section. Amazon Web Services (AWS) [7] provides us with cloud computing platforms and APIs and includes machine learning for SW developers to analysis data, construct mathematical model, and implement applications for their business. The platform has a layered structure with powerful features

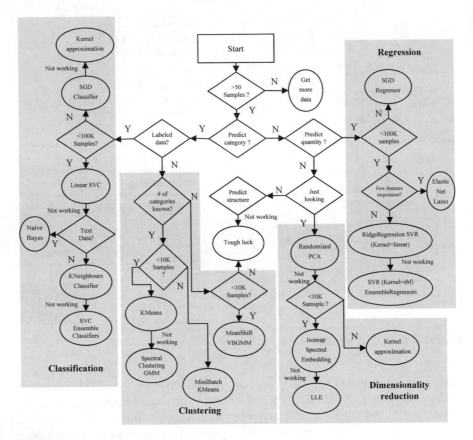

Fig. 2.5 Scikit-learn algorithm cheat sheet diagram

including machine learning. The machine learning service allows us to perform a binary classification as yes or no, multi-class classification to predict multiple conditions, and regression to predict the optimal values. In addition, deep learning amazon machine image (DLAMI) enables us to run the deep learning algorithm on AI HWs and engines. Apache MXNet is an open-source deep learning framework that handles the data processing. It supports a flexible model and multiple programming languages. AWS selected it as its deep learning frameworks. Apache MXNet is supported by many other AI companies including Facebook, Intel, Microsoft, and academies including MIT and CMU. Figure 2.6 illustrates Amazon AI platform.

As we can observe Fig. 2.6, Amazon AI platform enables us to develop AI services including (1) Rekognition: image recognition service, (2) Polly: text-to-speech service, and (3) Lex: conversational interfaces using voice and text. The Microsoft AI platform [8] is composed of AI services, infrastructure and tools. They can be implemented on the Microsoft Azure that is a cloud computing platform including open-source and standardized AI technologies. Figure 2.7 illustrates Microsoft AI platform.

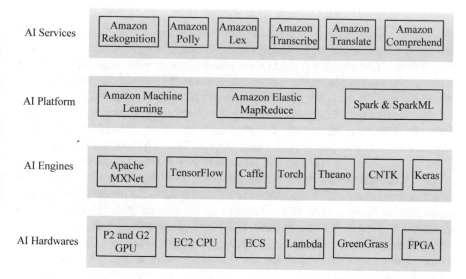

Fig. 2.6 Layered Amazon AI platform

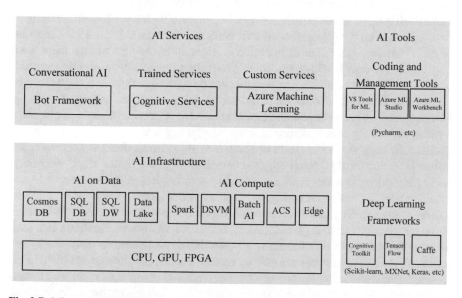

Fig. 2.7 Microsoft AI platform

As we can observe Fig. 2.7, AI services contain APIs for conversational interface, cognitive service to create functions about see, hear, speak, search understanding, and machine learning for training, deploying managing, and monitoring mathematical models. AI infrastructure provides us with AI computation, data storage, networking, and management. AI SW developers need a SW tool to implement AI ideas. AI Tools provide us with a comprehensive set for AI SW developments. They contain Azure machine learning, Caffe, Microsoft Cognitive Toolkit (CNTK), Kera, Scikit-learn, and so on. Caffe developed by University of California Berkeley is an open-source deep learning framework written in C++ with Python interface and supports many different types of deep learning models such as convolutional neural network (CNN), region-based CNN, and long short-term memory (LSTM). In addition, it supports AI acceleration computational kernel libraries. Theano is an open-source Python library for machine learning algorithms. Keras as an open-source deep learning library is designed to build and train deep learning models. This is written in Python.

As we discussed AI HW and SW platforms, they enable AI system designers to create a product and service efficiently. However, there are many challenges of AI system implementation. Firstly, as we briefly discussed in the previous section, data acquisition and management is a key part of AI implementation. The AI system we develop is as good as the data we have. It depends on the trained data. Bias is one of key challenges. For example, if one user group has favourite features and not use other features and we have the limited data set, AI system we develop could not learn from the other functions that are not used. In other ways, we can explain the bias problem as follows: People lie or have stereotypes, people create data, and data is biased. Thus, AI systems are not error-free. We could have an inaccurate model and AI systems can make a mistake. In addition, labelled data is more useful, but we face the scarcity of labelled data even if huge data are created every day. In order to use machine learning or deep learning, data labelling is required. Thus, in-house data management team may handle the data labelling and control over the process. However, this approach is not cost-efficient. Another challenge about data management is transparency of AI models. When AI models are built and AI systems run, the system becomes a block box and we receive a result without explanation. For example, a doctor receives a result about lung cancer detection from AI systems and does not agree with the decision. What if the doctor want to know how AI systems decide? Therefore, the rationale behind the decision is required to evaluate and clarify the result. Secondly, computing power is a key challenge of AI system implementation. In order to run machine learning or deep learning, we need to process massive volume of data and a powerful computing power is required. Cloud computing and massively parallel computing enables us to implement AI systems for a short term. However, in a long term, as the volume of data increases exponentially, bandwidths of connectivity are saturated, and AI algorithms are getting more complex, they would not be helpful. Obtaining enough computing power would be challenging to run AI systems. Thirdly, privacy protection and legal issues are important for running AI systems. Sometimes, data contain sensitive information like patient medical information and are stored in a local file server or shared with cloud servers. Privacy violation occurs when the sensitive data are stored in a server and illegally accessed by unauthorized persons. Thus, the company implementing AI systems should create an effective infrastructure to collect, store, protect, and manage the sensitive data. In addition, there are many legal issues connected to AI systems. For example, what data can be collected under General Data Protection Regulation (GDPR)? Who will take responsibility if AI systems cause damage? If AI systems collect sensitive data by themselves, it might be in violation of laws. Therefore, the lawmakers take these problems into account and create the legal system to keep up with the AI system developments. Lastly, there are challenges associated with AI business model developments and alignments. Many companies adopt AI systems in their own business. However, decision makers in a company have lack of AI knowledge, strategic approach, and business use cases. They may not know how much AI systems will be helpful for their business and solve business problems. More importantly, integrating AI system in their business is much more complex than adopting a plug-in solution. In-house AI experts should be involved in this process and development.

2.3 Artificial Intelligence Techniques and Selection

One algorithm can work well on some condition and environment, but the algorithm can't work well on other conditions and environments. This is very typical in many mathematical problems. It is very difficult to find a universal algorithm satisfying many conditions and environments. Thus, effective algorithm selection is an important topic. We should choose an algorithm from portfolio algorithms in a wide variety of environments and conditions. The problem of algorithm selection is formulated in [9]. The basic model is composed of problem stage, algorithm space, and performance measure space. Figure 2.8 illustrates the basic model of the algorithm selection problem. In Fig. 2.8, P, A, and R^n represent problem space, algorithm space, and n-dimensional real vector space of performance measures, respectively. x, A, and p are element of P problem to be solved, element of A algorithm applicable to problems from P, and mapping from $A \times P$ to R^n determining performance measure, respectively. S is mapping from P to A. $\|\|$ is norm on R^n to evaluate algorithm performance on a particular problem. The problem of algorithm selection can be described as follows: When all items in the basic model are given, the objective of this problem is to find $S(x)$ in order to have maximum performance of the algorithm. The best selection criteria can be expressed as follows: Choose $B(x)$ satisfying maximum performance for each problem:

$$\|p(B(x), x\| \geq \|p(A, x)\| \quad \text{for} \quad \forall A \in A. \tag{2.1}$$

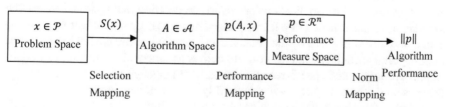

Fig. 2.8 Basic model of the algorithm selection problem

In machine learning, the main purpose of algorithm selection (also known as meta-learning) is to find a proper algorithm from the portfolio of algorithms (such as K means algorithm, support vector machine, deep neural networks, and others) satisfying minimum error on each data set. Algorithm selection is one important topic to study how it becomes flexible in solving machine learning problems. It is very challenging to select a proper machine learning algorithm and achieve the goals like solving a problem or improving a performance. Definitely, choosing the correct machine learning algorithm depends on many different variables such as training data, type of problem, accuracy, training time, linearity, number of parameters, number of features, and others.

As we can see Fig. 2.5, there is one possible guideline to help us algorithm selection. However, understanding key features, limitations, and applications of AI algorithms would be good approach to select a suitable algorithm for specific condition and environment. Machine learning algorithms can be classified as unsupervised learning, supervised learning, semi-supervised learning, reinforcement learning, and deep learning. Unsupervised learning algorithms are suitable for observing similarities between groups of data and clustering them. In addition, it is useful for reducing dimensionality such as the number of variables and features of data set. Key applications are customer segmentation and targeting in marketing, compression of data, big data visualization, feature elicitation, and others. Supervised learning algorithms are suitable for analysing the labelled training data, interpreting a function from them, and predicting unknown data on the data set. Thus, supervised learning algorithms are used for classification and regression. They both use the labelled data set to make predictions. The difference between them is that the output variable of regression is continuous and numerical, while the output variables of classification are discrete and categorical. Key applications are email filtering, image classification, forecasting, predictions, process optimization, and others. Semi-supervised learning algorithms are the mixed algorithms between unsupervised learning and supervised learning. They combine a small amount of labelled training data set with a large amount of unlabelled training data set. This approach can improve accuracy significantly. It is useful when enough labelled training data are not supplied. Key applications are speech analysis, lane finding on GPS data, text classification, and others. Reinforcement learning is different from both unsupervised learning and supervised learning. Reinforcement learning algorithms are goal-oriented algorithms to make a sequence of decisions by taking a suitable action to maximize cumulative rewards in a particular environment. Key applications are real-time decision, navigation, self-driving car, Game AI, and others. Deep learning also known as deep neural learning or deep neural network is inspired by the structure and function of the human brain. It uses multi-layered artificial neural networks to deliver accuracy of the model. Key applications are image and speech recognition, natural language processing, automated vehicles, and others. Figure 2.9 illustrates 3 types of machine learning.

In Part I of this book, these machine learning algorithms will be mainly discussed. The main purpose of Part I is to make wireless communications and networks system designers have enough mathematical background to use and apply AI algorithms

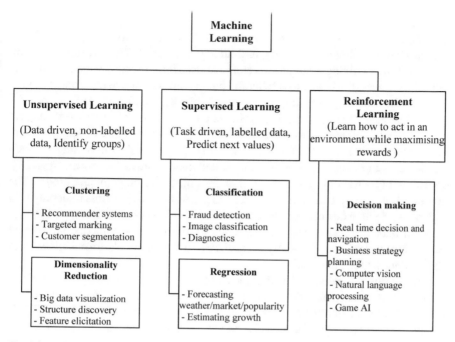

Fig. 2.9 Types of machine learning

for wireless communications and network systems. Thus, they select suitable AI algorithms and solve communications and networks systems problems.

2.4 Artificial Intelligence Workflow and Use Cases

As we discussed in the previous sections, AI platforms enable AI system designers to develop AI products or services efficiently. They have their own AI workflows to implement AI systems. The overall workflows for implementing AI systems are to create a model to learn a pattern in a training data set and then recognize the pattern presence in other different data sets. In general, AI workflows can be summarized as (1) accessing data, (2) creating a training model, (3) training a model, (4) validating a trained model, (5) deploying an inference model, (6) monitoring accuracy, (7) managing and updating the model. As we discussed the importance of data acquisition, we should access enough number of training data including suitable features. If needed, we can perform data pre-processing such as normalization, filtering, averaging, format alignment, and redundancy reduction. It is important to have a proper data set to increase the capability of AI systems. For example, if we need to interpolate pilot tones in OFDMA communication systems, we should collect enough training symbols of the OFDMA packets. The next step is to create a training model using established AI algorithms or designer's own AI algorithms. In this step, we define a set of models to identify which one solves the problem efficiently. Many AI platforms provide us with training, evaluating, and tuning AI models. The next step is to train a model. When training a model, we can feed data for which we already know the value. We can run the model to predict the target values, evaluate the trained model, and tune the model by adjusting its setting. Typically, small data set is trained for quick adjustment. The next step is to validate the trained model. In this step, we test the model with different known data sets. When the results are good enough, we can deploy an inference model without change. If needed, we adjust its setting for deep learning structure, learning rate or constant and then improve the model. Typically, we run multiple training processes to find the optimal settings for the model. The next step is to deploy an inference model. AI platforms provide tools to deploy it in desktop, embedded devices, enterprise systems, or cloud systems. Then, we monitor the performance of AI models. There are many tools to examine the operation of AI models in AI platforms. Typically, visualization tools of AI platforms allow us to analyse data set for the model.

Many industries adopt AI techniques and create new products and services. Communications and networks industries are not immune from AI techniques adoption. Especially, cellular communications and networks are evolving rapidly while adopting a new technique. AI techniques will be able to play an important role in efficiently managing huge network traffics over the cellular networks. For example, fault detection in cellular networks is still performed inefficiently by alarms and threshold violation after measuring network parameters. Thus, we have inefficient fault tolerance systems that enable a network to continue operating without interruption like power failures, human error, network equipment failures, and others. Anomaly detection using AI techniques will improve the fault tolerance of cellular networks significantly without human intervention. Using key features of machine learning algorithms, we can classify the faults and trace the components of networks causing the initial problem. In addition, the data about faults can be stored and analysed for future maintenance. We can find the patterns presence from the stored data and predict when and where faults are detected. If network components behaviours are predicted accurately, we can replace the components before the failures occur and reduce the operational cost significantly. Automation industries including factory automation, automated vehicles, robotics, intelligent control systems, and others are actively adopting AI technologies as well. Traditionally, they generate plenty of measurement data such as temperature, pressure, vibration, humidity, proximity, and others and should find an optimal point from them. In addition, autonomous systems are designed as model-based systems. However, data-driven systems are getting more important. Many autonomous systems are modelled as Markov Decision Process (MDP). Reinforcement learning as machine learning provides us with optimal solution. Machine learning becomes the key technique of automation application. In factory automation, it allows us to have predictive maintenance system detecting faults and preventing unwanted malfunctions. Thus, we can reduce operational costs and improve reliability. In automated vehicles, it allows us to reduce the number of accidents and re-define the driving itself. Machine learning makes auto-mated vehicles have an ability to collect data on their surroundings from camera, LiDAR, and radar and decide what action they take in real time. In order to process the image of surroundings, machine learning or deep learning will be key enablers of visual perception to detect and track objects such as neighbouring cars, pedestrians, traffic lights, cyclists, and others. We need a robust computational power because this task should be performed in real time and also required to process huge data. Thus, AI accelerators and distributed computing would be one of key technologies for automated vehicles. Finance industry is adopting AI techniques. Traditionally, many mathematical tools are widely used in finance industry to optimize the process such as pricing, performing risk management, stock markets prediction, credits decision, stock trading, and others. They are based on data-driven decision making on the time-varying data. AI techniques are well matched with this environments. AI techniques will help us with improving efficiency and productivity through reducing human errors by emotional factors. For example, algorithmic trading is one of key finan-cial services. Due to the demand for rapid decisions like portfolio risk calculations, high-performance computing, and deep learning are required.

When adopting AI algorithms to wireless communications and networks, we should understand different learning approaches of AI algorithms and consider to build AI enables wireless systems. One classification of AI algorithms is supervised learning, unsupervised learning, or semi-supervised learning in terms of the amount of knowledge or feedback for the learner. Supervised learning uses the labelled training data set of the known inputs and output in order to have a learning model. The trained supervised learning is able to find a patterns or predict the future output from new input data set. This approach is in line with the problems of wireless systems requiring a prior knowledge and predefined data set. Supervised learning can be applied to wireless systems extensively. For example, radio resource allo-cation, power control, channel estimation, localization, network traffic monitoring and activity recognition, fault detection, and so on. Unsupervised learning uses the training data set without a prior knowledge and is to find a pattern or reduce dimen-sionality from unlabelled data set. This approach is suitable for the problems of wireless systems without a prior knowledge about the outcomes or the problems with no labelling data set. In wireless systems, unsupervised learning can be applied to node clustering in mMTC application, data clustering, dimensionality reduction of data management, fault or event detection, and so on. Semi-supervised learning is a mixed form of supervised learning and unsupervised learning. It can be used when we have a mixed data set like small amount of labelled data and large amount of unlabelled data. It will be useful not to label all instances or reduce computa-tional cost when training a data set. For example, when wireless system collects not enough user data and then identify a user location, semi-supervised learning allows us to alleviate the tedious process of training data collection. Another classification of AI algorithms is offline or online learning in terms of how training is performed. The training step of offline learning is performed on the entire training data while the online learning performs training in a sequential order and update the representation of the model iteratively. The offline learning can be performed when the properties of the model is not dynamically changed. Wireless systems can be trained in the offline learning way if the channel status is not dynamically changed and the system parameters can't be changed significantly. On the other hands, online learning can be used when small training data set (not entire training data set) is processed one-by-one or the computational power is not enough to handle the entire training data set. This approach is useful for decentralized systems. For example, in IoT networks, each node has a limited computational power and the training data is limited. Thus, some task can be computed in the online learning way. We can update the model sequentially and iteratively. Active learning as a part of online learning collects a few useful data set and use them for training. This approach is meaningful when it is not easy to obtain data from all variables of interest. It has been one of key research topics in AI and ML. Figure 2.10 illustrates two classifications to adopt AI algorithms to wireless systems in terms of data set and training.

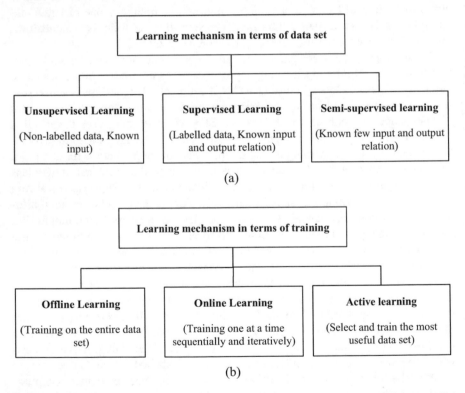

Fig. 2.10 Two classifications to adopt AI algorithms to wireless systems in terms of data set (**a**) and training (**b**)

References

1. S.M. McKinney, M. Sieniek, V. Godbole et al., International evaluation of an AI system for breast cancer screening. Nature **577**, 89–94 (2020). https://doi.org/10.1038/s41586-019-1799-6
2. J. Bughin, E. Hazan, S. Ramaswamy, M. Chui, T. Allas, P. Dahlstrom, N. Henke, M. Trench, *Artificial Intelligence: The Next Digital Frontier?* McKinsey Global Institute, Discussion paper (2017). https://www.mckinsey.com/~/media/McKinsey/Industries/Advanced%20Electronics/Our%20Insights/How%20artificial%20intelligence%20can%20deliver%20real%20value%20to%20companies/MGI-Artificial-Intelligence-Discussion-paper.ashx
3. https://www.tensorflow.org/
4. M. Fowler, *Refactoring: Improving the Design of Existing Code* (Pearson Education India, 1999)
5. https://www.ibm.com/se-en/cloud/watson-studio
6. https://scikit-learn.org/stable/index.html
7. https://aws.amazon.com/
8. https://www.microsoft.com/en-us/ai/ai-platform
9. J.R. Rice, The algorithm selection problem. Adv. Comput. **15**, 65–118

Chapter 3
Unsupervised Learning

There are two categories of classical machine learning: supervised and unsupervised learning according to system models and data types. The main difference is the type of data they use. Supervised learning uses the labelled input data, but the data used by unsupervised learning is not labelled. Supervised learning has the process learning from the training data, and it looks like a supervisor of the learning process. Supervised learning algorithms iteratively predict the training data set, and it is corrected by a supervisor until we reach an acceptable level of results. On the other hand, unsupervised learning does not have a feedback for predicting the results as well as does not know what the results should look like. Other differences are complexity, accuracy, reliability, and computation time. Supervised learning is more complex than unsupervised learning because it should understand the labelled data. Supervised learning is more accurate and reliable than unsupervised learning because supervised learning already knows the labelled data and it finds the hidden pattern or predicts the future output. Unsupervised learning can perform real-time data analysis, but supervised learning could not do in many applications. Due to those differences, supervised learning is useful to solve classification and regression problems and unsupervised learning is used to derive the structure by clustering the data. In this chapter, we discuss unsupervised learning techniques. Next chapter will cover supervised learning techniques.

3.1 Types and Performance Metrics of Unsupervised Learning

The typical steps of classical machine learning are problem definition, data acquisition, underlying model definition, underlying model improvement, and model building. In case of unsupervised learning, the data are not labelled and the model is developed to find any patterns on its own. Since it is easier to obtain unlabelled data from cellular networks and Internet, it is more useful for applying in exploratory

© The Author(s), under exclusive license to Springer Nature Switzerland AG 2022
H. Kim, *Artificial Intelligence for 6G*,
https://doi.org/10.1007/978-3-030-95041-5_3

analysis than supervised learning. In addition, it doesn't require the previous knowledge. It helps us to find useful features in a data set. We can reduce human error by intervention. It has relatively less complexity. It enables us to facilitate flexible and automated methods of machine learning in real time. The applications of unsupervised learning can be categorized by clustering, association, and dimension reduction. Clustering is a grouping with no pre-defined classes and finds a structure or pattern in an uncategorized data set. Therefore, a cluster is a collection of data that is similar to each other. Clustering is used to find out whether or not a data set is composed of a set of similar sub-groups or how much each group is similar. It is possible to classify clustering algorithms: exclusive, hierarchical, overlapping, and probabilistic clustering. As the name suggests, the exclusive clustering creates k mutually exclusive groups. Namely, one data set belongs to one cluster only. K-means clustering algorithm is one example of exclusive clustering. Hierarchical clustering enables us to build a hierarchy of clusters. There are two hierarchical clustering algorithms: agglomerative and divisive clustering. In agglomerative clustering, all data can be a cluster as an initial condition. The unions between two nearest clusters reduce the number of clusters iteratively. The iteration ends if it reaches the target number of clusters. On the other hand, divisive clustering starts at one cluster including all data. The cluster is successively divided into sub-clusters in terms of dissimilarity. In overlapping clustering, fuzzy sets are used for clustering data. It allows data points to belong to more than one cluster with different degrees of membership. Probabilistic clustering uses probability distribution to create clusters. The assignment of data points to clusters is based on the probability indicating strength of belief. Those clustering algorithms will be discussed from the next section in details. Association of unsupervised learning is to establish association or relation between variables in a large data set. This algorithm intends to discover regularities and find association rules. For example a group of lung cancer patients is highly related to a group of heavy smokers. Association of unsupervised learning is useful for forecasting sales and discount, analysing product purchasing pattern, placing products on the shelves, and so on. Many system designers want to avoid the curse of dimensionality. Working at high-dimensional spaces results in high complexity analysis. Dimension reduction is a key part of data analysis and signal processing. Dimension reduction of unsupervised learning is an algorithm to reduce the number of variables and represent data with lower dimension in order to find intrinsic patterns of the data set, have a better visualization, or learn in subsequent tasks.

It is one of the key issues in unsupervised learning to evaluate the quality of clustering results. Similarity measure is one important component of clustering analysis. The similarity is typically defined as the distance between data points. Different definition of the distance causes the different results. There are no general theoretical guideline to select a distance measure. It should be chosen according to applications and data feature. Therefore, it is an important starting point to define a suitable similarity measure. In order to evaluate high-dimensional data, Minkowski distance of order p between two points

$$X = (x_1, x_2, \ldots, x_n) \text{and} Y = (y_1, y_2, \ldots, y_n) \in \mathbb{R}^n \tag{3.1}$$

is defined as follows:

$$D(X, Y) = \left(\sum_{i=1}^{n} |x_i - y_i|^p \right)^{1/p} \tag{3.2}$$

where n and p are the dimensionality of data and an integer, respectively. They become Euclidean distance when $p = 2$ and Manhattan distance when $p = 1$. In [1], three properties of clustering performance measures are presented: scale invariance, richness, and consistency. A clustering function f is defined on a set S of $n \geq 2$ and the pairwise distances among data points. The set of data point is $S = (1, 2, \ldots, n)$, and the pairwise distances are given by a distance function $d(i, j)$ where $i, j \in S$. The distance function means the dissimilarity between data points. Minkowski distance can be used. The distance function satisfies the following properties: (1) $d(i, j) \geq 0$, (2) $d(i, j) = d(j, i)$, (3) $d(i, j) = 0$ if and only if $i = j$, and (4) $d(i, k) \leq d(i, j) + d(j, k)$ for all $i, j, k \in S$. The sets in a partition Γ of S are called a cluster. Three properties of clustering performance measures are as follows [1]:

Property 3.1. Scale Invariance *For any distance function d and any scaling factor* $\alpha > 0$,

$$f(d) = f(\alpha d).$$

Property 3.2. Richness *When* Range(f) *denotes the set of all partitions and* $f(d) = \Gamma$ *for some distance function d,* Range(f) *is equal to the set of all partitions of S.*

Property 3.3. Consistency *Let d and d' be two distance functions. d' is a Γ transformation of d if (1) for all i, j $\in$ S belonging to the same cluster of Γ, we have* $d'(i, k) \leq d(i, j)$ *and (2) for all i, j $\in$ S belonging to different cluster of Γ, we have* $d'(i, k) \geq d(i, j)$, *then* $f(d') = \Gamma$.

Property 3.1 indicates that the clustering function should not be sensitive to change when all distance are scaled by the scaling factor α. Property 3.2 indicates that the clustering function should be flexible to produce any partition of the data set. Property 3.3 indicates that the clustering process should be consistent and the clustering results should not be changed when shrinking distances inside a cluster and expanding distance in different cluster. In [1], the impossibility theorem for clustering is proposed: for every $n \geq 2$, there is no clustering function f satisfying all three properties. Therefore, the practical approach is to find a trade-off of desired features of clustering. For example we satisfy two properties while ignoring one property. In this point of view, three types of clustering stopping conditions are proposed as follows [1]:

Stopping condition 3.1. k cluster stopping condition.
Stopping clustering when we have k clusters.

Stopping condition 3.2. Distance r stopping condition.
Stopping clustering when the nearest pair of clusters is longer than r.
Stopping condition 3.3. Scale ϵ stopping condition.
Stopping clustering when the nearest pair of clusters is longer than a fraction ϵ of the maximum pairwise distance.

Stopping condition of clustering 3.1 violates the richness property because the number of clustering is fixed. Stopping condition of clustering 3.2 violates the scale invariance property because it builds in a fundamental length scale. Stopping condition of clustering 3.3 ignores consistency property. In many communication and network applications, we relax richness property (namely, the number of clusters is fixed priorly) and satisfy scale invariance and consistency properties. In [2], clustering algorithm evaluation is described in terms of clustering tendency, correct number of clusters, quality assessment without external data, comparison of external data, and comparison of different sets of clusters. The approaches are divided into expert analysis, relative analysis, internal data measure, and external data measure. The evaluation by an expert may provide us with good insight from data. However, the evaluation results may be not comparable and not standardized. The relative analysis is based on comparison of different clusters using different parameter settings under the same clustering algorithms. The advantage of internal data measure is not to require additional information. However, evaluation criteria should be optimized depending on data sets and algorithms. The external data measure requires an external ground truth and may be evaluated more accurately. However, it requires additional information. The internal evaluation enables us to establish the quality of clustering without external data using two metrics: cohesion and separation. Cohesion $M_{co}(C_i)$ represents how closely the elements of one cluster are placed each other. Separation $M_{se}(C_i, C_j)$ represents the level of separation between clusters. They are defined as follows:

$$M_{co}(C_i) = \sum_{x \in C_i, y \in C_i} f_{proximity}(x, y) \tag{3.3}$$

and

$$M_{se}(C_i, C_j) = \sum_{x \in C_i, y \in C_j} f_{proximity}(x, y) \tag{3.4}$$

where x and y are elements of clusters, C_i and C_j are clusters, and $f_{proximity}(x, y)$ is a proximity function determining how much similar of pairwise data is in terms of similarity, dissimilarity, distance, and others. As we can observe (3.3) and (3.4), cohesion is evaluated inside cluster and separation is evaluated between clusters. The cohesion metric (3.3) can be regarded as the sum of squared errors (SSE) in a cluster. The proximity function can be the squared Euclidean distance. The SSE is defined as follows:

$$SSE(C_i) = \sum_{x \in C_i} d(c, x)^2 \tag{3.5}$$

where c is a centroid of the cluster C_i. As we can observe (3.5), the small value of SSE represents good quality of clustering. The separation metric (3.4) can be regarded as the sum of squares between (SSB). The SSB is defined as follows:

$$SSB = \sum_{i=1}^{K} n_i d(m_i, m)^2 \tag{3.6}$$

where n_i is the number of elements in the cluster C_i, m_i is the mean of the individual cluster C_i, m is the overall mean, and K is the number of clusters. Since we should maximize the distance among clusters, the big value of SSB represents good quality of clustering. Those two metrics are useful for analysing clustering, but they do not work well when analysing density of data. Another approach is to find out correlation of actual data set and ideal data set. When evaluating hierarchical algorithms, we need to redefine the distance in hierarchical tree. The cophenetic distance represents how close data are grouped in a hierarchical tree. It is typically the height of the dendrogram (a diagram representing a tree) that two branches including two elements merge into single branch. It is widely used to evaluate cluster-based models of DNA sequences. The cophenetic correlation coefficient (CPCC) is used to evaluate hierarchical clustering algorithms [3]. We calculate the correlation between the elements of the cophenetic matrix P_c and the proximity matrix P. The elements of P_c and P are cophenetic distances and similarities, respectively. The CPCC is defined as follows [3]:

$$CPCC = \frac{\sum_{i<j}(d_{ij} - \overline{d})(d_{ij}^* - \overline{d^*})}{\sqrt{\sum_{i<j}(d_{ij} - \overline{d})^2 \sum_{i<j}(d_{ij}^* - \overline{d^*})^2}} \tag{3.7}$$

where d_{ij}, d_{ij}^*, $\overline{d}$, and $\overline{d}^*$ are the distance between elements (i, j), the cophenetic distance between them, the average distance in the proximity matrix, and the average cophenetic distance in the cophenetic matrix, respectively. The average distance in the proximity matrix and the average cophenetic distance in the cophenetic matrix can be calculated as follows: [4]

$$\overline{d} = \frac{\sum_{i<j} d_{ij}}{2(n^2 - n)} \tag{3.8}$$

and

$$\overline{d}^* = \sqrt{\frac{\sum_{i<j}\left(d_{ij} - d_{ij}^*\right)^2}{\sum_{i<j}\left(d_{ij}^*\right)^2}}. \tag{3.9}$$

where n is the number of elements in the cluster. If the value of the CCPC is high, the similarity of two matrix is high. The external evaluation enables us to assess a data set using external data that was not used for clustering. The external data is composed of a set of pre-defined data. The pre-defined data can be a standard data set or a ground truth by a data analyst. This approach is to evaluate how much close the results of clustering and the standard data set are. The external evaluation can be categorized by matching sets, peer-to-peer correlation, and information theory [4]. They depend on clustering algorithms and problems. The contingency matrix is used to define external evaluation metrics. The contingency matrix contains the number of true positive (TP), the number of true negatives (TN), the number of false positives (FP), and the number of false negatives (FN). TP represents the number of pairwise data found in the same cluster in both C and P where C is the results of clustering and P is the different partitions representing the pre-defined data, the standard data set, or the different results from other clustering algorithms. TN represents the number of pairwise data found in the different cluster in both C and P. FP represents the number of pairwise data found in the same cluster in C but in different clusters in P. FN represents the number of pairwise data found in different clusters in C but in the same cluster in P. Figure 3.1 illustrates the relationship among TP, TN, FP, and FN [5].

In the matching set, the similarity between clustering results and pre-defined data is compared. There are several external evaluation metrics: purity, precision, recall,

Fig. 3.1 Relationship among TP, TN, FP, and FN

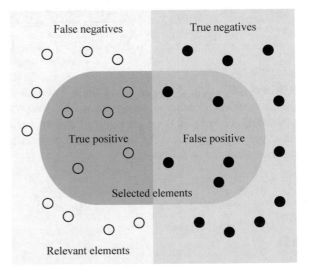

and F-measure. Purity is a measure to evaluate whether a cluster contains single class. It is defined as follows:

$$P_{\text{purity}} = \frac{1}{N} \sum_{m \in M} \max_{d \in D} |m \cap d| \tag{3.10}$$

where N, M, and D are the number of data, the number of clusters, and the number of classes, respectively. This metric checks whether or not a cluster contains data from same class. It doesn't work well in case of imbalanced data set. Precision is a measure to count how many data are classified in the same cluster as follows:

$$P_{\text{precision}} = \frac{TP}{TP + FP} = \frac{TP}{P}. \tag{3.11}$$

Recall is a measure to evaluate how much data are included in the same cluster as follows:

$$P_{\text{recall}} = \frac{TP}{TP + FN}. \tag{3.12}$$

The F-measure can be calculated by $P_{\text{precision}}$ and P_{recall} as follows:

$$F_\beta = \frac{(\beta^2 + 1) P_{\text{precision}} P_{\text{recall}}}{\beta^2 P_{\text{precision}} + P_{\text{recall}}} \tag{3.13}$$

where $\beta \geq 0$ is a real number. This metric is used for balancing the contribution of false negatives by weighting the recall. As we can observe (3.13), $F_0 = P_{\text{precision}}$. Namely, there is no impact on F_β from the recall when $\beta = 0$. The peer-to-peer correlation is based on the correlation between pairwise elements. The Jaccard coefficient is used to assess the similarity between C and P. It is defined as follows:

$$J = \frac{TP}{TP + FP + FN}. \tag{3.14}$$

The value of J is placed between 0 and 1. A coefficient of 0 represents that there is no common element. A coefficient of 1 denotes that two groups are identical. The Rand coefficient measures the percentage of the correct decision. It is defined as follows:

$$R = \frac{TP + TN}{TP + FP + TN + FN}. \tag{3.15}$$

The Fowlkes–Mallows coefficient measures the similarity between clusters found by clustering algorithms and ground truth. It is defined as follows:

$$FM = \sqrt{\frac{TP}{TP + FP} \frac{TP}{TP + FN}}. \tag{3.16}$$

The higher value of FM represents that they are more similar. Lastly, the method using information theory is based on the concept of entropy and mutual information. The entropy enables us to find a harmony in discrimination between data sets. The entropy of a cluster C is defined as follows:

$$H(C) = -\sum_i p_{C_i} \log p_{C_i} \tag{3.17}$$

where $p_{C_i} = n_i/n$ is the probability of the cluster C_i. The entropy of a class Y is defined as follows:

$$H(Y) = -\sum_j p_{Y_j} \log p_{Y_j} \tag{3.18}$$

where $p_{Y_j} = n_j/n$ is the probability of the class Y_j. The conditional entropy of the class Y for the cluster C_i is defined as follows:

$$H(Y|C_i) = -\sum_j \left(\frac{n_{ij}}{n_i}\right) \log\left(\frac{n_{ij}}{n_i}\right). \tag{3.19}$$

Given the cluster C, the conditional entropy of the class Y is defined as follows:

$$H(Y|C) = \sum_i \frac{n_i}{n} H(Y|C_i) = -\sum_i \sum_j \frac{n_{ij}}{n} \log\left(\frac{n_{ij}}{n_i}\right)$$
$$= -\sum_i \sum_j p_{ij} \log\left(\frac{p_{ij}}{p_{C_i}}\right), \tag{3.20}$$

$$H(Y|C) = -\sum_i \sum_j p_{ij}(\log p_{ij} - \log p_{C_i})$$
$$= -\left(\sum_i \sum_j p_{ij} \log p_{ij}\right) + \sum_i \left(\log p_{C_i} \sum_j p_{ij}\right)$$
$$= -\sum_i \sum_j p_{ij} \log p_{ij} + \sum_i p_{C_i} \log p_{C_i} = H(C, Y) - H(C) \tag{3.21}$$

where $p_{ij} = \frac{n_{ij}}{n}$ is the probability that an element in a cluster C_i belongs to a class Y_j and $H(C, Y)$ is the joint entropy of the cluster C and the class Y as follows:

$$H(C, Y) = -\sum_i \sum_j p_{ij} \log p_{ij}.$$ (3.22)

As we can observe (3.20) and (3.21), the conditional entropy approaches to zero for a good clustering. The conditional entropy $H(Y|C)$ means the remaining entropy of the class Y given the cluster C. If $H(Y|C) = 0$, a class Y is completely decided by a cluster C. When a class Y and a cluster C are independent, $H(Y|C) = H(Y)$. The mutual information $I(C, Y)$ measures how much information is shared between clusters and classes. It is defined as follows:

$$I(C, Y) = \sum_i \sum_j p_{ij} \log \frac{p_{ij}}{p_{C_i} p_{Y_j}}.$$ (3.23)

They represent the dependence between the observed joint probability and the expected joint probability. When a class Y and a cluster C are independent, $I(C, Y) = 0$. The mutual information is given as follows:

$$I(C, Y) = H(Y) - H(Y|C),$$ (3.24)

$$I(C, Y) = H(C) - H(C|Y).$$ (3.25)

Since $H(C|Y) \geq 0$ and $H(Y|C) \geq 0$, we have $I(C, Y) \leq H(C)$ and $I(C, Y) \leq H(Y)$. Normalized mutual information (NMI) is to determine the quality of clustering by reducing a bias for different cluster numbers. Considering $I(C, Y)/H(C)$ and $I(C, Y)/H(Y)$, the NMI is defined as follows:

$$\text{NMI}(C, Y) = \sqrt{\frac{I(C, Y)}{H(C)} \frac{I(C, Y)}{H(Y)}} = \frac{I(C, Y)}{\sqrt{H(C)H(Y)}}.$$ (3.26)

The range of NMI is between 0 and 1. It approaches to one for the good clustering.

Example 3.1 External Evaluation: Calculation of the Metric Purity Consider the clusters as shown in Fig. 3.2.
Calculate the purity in Fig. 3.2.

Solution

As we can observe Fig. 3.2, the majority class and the number of elements for each cluster are as follows:
Cluster A: circle ($\bigcirc$), 4
Cluster B: shaded circle ($\bullet$), 5
Cluster C: shaded triangle ($\blacktriangle$), 3
Using (3.10), we can calculate the purity as follows:

$$P_{\text{purity}} = \frac{1}{N} \sum_{m \in M} \max_{d \in D} |m \cap d| = \frac{1}{20}(4 + 5 + 3) = 0.6. \blacksquare$$

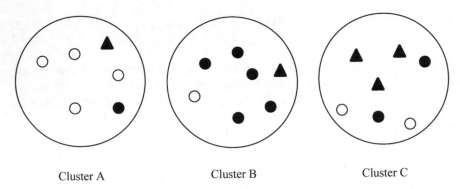

Cluster A   Cluster B   Cluster C

Fig. 3.2 Example of clusters

Table 3.1 Results of the medicine

	Better	No change
Medicine	43	7
Placebo	31	19

Example 3.2 External Evaluation: Precision, Recall, F-measure , Jaccard Coefficient, Rand Coefficient, and Fowlkes–Mallows Coefficient We have 100 patients and test a medicine. After giving the medicine, condition of the patients is observed and the following results are obtained [6, 7].

Calculate precision, recall, F-measure (when $\beta = 1$ and 2), Jaccard coefficient, Rand coefficient, and Fowlkes–Mallows coefficient of Table 3.1,

Solution

From Table 3.1, we can find TP, FP, FN, and TN as follows:
   TP $= 43$, FP $= 7$, FN $= 31$, TN $= 19$.
   Using (3.11), we can calculate the precision as follows:

$$P_{\text{precision}} = \frac{TP}{TP + FP} = \frac{43}{43 + 7} = 0.86.$$

Using (3.12), we can calculate the recall as follows:

$$P_{\text{recall}} = \frac{TP}{TP + FN} = \frac{43}{43 + 31} = 0.58.$$

Using (3.13), we can calculate the F-measure as follows:

$$F_{\beta} = \frac{\left(\beta^2 + 1\right) P_{\text{precision}} P_{\text{recall}}}{\beta^2 P_{\text{precision}} + P_{\text{recall}}},$$

$$F_1 = \frac{2P_{\text{precision}} P_{\text{recall}}}{P_{\text{precision}} + P_{\text{recall}}} = \frac{2 \cdot 0.86 \cdot 0.58}{0.86 + 0.58} = 0.69,$$

$$F_2 = \frac{5P_{\text{precision}} P_{\text{recall}}}{4P_{\text{precision}} + P_{\text{recall}}} = \frac{5 \cdot 0.86 \cdot 0.58}{4 \cdot 0.86 + 0.58} = 0.62$$

Using (3.14), we can calculate the Jaccard coefficient as follows:

$$J = \frac{\text{TP}}{\text{TP} + \text{FP} + \text{FN}} = \frac{43}{43 + 7 + 31} = 0.53.$$

Using (3.15), we can calculate the Rand coefficient as follows:

$$R = \frac{\text{TP} + \text{TN}}{\text{TP} + \text{FP} + \text{TN} + \text{FN}} = \frac{43 + 19}{43 + 7 + 19 + 31} = 0.62.$$

Using (3.16), we can calculate the Fowlkes–Mallows coefficient as follows:

$$\text{FM} = \sqrt{\frac{\text{TP}}{\text{TP} + \text{FP}} \frac{\text{TP}}{\text{TP} + \text{FN}}} = \sqrt{\frac{43}{43 + 7} \frac{43}{43 + 31}} = 0.5. \qquad \blacksquare$$

Example 3.3 External Evaluation: Entropy Consider a die toss with the following probability distribution:

	$X = 1$	$X = 2$	$X = 3$	$X = 4$	$X = 5$	$X = 6$
$P(X)$	1/2	0	1/4	1/8	0	1/8

Calculate the entropy.

Solution

Using (3.17), we can calculate as follows:

$$H(P) = - \sum_{X \in \{1,2,3,4,5,6\}} P(X) \log P(X)$$

$$= - \left(\frac{1}{2} \log \frac{1}{2} + \frac{1}{4} \log \frac{1}{4} + \frac{1}{8} \log \frac{1}{8} + \frac{1}{8} \log \frac{1}{8} \right) = 1.75. \qquad \blacksquare$$

Example 3.4 External Evaluation: Joint Entropy Consider the following joint distribution:

	$j = 1$	$j = 2$
$i = 1$	$p_{ij} = 1/2$	$p_{ij} = 1/4$
$i = 2$	$p_{ij} = 1/4$	$p_{ij} = 0$

Calculate the entropy of the joint distribution.

Solution

Using (3.22), we can calculate as follows:

$$H(C, Y) = -\sum_i \sum_j p_{ij} \log p_{ij}$$

$$= -\left(\frac{1}{2} \log \frac{1}{2} + \frac{1}{4} \log \frac{1}{4} + \frac{1}{4} \log \frac{1}{4} \right) = 1.5. \qquad \blacksquare$$

Example 3.5 External Evaluation: Normalized Mutual Information Consider the clusters as shown in Fig. 3.3.

Calculate the NMI in Fig. 3.3.

Solution

From Fig. 3.3, we have the Table 3.2:

The total number of the elements is 14. Using (3.17), we can calculate the entropy of the cluster C as follows:

$$H(C) = -\sum_i p_{C_i} \log p_{C_i} = -\sum_i \frac{n_i}{n} \log \frac{n_i}{n}$$

$$= -\left(\frac{6}{14} \log \frac{6}{14} + \frac{8}{14} \log \frac{8}{14} \right) = 0.9852$$

Using (3.17), we can calculate the entropy of the class Y as follows:

$$H(Y) = -\sum_j p_{Y_j} \log p_{Y_j} = -\sum_j \frac{n_j}{n} \log \frac{n_j}{n}$$

$$= -\left(\frac{7}{14} \log \frac{7}{14} + \frac{7}{14} \log \frac{7}{14} \right) = 1.$$

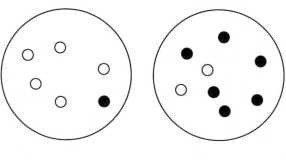

Cluster A   Cluster B

Fig. 3.3 Clusters

Table 3.2 Tables from Fig. 3.3

	n_i		n_j	n_{ij}	Class A	Class B
Cluster A	6	Class A	7	Cluster A	5	1
Cluster B	8	Class B	7	Cluster B	2	6

Using (3.23), we can calculate the mutual information as follows:

$$
I(C, Y) = \sum_i \sum_j p_{ij} \log \frac{p_{ij}}{p_{C_i} p_{Y_j}} = \sum_i \sum_j \frac{n_{ij}}{n} \log \frac{\frac{n_{ij}}{n}}{\left(\frac{n_i}{n}\right)\left(\frac{n_j}{n}\right)}
$$

$$
= \sum_i \sum_j \frac{n_{ij}}{n} \log \frac{n n_{ij}}{n_i n_j} = \left(\frac{5}{14} \log \frac{14 \cdot 5}{6 \cdot 7}\right) + \left(\frac{1}{14} \log \frac{14 \cdot 1}{6 \cdot 7}\right)
$$

$$
+ \left(\frac{2}{14} \log \frac{14 \cdot 2}{8 \cdot 7}\right) + \left(\frac{6}{14} \log \frac{14 \cdot 6}{8 \cdot 7}\right) = 0.2578.
$$

Using (3.26), we can calculate the normalized mutual information as follows:

$$
\text{NMI}(C, Y) = \frac{I(C, Y)}{\sqrt{H(C)H(Y)}} = 0.2598. \qquad \blacksquare
$$

Summary 3.1. Unsupervised Learning

1. In unsupervised learning, the data are not labelled and the model is developed to find any patterns on its own. It doesn't require the previous knowledge. It enables us to facilitate flexible and automated methods of machine learning in real time.
2. The applications of unsupervised learning can be categorised by clustering, association, and dimension reduction.
3. It is one of key issues in unsupervised learning to evaluate the quality of clustering results. Similarity measure is one important component of clustering analysis. The similarity is typically defined as the distance between data points.
4. Clustering algorithm evaluations are described in terms of clustering tendency, correct number of clusters, quality assessment without external data, comparison of external data, and comparison of different sets of clusters. The approaches are divided into expert analysis, relative analysis, internal data measure and external data measure.

3.2   Clustering Algorithms

The clustering problem can be simply formulated as follows: given a data set, we find clusters satisfying similar elements in the same cluster and dissimilar elements in different clusters. In the previous section, exclusive, hierarchical, overlapping, and probabilistic clustering are mentioned as types of clustering algorithms. It is possible to categorize clustering algorithms depending on different criteria. More simply, based on the properties of clusters, classifying them as hierarchical clustering and partitional clustering is widely used. Figure 3.4 illustrates the types of clustering algorithms [8].

The good quality of clustering can be achieved by maximizing interclusters distance (similarity or proximity) and minimizing intra-clusters distance. The performance of clustering depends on clustering algorithms, similarity metric, and the target applications. In addition, it can be measured by discovering some hidden patterns of a data set.

3.2.1   Hierarchical Clustering

Hierarchical clustering algorithms are grouping the data points with a sequence of partitions. They require only a measure of similarity or proximity among data points. The main difference between hierarchical clustering and partitional clustering is that hierarchical clustering does not assume a clustering number and only similarity or proximity is needed to calculate clustering algorithms. We can obtain any desired number of clusters by cutting a certain level of a tree-like diagram called a dendrogram. The tree structures from similarity of data points are built. The visualization

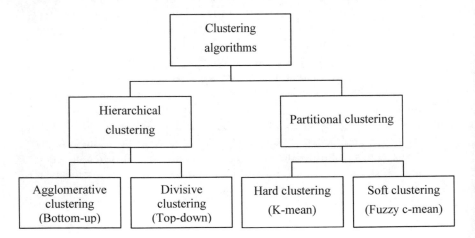

Fig. 3.4 Types of clustering algorithms

of the tree structures provides us with a useful summary of the data points. Two sub-categories are agglomerative algorithms and diverse algorithms. Agglomerative algorithms start grouping from each data point as a separate cluster and then merge into successively bigger groups. Diverse algorithms start from a whole data set and then divide into successively smaller groups. The key computation of hierarchical clustering is to measure the proximity between clusters. There are three cluster distance measures: single (or minimum) linkage, complete (or maximum) linkage, and average linkage. Figure 3.5 illustrates cluster distance measures. The single linkage measures the shortest distance between two elements in different clusters. It works well for non-elliptical shapes, but it is sensitive to noise and outliers. It is defined as follows:

$$d(A, B) = \min_{x_i \in A, x_j \in B} d(x_i, x_j). \tag{3.27}$$

The complete linkage measures the furthest distance between two elements in different clusters. It is less sensitive to noise and outliers than the single linkage. However, it tends to break large clusters and be biased towards globular clusters. It is defined as follows:

$$d(A, B) = \max_{x_i \in A, x_j \in B} d(x_i, x_j). \tag{3.28}$$

The average linkage measures the average distance between two elements in different clusters. This approach is a compromise between single linkage and complete linkage. It is defined as follows:

$$d(A, B) = \frac{1}{|A||B|} \sum_{x_i \in A, x_j \in B} d(x_i, x_j). \tag{3.29}$$

Example 3.5 Cluster Distance Measures Consider the two clusters $A = \{x_1, x_2, x_3\}$ and $B = \{x_4, x_5\}$ as shown in Fig. 3.6.

Calculate single linkage, complete linkage, and average linkage between two clusters.

Solution

From (3.27), we can calculate single linkage as follows:

$$d(A, B) = \min_{x_i \in A, x_j \in B} d(x_i, x_j)$$

$$= \min\{d(x_1, x_4), d(x_1, x_5), d(x_2, x_4), d(x_2, x_5), d(x_3, x_4), d(x_3, x_5)\}$$

$$= \min\{3, 4, 2, 3, 1, 2\} = 1.$$

From (3.28), we can calculate complete linkage as follows:

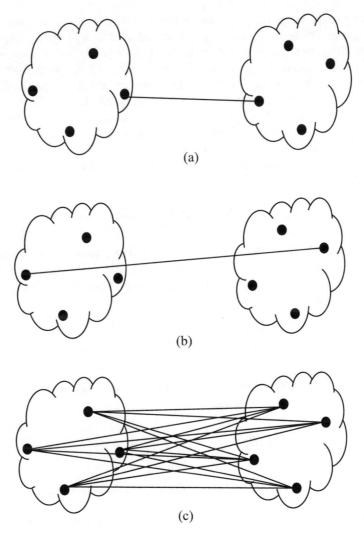

Fig. 3.5 Cluster distance measures. Cluster distance measures: single linkage (**a**), complete linkage (**b**), and average linkage (**c**)

Fig. 3.6 Clusters with one-dimensional feature.

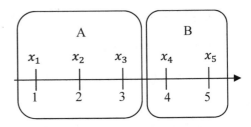

$$d(A, B) = \max_{x_i \in A, x_j \in B} d(x_i, x_j)$$
$$= \max\{d(x_1, x_4), d(x_1, x_5), d(x_2, x_4), d(x_2, x_5), d(x_3, x_4), d(x_3, x_5)\}$$
$$= \max\{3, 4, 2, 3, 1, 2\} = 4.$$

From (3.29), we can calculate average linkage as follows:

$$d(A, B) = \frac{1}{|A||B|} \sum_{x_i \in A, x_j \in B} d(x_i, x_j)$$

$$= \frac{3 + 4 + 2 + 3 + 1 + 2}{6} = 2.5.$$ ∎

The purpose of clustering algorithms is to find useful groups of elements that are in-line with data analysis. Thus, there are different types of clusters. The well-separated clusters are determined by the characteristics: The distance between two elements in different clusters is larger than the distance between any two elements within a cluster. The centre-based clusters are determined by the characteristics: each element is closer to the centre of its cluster than other clusters. When elements are denoted as graph including nodes and connections and the weight is given by the distance between nodes, a cluster is defined as a connected component. The graph-based clusters represent that two elements are connected only if they are within specific distance. The density-based clusters are defined as dense regions in the data space that is separated by regions of lower density. More generally, the conceptual clusters are defined as a group of elements sharing some property. Figure 3.7 illustrates different types of clusters.

Agglomerative clustering can be characterized as a greedy manner. The algorithm starts with individual elements as clusters, clusters are found by finding the maximum similarity (or minimum distance), and a pair of clusters is successively merged at each step. The proximity can be a distance between pairs of elements or clusters like Euclidian distance and single/complete/average linkage. The steps of the agglomerative clustering algorithm are summarized as follows:

Initial conditions: Compute the proximity (similarity or distance) matrix between the input elements, and each element is assigned as a cluster.

Repeat

- *Step 1*: Merge the closest two clusters
- *Step 2*: Update the proximity matrix.

Until: All elements are merged into a single cluster.

Divisive clustering starts with a single cluster including all elements and splits a cluster into two clusters with minimum similarity iteratively until each cluster contains single element. The steps of the divisive clustering algorithm are summarized as follows:

Initial conditions: Compute the proximity (similarity or distance) matrix between the input elements, and all elements are assigned in one cluster.

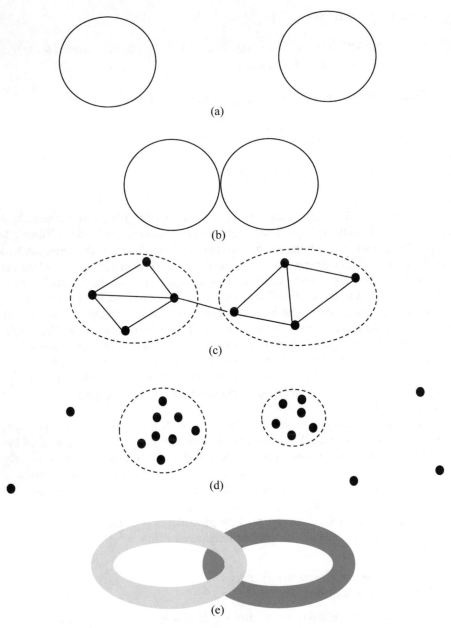

Fig. 3.7 Examples of different types of clusters: well-separated clusters (**a**), centre-based clusters (**b**), graph-based clusters (**c**), density-based clusters (**d**), and conceptual clusters (**e**)

Repeat

- *Step 1*: Select a cluster and split into two clusters.
- *Step 2*: Update the proximity matrix.

Until: Each cluster contains only one element.

Clusters of both algorithms can be obtained by cutting the dendrogram at the desired level. Agglomerative clustering is more widely used due to better search landscape and more flexibility. The limitation of hierarchical clustering can be summarized as follows: (1) an irreversible decision. If decision on merging or splitting clusters is made, it is not reversible. (2) High sensitivity to noise and outlier. (3) Difficulty for different size clusters and irregular shapes. (4) An objective function is not directly minimized or maximized. (5) Time complexity is high so that it is not suitable for large data. (6) Initial decision brings a big impact on the final results.

Example 3.6 Agglomerative Clustering algorithm Consider a data set with two-dimensional 8 elements as shown in Fig. 3.8.

Perform clustering using agglomerative clustering algorithm.

Solution

From Figure p3.4, we have the unlabelled 8 elements: (1.5, 2.5), (2, 4), (3, 3.5), (1, 2), (6, 5.5), (7, 7), (4, 6), (5, 5). Let each element be a cluster. The Euclidean distance is used as proximity. We compute the proximity matrix as shown in Table 3.3

From Table 3.3, we obtained the proximity among elements. In the first iteration, we are grouping the 1st element and the 4th element, the 5th element and the 8th

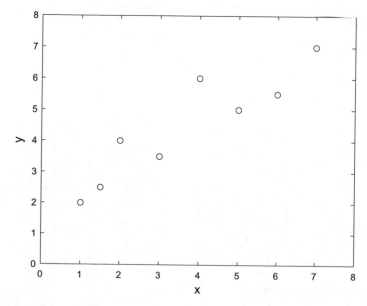

Fig. 3.8 Data set with 8 elements

Table 3.3 Proximity matrix

0	1.5811	1.8028	0.7071	5.4083	7.1063	4.3012	4.3012
1.5811	0	1.118	2.2361	4.272	5.831	2.8284	3.1623
1.8028	1.118	0	2.5	3.6056	5.3151	2.6926	2.5
0.7071	2.2361	2.5	0	6.1033	7.8102	5	5
5.4083	4.272	3.6056	6.1033	0	1.8028	2.0616	1.118
7.1063	5.831	5.3151	7.8102	1.8028	0	3.1623	2.8284
4.3012	2.8284	2.6926	5	2.0616	3.1623	0	1.4142
4.3012	3.1623	2.5	5	1.118	2.8284	1.4142	0

element, and the 2nd element and the 3rd element. In the second iteration, we are grouping the 7th element and a new cluster (including the 5th element and the 8th element). In the third iteration, we are grouping a new cluster (including the 1st element and the 4th element) and a new cluster (including the 2nd element and the 3rd element). In fourth iteration, we are grouping the 6th element and a new cluster (including the 5th element, the 8th element, and the 7th element). In the fifth iteration, we are grouping a new cluster (including the 1st element, the 4th element, the 2nd element, and the 3rd element) and a new cluster (including the 5th element, the 8th element, the 7th element, and the 6th element). Finally, we obtain the dendrogram as shown in Fig. 3.9. We can produce a set of nested clusters as shown in Fig. 3.10.

∎

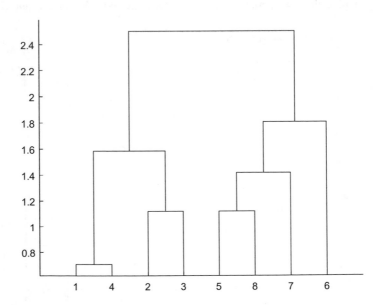

Fig. 3.9 Dendrogram of the data set

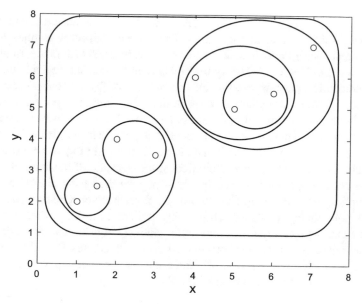

Fig. 3.10 Nested clusters for the data set

3.2.2 *Partitional Clustering*

Partitional clustering divides data points into multiple groups with certain similarity. There are two sub-categories: hard clustering and soft clustering. Hard clustering means that each data point belongs to a cluster or not. Soft clustering means that each data point belongs to a cluster with a certain degree like a certain probability. In the previous section, the evaluation of clustering was discussed. Selecting the best clustering algorithm is another challenge. Most of the clustering algorithms have the pros and cons and work well with certain environments and conditions. It is almost impossible to have a universal algorithm. The results of clustering algorithms are highly application dependent and often evaluated by an expert's own knowledge and insight subjectively. Thus, according to data structures, complexities, conditions, and applications, we should decide suitable similarity metrics, select algorithms, and set the parameter values. The most common practice of algorithm selection is to operate candidate algorithms with different similarity metrics and parameter setting, compare the results of algorithms, and then select the best one.

The approach of partitional clustering is to minimize the sum of squared distance between elements and the corresponding cluster centroids in each cluster and achieve optimal partition. K-means clustering algorithm as a hard partitioning clustering is one of the simplest unsupervised machine learning algorithms, where K represents the number of clusters. The approach of K-means clustering is to find a cluster in data by iteratively minimizing the measure between the cluster centre of the group and the given observation. It is easy to implement and provides us with fast computation if K is small. However, it is not easy to predict the number of clusters, and the initial conditions strongly affect to the results. Thus, the initialization is highly related to convergence speed and overall clustering performance. Candidate solutions to initialization problem are: (1) to try multiple initializations and find a better result, (2) use hierarchical clustering to find initial centroids, or (3) use post-processing like using results from other initialization algorithms. In addition, it provides us with a local optimum, but the global optimum is hard to achieve due to complexity. It is sensitive to outliers. Thus, some data elements that are further away from the centroids than others are sometimes removed to achieve efficient clustering. It is not suitable for finding clusters that are not hyperspherical shape. Despite of these disadvantages, K-means clustering algorithms are the most popular clustering algorithms due to their simplicity and efficiency. Sometimes, pre-processing of K-means clustering is used like normalizing the data or eliminating outliers. Post-processing of K-means clustering can (1) split/merge clusters with relatively high or low similarity or (2) eliminate some cluster representing outliers Fig. 3.11 illustrates the outliers problem.

Given n samples $x_1, \ldots, x_n \in \mathbb{R}^d$ and the cluster number k, we find k cluster centres $\mu_1, \ldots, \mu_k \in \mathbb{R}^d$ minimizing the following

$$f_k = \sum_i \min_j \|x_i - \mu_j\|^2, \quad i = 1, \ldots, n \text{ and } j = 1, \ldots, k. \tag{3.30}$$

As we can observe (3.30), the partition is achieved by minimizing the sum of the squared distance between all points to their closest cluster centre. Lloyd's algorithm [5] as a modified version of K-means clustering algorithm works well in practice. It is a simple iterative algorithm to find a local minimum. Here, (3.30) is modified as follows:

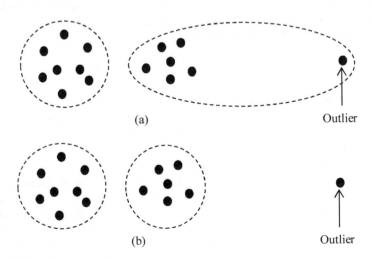

Fig. 3.11 Examples of outliers problem: including outlier (**a**) and removing outlier (**b**)

$$\arg\min_{C_j} \text{SSE} \quad \text{(3.31)}$$

where SSE is to sum of squared error

$$\text{SSE} = \sum_j \sum_i \|x_i - \mu_j\|^2, \quad i \in C_j \text{ and } j = 1, \ldots, k \quad \text{(3.32)}$$

and C_j is the sets of points to the closest cluster centre. K-means clustering algorithm is performed as the following steps:

Initial conditions: Set k cluster centres $\mu_1, \ldots, \mu_k \in \mathbb{R}^d$ randomly (or choose any k samples.).

Repeat.

– *Step 1*: Assign each x_i to its closest cluster centre as follows:

$$\arg\min_{C_j} \text{SSE}.$$

– *Step 2*: Compute new cluster centres μ_j as follows:

$$\mu_j = \frac{1}{|C_j|} \sum_{i \in C_j} x_i.$$

Until: It converges (when cluster centres do not change or objects in each cluster do not change).

As we can observe the K-means clustering algorithm, finding the optimum of the objective is NP-hard and K-means clustering is a heuristic approach. The algorithm terminates at a local optimum, but it is difficult to find a global optimum due to complexity. Kmeansclustering algorithm is a method for selecting the initial values. The initialization process of K-means $+ +$ clustering algorithm is helpful to get a better result than K-means clustering algorithm. The basic approach of K-means $+ +$ clustering algorithm is to put the initial centroids far away from each other. The progress of K-means $+ +$ clustering algorithm can be summarized as follows:

Initial conditions Choose an initial centre μ_1 uniformly at random among data points $\mathcal{X}$.

Repeat.

– *Step 1*: Assign each x_i to its closest cluster centre.
– *Step 2*: Choose the next centre μ_j, selecting $\mu_j = x' \in \mathcal{X}$ with the probability $D(x')^2 / \sum_{x' \in \mathcal{X}} D(x)^2$ where $D(x)$ represents the shortest distance between the data x_i and the closest centre μ_j.

Until: It converges.

As we can observe Figure 3.11, K-means clustering algorithm is sensitive to the outlier because a mean is significantly affected by outliers. K-medoids clustering algorithm also called as Partitioning Around Medoid (PAM) was proposed in 1987. This algorithm uses a medoid by picking actual data point to represent a cluster and assigns the remaining data points in the cluster. The basic idea is to minimize the sum of the dissimilarities between each data point and its corresponding medoid. The K-medoids clustering algorithm is more robust to outliers and noises than the K-means clustering algorithm. However, the complexity of the K-medoids clustering algorithm is higher than the K-means clustering algorithm. The absolute error criterion or the cost function is defined as follows:

$$E = \sum_{i=1}^{K} \sum_{p \in C_i} d(p, m_i) \tag{3.33}$$

where E is the sum of the absolute error for all data point p and m_i is the medoid of C_i. The K-medoids clustering algorithm is performed as the following steps:

Initial conditions: Set k data points randomly out of the n data points as the medoids.

– Associate each data point to the closest medoid by using a distance (or dissimilarity) metric.

Repeat: While the cost decreases:

– *Step 1*: For each medoid m and each data point p, consider the swap of m and p, associate each data point to the closest medoid, and compute the cost.
– *Step 2*: Perform the swap if decreasing the cost.

Until: It converges.

Example 3.7 K-means Clustering Algorithm Consider the following two dimensional data set:

Data set : (2, 2.5), (2, 3), (3, 4), (6, 6), (7, 8), (8, 8.5).

Find two clusters using K-means clustering algorithm.

Solution

As initial condition, we choose two cluster centres as follows:

$$\mu_1 = (2, 2.5) \text{and} \mu_2 = (6, 6).$$

We compute the Euclidean distance between the objects and cluster centres as follows:

$$d_1(x_i, \mu_1) = \sqrt{(x_{i,1} - \mu_{1,1})^2 + (x_{i,2} - \mu_{1,2})^2}$$

and $d_2(x_i, \mu_2) = \sqrt{(x_{i,1} - \mu_{2,1})^2 + (x_{i,2} - \mu_{2,2})^2}$.

and obtain the following distance matrix:

Based on the assigned group, we compute the new cluster centres as follows:
Table 3.4

$$\mu_1 = \left(\frac{2 + 2 + 3}{3}, \frac{2.5 + 3 + 4}{3} \right) = (2.33, 3.17),$$

$$\mu_2 = \left(\frac{6 + 7 + 8}{3}, \frac{6 + 8 + 8.5}{3} \right) = (7, 7.5).$$

We recompute the Euclidean distance and obtain the following distance matrix:
Table 3.5

In the second iteration, grouping is not changed and K-means clustering algorithm
terminates. We have two clusters as follows:

Cluster A : (2, 2.5), (2, 3), (3, 4),

Cluster B : (6, 6), (7, 8), (8, 8.5). ∎

Example 3.8 K-means and K-means + + Clustering Algorithm Consider a
randomly and uniformly distributed 300 IoT devices in a single cell. Perform clustering using both K-means clustering algorithm and K-means + + clustering
algorithm when we have 3 and 7 clusters in the single cell.

Table 3.4 First iteration

x_i	μ_1	μ_2	d_1	d_2	Closest cluster centre
(2,2.5)	(2,2.5)	(6,6)	0	5.3151	μ_1
(2,3)	(2,2.5)	(6,6)	0.5000	5.0000	μ_1
(3,4)	(2,2.5)	(6,6)	1.8028	3.6056	μ_1
(6,6)	(2,2.5)	(6,6)	5.3151	0	μ_2
(7,8)	(2,2.5)	(6,6)	7.4330	2.2361	μ_2
(8,8.5)	(2,2.5)	(6,6)	8.4853	3.2016	μ_2

Table 3.5 Second iteration

x_i	μ_1	μ_2	d_1	d_2	Closest cluster centre
(2,2.5)	(2.33,3.17)	(7,7.5)	0.7469	7.0711	μ_1
(2,3)	(2.33,3.17)	(7,7.5)	0.3712	6.7268	μ_1
(3,4)	(2.33,3.17)	(7,7.5)	1.0667	5.3151	μ_1
(6,6)	(2.33,3.17)	(7,7.5)	4.6344	1.8028	μ_2
(7,8)	(2.33,3.17)	(7,7.5)	6.7185	0.5000	μ_2
(8,8.5)	(2.33,3.17)	(7,7.5)	7.7819	1.4142	μ_2

Solution

Using computer simulation, we perform two algorithms under the given system model. In the first simulation, we consider 3 clusters in the single cell. Figure 3.12

Fig. 3.12 K-means clustering (**a**) and K-means + + clustering (**b**) with 3 clusters in a single cell

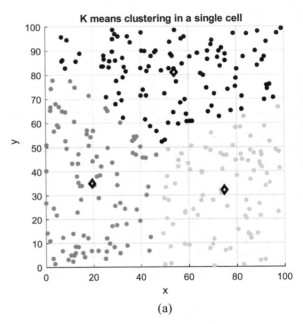

(a)

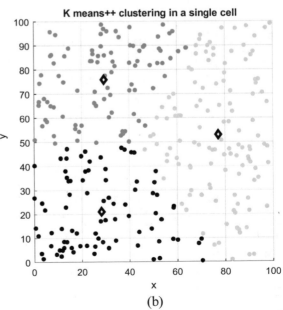

(b)

illustrates the clustering results. In the figure, diamonds and different coloured circles represent cluster centres and clustered IoT devices, respectively. K-means clustering and K-mean + + clustering performed 11 iterations and 8 iterations to find the cluster centres, respectively. Figure 3.13 illustrates how many IoT devices are grouped in each cluster. When using K-means clustering, each cluster includes 111, 99, 90 IoT devices. When using K-means + + clustering, each cluster includes 83, 92, and 125 IoT devices.

In the second simulation, we consider 7 clusters in the single cell. Figure 3.14 illustrates the clustering results. K-means clustering and K-mean + + clustering performed 18 iterations and 9 iterations to find the cluster centres, respectively. Figure 3.15 illustrates how many IoT devices are grouped in each cluster. When using K-means clustering, each cluster includes 53, 35, 32, 42, 48, 50, and 40 IoT devices. When using K-means + + clustering, each cluster includes 36, 47, 46, 30, 43, 51, and 47 IoT devices.

Example 3.9 K-Medoids Clustering Algorithm Consider a randomly distributed 100 and 300 IoT devices in a single cell. Perform clustering using K-medoids clustering algorithm.

Solution

Using computer simulation, we perform clustering. Firstly, we generate random 100 and 300 data of IoT devices location in two-dimensional domain as shown in Fig. 3.16

Secondly, we perform K-medoids clustering algorithm for the data and obtain the results. Figure 3.17 illustrates the clustering results. In the figure, diamonds and circles/asterisks represent cluster centres and clustered IoT devices, respectively.

In this simulation, we used Euclidean distance as dissimilarity metric. The total sums of distances are 89.6592 and 257.581 for 100 and 300 IoT devices, respectively. ∎

Soft clustering such as fuzzy C-means (FCM) clustering is a grouping method that allows each data point to belong to more than one cluster. FCM clustering is now widely used, and the data points on boundaries among clusters can be assigned with some weight. It is based on a fuzzy logic and fuzzy set theory. Fuzzy set theory enables each data point to belong to a set or group with a degree of membership between 0 and 1. We assume a set of n data points $X = \{x_1, x_2, \ldots, x_n\}$, and x_i is n-dimensional data point. We have K clusters C_j and a partition matrix $W = w_{ij} \in [0, 1]$, for $i = 1, \ldots, n$ and $j = 1$ where w_{ij} is a weight representing the degree of membership of an data point i in a cluster C_j. All w_{ij} satisfies the following:

$$\sum_{j=1}^{K} w_{ij} = 1 \tag{3.34}$$

and

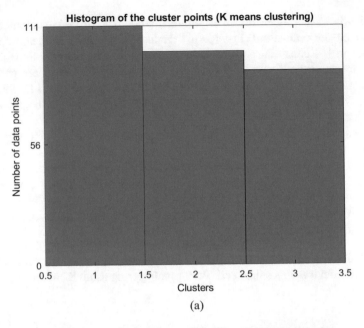

(a)

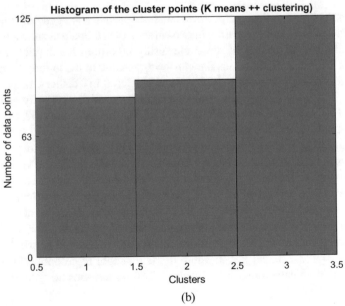

(b)

Fig. 3.13 Histogram of K-means clustering (each cluster includes 111, 99, 90 IoT devices) (**a**) and K-means + + clustering (each cluster includes 83, 92, and 125 IoT devices) (**b**) with 3 clusters in a single cell

Fig. 3.14 K-means clustering (**a**) and K-means ++ clustering (**b**) with 7 clusters in a single cell

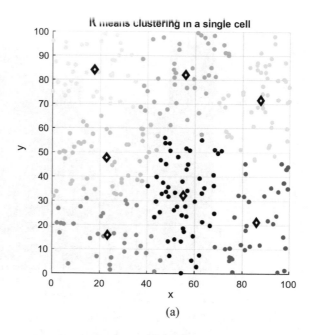

(a)

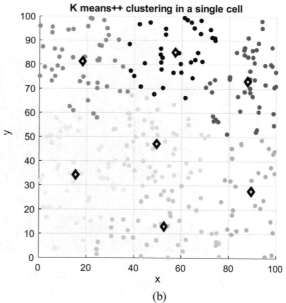

(b)

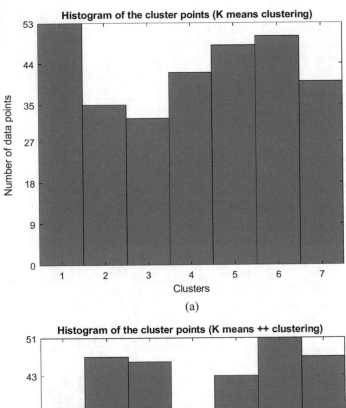

(a)

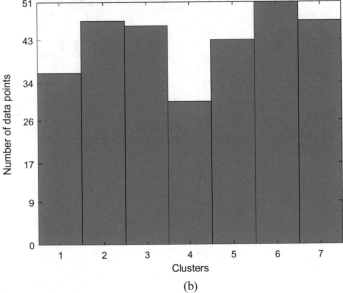

(b)

Fig. 3.15 Histogram of K-means clustering (each cluster includes 53, 35, 32, 42, 48, 50, and 40 IoT devices) (**a**) and K-means + + clustering (each cluster includes 36, 47, 46, 30, 43, 51, and 47 IoT devices) (**b**) with 7 clusters in a single cell ∎

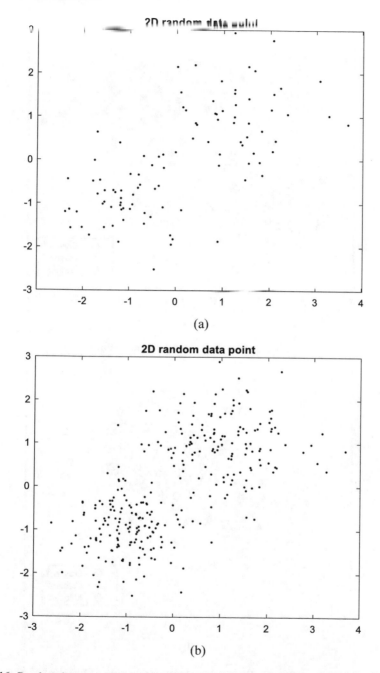

Fig. 3.16 Random data generation with 100 IoT devices (**a**) and 300 IoT devices (**b**)

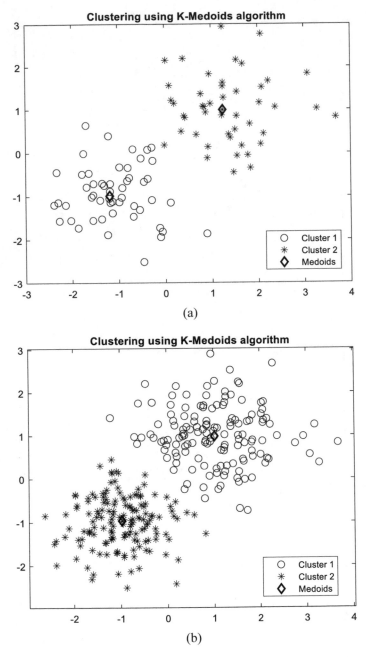

Fig. 3.17 Clustering results using K- medoids with 100 IoT devices (53 circles and 47 asterisks) (**a**) and 300 IoT devices (146 circles and 154 asterisks) (**b**)

$$0 < \sum_{i=1}^{n} w_{ij} < n. \tag{3.35}$$

Similar to K-means clustering, FCM is to minimize the sum of the squared error as follows:

$$\mathrm{SSE} = \sum_{j=1}^{K} \sum_{i=1}^{n} w_{ij}^{m} d(x_i, c_j) \tag{3.36}$$

where m is a parameter representing the level of cluster fuzziness and a real number greater than 1. If m is close to 1, it becomes a hard clustering like K-means clustering. m approaches infinity, and it leads to complete fuzziness. c_j is the corresponding centroid of a cluster C_j. It is defined as follows:

$$c_j = \frac{\sum_{i=1}^{n} w_{ij}^{m} x_i}{\sum_{i=1}^{n} w_{ij}^{m}}. \tag{3.37}$$

w_{ij} is defined as follows:

$$w_{ij} = \frac{\left(1/d(x_i, c_j)^2\right)^{\frac{1}{m-1}}}{\sum_{l=1}^{k} \left(1/d(x_i, c_l)^2\right)^{\frac{1}{m-1}}}. \tag{3.38}$$

As we can observe (3.38), w_{ij} should be a high value if x_i is close to c_j. The FCM clustering algorithm can be summarized as follows:

Initial conditions Choose a number of clusters K.

- Assign initial values to all w_{ij}.

Repeat

- *Step 1*: Compute the centroid of each cluster using the fuzzy partition.
- *Step 2*: Update the fuzzy partition.

Until: It converges.

Example 3.10 FCM Clustering Algorithm Consider a randomly distributed 50 and 100 IoT devices in a single cell. Perform clustering using FCM clustering algorithm and observe the effect of m when $K = 2$.

Solution

Using computer simulation, we perform FCM clustering algorithm. Firstly, we generate 50 and 100 IoT devices location data in two-dimensional domain. Secondly, we perform FCM clustering algorithm for the data and obtain the results. We set the level of cluster fuzziness $m = 1.1$, 1.5, 2, and 4. Figures 3.18 and 3.19 illustrate

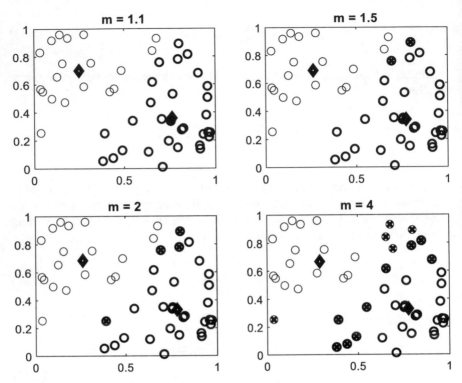

Fig. 3.18 Clustering results using FCM clustering when single cell includes 50 IoT devices

the clustering results. In the figures, diamonds, circles, and crosses represent cluster centres, clustered IoT devices, and overlapped clustering IoT devices, respectively.

As we can observe Figs. 3.18 and 3.19 as m is close to 1, it becomes a hard clustering like K-means clustering.   ∎

Summary 3.2. Clustering Algortihms

1. The clustering problem can be simply formulated as follows: Given a dataset, we find clusters satisfying similar elements in the same cluster and dissimilar elements in different clusters.
2. The purpose of clustering algorithms is to find useful groups of elements that is in-line with data analysis.
3. The good quality of clustering can be achieved by maximizing inter-clusters distance (similarity or proximity) and minimizing intra-clusters distance.

4. Hierarchical clustering algorithms are grouping the data point with a sequence of partitions. We can obtain any desired number of clusters by cutting a certain level of a tree like diagram called a dendrogram.

5. The approach of partitional clustering is to minimize the sum of squared distance between elements and the corresponding cluster centroids in each cluster and achieve optimal partition.

6. The approach of K-means clustering is to find a cluster in data by iteratively minimizing the measure between the cluster centre of the group and the given observation.

7. Soft clustering such as fuzzy c-mean (FCM) clustering is a grouping method that allows each data point to belong to more than one cluster. Fuzzy set theory enables each data point to belong to a set or group with a degree of membership between 0 and 1

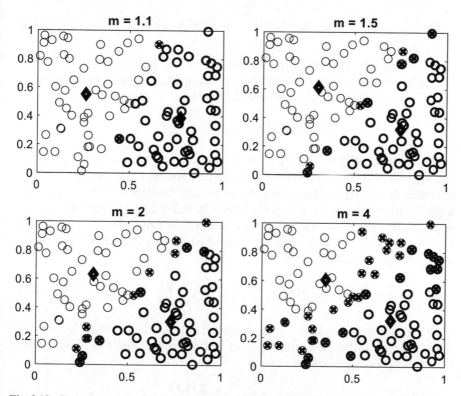

Fig. 3.19 Clustering results using FCM clustering when single cell includes 100 IoT devices

3.3   Association Rule Mining

As we discussed in the previous section, clustering is a method of grouping a set of data elements in that data in the same group is more similar to each other than other data in other groups, whereas association rule is a method for discovering relationships between the attributes of large data. Finding inherent patterns in data is very useful tool for many applications. This method provides us with a new information such as how frequently data transaction happens. Association analysis is widely used in many applications such as purchase patterns analysis and decision-making for efficient marketing. The basic concept of association rule was introduced to discover the regularities among product items in supermarkets [6]. The problem of association rule mining is defined as follows [6]: Let $I = \{i_1, i_2, \ldots, i_m\}$ be a set of items. Let $T = \{t_1, t_2, \ldots, t_n\}$ be a set of transactions (or database). Each transaction t_i in T has an unique transaction ID and contains a set of items in I such that $t_i \subseteq T$. An association rule as an implication of the form is defined as follows:

$$X \rightarrow Y, \text{ where } X, Y \subseteq I \text{ and } X \cap Y = \emptyset \tag{3.39}$$

where X and Y are a set of items or attributes. X and Y are called left-hand side (LHS) or antecedent and right-hand side (RHS) or consequent, respectively. They are also called itemsets. One example of an association rule is as follows:

$$\{\text{milk, bread, apple}\} \rightarrow \{\text{egg}\} \tag{3.40}$$

We can interpret (3.40) as follows: if one person buys milk, bread, and apple, then egg will be highly likely brought too. The association rules are presenting association or correlation among itemsets. In order to select rules from the database, two metrics support and confidence are widely used. Support indicates how frequently itemsets appear in the transaction set T. One important property of an itemset is its support count representing frequency of occurrence of an itemset. The support count $\sigma(X)$ of X in T can be expressed as follows:

$$\sigma(X) = |\{t_i | X \subseteq t_i, t_i \in T\}| \tag{3.41}$$

where $|\ |$ represents the number of elements in a set. The support of a rule $X \rightarrow Y$ is the transaction percentage containing $X \cup Y$ in T. Let N be the number of transactions in T. It is defined as follows:

$$s(X \rightarrow Y) = \frac{\sigma(X \cup Y)}{N}. \tag{3.42}$$

Confidence indicates how often items in Y appear in transaction T containing X. The confidence of a rule $X \rightarrow Y$ is the transaction percentage in T that contains X which also contains Y. It is defined as follows:

$$c(X \rightarrow Y) = \frac{\sigma(X \cup Y)}{\sigma(X)}. \tag{3.43}$$

Support is a useful measure. If it is very low, the rule may occur by change. In addition, a support rule covering few transactions may be uninteresting because it means people seldom buy them together and it may not be profitable to promote. Thus, support is used to remove uninteresting rules. On the other hand, confidence determines the predictability of the rule. If it is very low, we cannot reliably predict Y from X. Confidence provides us with an estimate of the conditional probability of Y given X. Another important concept is lift. The lift represents how much attributes happen independently. It is defined as follows:

$$l(X \rightarrow Y) = \frac{s(X \cup Y)}{s(X)s(Y)}. \tag{3.44}$$

If the lift >1, two occurrences are dependent on each other. If the lift $= 1$, they are independent and no rule can be created between them. Now, we can formulate association rule mining problem as follows: given a transaction set T, the problem of association rule mining is to find all association rules having (1) support $\geq$ the user-specified minimum support (minsup threshold) and (2) confidence $\geq$ the user-specified minimum confidence (minconf threshold).

Example 3.11 Association Rule Mining: Support Count, Support, and Confidence
Consider a set of 7 transactions as shown in Table 3.6. Each transaction t_i is a set of purchased items in a store by a customer.

Calculate   support   count $\sigma(\{Chicken, Bread\})$,   support $s(\{Chicken, Bread\} \rightarrow \{Milk\})$, and confidence $c(\{Chicken, Bread\} \rightarrow \{Milk\})$.

Solution

From (3.41), (3.42), and (3.43), the support count is computed as follows:

$$\sigma(\{Chicken, Bread\}) = 3$$

Table 3.6 Transaction data

Transaction ID	Items
t_1	Sugar, Chicken, Milk
t_2	Beef, Egg
t_3	Egg, Sugar
t_4	Beef, Chicken, Egg
t_5	Beef, Chicken, Bread, Egg, Milk
t_6	Chicken, Bread, Milk
t_7	Chicken, Sugar, Bread

because there are only three transactions containing all two items in Table 3.6. The total number of transaction is 7. The support is computed as follows:

$$s(\{\text{Chicken, Bread}\} \rightarrow \{\text{Milk}\}) = \frac{\sigma(\{\text{Chikcken, Bread, Milk}\})}{N}$$

$$= \frac{2}{7} = 0.286.$$

The confidence is computed as follows:

$$c(\{\text{Chicken, Bread}\} \rightarrow \{\text{Milk}\}) = \frac{\sigma(\{\text{Chikcken, Bread, Milk}\})}{\sigma(\{\text{Chikcken, Bread}\})}$$

$$= \frac{2}{3} = 0.667.$$ ∎

Example 3.12 Association Rule Mining : minsup and minconf Threshold Consider a set of 7 transactions as shown in Table 3.6. Given the user-specified minimum support = 30% and minimum confidence = 70%, check whether or not the association rule {Chicken, Bread} → {Milk} is valid.

Solution

From the result of Example 3.11, $s(\{\text{Chicken, Bread}\} \rightarrow \{\text{Milk}\})$ is 28% <30% and $c(\{\text{Chicken, Bread}\} \rightarrow \{\text{Milk}\})$ is 66% <70%. Thus, the association rule {Chicken, Bread} → {Milk} is not valid.

∎

Association rule generation is normally composed of two steps: (1) frequency itemset generation and (2) rule generation. It is required to satisfy both minsup and minconf at the same time. The first step is to generate all itemsets whose supports are no less than minsup. These itemsets are called frequent itemsets. It requires an intensive computation. The second step is to generate confidence rules above minconf. The main purpose of this step is to extract all high confidence rules from frequent itemsets. These rules are called strong rules. It is not simple task to find all frequent itemsets from a database because we should search all possible combinations. The number of possible itemsets is 2^n-1 where n is the number of unique items. In order to search efficiently, we can use the downward-closure property of support that is also called anti-monotonicity. For a frequent itemset, all its subsets are frequent as well. For example if {Chicken, Bread} is frequent, both {Chicken} and {Bread} are frequent. For an infrequent itemset, all its supersets are infrequent. For example if {Chicken} is infrequent, {Chicken, Bread}, {Chicken, Milk}, and {Chicken, Bread, Milk} are infrequent. This property is useful for sorting candidate itemsets. In order to find frequent itemsets, a brute force approach is to list all possible association rules, calculate the support and confidence for every possible rules, and prune rules failing the minsup and minconf thresholds. Each itemset can be a candidate frequent itemset. We determine the support count of each candidate by searching the database. We need to match each transaction against every candidate as shown in Fig. 3.20

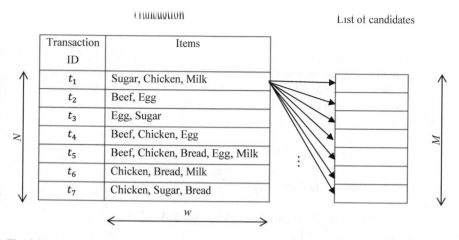

Fig. 3.20 Examples of the support count of candidate itemsets

This approach requires a high complexity because we need $O(NMw)$ comparison where N, M, and w are the number of transactions, the number of candidate itemsets, and the maximum transactions width, respectively. The total number of possible association rules containing d items is

$$R = \sum_{k=1}^{d-1} \left(\binom{d}{k} \sum_{j-1}^{d-k} \binom{d-k}{j} \right) = 3^d - 2^{d+1} + 1. \qquad (3.45)$$

If we have 5 items ($d = 5$) or 6 items ($d = 6$), this approach is required to compute the support and confidence for 178 or 602 association rules. In order to reduce the complexity, we need to prune rules before computation. There are several ways to reduce the computational complexity: (1) reduce the number of candidate itemsets (M), (2) reduce the number of transactions (N), and (3) reduce the number of comparisons (NM). The first way is Apriori algorithm. It is effective to remove some candidate itemsets without computing their supports. The second way is to reduce N size as the size of items increases. The third way is to use efficient data structure to store candidates or transaction instead of all comparisons and reduce the number of comparisons. Apriori algorithm is the best-known association rule algorithm. Apriori algorithm is to count transactions and generate candidate itemsets by exploiting downward-closure property of support. It is based on Apriori principle [7]: If an itemset is frequent, then all of its subsets must also be frequent. Apriori algorithms are basically a bottom-up approach and a level-wise and breadth-first algorithm. The frequent sets are extended one item at a time and then candidates groups are tested against the data. The algorithm ends until there are no further extensions. The algorithm is composed of two steps: the first step is to find all sets of items having support values greater than the support threshold. These items are called large itemsets. The second step is to generate the desired rules using the large itemsets. Apriori algorithm pseudocode is given for a transaction database T and a support threshold $\in$ as follows:

> ***Procedure*** Apriori (T, ϵ)
> C_k: Candidate itemset of size k
> L_k: Frequent itemset of size k
> $L_1 = \{$frequence items$\}$;
> ***for*** $(k = 2; L_1! = \emptyset; k + +)$ do
> ***begin***
> $C_k =$ candidates generated from L_{k-1}
> // New candidates, Eliminating candidates that are infrequent
> ***for*** each transaction t in database go
> ***begin***
> increment the count of all candidates in C_k containing in t
> $L_k =$ candidates in C_k with ϵ
> ***end***
> ***end***
> ***end***
> ***end***
> ***return*** $\cup_k L_k$

As we can observe the above pseudocode, there are two sets: C_k: candidate itemset of size k and L_k: frequent itemset of size k. Apriori algorithm generates candidate itemsets of length k from itemset of length k-1 and then prunes candidates having an infrequent sub-pattern.

Example 3.13 Apriori Algorithm Consider a set of 5 transactions as shown in Table 3.7. Each transaction t_i is a set of purchased items in a store by a customer.
    Find frequent itemsets as setting minimum support $= 2$ using Apriori algorithm.

Solution

In order to count candidates, we need to scan the database of transactions to determine the support of each candidate. Thus, the first step of Apriori algorithm is to count up the supports (also called the frequencies) of each item as follows:

Table 3.7 Transaction data

Transaction ID	Items
t_1	Milk, Chicken, Sugar
t_2	Egg, Chicken, Bread
t_3	Milk, Egg, Chicken, Bread
t_4	Egg, Bread
t_5	Beef

Items	Support
Milk	2
Egg	3
Chicken	3
Bread	3
Sugar	1
Beef	1

We don't need to generate candidates involving sugar and beef. The next step is to discard the items with minimum support less than 2 as follows:

Items	Support
Milk	2
Egg	3
Chicken	3
Bread	3

In this way, Apriori algorithm prunes the tree of all possible itemsets. The next step is to select only 2 itemsets which are frequent and generate a list of all 2 itemsets of the frequent items as follows:

Items	Support
Milk, Egg	1
Milk, Chicken	2
Milk, Bread	1
Egg, Chicken	2
Egg, Bread	3
Chicken, Bread	2

The next step is to count up the frequencies and discard the items with minimum support less than 2 as follows:

Items	Support
Milk, Chicken	2
Egg, Chicken	2
Egg, Bread	3
Chicken, Bread	2

The next step is to generate a list of all 3 itemsets of the frequent items as follows:

Items	Support
Egg, Chicken, Bread	2

Now, the algorithm ends because we have only one frequent itemset.   ∎

As we can observe Example 3.13, the procedure of the algorithm is focused on reducing the computational complexity. There are several parameters affecting computational complexity. Firstly, the minimum support is highly related to select the frequent itemsets. If we have a lower minimum support, more frequent itemsets are selected and complexity increases. Secondly, the dimensionality or the number of items of data affects to store support of each itemset. If the dimensionality increases, computational complexity increases as well. Thirdly, size of database affects to increase with the number of transactions. Fourthly, the transaction width may increase length of frequent itemsets. Thus, the main disadvantages of Apriori algorithm are summarized as follows: (1) a large number of candidates are found, and this is computationally expensive. (2) It requires to scan an entire database. On the other hand, the main advantages are summarized as follows: (1) it can be used for large itemsets. (2) It is easy to implement.

Association rule mining sometimes produces too many association rules. Many of them might be redundant. Interestingness measures play an important role in data mining in order to prune or rank the rules or patterns. The basic idea is to select or rank rules or patterns in terms of user's interest. Figure 3.21 illustrates the role of interestingness measures in the general data mining process.

As we can observe Figure 3.21, the interestingness measures can be used for pruning patterns in the data mining. Thus, it enables us to improve the efficiency. For example the minimum support threshold can be defined for pruning uninteresting patterns. In addition, it can be used for ranking patterns or selecting specific patterns in terms of user's interests or statistical methods. The statistical methods as objective interestingness measure how exceptional the patterns are. It can be derived from

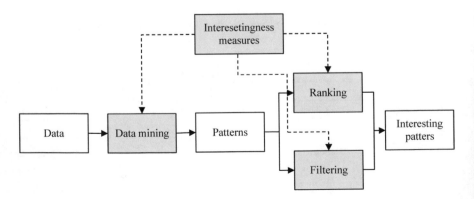

Fig. 3.21 Interestingness measure in the data mining process

the data and applied of specific applications independently. The user's interests as subjective interestingness are based on the expert's knowledge.

Summary 3.3. Association Rule Mining

1. Association rule is a method for discovering relationships between the attributes of large data. This method provides us with a new information such as how frequently data transaction happens.
2. Association analysis is widely used in many applications such as purchase patterns analysis and decision making for efficient marketing.
3. Association rule generation is normally composed of two steps: (1) frequency itemset generation and (2) rule generation.
4. Apriori algorithm is the best known association rule algorithm. Apriori algorithm is to count transactions and generate candidate itemsets by exploiting downward-closure property of support. The algorithm is composed of two steps: The first step is to find all sets of items having support values greater than the support threshold. These items is called large itemsets. The second step is to generate the desired rules using the large itemsets.
5. Association rule mining sometimes produces too many association rules. Many of them might be redundant. Interestingness measures play an important role in data mining in order to prune or rank the rules or patterns. The basic idea is to select or rank rules or patterns in terms of user's interest.

3.4 Dimensionality Reduction

High-dimensional data sets are often found in many problems of machine learning, but all the measured variables are not important for performing data analysis or understanding phenomena. Most of the machine learning or data mining techniques on high dimension is not so much effective. The intrinsic dimension may be small. If we transform high-dimensional data sets to low-dimensional data sets, it would be helpful for finding a pattern or information from the data sets. Dimensionality reduction techniques can be categorized into feature selection and feature extraction/reduction. The feature selection is to select an optimal subset of features based on some criteria or objective function. The objective is to reduce dimensionality, remove noise, or improve the performance of data mining. The feature extraction is to use the mapping of high-dimensional data onto a low-dimensional space. The criterion of the feature extract can be different in terms of the problem formulation. For example it can minimize data loss for unsupervised setting.

Given the p-dimensional random variable $\mathbf{x} = \left[x_1, x_2, \ldots, x_p\right]^T$, we find a lower dimensional representation $\mathbf{u} = [u_1, u_2, \ldots, u_k]^T$ where $k \leq p$. Variable, feature, and attribute can be used interchangeably. Assume that there are n observations with a realization of the p-dimensional random variable $\mathbf{x}$ as follows:

$$\mathbf{X} = \{x_{ij} | 1 \leq i \leq p, 1 \leq j \leq n\}. \tag{3.46}$$

Its mean and covariance are as follows:

$$E(\mathbf{x}) = \boldsymbol{\mu} = \left[\mu_1, \mu_2, \ldots, \mu_p\right]^T \tag{3.47}$$

and

$$E\left((\mathbf{x} - \boldsymbol{\mu})(\mathbf{x} - \boldsymbol{\mu})^T\right) = \boldsymbol{\Sigma}_{p \times p} \tag{3.48}$$

respectively. We can standardize x_{ij} by $\left(x_{ij} - \widehat{\mu_i}\right)/\widehat{\sigma_i}$ where

$$\widehat{\mu_i} = \frac{\sum_{j=1}^{n} x_{ij}}{n} \tag{3.49}$$

and

$$\widehat{\sigma_i} = \frac{\sum_{j=1}^{n} \left(x_{ij} - \widehat{\mu_i}\right)^2}{n}. \tag{3.50}$$

Now, we can express a lower dimensional variable $\mathbf{u}$ as a linear combination of the the p-dimensional random variable $\mathbf{x}$ as follows:

$$u_i = w_{i1}x_1 + w_{i2}x_2 + \ldots + w_{ip}x_p, \quad i = 1, 2, \ldots, k. \tag{3.51}$$

Its matrix form is as follows:

$$\mathbf{u} = \mathbf{W}\mathbf{x} \tag{3.52}$$

where $\mathbf{W}$ is a $k \times p$ linear transformation weight matrix. We can express it in a different form as follows:

$$\mathbf{x} = \mathbf{V}\mathbf{u} \tag{3.53}$$

where $\mathbf{V}$ is a $p \times k$ linear transformation weight matrix. $\mathbf{u}$ is called the hidden variable. In addition, we can express it in terms of $n \times p$ observation $\mathbf{X}$ as follows:

$$u_{ij} = w_{i1}x_{1j} + w_{i2}x_{2j} + \ldots + w_{ip}x_{pj}, \quad i = 1, 2, \ldots, k \text{ and } j = 1, 2, \ldots, n$$
$$(3.54)$$

where j denotes the jth realization. Its matrix forms are as follows:

$$\mathbf{U}_{k \times n} = \mathbf{W}_{k \times p} \mathbf{X}_{p \times n} \tag{3.55}$$

and

$$\mathbf{X}_{p \times n} = \mathbf{V}_{p \times k} \mathbf{U}_{k \times n}. \tag{3.56}$$

As we can observe above equations, the transformed variable can be expressed by linear combination of the original variable. This linear feature extraction technique is simple and easy to implement. Principal components analysis (PCA) is a popular technique for dimensionality reduction. The basic idea is to reduce the dimensionality of a data set by finding a new set of variables with a lower dimension than the original set of variables. The linear sub-space can be specified by k orthogonal vectors forming a new reference (or coordinate) system. The new variables are called principal components and uncorrelated each other. The principal components are linear transformation of the original variables. The objective of this algorithm is to minimize the projection error or maximize the variance of the projected data. Both approaches are theoretically equivalent. The outputs of PCA are the principal components, that is less than or equal to the number of original variables. The applications of PCA are a dimensionality reduction or data compression. In addition, PCA can be regarded as a symmetric regression because the hyperplane spanned by the principal components is the regression hyperplane minimizing the orthogonal distances to a data set. We call it principal component regression (PCR). In PCR, the principal components are used as regressors. It creates the principal components without the response variables. There is no guarantee that the outcomes are interpretable. Let the first principal component u_1 be a linear combination with the largest variance. In matrix form, we have $\mathbf{u}_1 = \mathbf{x}^T \mathbf{w}_1$ where the first weight vector $\mathbf{w}_1 = \begin{bmatrix} w_{11}, w_{12}, \ldots, w_{1p} \end{bmatrix}^T$ should satisfy

$$\mathbf{w}_1 = \arg\max_{\mathbf{w}=1} \operatorname{var}(\mathbf{x}^T \mathbf{w}) \tag{3.57}$$

in order to maximize the variance. The second principal component u_2 is a linear combination with the second largest variance and orthogonal to the first principal component. We continue to find the remaining principal components in this way. The first several principal components are key components, and the remaining can be ignored with minimal loss of information. For example when we have two-dimensional data set as shown in Figure 3.22a, the first principal component u_1 is a minimum distance fit to a line in the original space, and the second principal component u_2 is a minimum distance fit to a line in the perpendicular plane to the first

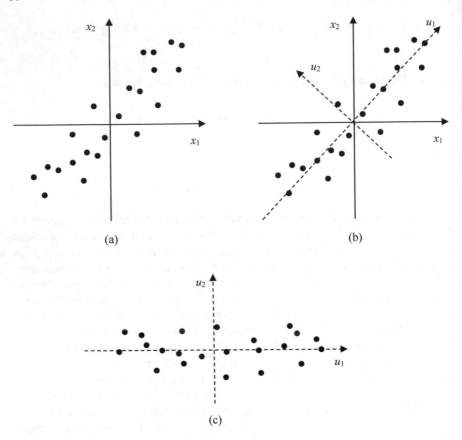

Fig. 3.22 Example of geometric expression of data set (**a**), principal components (**b**), and original data projected into the principal component space (**c**)

principal component as shown in Fig. 3.22b. We can have a new reference system as the projected data set as shown in Fig. 3.22c. The principal components are a series of linear least squares fits to a data set, and each component is orthogonal to the previous component. As we can observe Fig. 3.22b, the variance is greater along axis u_1 than axis u_2. The spatial relationships are unchanged, it has rotated the data points, and the principal components are uncorrelated. The principal components u_1 and u_2 are regarded as a size measure and a measure of shape, respectively. For data set with many variables, the variance of axes may be greater, whereas the other may be smaller Fig. 3.22.

PCA is traditionally performed on a square symmetric matrix such as sums of squares and cross-products (SSCP) matrix, covariance matrix, or correlation matrix. Basically, the results of SSCP and covariance matrix are same because the difference of them is a global scaling factor. A correlation matrix can be used if the variances are on different scales. Let us consider a standardized data with covariance matrix

$$\mathbf{\Sigma}_{p \times p} = \frac{1}{n} \mathbf{X}\mathbf{X}^T \tag{3.58}$$

and the matrix $\mathbf{\Sigma}$ is symmetric semidefinite positive and admits a spectral decomposition as follows:

$$\mathbf{\Sigma} = \mathbf{Q}\mathbf{\Lambda}\mathbf{Q}^T \tag{3.59}$$

where $\mathbf{\Lambda} = \mathrm{diag}(\lambda_1, \ldots, \lambda_p)$ is the diagonal matrix of the eigenvalues $\lambda_1 \leq \ldots \leq \lambda_p$ and $\mathbf{Q} = [\mathbf{q}_1, \mathbf{q}_2, \ldots, \mathbf{q}_p]$ is a $p \times p$ orthogonal matrix containing the normalized eigenvector $\mathbf{q}_i$ of $\mathbf{\Sigma}$ associated with eigenvalue λ_i. The eigenvectors and eigenvalues are key elements of PCA. The eigenvectors decide the directions of a new reference system. The eigenvalues determine their magnitude representing the variance of the data set along the new reference system. The principal component transformation results in a reference system in that the data set has zero mean and the variables are uncorrelated. The principal components are given by the p rows of $\mathbf{U}$ matrix [8]. We have the following matrix:

$$\mathbf{U} = \mathbf{Q}^T\mathbf{X} \tag{3.60}$$

and the transformation weight matrix is given by the orthogonal matrix $\mathbf{Q}^T$. In a new reference system, we can discard the variables with small variable by projecting on the sub-space spanned by the first k principal components. When we have the dimensionality reduction linear map $\mathbf{x} \in \mathbb{R}^p \to \mathbf{u} \in \mathbb{R}^k$, the key properties of PCA can be summarized as follows:

Property 3.4 Maximal Variance in the Projected Space

$$\max_{\mathbf{A}^T\mathbf{A}=\mathbf{I}} \left(\mathrm{tr}\left(\frac{1}{n} \sum_{j=1}^{n} \mathbf{u}\mathbf{u}^T \right) \right) = \sum_{i=1}^{k} \lambda_i \tag{3.61}$$

where tr() is a trace matrix, $\mathbf{u} = \mathbf{A}^T(\mathbf{x} - \mathbf{\mu})$, $\mathbf{A} = \mathbf{Q}_k$, and λ_i is the first k eigenvalues of the covariance matrix.

Property 3.5 Least Squared Sum of Errors

$$\min_{\mathbf{A}} \left(\sum_{j=1}^{n} \left\| \mathbf{x} - \mathbf{x}^* \right\|^2 \right) = n \sum_{i=1}^{k} \lambda_i \tag{3.62}$$

where $\mathbf{x} = \mathbf{A}\mathbf{A}^T(\mathbf{x} - \mathbf{\mu})$, $\mathbf{A} = \mathbf{Q}_k$, and λ_i is the first k eigenvalues of the covariance matrix.

Property 3.6 Maximal Mutual Information

$$\min_{A}(I(\mathbf{x}; \mathbf{u})) = \frac{1}{2} \ln \left(\prod_{i=1}^{k} 2\pi e \lambda_i \right) \tag{3.63}$$

where $\mathbf{u} = \mathbf{A}^T(\mathbf{x} - \mathbf{\mu})$, $\mathbf{A} = \mathbf{Q}_k$, and λ_i is the first k eigenvalues of the covariance matrix.

PCA is to find orthogonal projections of the data set containing the highest variance. This approach enables PCA to find directions of the data set if some of the variables in the data set are linearly correlated. However, if the data set is not linearly correlated, PCA doesn't work well. PCA is based on orthogonal transformations. Principal components are orthogonal each other. Thus, it can restrict to find projections with the highest variance. Principal components may not be readable or interpretable as original features. Another limitation is a high computational complexity. The covariance matrix requires an intensive computation for data set with a high number of dimensions. Data standardization is required in order to implement PCA. Outliers in the data set can affect the performance of PCA analysis. The steps of PCA can be summarized as follows:

Step 1:   Standardize a data set.
Step 2:   Calculate a square symmetric matrix (SSCP matrix, covariance matrix, or correlation matrix) for the features in the data set.
Step 3:   Calculate the eigenvalues and eigenvectors for the matrix to identify the principal components.
Step 4:   Sort eigenvalues and corresponding eigenvectors, pick k eigenvalues, and form a matrix of eigenvectors.
Step 5:   Transform the data set along the new reference system.

Example 3.14 Principal Component Analysis Consider the following data set

$$x = \begin{bmatrix} -5 & 7 & 0 & -1 & 1 & 0 & 5 & 0 & -4 & 5 \\ 9 & -1 & -2 & -2 & 1 & -1 & -3 & 1 & 0 & -3 \\ 0 & 8 & -5 & 1 & -2 & 0 & -2 & 0 & -3 & -1 \\ 3 & 0 & 3 & 0 & -1 & 5 & -3 & -2 & 6 & 0 \\ 4 & 2 & -9 & 1 & -3 & -6 & 0 & -2 & 1 & -3 \end{bmatrix}$$

and perform the dimensionality reduction using PCA.

Solution

In the matrix $\mathbf{X}$, each row and each column are one variable and one observation, respectively. Using computer simulation, the eigenvalues of the covariance matrix are found as follows:

[Ი7 Ი1Ი4 1Ი.Ი0Ი) Ი.Ი0Ი5 4.Ი0Ი0 1.Ი16Ი].

We can find the first two principal components explaining 80% of the variance. The corresponding two eigenvectors of the covariance matrix are

$$\mathbf{u}_1 = [0.695 \ -0.5538 \ 0.171 \ -0.3383 \ -0.2582],$$

$$\mathbf{u}_2 = [0.1999 \ 0.2443 \ 0.5867 \ -0.2332 \ 0.7084].$$

Each observation by projecting on the reference system is given by two eigenvectors. The original variables can be expressed by using the values in the eigenvectors Fig. 3.23 illustrates the variables and observations.

Summary 3.4. Dimensionality Reduction

1. Most of machine learning or data mining techniques on high dimension is not so much effective. The intrinsic dimension may be small.
2. Dimensionality reduction techniques can be categorized into feature selection and feature extraction/reduction. The feature selection is to select an optimal subset of features based on some. The feature extraction is to use the mapping of high dimensional data onto a low dimensional space.

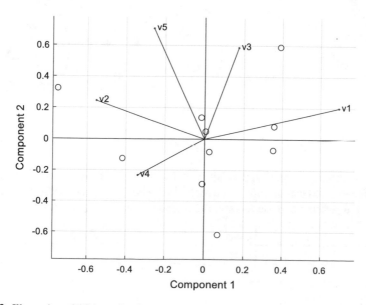

Fig. 3.23 Illustration of PCA application

3.   The basic idea of PCA is to reduce the dimensionality of a dataset by finding a new set of variables with a lower dimension than the original set of variables. The linear subspace can be specified by k orthogonal vectors forming a new coordinate system. The new variables are called principal components and uncorrelated each other.

3.5   Problems

3.1   Describe difference between supervised learning and unsupervised learning.

3.2   Describe applications using unsupervised learning and explain why it is useful.

3.3   Explain why unsupervised learning is used for clustering, association, and dimension reduction.

3.4   Describe how to evaluate the results of unsupervised learning.

3.5   Describe the pros and cons of the evaluation methods: expert analysis, relative analysis, internal data measure, and external data measure.

3.6   Describe meaning and difference about the contingency matrix: the number of true positive (TP), the number of true negatives (TN), the number of false positives (FP), and the number of false negatives (FN).

3.7   Consider the following clusters:

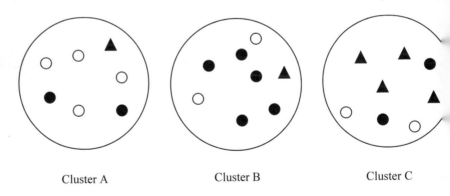

Cluster A   Cluster B   Cluster C

Calculate the purity, precision, recall, and F-measure.

3.8   We have 200 patients and test a medicine. After giving the medicine, condition of the patients is observed and the following results are obtained.

	Better	No change
Medicine	123	13
Placebo	41	23

Calculate precision, recall, F-measure (when $\beta = 1$ and 2), Jaccard coefficient, Rand coefficient, and Fowlkes–Mallows coefficient.

3.9 Describe types, pros and cons, and key applications of clustering algorithms.

3.10 Consider the following two-dimensional data set:

Data set: (1, 2), (2.5, 2.5), (3.5, 3.5), (4, 3.8), (7, 6), (7.5, 8), (8, 8.5), (10.9)

Find two clusters using K-means clustering algorithm.

3.11 Consider a randomly and uniformly distributed 100 IoT devices in a single cell. Perform clustering using both K-means clustering algorithm and K-means + +clustering algorithm when we have 3, 5, and 7 clusters in the single cell.

3.12 Consider a randomly distributed 100 and 200 IoT devices in a single cell. Perform clustering using FCM clustering algorithm and observe the effect of m when $K = 3$.

3.13 Describe key application of association rule mining

3.14 Describe key components affecting computational complexity of Apriori algorithm.

3.15 Consider a set of 7 transactions as shown in the below table. Each transaction t_i is a set of purchased items in a store by a customer.

Transaction ID	Items
t_1	Egg, Milk, Chicken, Sugar
t_2	Sugar, Egg, Chicken, Bread
t_3	Milk, Egg, Chicken, Bread, Egg
t_4	Egg, Bread
t_5	Beef, Sugar
t_6	Egg, Milk
t_7	Milk

Find frequent itemsets as setting minimum support $= 2$ using Apriori algorithm.

3.16 Describe pros and cons of feature select and feature extraction.

3.17 Investigate the properties of PCA except Properties 3.4, 3.5, and 3.6.

3.18 Consider the following data set

$$
\mathbf{X} = \begin{bmatrix}
7 & 4 & 0 & -1 & -1 & 0 & 5 & 0 & 4 & -5 \\
2 & -1 & 2 & 7 & 1 & 1 & -3 & -1 & 0 & 3 \\
0 & 4 & 5 & 10 & -2 & 0 & 7 & 0 & -3 & -1 \\
-1 & 0 & 3 & 0 & -8 & 2 & -3 & -2 & 6 & 0 \\
4 & 2 & -8 & 1 & -3 & -6 & 0 & -2 & 1 & 0
\end{bmatrix}
$$

and perform the dimensionality reduction using PCA.

References

1. J. Kleinberg, "An impossibility theorem for clustering," *The 15th International Conference on Neural Information Processing Systems (NIPS'02)* (MIT Press, 2002), pp. 463–470
2. P.-N. Tan, M. Steinbach, V. Kumar, *Introduction to Data Mining*. Pearson, Inc., 2005. ISBN-13: 978–0321321367
3. G. Gan, C. Ma, J. Wu, *Data Clustering: Theory, Algorithms, and Applications* (Society for Industrial and Applied Mathematics, ASA-SIAM Series on Statistics and Applied Mathematics, 2007)
4. M. Cord, P. Cunningham (eds), *Machine Learning Techniques for Multimedia: Case Studies on Organization and Retrieval* (Springer, 2008). ISBN 978–3–540–75171–7
5. S.P. Lloyd, Least squares quantization in PCM. IEEE Trans. Inf. Theory **28**, 129–137 (1982)
6. R. Agrawal, T. Imieliński, A. Swami, "Mining association rules between sets of items in large databases". Proceedings of the ACM International Conference on Management of Data (SIGMOD '93), (1993)
7. R. Agrawal, R. Srikant, "Fast algorithms for mining association rules in large databases," In Proc. of the 20th International Conference on Very Large Data Bases (Santiago, Chile, 1994), pp. 487–499
8. K.V. Mardia, J.T. Kent, J.M. Bibby, *Multivariate Analysis* (Academic Press, Probability and Mathematical Statistics, 1995)

Chapter 4
Supervised Learning

A human learns from past experience but a machine does not have experience. However, a machine can learn from past data. A machine as a learner has a training data set and a test data set and develops a rule or a procedure to perform classification or regression by analysing samples in the test set. For example, the training data set can be different types of dogs such as poodle, bulldog, golden retriever, and so on. The identity of them is given to the learner. The test data set can be unidentified dogs. A machine develops a rule identifying them in the test data set. The main purpose of machine learning is to infer from sample data in statistics point of view. In terms of computer science or electrical engineering, machining learning algorithms solve optimization problems or build a mathematical model for inference. The main goal of supervised learning is to learn a target function that can be used to predict the value of a class. Supervised learning techniques construct a predictive model by learning from a labelled training data. Supervised learning enables us to make a decision such as approval or no approval, high risk or low risk, and so on. In this chapter, we discuss supervised learning techniques.

4.1 Supervised Learning Workflow, Metrics, and Ensemble Methods

Machine learning is continuously evolving. Supervised learning is a core sub-field of machine learning. It turns raw data into a revelatory insight and allows us to improve the performance of a machine or make business decision faster and more accurately. It has been used in many applications such as automated vehicles, medical diagnosis, and so on. Machine learning helps us to analyse huge data and has an automated process. It has changed the way about data analysis including data extraction and interpretation works. In supervised learning, the inputs are independent variables, predictors, and features. The outputs are dependent variables and responses. The goal of supervised learning is to predict output values. We can formulate supervised learning problems as follows: Given n paired data points $[(x_1, y_1), \ldots, (x_n, y_n)]$, we develop a model with a function $f()$, $\mathbf{y} = f(\mathbf{x})$ and find the response value predicted by the function $f()$ on $\mathbf{x}$, $\hat{\mathbf{y}} = f(\mathbf{x})$. The function types would be linear, polynomial, logistic, and so on. The functions as a prior knowledge would be suggested by the nature of the problems. The supervised learning process is composed of two phases: training and testing. The training phase is to learn a model using a training data set. The testing phase is to test a model using a test data set to evaluate the accuracy of the model. Figure 4.1 illustrates the supervised learning workflow. In the first phase, we collect a data set and perform training on the data set. If there are experts who can suggest which attribute or feature is the most informative, we can obtain a better attribute or feature of the data set. Another approach is brute-force method [1]. It measures everything available that the informative or relevant features can be isolated [1]. However, the data set by the brute-force method is not directly suitable because in many cases it contains noises and missing feature values. Therefore, pre-processing is required. In the second phase, data preparation and pre-processing are performed. Many techniques have been developed to deal with missing data, detect outliers, remove irrelevant feature, reduce the dimensionality of a data set, or manage noises.

There are three types of generalization errors in machine learning: approximation error, estimation error, and optimization error. The approximation error represents

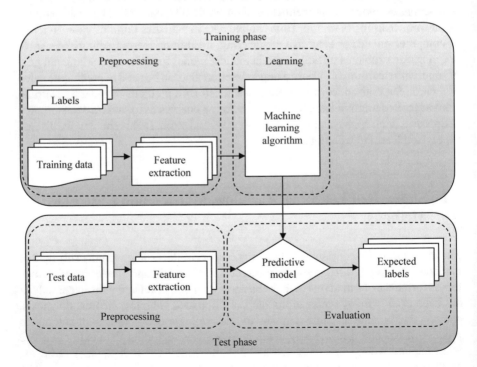

Fig. 4.1 Supervised learning workflow

how the solution we found can be approximated and implementable in the machine learning system. The estimation error represents how much accurately we can find a model or function using a finite training data set. In general, the estimation error depends on the volume of a training data set and the capacity of functions. The optimization error represents how closely we can calculate the function satisfying the information in a finite training data set. There are several more evaluation issues in machine learning. For example, predictive accuracy, efficiency about model construction, robustness of noise and missing data, scalability, interpretability, and others. Key questions about them can be summarized as follows: (1) What metrics should we use? (2) How much reliable are the results? And (3) How do we manage an error on a training data set? The model is generally evaluated by accuracy. In order to access the model, the following metric "accuracy" as overall recognition rate of the classifier is used:

$$\text{Accuracy} = \frac{\text{Number of correct classfications/predictions}}{\text{Total number of test cases}}. \tag{4.1}$$

The misclassification rate can be represented as accuracy. As we discussed in Chap. 3, we can express it using contingency matrix. It can be calculated as well in terms of true positives (TP), true negatives (TN), false positives (FP), and false negatives (FN) as follows:

$$P_{\text{accuracy}} = \frac{\text{TP} + \text{TN}}{\text{TP} + \text{TN} + \text{FP} + \text{FN}}. \tag{4.2}$$

For example, we have 100 patients and test a medicine. After giving the medicine, the condition of the patients is observed and the following results are obtained: TP $= 90$, TN $= 1$, FP $= 2$, and FN $= 7$. From (4.2), the accuracy is $(90 + 1)/(90 + 1 + 2 + 7) = 90\%$. This metric works well only if we have equal number of samples belonging to each class. If we deal with class imbalanced data set, it doesn't work well. Other metrics we discussed in the previous chapter can be used when accuracy doesn't work well. One important assumption of supervised learning model is that the distribution of a training data set is same as the distribution of a test data set including future unknown samples. Namely, the training data set should be representative of the test data set. If this assumption is violated, the accuracy of the classification or prediction will be poor. In order to evaluate how close the predicted values are from the ground true, we can compute loss functions as follows:

$$\text{Absolute error: } \Delta x_a = \left| x_i - x_i' \right| \tag{4.3}$$

and

$$\text{Squared error: } \Delta x_s = \left(x_i - x_i' \right)^2 \tag{4.4}$$

where x_i and x_i' are the ground true and the predicted value, respectively. The average loss can be expressed as follows:

$$\text{Mean absolute error: MAE} = \frac{1}{N} \sum_{i=1}^{N} \left| x_i - x_i' \right| \tag{4.5}$$

and

$$\text{Mean-squared error: MSE} = \frac{1}{N} \sum_{i=1}^{N} \left(x_i - x_i' \right)^2 \tag{4.6}$$

where N is the test data size. We can normalize them by dividing by the total squared error of the simple predictor as follows:

$$\text{Relative absolute error: RAE} = \frac{\sum_{i=1}^{N} \left| x_i - x_i' \right|}{\sum_{i=1}^{N} \left| x_i - \mu \right|} \tag{4.7}$$

and

$$\text{Relative squared error: RSE} = \frac{\sum_{i=1}^{N} \left(x_i - x_i' \right)^2}{\sum_{i=1}^{N} \left(x_i - \mu \right)^2} \tag{4.8}$$

where μ is the mean. The performance of supervised learning depends on multiple factors including supervised learning algorithms, size of a data set, data distribution, misclassification cost, and others. The size of training data and test data is highly related to the accuracy of classifiers. The holdout is the simplest of cross-validation. The data set is divided into two sets: training data set and test data set. Typically, two-thirds of a data set are allocated to a training data set and one-thirds is allocated to a test data set. This method depends on which data points end up in training data set or test data set. Therefore, this evaluation may have a high variance. Another approach is k-fold cross-validation. A data set is split into k equal-sized and mutually exclusive subset (or subsamples, folds): $d_1, d_2, ..., d_k$. Each subset is in turn used for testing and the remaining $k - 1$ subsets are used for training. For example, in the first iteration, d_1 is used for testing and $d_2, ..., d_k$ is used for training. In the second iteration, d_2 is used for testing and $d_1, d_3, ..., d_k$ is used for training. This process is repeated k times. The k results can be averaged to produce an estimation. The subsets are stratified before it is performed in order to reduce the variance of the estimate. Typically, tenfold cross-validation is widely used because extensive experiments represent good trade-off between computational complexity and accurate estimate. The leave-one-out cross-validation is a special case of k-fold cross-validation when the number of folds is equal to the number of data points in the data set. Thus, only one data point is left out at a time for test. It provides us with a reliable and unbiased estimate but it is computationally expensive. When dealing with a data set, we face three types of prediction errors: bias, variance, and noise. The noise is difficult to reduce by algorithm selection or other methods. However, both bias and variance can be reduced. The bias error occurs due to erroneous assumption. That is to say. When a machine learning algorithm has limited flexibility to learn from a data set, the bias errors appear. It is the difference between the prediction and the expectation. High bias can cause less assumptions about the target function form. The variance error occurs due to algorithm sensitivity in the training data set. High variance means large fluctuation about the estimate of the target function in the training data set. In machine learning, the bias variance trade-off is key aspect to achieve low bias and variance. Increasing the bias or the variance decreases the variance and the bias, respectively. If we find a trade-off between bias and variance, it implies that a model should balance under-fitting and overfitting. In machine learning, ensemble methods create multiple models and then combine them to produce a better result. They usually provide us with more accurate results. In ensemble methods, we call base learners or weak learners that perform not so well by themselves due to bias or variance errors. When setting up an ensemble method, we need to select a base model to be aggregated. If a single base learning algorithm is used and base learners are trained in different way, we call it homogeneous ensemble. If different base learning algorithms are used, we call it heterogeneous ensemble. One important aspect about a base learner is to be coherent with the aggregation models. If we have a base model with a low bias and high variance, the aggregation will reduce the variance. On the other hand, if we have a base model with high bias and low variance, it will reduce the bias. This is about how to set a target and combine models. The ensemble methods enable us to decrease the variance by bagging, reduce the bias by boosting, or improve predictions by stacking. Ensemble methods can be divided into parallel ensemble methods and sequential ensemble methods. The basic concept of parallel ensemble methods is to exploit independence between base learners. In the parallel ensemble methods, base learners are generated in parallel and the variance can be reduced by averaging. The basic concept of sequential ensemble methods is to exploit dependence between base learners. In the sequential ensemble methods, base learners are generated in a sequence. The bias error can be reduced, and the performance of the model can be improved by assigning different weights to previously misrepresented learners. Stacking uses the concept of heterogeneous ensemble and combines multiple classification or regression models by training a meta-model. The base models are composed of different learning algorithms. They are trained using a complete training set, and then a meta-model is trained on the output of the base models.

Bagging as a parallel ensemble method stands for bootstrap aggregation. It generates additional data in the training and reduces the variance of the prediction model. It is useful to avoid overfitting. Bagging is composed of bootstrapping and aggregating. Figure 4.2 illustrates the process of bagging.

We firstly define bootstrapping. It is a statistical technique using random sample with replacement. The bootstrapping method divides a data set with size N into B samples of identical size with replacement. Figure 4.3 illustrates an example of bootstrapping process.

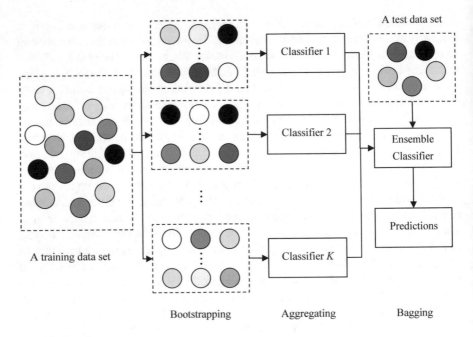

Fig. 4.2 Bagging process

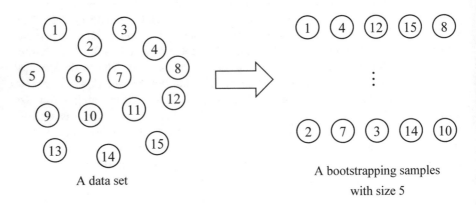

Fig. 4.3 Example of bootstrapping process

These samples are useful because they are regarded as representative and independent samples of the data distribution. Sometimes, they are used for evaluating variance and finding confidence intervals of estimators. The bootstrapping process can be summarized as follow: Given a data set D with N training samples, we create a bootstrapping training set D_k, $k = 1,\ldots, K$ that contains N training samples drawn randomly from D with replacement. We perform machine learning algorithm and construct a hypothesis h_k on D_k. We use h_k to make predictions. Then, we repeat this process with K iterations. Aggregation is a process to combine each prediction and produces the final prediction. It can be done by picking the majority for classification or averaging all predictions for regression. The bagging process is similar to the bootstrapping process. The data sets are different from each other. We train models separately on each of data sets $D_1, D_2, \ldots D_k$, to obtain classifier $h_1, h_2, \ldots, h_k$. We use average model $h = \frac{1}{K} \sum_{i=1}^{K} h_i$ as the final model. This model is useful for high variance and noisy data. For a test data set, we use the final model and make a prediction. Each data point has the probability $P(\text{not selected}) = (1 - 1/N)^N$ of being selected as test data. As N goes to infinity, this probability goes to $e^{-1} = 0.368$. Thus, we can compute the probability that a sample is selected as $P(\text{selected}) = 1 - (1 - 1/N)^N = 0.632$ for reasonably large data sets. This means that the training data contains approximately 63.2% of the training samples in any given bootstrapping set. Typically, bagging reduces variance of the base models and eliminates the overfitting of base models. However, it has high computational complexity and it doesn't affect to stable machine learning algorithms. In addition, it introduces a loss of interpretability of a model. Interpretability also known as human-interpretable interpretations is the measure to which a human can understand the cause of choices in their decision-making process. The higher interpretability of a machine learning model means the easier understanding of why certain decisions have been made. Basically, there is a trade-off relationship between interpretability and accuracy. Figure 4.4 illustrates the relationship between interpretability and accuracy for machine learning algorithms.

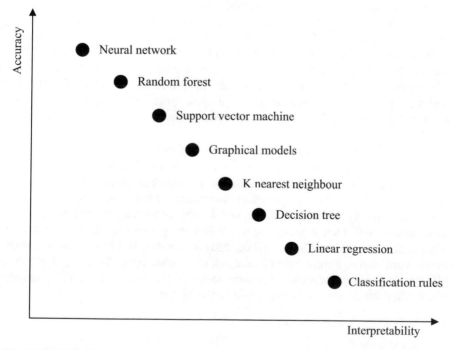

Fig. 4.4 Relationship between interpretability and accuracy for machine learning algorithms

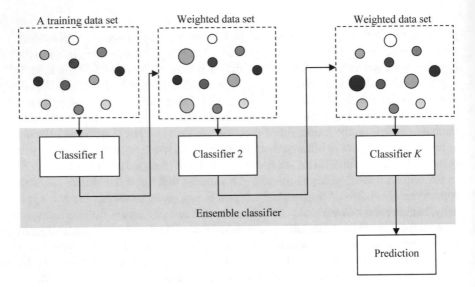

Fig. 4.5 Boosting process

Typically, high accurate models have nonlinear and non-smooth relationship and high computational complexity. High interpretable models have linear and smooth relationship and low computational complexity.

Boosting is a sequential ensemble method converting weak learners to strong learner by getting multiple weak learners and weighting samples. Unlike bagging, boosting does not construct the hypotheses independently but focuses on problematic samples for exiting hypotheses. In boosting, each classifier is trained on a data set while considering the previous decision. The mis-predicted data points in the previous round are identified, and more weight is given to them. Therefore, the subsequent learner pays more attention to get them correct. Figure 4.5 illustrates the process of boosting.

The boosting process can be summarized as follows: Firstly, a base learner uses a training data set and assigns equal weight to each observation. Secondly, if false predictions are identified, we pay more attention to the false prediction and assign to the next base learner with higher weight on the incorrect prediction. Thirdly, we repeat these steps until higher accuracy is achieved. Lastly, we combine the hypotheses from base learners and create a strong learner. When comparing bagging with boosting, there is no a clear winner because it depends on the data and circumstances. Both of them are used to improve the accuracy of classifiers. Typically, bagging is more efficient than boosting because bagging can train multiple models in parallel. Bagging can't reduce the bias but boosting can decrease the bias.

Supervised learning can be categorized as classification and regression [2]. Classification is a technique for predicting where we categorize data into classes. The main purpose of classification is to identify which class the data will belong to. Key supervised learning algorithms for classification are naive Bayes, K-nearest neighbours, decision tree, random forest, support vector machine, and so on. Key applications of classification are spam mail detection, image detection, text recognition, and others. Regression is a technique for finding the correlation between target variable and prediction variable and predicting the continuous values. Key supervised learning algorithms for regression are simple linear regression, Lasso regression, logistic regression, support vector machines, multivariate regression algorithm, multiple regression algorithm, and so on. Key applications of regression are stock price prediction, forecasting market, financial analysis, and so on. The main difference between classification and regression is the output variable. Regression predicts a continuous variable while classification predicts discrete class labels. The performances of classification and regression are evaluated by accuracy and root mean-squared (RMS) error, respectively. Some machine learning algorithms such as neural networks can be used for both classification and regression. Sometimes, classification algorithms can predict a continuous variable in the form of probability or regression can predict discrete class labels in the form of an integer quantity. In addition, it is possible to transform a classification problem into a regression problem. Figure 4.6 illustrates the comparison of classification and regression.

From the next section, we will discuss supervised learning algorithms for classification and regression.

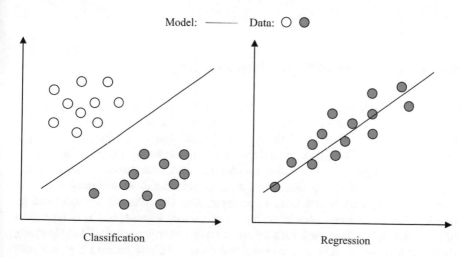

Fig. 4.6 Classification versus regression

Summary 4.1 Supervised Learning

1. Supervised learning turns raw data into a revelatory insight and allows us to improve the performance of a machine or make business decision faster and more accurately.
2. The supervised learning process is composed of two phase: (1) The training phase is to learn a model using a training data set. (2) The testing phase is to test a model using a test data set to evaluate the accuracy of the model.
3. The performance of supervised learning depends on multiple factors including supervised learning algorithms, size of a data set, data distribution, misclassification cost and others.
4. In machine learning, ensemble methods create multiple models and then combine them to produce a better result. They usually provides us with more accurate results.
5. The ensemble methods enables us to decrease variance by bagging, reduce bias by boosting, or improve predictions by stacking.
6. Bagging as a parallel ensemble method stands for bootstrap aggregation. It generates additional data in the training and reduce the variance of the prediction model. It is useful to avoid overfitting.
7. Boosting is a sequential ensemble method converting weak learners to strong learner by getting multiple weak learners and weighting samples. Unlike bagging, boosting does not construct the hypotheses independently but focuses on problematic samples for exiting hypotheses.

4.2   Classification of Supervised Learning

In general, classification steps are composed of training, testing, and usage. In the step of training, a model is created by training data. Classification algorithms find relationships from a data set. The relationships are formulated by constructing a model. In the step of testing, we test the model on a data set with known class labels. In the step of usage, we perform classification on a new data set with unknown class labels using the model. Depending on a data set and classification algorithms, those steps may be merged. There is no clear separation. For example, a model can be developed before classification or during classification. A model can be independent of a test data set or dependent on a test data set. In machine learning, lazy learning simply stores the training data and waits for query before generalizing, whereas eager learning constructs a classification model before receiving a data set. Thus, lazy learning requires less time in the step of training but more time in predicting. Eager learning learns slowly but classifies quickly. Consider a classification task of mapping an attribute data set $\mathbf{x}$ into a class label y. Figure 4.7 illustrates a general

Fig. 4.7 General
classification system model

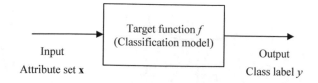

classification system model.

The data point can be characterized by a tuple $(\mathbf{x}, y)$ where $\mathbf{x}$ and y are the attribute set and the class label, respectively. The attribute set includes properties of a variable and can contain discrete or continuous values. The class label must be a discrete value. The target function also known as a classification model distinguishes between data points of different class labels or predicts the class label of unknown input data sets. Classification techniques are working well at input data sets with binary or nominal categories. The classification is defined as follows: Classification is a learning task with a target function f mapping an attribute data set $\mathbf{x}$ into one of the predetermined class label y. A classification model is used for either a predictive model or a descriptive model. The predictive model as the name said is to predict a class of previously unseen data points. It requires a high accuracy and a fast response time. The descriptive model is an explanatory tool to distinguish data from different classes. It determines characteristic of a data set. A decision tree model is relatively more descriptive than other classifier models. As we discussed the supervised learning flow in the previous section, a training set is composed of data points with attributes values and known class labels and is used for building a classification model. From the training data, building or learning a classification model is known as induction. The classification model is applied for a test data set with unknown class labels. This process on the test data is known as deduction. Thus, we can say that classification is composed of two steps: learning a model from a training data set and applying a model to assign labels to a test data set.

4.2.1 Decision Tree

Decision tree is a technique for approximating target functions with discrete variables. It is widely used for inductive inference because it is robust to noise and the algorithm is simple. Decision trees have a hierarchical data structure and are built using recursive partitioning based on a divide-and-conquer strategy. The decision tree is composed of a root node, internal (or test) nodes, leaf (or terminal, decision) nodes, and edges (or branches). The root node as a topmost node of a decision tree is the starting point of the tree. It has no incoming edges, and outgoing edges lead to an internal node or a leaf node. Internal nodes have one incoming edge and at least two outgoing edges. They represent a test on an attribute. Leaf nodes are the bottommost nodes of the decision tree. They represent a numeric prediction of the

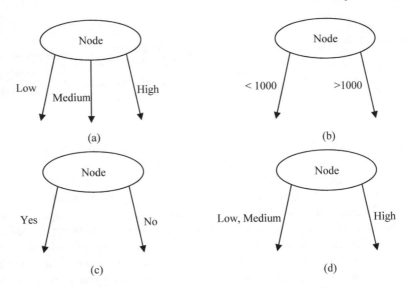

Fig. 4.8 Example of edge types: Discrete value (**a**), continuous value (**b**), binary value (**c**), and attribute value (**d**)

decision tree. Each leaf node has one incoming edge and one class label or distribution. Edges correspond to the outcome of the test on the training data set and connect to the next nodes. There are four types of the edge: discrete value, continuous value, binary value, and attribute value [3]. The discrete valued edges have only one edge for each attribute value. Each node makes an n splits depending on the attribute of the node. The continuous valued edge has continuous value. It divides the value into two or three intervals. For example, greater or less than the specific value. The binary discrete valued edge has the yes or no (1 or 0) value. The attribute valued edge can be merged in one edge to get more accurate decision. Figure 4.8 illustrates an example of edge types.

The decision tree techniques are useful when the function and data set have the following properties: (1) Disjunctive expressions are required. The Boolean function can be fully expressed in the decision tree. Many other functions can be approximated by Boolean functions. (2) A data set is expressed by attribute value pairs. (3) The output of the target functions is discrete. The hypothesis space is the set of all possible finite discrete functions. The finite discrete functions can be expressed by a decision tree. The decision tree techniques are robust to classification errors and can be applied to classification and regression problems. There are two types of decision trees: classification trees and regression trees. If the decision variable is discrete, the decision tree is built by a finite number of class values. We call it classification tree. The attributes are numerical or categorical values. Leaf nodes represent class labels, and edges represent conjunctions of features. The main goal of classification trees is to classify a data set. The attribute with the highest purity (or the lowest impurity) is chosen. Purity and impurity represent homogeneity and heterogeneity, respectively. The entropy value is widely used as impurity measure. The basic idea of the classification tree is a top-down approach. We recursively split from the root node and construct continuously until there are no more splits while each node is branched in terms of the highest purity (or the lowest impurity). The concept learning system (CLS) constructs a decision tree to minimize the costs of classification [4]. The basic idea of the CLS affects to interactive dichotomizer 3 (ID3) [5], C4.5 [6] and classification and regression tree (CART) [7]. C4.5 and CART are widely used in machine learning area. The CLS begins with an empty decision tree and iteratively builds the tree by adding nodes until the tree classifies all the training data correctly [4]. Regression trees are very similar to classification trees. If the decision variable can take continuous values, entropy as an impurity measure cannot be used and leaf nodes typically set to the mean values. There are no class labels. The main difference is leaf node generation by averaging over the target values. The nodes are branched into two partitions like greater or less than the specific value. Automatic interaction detection (AID) as the first regression tree was developed for fitting trees to predict a quantitative variable [8]. The AID performs stepwise splitting.

Once a decision tree is constructed, classification of a data set is straightforward. Starting from the root node, specific attribute is applied to a test data set. Based on the results, appropriate edges are selected. It will lead to an internal node or a leaf node. If it reaches to a leaf node, the class label associated with the leaf node is allocated to a data point. Figure 4.9 illustrates an example of decision tree.

As we can observe Fig. 4.9, the decision about car change is made in terms of two attributes: car condition and financial status. The root node is the problem statement "Old car over 20 years". It has three outgoing edges: not working, working and less trouble, and working but a lot of trouble. They represent the values of the attribute

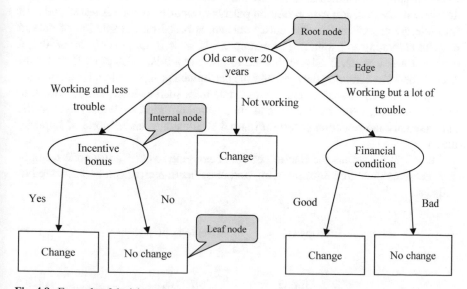

Fig. 4.9 Example of decision tree

about car condition. After that, we have two internal nodes (incentive bonus and financial condition) and one leaf node (change). The leaf node is decided as car change when a car doesn't work. Two internal nodes represent the new attribute financial status. According to the financial status, the decision on car change is made. This is a top-down approach. This example has only discrete and categorical attributes. How can we build a decision tree? Plenty of decision trees can be constructed from a set of attributes. It is computationally infeasible to construct the optimal tree because finding accurate tree is NP-hard. Usually, a greedy strategy is employed and Hunt's algorithm [4] is the basis of many decision tree induction algorithms. It enables us to construct a decision tree in an iterative way by partitioning the training data into subsets until each path ends in a pure subset. Let D_t be the set of training data associated with a node t. Let y be the class label. The recursive procedure of Hunt's algorithm can be summarized as follows: (1) If all data in D_t belong to the same class, t is a leaf node with class label y_t. (2) If D_t is an empty set, t is a leaf node. (3) If D_t contains data belonging to more than one class, we select attribute test condition to split data into smaller subsets, create a child node for each output of test condition, and apply this procedure to each child node iteratively. This algorithm is useful if every combination of attribute values is present in the training data and each combination has a unique class label [9].

There are two design issues of decision tree induction: How to split a training data? And how to stop the splitting procedure? Based on attribute tests, the training data can be split. The first question can be transformed to the question: What attribute is the best one? We need to evaluate the goodness of each attribute, specify the test conditions, and select the best one. Different attribute types have been described in Fig. 4.8. In order to split a data set at each node of the decision tree, the selection of an attribute is fundamental to construct a decision tree. It affects to efficiency and accuracy of the decision tree. Based on purity or impurity, a node is tested and split into internal or leaf nodes. The purity concept is based on the fraction of data in a group belonging to a subset. For example, if a node is split evenly 50%/50%, a node is 100% impure. If all data belong to one class, it is 100% pure. It partitions data elements in a training data set, and the partitioned group would have all or most of data elements in the same class. Univariate trees such as information gain, gain ratio, Gini index, twoing criterion, and chi-squared criterion are widely used as the heuristic attribute selection criteria. Figure 4.10 illustrates an example of impurity measure.

Let $p(i|t)$ represents the fraction of data belonging to class i at a node t. Impurity measures including entropy, Gini, and classification error can be expressed as follows:

$$\text{Entropy } (t) = -\sum_{i=0}^{c-1} p(i|t) \log_2 p(i|t), \tag{4.9}$$

$$\text{Gini } (t) = 1 - \sum_{i=0}^{c-1} (p(i|t))^2, \tag{4.10}$$

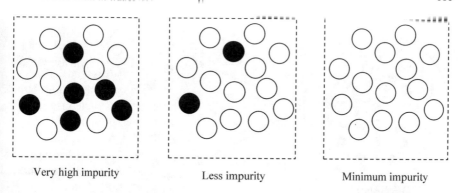

| Very high impurity | Less impurity | Minimum impurity |

Fig. 4.10 Example of impurity measure

$$\text{Classification error } (t) = 1 - \max_i p(i|t) \tag{4.11}$$

where c is the number of classes. Information gain is one of common ways to measure impurity. The entropy plays an important role in calculating information gain. Based on information theory, entropy is defined as information-theoretic measure of randomness. The higher the impurity means the higher the entropy. In terms of machine learning, it helps us to find the suitable attribute for a learning model. If a data set belongs to one class, the entropy is calculated as follows: $-1 \log_2 1 = 0$. This is not good for training. If a data set evenly belongs to two classes, the entropy is calculated as follows: $-0.5 \log_2 0.5 - 0.5 \log_2 0.5 = 1$. This is a good training set for learning. The information gain allows us to determine which attribute in a given training data set is most useful. It is based on reduction in entropy after a data set is split on an attribute. It is useful for measuring the effectiveness of the attribute in classifying the training data points. The information gain IG is defined as follows:

$$\text{IG}(T, a) = \text{H}(T) - \text{H}(T|a) \tag{4.12}$$

where T and a are a set of training data and attribute, respectively. $\text{H}(T)$ is an entropy of parent. $\text{H}(T|a)$ as the conditional entropy of T given the value of attribute a is entropy sum of children. We would simply choose the attribute with higher information gain. In machine learning, (4.12) can be rewritten in order to determine the goodness of split as follows:

$$\text{IG} = \text{H}(\text{parent}) - \sum_{j=1}^{k} \frac{N(v_j)}{N} \text{H}(v_j) \tag{4.13}$$

where N is the total number of data, k is the number of attributes, and $N(v_j)$ is the number of data corresponding to the child node v_j.

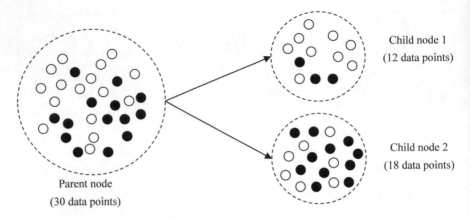

Fig. 4.11 Data set

Example 4.1 Information Gain Consider the decision tree with 16 white circles and 14 black circles at the parent node as shown in Fig. 4.11.

Calculate the information gain of the data set.

Solution

From (4.13), we calculate information gain as follows:

$$IG = H(\text{parent}) - \sum_{j=1}^{k} \frac{N(v_j)}{N} H(v_j),$$

$$H(\text{parent}) = -\left(\frac{16}{30} \log_2 \frac{16}{30}\right) - \left(\frac{14}{30} \log_2 \frac{14}{30}\right) = 0.997,$$

$$H(\text{child } 1) = \frac{N(v_1)}{N} H(v_1) = \frac{12}{30}\left(-\left(\frac{9}{12} \log_2 \frac{9}{12}\right) - \left(\frac{3}{12} \log_2 \frac{3}{12}\right)\right) = 0.323,$$

$$H(\text{child } 2) = \frac{N(v_2)}{N} H(v_2) = \frac{18}{30}\left(-\left(\frac{7}{18} \log_2 \frac{7}{18}\right) - \left(\frac{11}{18} \log_2 \frac{11}{18}\right)\right) = 0.578,$$

$$IG = 0.997 - (0.323 + 0.578) = 0.094.$$

■

Gini as an index of unequal distribution measures how often a randomly selected element would be incorrectly labelled. It is pure if all elements are in a single class. Gini index varies between 0 and 1. The value close to 0 means high purity. Namely, most of the elements belong to a specific class. The value close to 1 means high impurity. Namely, most of the elements have a random distribution across multiple classes. The value of 0.5 means an equal distribution of elements. The Gini index is useful for the wider distributions but the information gain works well at narrow distributions with small counts and multiple values. The Gini index performs well at binary split and the categorical target variables. Equations (4.9) and (4.10) of entropy and Gini can be rewritten in case of two class variables as follows:

$$\text{Entropy} = -\big(p \log_2 p + (1-p) \log_2 (1-p)\big), \tag{4.14}$$

and

$$\text{Gini} = 1 - \big(p^2 + (1-p)^2\big) \tag{4.15}$$

where p is the probability of one class. Figure 4.12 illustrates the comparison between entropy index and Gini index for two class variables.

Example 4.2 Gini Index Consider the decision tree with 12 white circles and 12 black circles at the parent node and two different splits by both attribute a and b as shown in Fig. 4.13.

Calculate and compare the Gini indexes of two splits by attribute a and b.

Solution

The Gini index of the parent node is 0.5 due to equal distribution. In the first split by attribute a, the Gini indexes of the children nodes are calculated as follows:

$$\text{Gini (child node } a1) = 1 - \left(\left(\frac{8}{14}\right)^2 + \left(\frac{6}{14}\right)^2\right) = 0.49,$$

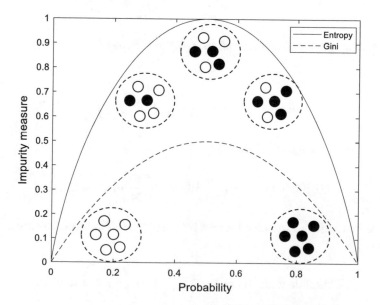

Fig. 4.12 Comparison between entropy index and Gini index for two class variables

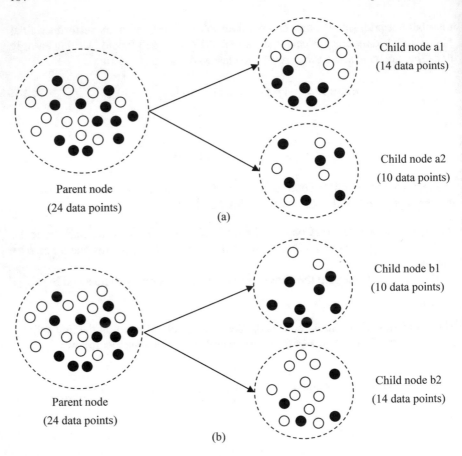

Fig. 4.13 Data set split a by attribute a (**a**) and split b by attribute b (**b**)

$$\text{Gini (child node } a2) = 1 - \left(\left(\frac{4}{10} \right)^2 + \left(\frac{6}{10} \right)^2 \right) = 0.48,$$

and the weight averaged Gini index is

$$\text{Gini(attibute } a) = \left(\frac{14}{24} \right) 0.49 + \left(\frac{10}{24} \right) 0.48 = 0.49.$$

In the first split by attribute b, the Gini indexes of the children nodes are calculated as follows:

$$\text{Gini(child node } b1) = 1 - \left(\left(\frac{2}{10} \right)^2 + \left(\frac{8}{10} \right)^2 \right) = 0.32,$$

$$\text{Gini(child node } b2) = 1 - \left(\left(\frac{10}{14} \right)^2 + \left(\frac{4}{14} \right)^2 \right) = 0.41,$$

and the weight averaged Gini index is

$$\text{Gini(attibute } b) = \left(\frac{10}{24} \right) 0.32 + \left(\frac{14}{24} \right) 0.41 = 0.37.$$

The split b by attribute b has smaller Gini index value. Therefore, the attribute b is selected for the classifier. Typically, for classification, we choose the test with higher information gain. For regression, we select the test with the lowest mean-squared error.

■

The above attribute selection criteria can be biased. They tend to have attributes with a large number of distinct values. It causes a poor accuracy. Thus, information gain ratio as a ratio of information gain to the intrinsic information is widely used [10]. The intrinsic information means the entropy of data distribution into branches. It reduces a bias towards attributes with many values by considering the number and size of branches when choosing an attribute [10]. The normalization of information gain is defined as follows:

$$\text{Gain Ratio } (T, a) = \frac{IG(T, a)}{H(T)}. \tag{4.16}$$

The gain ratio enables us to outperform the information gain in terms of accuracy and complexity.

There are multiple stop conditions. One of them is spilt until we can't split anymore. Another approach is to set a threshold to find the best candidate split and reduce the impurity. Common stop conditions are when (1) the maximum tree depth has arrived, (2) all training data belong to a single value, (3) the minimum number of children nodes has been reached, or (4) the best split condition has reached at a certain threshold. If the tree is grown until each lead node has the lowest impurity measure, the data can overfit. It is quite serious problem in decision tree classification. We should minimize error and avoid the overfit. The main reason of the overfit is too much noise or variance in the training data and lack of representative samples. Figure 4.14 illustrates an example of decision tree overfitting and balance of bias and variance of the model. If the decision tree is overfitted, the accuracy of the training data increases as the decision tree grows. However, the accuracy of the test data decreases at some point. This means that a large enough tree cannot generalize to test data.

Pruning is a key technique to avoid overfitting. There are two different pruning techniques: pre-pruning and post-pruning. In the pre-pruning, the decision tree stops growing at some point when there is unreliable information. If growing decision tree stops too early, the performance might be poor. However, if we stop before it

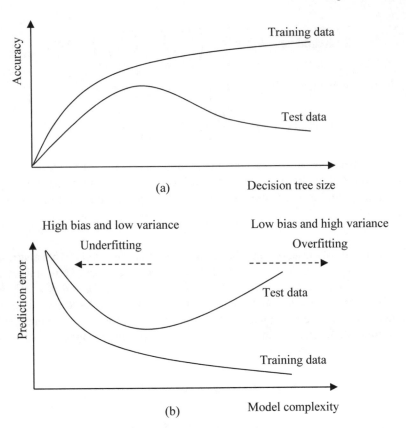

Fig. 4.14 Overfit of decision tree (**a**) and balance of bias and variance (**b**)

perfectly classifies the training data, we can avoid overfitting. Typically, the stop condition is 5% of total training data. It provides us with good trade-off between complexity and accuracy. The main advantage of this pre-pruning is to save time because there are no wasted growing subtrees. In the post-pruning, growing the decision tree is not restricted by a threshold. The decision tree is grown fully until all training data are classified and then we remove nodes with insufficient evidence. In practice, post-pruning is preferred because pre-pruning can stop too early. After growing the tree fully, we perform pruning by eliminating lower nodes with low information gain and assigning the majority label of the parent to them. There are two operations of post-pruning: subtree replacement and subtree raising. The subtree replacement is a bottom-up approach. It considers to replace subtree with a leaf node. It might decrease accuracy of the training data but increase accuracy of the test data. The subtree raising is more complex than the subtree replacement. In the subtree raising, a subtree is pruned but replaced by another subtree with a better accuracy. The criteria of labelling are frequency. After pruning, the most frequent class is labelled as the leaf node class. This approach is widely used in practice. One

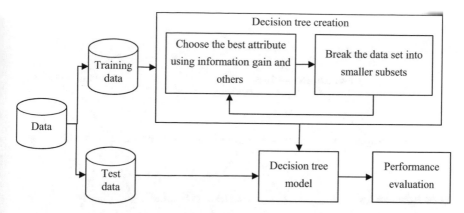

Fig. 4.15 Process of decision tree algorithms

application is Microsoft xBox. In the body part recognition algorithm of Microsoft xBox, the decision tree algorithm is used to classify body parts. The advantages of decision trees can be summarized as follows: (1) decision trees are self-explanatory and the representation of the decision tree is easy to understand. (2) It deals with both nominal and numerical input attributes. (3) It is robust to missing data or errors. (4) We don't need assumption about data distribution or structure. (5) It is easy to use and implement and also computationally cheap for small size trees. (6) It allows us to handle discrete and continuous data, classification and regression, and variable size of data. On the other hand, the disadvantages are as follows: (1) In many decision tree algorithms such as ID3 and C4.5, the target attributes have discrete values. (2) It doesn't work well if many highly relevant attributes exist. (3) It is sensitive to training data with irrelevant data and noise. (4) It can overfit. (5) It doesn't have good performance for predicting the value of continuous class attributes. Figure 4.15 illustrates the process of decision tree algorithms.

Example 4.3 Decision Tree for Classification Consider the random three classes (class 1, class 2, and class 3) data with target matrix 18 observations with 3 features ($x1$, $x2$, and $x3$) as a training data set as follows:

$$
\text{Data} =
\begin{bmatrix}
20 & 10 & 10 & 2 & 22 & 12 & 5 & 5 & 5 & 19 & 10 & 10 & 22 & 21 & 23 & 5 & 5 & 15 \\
0 & 0 & 11 & 15 & 10 & 11 & 0 & 0 & 1 & 3 & 4 & 1 & 10 & 10 & 11 & 0 & 0 & 5 \\
5 & 1 & 2 & 0 & 1 & 0 & 10 & 11 & 10 & 11 & 10 & 0 & 0 & 1 & 0 & 12 & 12 & 10
\end{bmatrix}
$$

and a test data set is

$$
\text{Data} =
\begin{bmatrix}
10 & 0 & 0 & 11 & 3 & 0 \\
0 & 2 & 0 & 12 & 4 & 12 \\
0 & 0 & 3 & 0 & 5 & 34
\end{bmatrix}.
$$

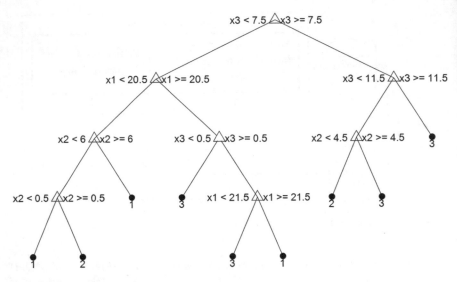

Fig. 4.16 Decision tree generation

Create the decision tree and then perform classification for the test data set.

Solution

Using computer simulation, we calculate impurity measures and obtain the decision tree as shown in Fig. 4.16.

Using this decision tree model, the test data is applied. The test data is classified as class 1, class 2, class 1, class 1, class 2, and class 3.

                                                                              ■

Summary 4.2 Decision Tree

1. Decision trees have a hierarchical data structure and are built using recursive partitioning based on a divide-and-conquer strategy.
2. The decision tree techniques are useful when the function and data set have the following properties: (1) Disjunctive expressions are required. The Boolean function can be fully expressed in the decision tree. Many other functions can be approximated by Boolean functions. (2) A data set is expressed by attribute value pairs. (3) The output of the target functions are discrete.
3. There are two types of decision trees: classification trees and regression trees. If the decision variable is discrete, the decision tree is built by a finite number of class values. We call it classification tree. Regression trees are very similar to classification trees. If the decision variable can

take continuous values, entropy as an impurity measure cannot be used and leaf nodes typically set to the mean values. There are no class labels.

4. Pruning is a key technique to avoid overfitting. In the pre-pruning, the decision tree stops growing at some point when there is unreliable information. In the post-pruning, growing the decision tree is not restricted by a threshold. The decision tree is grown fully until all training data is classified and then we remove nodes with insufficient evidence.

4.2.2 K-Nearest Neighbours

The K-nearest neighbours (KNN) algorithm is widely used because it is simple but powerful for both classification and regression problems. Classification accuracy is quite good. It has flexible decision boundaries. It is robust to noisy training data and effective when the training data is large enough. The disadvantages can be summarized as (1) run time is long when we have large training data, (2) it is sensitive to irrelevant features and they should be eliminated, (3) computational cost is high and a large memory is required, and (4) it is difficult to handle high dimensionality. Typically, it cannot handle more than 30 features. In order to solve a long run time problem, we can use subset of dimensions, calculate an approximate distance, remove redundancy, or sort training data priorly. In order to solve an irrelevant feature problem, we can eliminate some features or adapt weight of features. In order to solve a computational cost problem, we can remove redundancy or increase the storage requirements by pre-sorting. In order to deal with a high-dimensional data, we can increase the required training data and computational power. The nearest neighbour (NN) algorithm is a special case of the KNN algorithm when $k = 1$. It classifies data points with attributes to its nearest neighbours in a data set, where "nearest" is computed as a similarity measure. For regression, the value for the test data set becomes the average or the weighted average of the values of the k neighbours. It is mainly used for classification problems. As we discussed in Chap. 3, K-mean clustering algorithm is unsupervised learning algorithm, where k represents the number of clusters. However, in KNN algorithms as supervised learning, k represents the number of closest neighbours we will consider. The basic concept of KNN algorithm is to classify a data set using the majority vote of the k closest data points. Figure 4.17 illustrates an example of K-nearest neighbours. In this figure, the cross-query point is the point to classify. The algorithm performs to find which class it belongs to circle or dot. When 3 nearest neighbours are considered, the class is assigned to the cross query point by the majority vote. That is to say. There are 1 circle and 2 dots in the dashed circle. The majority vote determines that it is of the dot class. If the k value is even number, the majority vote doesn't work. Therefore, an odd or even number of k value should be chosen when the data has an even or odd number of classes, respectively.

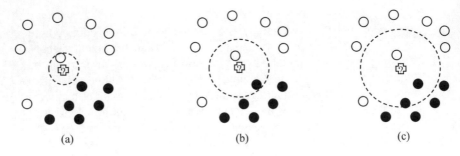

Fig. 4.17 Example of K-nearest neighbours: 1 nearest neighbour (**a**), 2 nearest neighbour (**b**), and 3 nearest neighbour (**c**)

This algorithm was proposed by Cover [11]. In this algorithm, the nearest neighbour is computed on the basis of value of k that specifies how many nearest neighbours are to be considered to define a class of a sample data point [11]. Two key properties of KNN algorithms are lazy and nonparametric learning. KNN algorithms do not need a training phase and use all data sets for training during classification. Therefore, it is lazy learning (also called instance-based learning). In addition, it doesn't assume anything on the underlying data distribution. There is no explicit training or model. Therefore, it is nonparametric. In order to make KNN algorithms work well, it is important to determine k values properly. If it is too small, the model is noisy and less stable. If it is too large, the model is computationally expensive and less precise. In principle, when we have the infinite number of data, larger k value enables us to have better classification. Therefore, it is important to find a proper k value.

Consider the labelled data set $\mathcal{D}_n = \{(\mathbf{x}_1, \theta_1), \ldots, (\mathbf{x}_n, \theta_n)\}$ where θ_i is the class label for $\mathbf{x}_i$. We assume the number of classes as C: $\theta_i \in \{1, 2, \ldots, C\}$. The value of the target function for a new data set is estimated from the known value of the nearest training data. The input $\mathbf{x}$ and its nearest neighbour $\mathbf{x}'$ satisfy

$$\mathbf{x}' = \arg \min_{\mathbf{x_i} \in \mathcal{D}_n} \text{dist}(\mathbf{x}, \mathbf{x_i}) \qquad (4.17)$$

where dist() is Euclidean distance. The 1-nearest neighbour decision rule is to find $(\mathbf{x}', \theta')$ from the training data set to the test data $\mathbf{x}$. In the NN classification model, we consider a Euclidean distance and compute the decision boundaries. The decision boundary between any two data points is a straight line. It shows us how input space is divided into classes. Each line segment is equidistant between two data points of classes. It is clear that the decision boundary of 1-NN is part of Voronoi diagram. Figure 4.18 illustrates an example of the nearest neighbour decision boundary as Voronoi diagram. In this figure, the circles represent the data points and the decision boundaries are determined by the straight line between two data points of opposite classes.

If data points have two different labels (circles and triangles) as shown in Fig. 4.19a, the classification can be performed as the bold line of Fig. 4.19b.

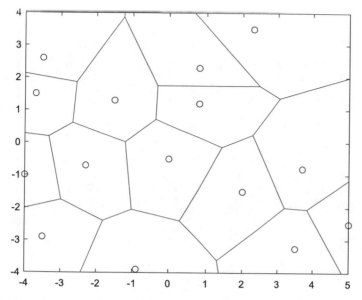

Fig. 4.18 Example of the nearest neighbour decision boundaries

In KNN algorithms, the class label can be considered as the outcome of majority voting. For the input $\mathbf{x}$, we find its K-nearest neighbours $\mathcal{D}_k = \{(\mathbf{x}_1, \theta_1), \ldots, (\mathbf{x}_k, \theta_k)\}$ of a query point $\mathbf{x_q}$. The K-nearest neighbour decision rule is to find k data points closest to the test data point $\mathbf{x}$ and the output of classification is as follows:

$$\theta' = \arg\max_{\theta_j} \sum_{i=1}^{k} \delta(\theta_j, \theta_i) \tag{4.18}$$

where $\delta()$ is the Kronecker delta function as follows:

$$\delta(x, y) = \begin{cases} 1, & \text{if } x = y \\ 0, & \text{if } x \neq y \end{cases} . \tag{4.19}$$

The pseudocode of KNN algorithm can be summarized as follows:

Procedure KNN $(\mathcal{D}_n)$

Load the training and test data;

Choose the value of k;

for Each data point in test data

- Compute the Euclidean distances between the query data points and all training data points
- Store and sort the Euclidean distances

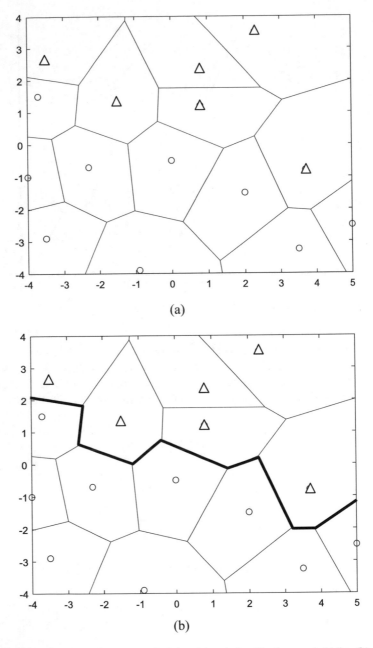

Fig. 4.19 Example of two different labelled data (**a**) and classification as a bold line (**b**)

- Choose the first k points
- Assign a class to the test data point by a majority vote

end

return θ_i

There are many variations of KNN algorithms to improve performance. The weighted KNN algorithm [12] gives weights to neighbours based on distance calculation and improves the performance. The condensed nearest neighbour algorithm [13] eliminates data sets with similarity and improves run time and memory requirements. The reduced nearest neighbour algorithm [14] eliminates patterns affecting training data set results, reduces the training data size, and improves the performance. The rank nearest neighbour algorithm [15] assigns ranks to training data and improves performance when we have many variations among features. The clustered K-nearest neighbour algorithm [16] enables us to form clusters to select nearest neighbours and overcome uneven distribution of training data.

Example 4.4 KNN Algorithm 1 Consider the following labelled data set:

i	$\mathbf{x}_i$	θ_i
1	(14, 14)	Class A
2	(14, 8)	Class A
3	(6, 8)	Class B
4	(2, 8)	Class B

When we have a new data point (6,14) and $k = 3$, find the classification of the data point.

Solution

Firstly, we compute the Euclidean distances between the new data point and all training data points as follows:

i	Euclidean distances
1	$\sqrt{(14 - 6)^2 + (14 - 14)^2} = 8$
2	$\sqrt{(14 - 6)^2 + (8 - 14)^2} = 10$
3	$\sqrt{(6 - 6)^2 + (8 - 14)^2} = 6$
4	$\sqrt{(2 - 6)^2 + (8 - 14)^2} = 7.2$

Secondly, we sort the Euclidean distances and determine the nearest neighbours as follows:

i	Euclidean distances	Rank and 3 nearest neighbours
1	8	3 and Yes
2	10	4 and No
3	6	1 and Yes
4	7.2	2 and Yes

Thirdly, we choose the first k points and assign a class to the test data point by the majority vote:

i	Euclidean distances	Rank and 3 nearest neighbours	Classification
1	8	3 and Yes	Class A
2	10	4 and No	
3	6	1 and Yes	Class B
4	7.2	2 and Yes	Class B

From the above table, we have 1 class A and 2 class B. By the majority vote, we classify the new data point (6,14) in class B.   ∎

Example 4.5 KNN Algorithm 2 Consider the labelled data set with three classes (Class A, B, and C) as shown in Fig. 4.20.

We have new data points (2.3, 0.7), (5.4, 1.5), and (6.5, 2.2). Compare the classification of the data points when and $k = 4$ and 8.

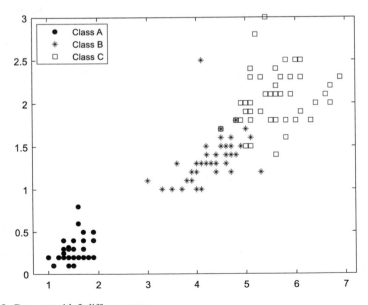

Fig. 4.20 Data set with 3 different types

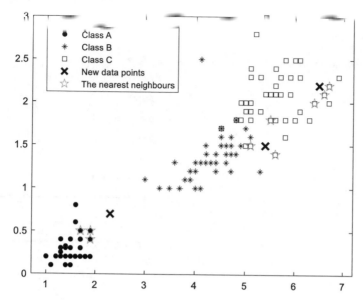

Fig. 4.21 Nearest neighbours when $k = 4$

Solution

Using computer simulation, when $k = 4$, we have the nearest neighbours as shown in Figs. 4.21 and 4.22.

As we can observe Figs. 4.21 and 4.22, we can classify the new point (2.3, 0.7) as class A, the new point (5.4, 1.5) as class C, and the new point (6.5, 2.2) as class C by the majority vote. When $k = 8$, we have the nearest neighbours as shown in Figs. 4.23 and 4.24.

As we can observe Figs. 4.23 and 4.24, we can classify the new point (2.3, 0.7) as class A, the new point (5.4, 1.5) as class C, and the new point (6.5, 2.2) as class C by the majority vote. In this example, the results of both cases are same. In general, choosing k value is important to improve accuracy. We call it parameter tuning. One possible approach is to try and error when the training data is unknown. Another approach is to set $k = \sqrt{N}$ where N is the number of the training data. Another approach is to choose k by cross-validation. That is to say. We take small part of the training data, validate data, and evaluate different k values. In this way, we minimize the validation error and obtain the final setting.

■

Fig. 4.22 Zoomed nearest
neighbours when $k = 4$ at the
new points $(2.3, 0.7)$ (**a**),
$(5.4, 1.5)$ (**b**), and $(6.5, 2.2)$
(**c**)

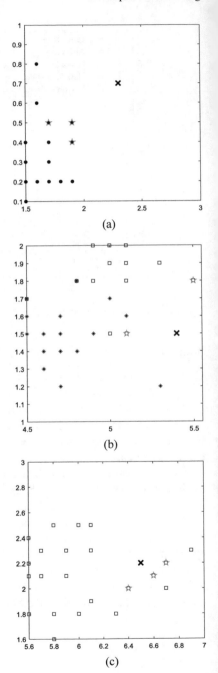

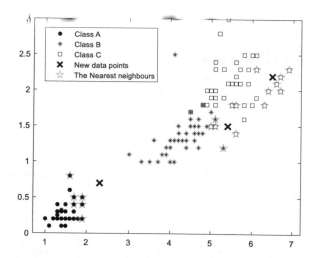

Fig 4.23 Nearest neighbours when $k = 8$

Summary 4.3 K Nearest Neighbours

1. KNN algorithms classify data points with attributes to its nearest neighbours in a data set, where "nearest" is computed as a similarity measure.
2. K-mean clustering algorithm is unsupervised learning algorithm, where k represents the number of clusters. However, in KNN algorithms as supervised learning, k represents the number of closest neighbours we will consider.
3. The pseudocode of KNN algorithm can be summarized as follows:

Procedure KNN $(\mathcal{D}_n)$

Load the training and test data;

Choose the value of k;
    for Each data point in test data

– Compute the Euclidean distances between the query data points and all training data points
– Store and sort the Euclidean distances
– Choose the first k points
– Assign a class to the test data point by majority vote

end

return θ_i.

Fig. 4.24 Zoomed nearest
neighbours when $k = 8$ at the
new points (2.3, 0.7) (**a**),
(5.4, 1.5) (**b**), and (6.5, 2.2)
(**c**)

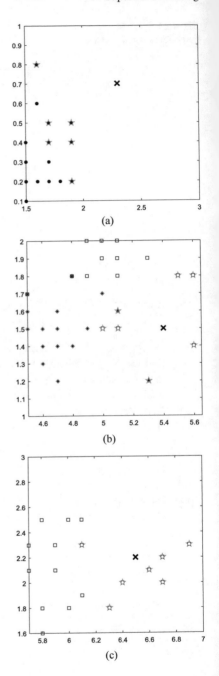

(a)

(b)

(c)

Fig. 4.25 Example of a linear classifier

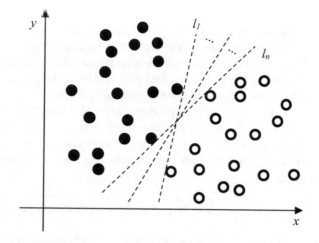

Fig. 4.25 Example of a linear classifier

4.2.3 Support Vector Machine[1]

The support vector machine (SVM) as one of the key machine learning techniques is a classification and regression tool. It was theoretically motived by statistical learning theory [17]. The SVMs have been recently developed [18] and have been applied to many applications including pattern recognition, medical diagnosis, and others. The SVM technique became famous because it provides us with accuracy in handwriting recognition tasks [19]. The advantages of SVMs can be summarized as follows: (1) It is robust to large number of variables and small samples. (2) It works very well when there is a margin between classes and an unstructured data set. (3) It is scalable and works well at high-dimensional data. (4) Overfitting is not common. (5) It doesn't trap in local minima. On the other hand, SVMs have the following disadvantages: (1) It is not suitable for large data. (2) It doesn't work well when data has more noise. (3) It doesn't work well when the number of features for data points exceeds the number of training data. (4) It takes a long time for learning due to quadratic programming optimization. The basic concept of SVMs is to determine an optimal separating hyperplane maximizing the gap between classes and decision boundary and minimizing the misclassified training data and unseen test data. Let's consider linearly separable training samples in order to understand the basic concept of SVMs. Figure 4.25 illustrates an example of a linear classifier.

As we can observe Fig. 4.25, there are multiple linear separating lines between two classes. It is possible to classify them by many lines (l_1, ..., l_n). A good margin is to satisfy larger separation for both classes, and a bad margin is the one that is very close to one class. The SVM techniques find a linear decision surface separating classes as maximizing the distance (or gap, margin) between classes. Some points that are the closest points to the optimal line (or hyperplane) are important, and

---

[1] Refer to the book "Haesik Kim, *Design and Optimization for 5G Wireless Communications*, Wiley, April 2020, ISBN: 978-1-119-49452-2.", the Sect. 4.2.3 is written and new contents are added.

other samples do not affect to find the optimal line. We call them support vectors. They are the critical elements of the training samples. The support vectors would change the position of the separating hyperplane if it is removed. If there is no a linear decision surface, we can map the data into a higher-dimensional space where separating decision surface can be found as shown in Fig. 4.26a, b. The feature space can be constructed by mathematical projection.

The mapping of the input data space into feature space involves increasing dimension of the feature space. After transforming, we can obtain simpler mathematical function describing the separating boundary and find a solution by examining simpler feature space. The data transformation as transform function $\varphi()$ can be expressed as follows:

$$\mathbf{x} = (x_1, \ldots, x_n) \rightarrow \varphi(\mathbf{x}) = (\varphi(x_1), \ldots, \varphi(x_n)). \tag{4.20}$$

Figure 4.27 illustrates an example of data transformation.

As we can observe Fig. 4.27a, the circles and dots are randomly distributed and cannot be separated by a straight line. In machine learning, a kernel method is used to solve a nonlinear problem by a linear classifier. The kernel function transforming linearly inseparable data to linearly separable data is computationally cheaper than the explicit computation. The optimal hyperplanes may be useless if the training data set is not linearly separable. The kernel function enables us to find complicated decision boundaries accommodating a training data set. However, this approach may not work well if the model is noisy. The SVM is one of the well-known kernel methods. The decision function is specified by a subset of training samples (support vectors). Finding the optimal separating hyperplane is an optimization problem. This becomes a quadratic programming problem that is easy to solve using Lagrange multipliers in a standard form. It would be optimal choice when maximizing the margin. We can find the optimal separating hyperplane. Figure 4.28 illustrates an example of the support vectors and maximum margin. The position of the optimal hyperplane is determined by the few samples closest to the hyperplane.

In general, there is a trade-off relationship between the margin and generalization error (or misclassified points). We consider n training samples of the form: $\{\mathbf{x}_i, y_i\}$ where input vector $\mathbf{x}_i \in \mathbb{R}^d$ with $i = 1, \ldots, n$ is of dimensionality d and belongs to the response variables $y_i \in \{-1, +1\}$. The training data is a subset of all possible data. Figure 4.29 illustrates example of poor and good generalization. As we can observe Fig. 4.29a, new data is on the wrong side of the hyperplane. This is poor generalization. In Fig. 4.29b, new data points are close to the training data and are on the right side of the hyperplane. This is good generalization.

We assume that data is linearly separable and finds a linear classifier for a binary classification problem. The line $f(\mathbf{x}) = \mathbf{w}^T \mathbf{x} - b$ classifies samples with $y_i = -1$ on one side ($f(\mathbf{x}_i) < 0$) and samples with $y_i = +1$ on the other side ($f(\mathbf{x}_i) > 0$). The separating line (or hyperplane) can be described as follows:

$$\mathbf{w}^T \mathbf{x} = b \tag{4.21}$$

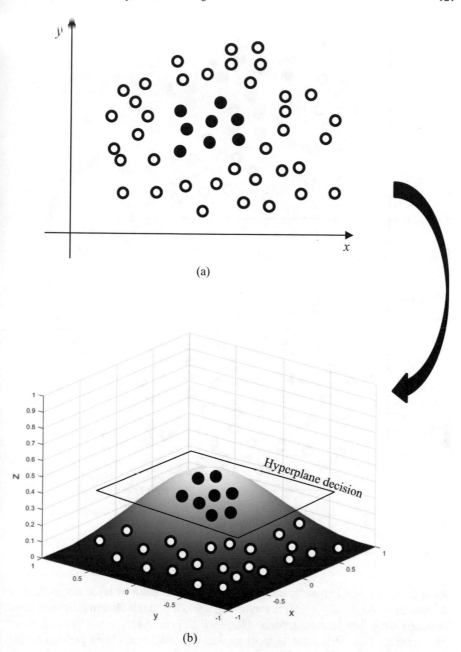

Fig. 4.26 Example of hyperplane decision in two-dimensional space (**a**) and multidimensional space (**b**)

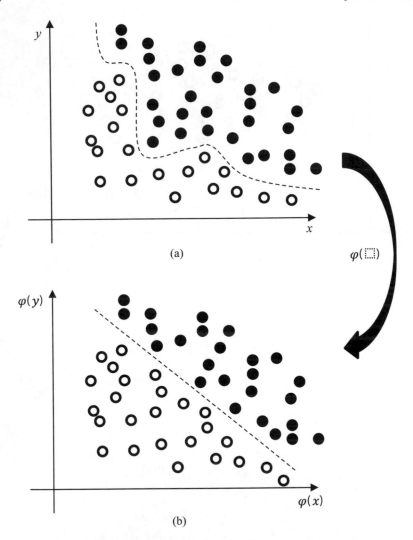

Fig. 4.27 Example of input data space (**a**) and feature space (**b**)

where the vector **w** is perpendicular to the line $\mathbf{w}^T\mathbf{x} = b$ and $b/\mathbf{w}$ is the perpendicular distance between the origin and the separating line. We call this discriminant function. The separating line acts as a linear classifier. Support vectors are the closest to the separating line. Using the support vectors, we define two lines parallel to the separating line. Two lines cut through the closest training samples on each side. We call those support lines. The distances between them are d_1 and d_2. The margin γ is defined as addition of d_1 and d_2. Two support lines are described as follows:

$$\mathbf{w}^T\mathbf{x} = b + \delta \text{ and } \mathbf{w}^T\mathbf{x} = b - \delta \tag{4.22}$$

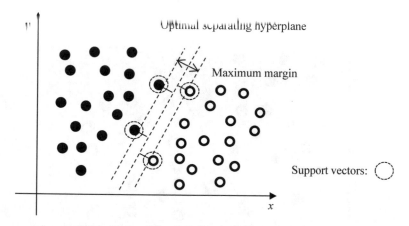

Fig. 4.28 Example of maximum margin and support vectors

where δ is the offset on both sides. The SVM finds the optimal $\mathbf{w}$ for any δ. Thus, any nonzero offset establishes a margin optimization. The margin γ is twice distance d ($=d_1$ or d_2) of the closest samples to the separating line because of equidistant from both support lines. Thus, the margin γ is $2/\|\mathbf{w}\|$. Given a vector $\mathbf{x}_i$, the following equations should be satisfied:

$$\mathbf{w}^T\mathbf{x}_i - b \leq -1 \text{ for } y_i = -1 \tag{4.23}$$

$$\mathbf{w}^T\mathbf{x}_i - b \geq +1 \text{ for } y_i = +1 \tag{4.24}$$

and (4.23) and (4.24) can be written as follows:

$$y_i\left(\mathbf{w}^T\mathbf{x}_i - b\right) - 1 \geq 0, 1 \leq i \leq n. \tag{4.25}$$

(4.25) becomes the constraint of the optimization problem. Now, we formulate the primal problem of linear SVMs as follows:

$$\min \frac{1}{2}\|\mathbf{w}\|^2 \tag{4.26}$$

subject to

$$y_i\left(\mathbf{w}^T\mathbf{x}_i - b\right) - 1 \geq 0, 1 \leq i \leq n. \tag{4.27}$$

This is a constrained quadratic programming problem. We maximize the margin subject to the constraint that all data falls on one side or the other. It cannot be solved by quadratic optimization when the number of training samples is large. Thus, we transform to the dual formulation and use the method of Lagrange multipliers α as

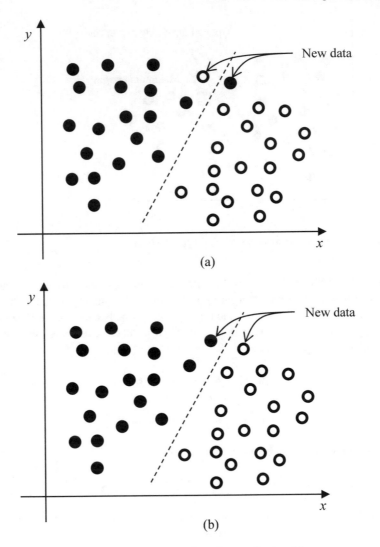

Fig. 4.29 Example of poor generalization (**a**) and good generalization (**b**)

follows:

$$L(\mathbf{w}, b, \boldsymbol{\alpha}) = \frac{1}{2}\|\mathbf{w}\|^2 - \boldsymbol{\alpha}\big(y_i\big(\mathbf{w}^T\mathbf{x}_i - b\big) - 1\big), \quad 1 \leq i \leq n \qquad (4.28)$$

$$= \frac{1}{2}\|\mathbf{w}\|^2 - \sum_{i=1}^{n}\alpha_i\big(y_i\big(\mathbf{w}^T\mathbf{x}_i - b\big) - 1\big), \qquad (4.29)$$

$$-\frac{1}{2}\|\mathbf{w}\|^{1}-\sum_{i=1}^{n}\alpha_{i}y_{i}\left(\mathbf{w}^{1}\mathbf{x}_{i}-b\right)+\sum_{i=1}^{n}\alpha_{i}. \tag{4.30}$$

The dual formulation doesn't require to access the original data and allows us to compute a simple equation. Finding the minimum of $L(\mathbf{w}, b, \boldsymbol{\alpha})$ means that we differentiate it with respect to $\mathbf{w}$ and b and set the derivatives to zero as follows:

$$\frac{\partial L}{\partial \mathbf{w}} = 0 \rightarrow \mathbf{w} = \sum_{i=1}^{n}\alpha_{i}y_{i}\mathbf{x}_{i}, \tag{4.31}$$

$$\frac{\partial L}{\partial b} = 0 \rightarrow 0 = \sum_{i=1}^{n}\alpha_{i}y_{i}. \tag{4.32}$$

In addition, inequality constraint and complementary slackness should be satisfied as follows: $y_{i}\left(\mathbf{w}^{T}\mathbf{x}_{i}-b\right) - 1 \geq 0$ and $\alpha\left(y_{i}\left(\mathbf{w}^{T}\mathbf{x}_{i}-b\right) - 1\right) = 0$. This condition means that the inequality constraint is satisfied when $\alpha_{i} \geq 0$ and the inequality is saturated when $\alpha_{i} = 0$. By substituting (4.31) and (4.32) into (4.30), we have its dual form as follows:

$$\max_{\boldsymbol{\alpha}} L = \sum_{i=1}^{n}\alpha_{i} - \frac{1}{2}\sum_{i=1}^{n}\sum_{j=1}^{n}\alpha_{i}\alpha_{j}y_{i}y_{j}\mathbf{x}_{i}^{T}\mathbf{x}_{j} \tag{4.33}$$

subject to

$$\sum_{i=1}^{n}\alpha_{i}y_{i} = 0, \quad \alpha_{i} \geq 0, 1 \leq i \leq n. \tag{4.34}$$

This dual form is a constrained quadratic programming problem as well but the number of new variables $\boldsymbol{\alpha}$ is same as the number of training samples. The term L of (4.33) can be rewritten as follows:

$$L = \sum_{i=1}^{n}\alpha_{i} - \frac{1}{2}\sum_{i=1}^{n}\sum_{j=1}^{n}\alpha_{i}\alpha_{j}y_{i}y_{j}\mathbf{x}_{i}^{T}\mathbf{x}_{j} = \sum_{i=1}^{n}\alpha_{i} - \frac{1}{2}\sum_{i=1}^{n}\sum_{j=1}^{n}\alpha_{i}H_{ij}\alpha_{j}, \tag{4.35}$$

$$L = \sum_{i=1}^{n}\alpha_{i} - \frac{1}{2}\boldsymbol{\alpha}^{T}\mathbf{H}\boldsymbol{\alpha} \tag{4.36}$$

where $H_{ij} = y_{i}y_{j}\mathbf{x}_{i}^{T}\mathbf{x}_{j}$. The dual form is rewritten as follows:

$$\max_{\boldsymbol{\alpha}} \sum_{i=1}^{n}\alpha_{i} - \frac{1}{2}\boldsymbol{\alpha}^{T}\mathbf{H}\boldsymbol{\alpha} \tag{4.37}$$

subject to

$$\sum_{i=1}^{n} \alpha_i y_i = 0, \quad \alpha_i \geq 0, 1 \leq i \leq n. \tag{4.38}$$

Figure 4.30 illustrates example of non-separable case and slack variables. The slack variables are introduced to allow certain constraints to be violated. The slack variable is a measure of deviation. The slack variables $s_i > 1$ represent that sample is located in the wrong side of separating hyperplane. The slack variables $0 < s_i < 1$ represent that sample is located in the right side of separating hyperplane but within

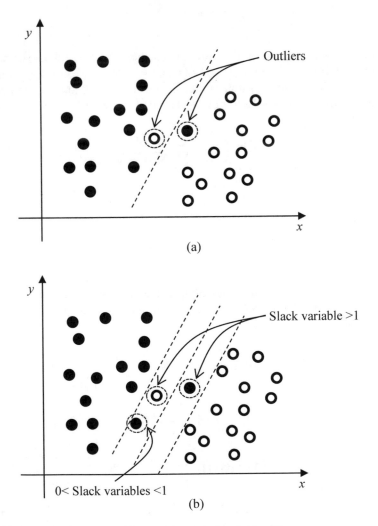

Fig. 4.30 Example of non-separable case (**a**) and slack variables (**b**)

the maximum margin region. The slack variables $s_i < 1$ represent that sample is located in the right side of separating hyperplane.

In some cases, the training samples may not be separable, some samples may lie within the margin, and they couldn't be classified correctly. We want to handle this case without a big change. Thus, a cost penalty C is used. If $C \to 0$, it maximizes the margin, and misclassified points are allowed. If $C \to \infty$, it minimizes the number of misclassified points and no errors are allowed. The cost penalty C acts like a regularizing parameter. We can generalize linear SVM. The classifier (4.25) can be rewritten by n non-negative slack variables s_i as follows:

$$y_i\left(\mathbf{w}^T \mathbf{x}_i - b\right) \geq 1 - s_i,\ 1 \leq i \leq n \tag{4.39}$$

and the cost function becomes

$$\|\mathbf{w}\|^2 + C \sum_{i=1}^{n} s_i^m \tag{4.40}$$

where m is an integer. The cost penalty C represents the effect of the slack variables, and the slack variables s_i affect to misclassified points, generalization, or computational efficiency. Thus, the generalized optimal separating hyperplane is formulated as follows:

$$\min \frac{1}{2}\|\mathbf{w}\|^2 + C \sum_{i=1}^{n} s_i^m \tag{4.41}$$

subject to

$$y_i\left(\mathbf{w}^T \mathbf{x}_i - b\right) \geq 1 - s_i,\quad s_i \geq 0, 1 \leq i \leq n. \tag{4.42}$$

After including the slack variables, the associated dual form is

$$\max_{\alpha} \sum_{i=1}^{n} \alpha_i - \frac{1}{2}\alpha^T \mathbf{H}\alpha \tag{4.43}$$

subject to

$$\sum_{i=1}^{n} \alpha_i y_i = 0,\quad 0 \leq \alpha_i \leq C, 1 \leq i \leq n. \tag{4.44}$$

As we can observe (4.37), (4.38), (4.43), and (4.44), the only difference is the constraint of the Lagrange multipliers. It includes an upper bound C. If $C \to 0$, misclassified points are not allowed. If $C \to \infty$, it is same as the original problem and implies smaller errors on the training samples. They are a trade-off relationship.

A support vector $\mathbf{x}_s$ satisfies (4.32), and we have the following equation:

$$y_s\left(\mathbf{w}^T\mathbf{x}_s - b\right) = 1. \tag{4.45}$$

By substituting (4.45) into (4.31), we have

$$y_s\left(\sum_{l \in S} \alpha_l y_l \mathbf{x}_l\, \mathbf{x}_s - b\right) = 1 \tag{4.46}$$

where S is the set of support vectors. If a sample is not a support vector, $\alpha_i = 0$. If a sample is a support vector, $\alpha_i > 0$ and it satisfies (4.46). From (4.23) and (4.24), $y_s^2 = 1$. Thus, (4.46) can be rewritten as follows:

$$y_s^2\left(\sum_{l \in S} \alpha_l y_l \mathbf{x}_l\, \mathbf{x}_s - b\right) = y_s, \tag{4.47}$$

$$\left(\sum_{l \in S} \alpha_l y_l \mathbf{x}_l\, \mathbf{x}_s - b\right) = y_s, \tag{4.48}$$

$$b = \sum_{l \in S} \alpha_l y_l \mathbf{x}_l\, \mathbf{x}_s - y_s. \tag{4.49}$$

Using (4.49) and support vectors, we can determine b. In order to achieve numerical stability, we can take an average over all of support vectors. Alternatively, we can find it from complementary slackness condition as follows:

$$0 = y_i\left(\mathbf{w}^T\mathbf{x}_i - b\right) - 1 \text{ when } \alpha_i > 0 \tag{4.50}$$

$$b = -\frac{1}{y_i} + \mathbf{w}^T\mathbf{x}_i, \tag{4.51}$$

In addition, we can find $\mathbf{w}$ using (4.31). Thus, we have both variables b and $\mathbf{w}$ to define the separating line. The important thing is that samples appear only through the inner products of $\mathbf{x}_i^T\mathbf{x}_j$. When dealing with samples in a higher-dimensional space, we can replace it with kernel function (or kernel matrices). The kernel function is $K\left(\mathbf{x}_i, \mathbf{x}_j\right) = \mathbf{x}_i^T\mathbf{x}_j$. The vector $\boldsymbol{\alpha}$ of the dual form contains many zeros. The sparsity provides us with the reduction of the memory and computational time in practice.

The data set we will face in practice will be most likely not linearly separable. However, a linear classifier may still be useful. Thus, we need to change to nonlinear support vector machines and apply it for non-separable cases. We relax the constraints for (4.23) and (4.24) by introducing slack variables s_i as follows:

$$\mathbf{w}^T\mathbf{x}_i - b \leq -1 + s_i \text{ for } y_i = -1 \tag{4.52}$$

$$\mathbf{w}^T \mathbf{x}_i - b \geq +1 - s_i \text{ for } y_i = +1 \tag{4.53}$$

$$s_i \geq 0, \ 1 \leq i \leq n \tag{4.54}$$

where the slack variables allow for misclassified points. The primal problem has been reformulated to (4.41) and (4.42) using the penalty function $C \sum_{i=1}^{n} s_i^m$ of (4.40). The penalty function becomes a part of the objective function. When m is small integer (For example, $m = 1$ or 2), it is still a quadratic programming problem. The penalty function represents how much samples are misclassified. Now, we reformulate (4.41) and (4.42) using a Lagrangian as follows:

$$L(\mathbf{w}, b, s, \boldsymbol{\alpha}, \boldsymbol{\mu}) = \frac{1}{2}\|\mathbf{w}\|^2 + C \sum_{i=1}^{n} s_i - \sum_{i=1}^{n} \alpha_i \left(y_i \left(\mathbf{w}^T \mathbf{x}_i - b\right) - 1 + s_i \right) - \sum_{i=1}^{n} \mu_i s_i. \tag{4.55}$$

In (4.55), the term $\frac{1}{2}\|\mathbf{w}\|^2 + C \sum_{i=1}^{n} s_i$ is minimization function, the term $C \sum_{i=1}^{n} s_i - \sum_{i=1}^{n} \alpha_i \left(y_i \left(\mathbf{w}^T \mathbf{x}_i - b\right) - 1 + s_i \right)$ is margin constraint, and the term $\sum_{i=1}^{n} \mu_i s_i$ is error constraint. We drive the KKT conditions as follows:

$$\frac{\partial L(\mathbf{w}, b, s, \boldsymbol{\alpha}, \boldsymbol{\mu})}{\partial \mathbf{w}} = 0 \rightarrow \mathbf{w} - \sum_{i=1}^{n} \alpha_i y_i \mathbf{x}_i = 0, \tag{4.56}$$

$$\frac{\partial L(\mathbf{w}, b, s, \boldsymbol{\alpha}, \boldsymbol{\mu})}{\partial b} = 0 \rightarrow \sum_{i=1}^{n} \alpha_i y_i = 0, \tag{4.57}$$

$$\frac{\partial L(\mathbf{w}, b, s, \boldsymbol{\alpha}, \boldsymbol{\mu})}{\partial s} = 0 \rightarrow C - \alpha_i - \mu_i = 0, \tag{4.58}$$

$$y_i \left(\mathbf{w}^T \mathbf{x}_i - b\right) - 1 + s_i \geq 0, \tag{4.59}$$

$$s_i \geq 0, \ \alpha_i \geq 0, \ \mu_i \geq 0, \tag{4.60}$$

$$\alpha_i \left(y_i \left(\mathbf{w}^T \mathbf{x}_i - b\right) - 1 + s_i \right) = 0, \tag{4.61}$$

$$\mu_i s_i = 0. \tag{4.62}$$

In non-separable case, we can use a linear classifier by projecting data set into a higher dimension using a function $\varphi()$. In order to solve nonlinear classification problems with a linear classifier, we firstly project a data set $\mathbf{x}$ to high dimension with mapping function $\varphi(\mathbf{x})$. Secondly, we find a linear discriminant function for the transformed data. The discriminant function on the original data set would not be linear but the discriminant function on the transformed data set would be linear. This method is very useful when dealing with nonlinear classification problems. However, we should carefully deal with this projection due to the curse of dimensionality. As the number of dimensions increases, the amount of data we need to generalize correctly increases exponentially and there is a risk of overfitting the data. In addition, computational complexity increases with the dimension of the space. We call this the curse of dimensionality. In order to avoid this problem, we define an implicit mapping to a higher-dimensional feature space and computation in a high dimension is performed implicitly using the kernel functions. In this case, the kernel function is $K\left(\mathbf{x}_i, \mathbf{x}_j\right) = \varphi(\mathbf{x}_i)^T \varphi(\mathbf{x}_j)$. It enables us to avoid an explicit mapping. The kernel function is very useful because it allows us to compute the separating hyperplane without mapping to a high-dimensional space. The required calculation of the kernel function is in $\mathbf{x}_i \in \mathbb{R}^d$, and the complexity is not so high. There are multiple forms of the kernel functions as follows:

$$\text{Linear: } K\left(\mathbf{x}_i, \mathbf{x}_j\right) = \mathbf{x}_i^T \mathbf{x}_j, \tag{4.63}$$

$$\text{Polynomial: } K\left(\mathbf{x}_i, \mathbf{x}_j\right) = \left(\mathbf{x}_i \mathbf{x}_j + c\right)^q, \tag{4.64}$$

$$\text{Gaussian radial basis function: } K\left(\mathbf{x}_i, \mathbf{x}_j\right) = \exp\left(-\frac{\|\mathbf{x}_i - \mathbf{x}_j\|^2}{2\sigma^2}\right), \tag{4.65}$$

$$\text{Sigmoide: } K\left(\mathbf{x}_i, \mathbf{x}_j\right) = \tanh\left(a\mathbf{x}_i \mathbf{x}_j - b\right) \tag{4.66}$$

where $\sigma > 0$ is a scale parameter, $a, b, c > 0$, and degree q is an integer. When training data in SVM, the key parameters are the kernel function K and the cost penalty C. In general, there is no optimal way for choosing a kernel function for a given problem. Thus, one popular approach is to try Gaussian radial basis function because it requires only one parameter σ, or polynomial kernel with low degree ($q = 1$ or 2). The kernel function is also known as similarity function because it evaluates the similarity between data points. The output of the kernel function is maximized as two data points are equivalent. The SVMs work well at binary classification. When dealing with multiple classification, there are two approaches: the one-against-the-rest and the one-against-the-one. In the one-against-the-rest approach, we divide the K-class problem into K binary classification sub-problems such as "kth class" vs. "not kth class", k = 1, 2,..., K. We construct a hyperplane between them. In the one-against-the-one approach, we divide the K-class problem into comparisons of all pairs of classes. We also control a measure of complexity by maximizing the margin of the hyperplane. The complexity of nonlinear SVM is generally between $\mathcal{O}(n^2)$ and $\mathcal{O}(n^3)$ where n is a training data set size.

Example 4.6 *Support Vector Machine 1* Consider the following data set with 6 samples:

Class A: (1,4), (2,10), (4,9),
Class B: (6,2), (7,8), (8,3).

Find a classifier using SVM.

Solution

Firstly, we create two matrices from the given data set as follows:

$$\mathbf{x} = \begin{bmatrix} 1 & 4 \\ 2 & 10 \\ 4 & 9 \\ 6 & 2 \\ 7 & 8 \\ 8 & 3 \end{bmatrix} \text{ and } \mathbf{y} = \begin{bmatrix} 1 \\ 1 \\ 1 \\ -1 \\ -1 \\ -1 \end{bmatrix}.$$

Figure 4.31 illustrates the data set.
Secondly, we find the matrix $\mathbf{H}$ using $H_{ij} = y_i y_j \mathbf{x}_i^T \mathbf{x}_j$ as follows:

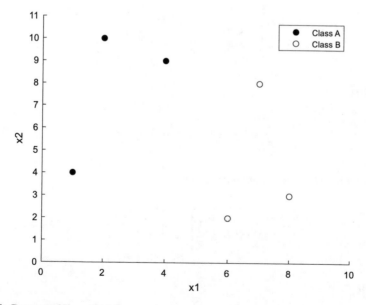

Fig. 4.31 Data set of Example 4.6

$$H = \begin{bmatrix} 17 & 42 & 40 & -14 & -39 & -20 \\ 42 & 104 & 98 & -32 & -94 & -46 \\ 40 & 98 & 97 & -42 & -100 & -59 \\ -14 & -32 & -42 & 40 & 58 & 54 \\ -39 & -94 & -100 & 58 & 113 & 80 \\ -20 & -46 & -59 & 54 & 80 & 73 \end{bmatrix}.$$

Thirdly, we solve a constrained quadratic programming problem:

$$\max_{\alpha} \sum_{i=1}^{n} \alpha_i - \frac{1}{2}\alpha^T H\alpha$$

subject to

$$\sum_{i=1}^{n} \alpha_i y_i = 0, \quad \alpha_i \geq 0, \ 1 \leq i \leq n.$$

We need to transform the minimization optimization problem to standard form as follows:

$$\min_{\alpha} - \sum_{i=1}^{6} \alpha_i + \frac{1}{2} \begin{bmatrix} \alpha_1 \\ \vdots \\ \alpha_6 \end{bmatrix}^T H \begin{bmatrix} \alpha_1 \\ \vdots \\ \alpha_6 \end{bmatrix}$$

subject to

$$\sum_{i=1}^{6} \alpha_i y_i = 0, \alpha_i \geq 0, 1 \leq i \leq 6.$$

We rewrite the above problem as follows:

$$\min_{\alpha} f^T\alpha + \frac{1}{2} \begin{bmatrix} \alpha_1 \\ \vdots \\ \alpha_6 \end{bmatrix}^T H \begin{bmatrix} \alpha_1 \\ \vdots \\ \alpha_6 \end{bmatrix}$$

subject to

$$B\alpha = b, A\alpha \leq a, 1 \leq i \leq 6$$

where $\mathbf{f}$, $\mathbf{B}$, $\mathbf{b}$, $\mathbf{A}$, and $\mathbf{a}$ are $\mathbf{f} = \begin{bmatrix} -1 \\ \vdots \\ -1 \end{bmatrix}$, $\mathbf{B} = \begin{bmatrix} y_1 & & y_6 \\ 0 & \cdots & 0 \\ \vdots & \ddots & \vdots \\ 0 & \cdots & 0 \end{bmatrix}$, $\mathbf{b} = \begin{bmatrix} 0 \\ \vdots \\ 0 \end{bmatrix}$, $\mathbf{A} =$

$\begin{bmatrix} -1 & \cdots & 0 \\ \vdots & \ddots & \vdots \\ 0 & \cdots & -1 \end{bmatrix}$, and $\mathbf{a} = \begin{bmatrix} 0 \\ \vdots \\ 0 \end{bmatrix}$.

Using a quadratic programming solver, we find α as follows:

$$\alpha = \begin{bmatrix} 0 \\ 0 \\ 0.2 \\ 0 \\ 0.2 \\ 0 \end{bmatrix}.$$

This means the support vectors are the data points (4, 9) and (7, 8).
Fourthly, we find $\mathbf{w}$ as follows:

$$\mathbf{w} = \sum_{i=1}^{6} \alpha_i y_i \mathbf{x}_i = \begin{bmatrix} -0.6 \\ 0.2 \end{bmatrix}.$$

Fifthly, since $\alpha_3 > 0$, we find b as follows:

$$b = -\frac{1}{y_3} + \mathbf{w}^T \mathbf{x}_3 = -1.6.$$

Figure 4.32 illustrates the classifier using SVM.

∎

Example 4.7 Support Vector Machine 2 Consider the following data set with 50 samples:

Class A: (8, 10), (6, 2), (2,1), (5, 4), (6, 6), (11, 6), (2,1), (1, 2), (2, 3), (5,5), (8,5), (7,7), (8,6), (7, 5), (5,6), (10,8), (5,1), (3,9), (4, 2), (8,5), (6,3), (8,4), (10,7), (10, 2), (6,3).
Class B: (12,19), (10,17), (13,14), (19,18), (15,20), (14,15), (11,16), (10,11), (15,17), (12,19), (10,13), (11,15), (20,13), (14,10), (19,12), (16,11), (14,15), (19,10), (18,20), (15,17), (18,10), (19,10), (14,12), (13,15), (16,11).

Find a classifier using SVM.

Solution

In the same manner as Example 4.6, we find a classifier. Figure 4.33 illustrates the

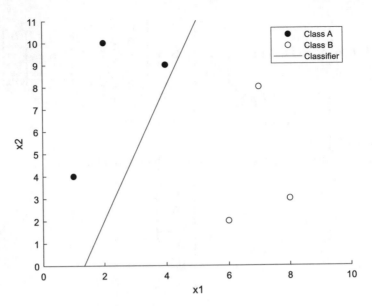

Fig. 4.32 Classifier for the Example 4.6 data set

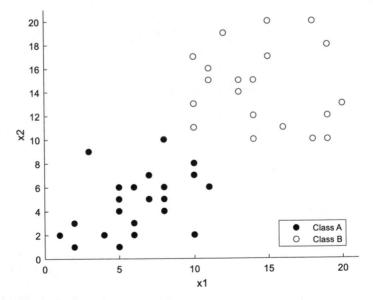

Fig. 4.33 Data set of Example 4.7

data set

After finding the matrix $\mathbf{H}$ using $H_{ij} = y_i y_j \mathbf{x}_i^T \mathbf{x}_j$ and solving a constrained quadratic programming problem, we find α as follows:

$$\alpha = [0.33\,0\,0\,0\,0\,0\,0\,0\,0\,0\,0\,0\,0\,0\,0.11\,0\,0\,0\,0\,0\,0\,0\,0\,0\,0\,0\,0\,0\,0\,0\,0.44\,0\,0\,0\,0\,0\,0\,0\,0\,0\,0\,0\,0\,0\,0\,0\,0\,0]^T$$

Thus, $\alpha(1)$, $\alpha(16)$, and $\alpha(33)$ are nonzero. The support vectors are $(8, 10)$, $(10,8)$, and $(10,11)$. Then, we find w as follows:

$$\mathbf{w} = \sum_{i=1}^{50} \alpha_i y_i \mathbf{x}_i = \begin{bmatrix} -0.67 \\ -0.67 \end{bmatrix}.$$

Since $\alpha_1 > 0$, we find b as follows:

$$b = -\frac{1}{y_1} + \mathbf{w}^T \mathbf{x}_1 = 13.$$

Figure 4.34 illustrates the classifier using SVM.

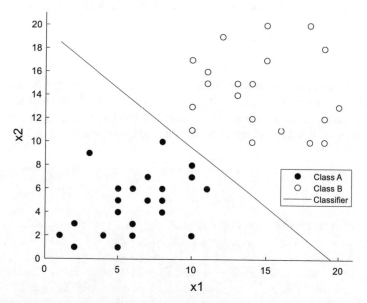

Fig. 4.34 Classifier for the Example 4.7 data set ∎

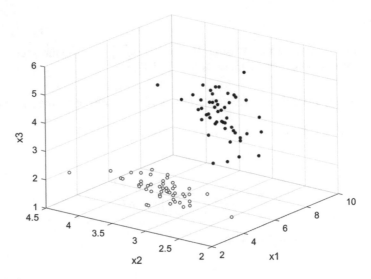

Fig. 4.35 Data set

Example 4.8 Support Vector Machine 3 Consider the following data set with three-dimensional 150 samples and two classes (Class A (Circle) and B (Dot)) as shown in Fig. 4.35.

Find a classifier using SVM.

Solution

We use a Gaussian radial basis function as kernel function and apply SVM method. Using computer simulation, we find the support vectors. The support vectors in each class are represented as a large circle. Figure 4.36 illustrates the support vectors.

The classifier as a hyperplane is expressed as grey surface in Fig. 4.37.

Summary 4.4. Support Vector Machines

1. The basic concept of SVMs is to determine an optimal separating hyperplane maximizing the gap between classes and decision boundary and minimizing the misclassified training data and unseen test data.
2. The kernel function enables us to find complicated decision boundaries accommodating a training data set. It is very useful because it allows us to compute the separating hyperplane without mapping to a high dimensional space.
3. Finding the optimal separating hyperplane is an optimization problem. This becomes a quadratic programming problem which is easy to solve

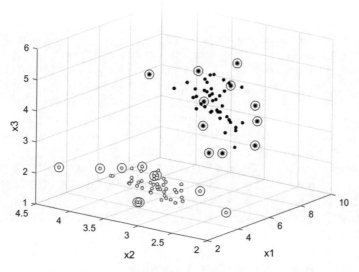

Fig. 4.36 Support vectors for data set

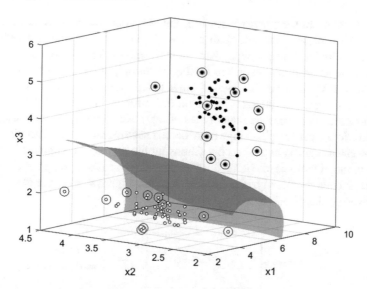

Fig. 4.37 Classifier for data set ■

using Lagrange multipliers in a standard form. It would be optimal choice when maximizing the margin.

4.   The advantages of SVMs can be summarized as follows: (1) It is robust to large number of variables and small samples. (2) It works very well when there is a margin between classes and an unstructured data set. (3) It is scalable and works well at high dimensional data. (4) Overfitting is not common. (5) It doesn't trap in local minima.

5.   SVMs have the following disadvantages: (1) It is not suitable for large data. (2) It doesn't work well when data has more noise. (3) It doesn't work well when the number of features for data points exceeds the number of training data. (4) It takes a long time for learning due to quadratic programming optimisation.

4.3   Regression of Supervised Learning

As we discussed in the previous section, the classification enables us to identify the categorical class associated with a given input. In regression, we estimate a continuous variable using independent input variables. The regression is defined as follows: regression is a statistical method determining the relationship between one dependent variable and a series of other independent variables and enables us to predict a continuous outcome based on the value of one or multiple variables. In machine learning, there are many types of regression techniques including linear regression, logistic regression, ridge regression, Lasso regression, polynomial regression, Bayesian regression, and so on. Choosing the regression techniques depends on type of target variable, the number of independent variables, shape of the regression line, model complexity, accuracy, and so on. Figure 4.38 illustrates a simple comparison of classification and regression.

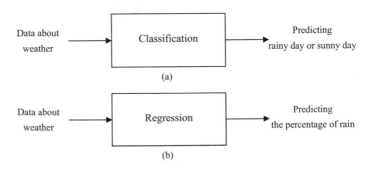

Fig. 4.38 Comparison of classification (**a**) and regression (**b**)

4.3.1 Linear Regression[2]

The basic concept of regression was used to find the orbits of comets and planets around the sun by French mathematician A. M. Legendre in 1805. In 1806, Gauss's derivation of the normal distribution for error term was introduced. F. Galton coined the term regression to describe biological phenomena and created the statistical concept of regression in 1870s. Mathematical background was developed in the field of statistics and studied to describe the relationship between inputs and outputs variables. However, machine learning adopts this concept and uses it for many applications. The main difference between statistical methods and machine learning methods can be summarized as follows: statistical methods are developed for inference to describe the relationship between inputs and output variables. They focus on model structure and statistical distribution. Machine learning models are developed for accurate predictions while they require more training data and take a risk like overfitting. They focus on model flexibility and predictive accuracy. However, their differences are getting more blurred. Linear regression can be regarded as a type of supervised learning. Like classification algorithms, we use a training data set to develop a model and then make a prediction. In linear regression, we assume the underlying behaviour is linear and predicts continuous values. If there is only one variable, it is a straight line. Linear regression provides us with good introduction to many regression concepts of machine learning. We can expand this concept to solve other complicated problems. For example, we can have a generalized linear model and they learn nonlinear mappings but their properties are maintained as a linear model. In addition, using kernel methods, they can apply to nonlinear functions or avoid overfitting.

Firstly, we consider simple linear regression model. The one-dimensional independent variable (or predictor variable, explanatory variable) is represented as x along the horizontal axis and the dependent variable (or response variable, target variable) is represented as y along the vertical axis. The purpose of the simple linear regression is to determine the best bit line that connects x and y where x is the observed values and y is what we are trying to predict. If we assume the relationship between x and y is linear, we have a linear model as follows:

$$y = \beta_0 + \beta_1 x \tag{4.67}$$

where β_0 is an intercept (also called bias) term and β_1 is a slope coefficient. This is a linear regression problem. Consider there are n paired data points $(x_1, y_1), (x_2, y_2)$, ..., (x_n, y_n) and the paired observations satisfy a linear model. We can rewrite (4.67) as expressing y_i as a function of x_i as follows:

$$y_i = \beta_0 + \beta_1 x_i + \varepsilon_i \tag{4.68}$$

---

[2] Refer to the book "Haesik Kim, *Design and Optimization for 5G Wireless Communications*, Wiley, April 2020, ISBN: 978-1-119-49452-2.", the Sect. 4.3.1 is written and new contents are added.

Fig. 4.39 Example of
residues

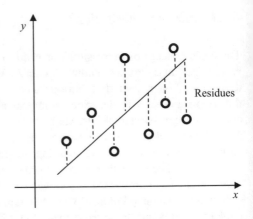

where ε_i is an error term. This term represents that the data does not fit the model.
The error term can be expressed as Gaussian noise: $\varepsilon \sim \mathcal{N}(0, \sigma^2)$. We assume that
it is observed as independent and identically distributed random variables with zero
mean and constant variance σ^2. Namely, we assume that it does not depend on x_i.
We call this model a probabilistic model because the error term is a random variable.
The variance σ^2 means how spread out the data are from the straight line. What we
have to do is finding β_0, β_1, and σ^2. If we find them, we can determine the best bit
line. In order to determine the values of the parameters, n paired observations are
used. In principle, there are many different methods to estimate $\hat{\beta}_0$, $\hat{\beta}_1$, and $\hat{\sigma}^2$ from a
data set. Among them, least squares methods and maximum likelihood methods are
widely used to estimate them. The difference between the predicted data point (or
fitted value) and the actual observation is called the residues. Figure 4.39 illustrates
an example of residues.

In a simple linear regression, the sum of the squared errors (SSE) can be expressed
as $\sum(\text{predicted point} - \text{actual observation})^2 = \sum(\text{residues})^2$. Since we don't want
residuals to cancel each other, SSE is widely used and leads a parabolic error surface.
This is useful for gradient descent. The least squares approach is to minimize the
SSE and find a best fit line. We call this the least squared criterion. The least squares
estimation of β_0 and β_1 is obtained by solving the following optimization problem:

$$\min_{\beta_0,\beta_1}\left(\sum_{i=1}^{n}\varepsilon_i^2\right) = \min_{\beta_0,\beta_1}\left(\sum_{i=1}^{n}(y_i - (\beta_0 + \beta_1 x_i))^2\right). \tag{4.69}$$

This problem is known as the least squares linear regression problem. The sum
of the squared errors term of (4.69) is rewritten as follows:

$$\text{SSE} = \sum_{i=1}^{n}\varepsilon_i^2 = \sum_{i=1}^{n}\left(y_i^2 - 2y_i(\beta_0 + \beta_1 x_i) + \beta_0^2 + 2\beta_0\beta_1 x_i + \beta_1^2 x_i^2\right). \tag{4.70}$$

The partial derivatives of SSE with respect to β_0 are

$$\frac{\partial \text{SSE}}{\partial \beta_0} = \sum_{i=1}^{n}(-2y_i + 2\beta_0 + 2\beta_1 x_i) \tag{4.71}$$

and the partial derivatives of SSE with respect to β_1 are

$$\frac{\partial \text{SSE}}{\partial \beta_1} = \sum_{i=1}^{n}(-2x_i y_i + 2\beta_0 x_i + 2\beta_1 x_i^2). \tag{4.72}$$

The values of β_0 and β_1 are obtained by setting

$$\frac{\partial \text{SSE}}{\partial \beta_0} = 0, \frac{\partial \text{SSE}}{\partial \beta_1} = 0. \tag{4.73}$$

Thus, Eq. (4.71) is rewritten as follows:

$$\sum_{i=1}^{n}(-2y_i + 2\beta_0 + 2\beta_1 x_i) = 0, \tag{4.74}$$

$$\sum_{i=1}^{n}\left(-y_i + \hat{\beta}_0 + \hat{\beta}_1 x_i\right) = 0, \tag{4.75}$$

$$-n\bar{y} + n\hat{\beta}_0 + \hat{\beta}_1 n\bar{x} = 0, \tag{4.76}$$

$$\hat{\beta}_0 = \bar{y} - \hat{\beta}_1 \bar{x} \tag{4.77}$$

where $\hat{\beta}_i$ is the predicted value of β_i and $\bar{y}$ and $\bar{x}$ are the sample means as follows:

$$\bar{y} = \frac{1}{n}\sum_{i=1}^{n} y_i \text{ and } \bar{x} = \frac{1}{n}\sum_{i=1}^{n} x_i. \tag{4.78}$$

Likewise, Eq. (4.72) is rewritten as follows:

$$\sum_{i=1}^{n}(-2x_i y_i + 2\beta_0 x_i + 2\beta_1 x_i^2) = 0, \tag{4.79}$$

$$\sum_{i=1}^{n} -x_i y_i + \hat{\beta}_0 \sum_{i=1}^{n} x_i + \hat{\beta}_1 \sum_{i=1}^{n} x_i^2 = 0, \tag{4.80}$$

$$-\sum_{i=1}^{n} x_i y_i + \left(\bar{y} - \hat{\beta}_1 \bar{x}\right)\sum_{i=1}^{n} x_i + \hat{\beta}_1 \sum_{i=1}^{n} x_i^2 = 0, \tag{4.81}$$

$$\hat{\beta}_1 = \frac{\sum_{i=1}^{n} x_i (y_i - \overline{y})}{\sum_{i=1}^{n} x_i (x_i - \overline{x})}, \tag{4.82}$$

$$\hat{\beta}_1 = \frac{\sum_{i=1}^{n} (x_i - \overline{x})(y_i - \overline{y})}{\sum_{i=1}^{n} (x_i - \overline{x})^2} = \frac{S_{xy}}{S_{xx}} \tag{4.83}$$

where S_{xy} and S_{xx} are defined as follows:

$$S_{xx} = \sum_{i=1}^{n} (x_i - \overline{x})^2 = \sum_{i=1}^{n} (x_i)^2 + \frac{\left(\sum_{i=1}^{n} x_i\right)^2}{n}, \tag{4.84}$$

$$S_{xy} = \sum_{i=1}^{n} (x_i - \overline{x})(y_i - \overline{y}) = \sum_{i=1}^{n} x_i y_i + \frac{\left(\sum_{i=1}^{n} x_i\right)\left(\sum_{i=1}^{n} y_i\right)}{n}. \tag{4.85}$$

In addition, we can express (4.83) as follows:

$$\hat{\beta}_1 = \frac{\sum_{i=1}^{n} (x_i - \overline{x})(y_i - \overline{y})}{\sum_{i=1}^{n} (x_i - \overline{x})^2} = r \frac{s_y}{s_x} \tag{4.86}$$

where s_y and s_x are standard deviations of x and y, respectively. r is the correlation coefficient as follows:

$$r = \frac{1}{n-1} \sum_{i=1}^{n} \left(\frac{x_i - \overline{x}}{s_x}\right)\left(\frac{y_i - \overline{y}}{s_y}\right). \tag{4.87}$$

It means how much x is related to y and r^2 is called the coefficient of determination. In order to assess accuracy of the estimated coefficients, the standard error (SE) is widely used and we can define them as follows:

$$SE\left(\hat{\beta}_1\right) = \frac{\sigma}{\sqrt{\sum_{i=1}^{n} (x_i - \overline{x})^2}}, \tag{4.88}$$

$$SE\left(\hat{\beta}_0\right) = \sigma \sqrt{\frac{1}{n} + \frac{\overline{x}^2}{\sum_{i=1}^{n} (x_i - \overline{x})^2}}, \tag{4.89}$$

where the mean-squared error σ^2 is

$$\sigma^2 = \frac{\sum_{i=1}^{n} \left(\hat{y}_i - y_i\right)^2}{n - 2} \tag{4.90}$$

where $\hat{y}_i$ is the predicted value of y_i. The standard error means how close x values are and how large the errors are. It reflects how much it varies under repeated sampling.

If a value is very close, the standard error $SE(\hat{\beta}_1)$ is large and the slope is less confident. Thus, the standard error gets larger if the error gets bigger. The bigger error indicates a worse fit. In addition, the standard errors are used to calculate confidence intervals. The confidence interval tells us about how stable the estimation is. If we define 95% confidence interval, the 95% confidence interval β_1 will be $\left[\hat{\beta}_1 - 2\,SE(\hat{\beta}_1), \hat{\beta}_1 + 2\,SE(\hat{\beta}_1)\right]$.

Many applications we will face contain more than one independent variable. We need to create a multiple regression model and predict the dependent variable y as a linear model of the different x independent variables as follows:

$$y = \beta_0 + \beta_1 x_1 + \beta_2 x_2 + \cdots + + \beta_n x_n. \tag{4.91}$$

As we can observe (4.91), it is still linear in the coefficient β_i. However, this model is more versatile. For example, if we want to predict a used car price (y), we should consider a new car price ($\beta_1 x_1$), a similar model car price ($\beta_2 x_2$) and so on. We can predict one dependent variable y as a function of multiple x independent variables. In addition, we can easily transform to any nonlinear function by replacing one independent variable x_i with x_i^2. The coefficients can be estimated by least squares approaches. Assume there are n dependent variables y_i ($i = 1, 2, ..., n$) and k independent variable x_j ($j = 0, 1, 2, ..., k$ and $x_0 = 1$). We can write the linear regression model with multiple independent variables as follows:

$$y_i = \beta_0 + \beta_1 x_{i1} + \beta_2 x_{i2} + \cdots + \beta_k x_{ik} + \varepsilon_i \tag{4.92}$$

where ε_i is the error term. We can rewrite (4.92) in the following matrix form:

$$\mathbf{y} = \mathbf{X}\boldsymbol{\beta} + \boldsymbol{\varepsilon} \tag{4.93}$$

where

$$\mathbf{y} = \begin{bmatrix} y_1 \\ y_2 \\ \vdots \\ y_n \end{bmatrix}, \ \mathbf{X} = \begin{bmatrix} 1 & x_{11} & \cdots & x_{1k} \\ 1 & x_{21} & \cdots & x_{2k} \\ \vdots & \vdots & \ddots & \vdots \\ 1 & x_{n1} & \cdots & x_{nk} \end{bmatrix}, \ \boldsymbol{\beta} = \begin{bmatrix} \beta_0 \\ \beta_1 \\ \vdots \\ \beta_k \end{bmatrix}, \ \boldsymbol{\varepsilon} = \begin{bmatrix} \varepsilon_1 \\ \varepsilon_2 \\ \vdots \\ \varepsilon_n \end{bmatrix}. \tag{4.94}$$

Similar to a single regression model, the least squares approach is to minimize the sum of the squared errors (SSE). The least squares estimation of $\boldsymbol{\beta}$ is obtained by solving the following optimization problem:

$$\min_{\boldsymbol{\beta}} \left(\sum_{i=1}^{n} \varepsilon_i^2 \right) = \min_{\boldsymbol{\beta}} \left(\sum_{i=1}^{n} (y_i - \mathbf{X}_i \boldsymbol{\beta})^2 \right) \tag{4.95}$$

where $\mathbf{X}_i$ is a row vector of the matrix X. The least squares estimation of $\boldsymbol{\beta}$ is

$$\hat{\boldsymbol{\beta}} = \left(\mathbf{X}^T\mathbf{X}\right)^{-1}\mathbf{X}^T\mathbf{y}. \tag{4.96}$$

In order to solve this equation, it requires that k is smaller than or equal to n ($n > k$). If $\left(\mathbf{X}^T X\right)^{-1}$ does not exist, the solution may not be unique. If it exists, the columns of $\mathbf{X}$ are linearly independent. We can express the predicted value of $\hat{\mathbf{y}}$ as follows:

$$\hat{\mathbf{y}} = \mathbf{X}\hat{\boldsymbol{\beta}} = \mathbf{X}\left(\mathbf{X}^T\mathbf{X}\right)^{-1}\mathbf{X}^T\mathbf{y} = \mathbf{H}\mathbf{y} \tag{4.97}$$

where $\mathbf{H}$ is a $n \times n$ projection matrix and it projects $\mathbf{y}$ onto the space spanned by the column of $\mathbf{X}$. The matrix $\mathbf{H}$ is often called the hat matrix. It plays an important role in diagnostics of a regression model.

Example 4.9 Linear Least Square Regression Consider 5 paired data points (2, 1), (4, 5), (6, 7), (8, 11), (10, 15) and the paired data satisfy a linear model. Find β_0 and β_1 of the linear model and compare the paired data and the linear model.

Solution

In order to calculate (4.77) and (4.83), the first step is to compute the following:

i	x_i	y_i	x_i^2	$x_i y_i$
1	2	1	4	2
2	4	5	16	20
3	6	7	36	42
4	8	11	64	88
5	10	15	100	150

and then we calculate sum of x, y, x^2, and xy as follows:

$$\sum_{i=1}^{5} x_i = 30, \ \sum_{i=1}^{5} y_i = 39, \ \sum_{i=1}^{5} x_i^2 = 220, \ \sum_{i=1}^{5} x_i y_i = 302.$$

By (4.84) and (4.85),

$$S_{xx} = \sum_{i=1}^{5}(x_i)^2 + \frac{\left(\sum_{i=1}^{5} x_i\right)^2}{n} = 220 + \frac{30^2}{5} = 400,$$

$$S_{xy} = \sum_{i=1}^{5} x_i y_i + \frac{\left(\sum_{i=1}^{5} x_i\right)\left(\sum_{i=1}^{5} y_i\right)}{n} = 302 + \frac{30 \cdot 39}{5} = 536.$$

Now, we can obtain the slope and intercept as follows:

$$\hat{\beta}_1 = \frac{S_{xy}}{S_{xx}} = \frac{536}{400} = 1.34,$$

$$\hat{\beta}_0 = \bar{y} - \hat{\beta}_1\bar{x} = \frac{39}{5} - 1.34 \cdot \frac{30}{5} = -0.24.$$

Thus, we obtain the following linear model:

$$\hat{y} = -0.24 + 1.34x.$$

We can compare the pair data and estimation as follows:

i	X	Y	Estimation ($\hat{y}$)	Error (ε)
1	2	1	2.44	1.44
2	4	5	5.12	0.12
3	6	7	7.8	0.8
4	8	11	10.48	-0.52
5	10	15	13.16	-1.84

Figure 4.40 illustrates comparison of the linear model and the paired data points. ∎

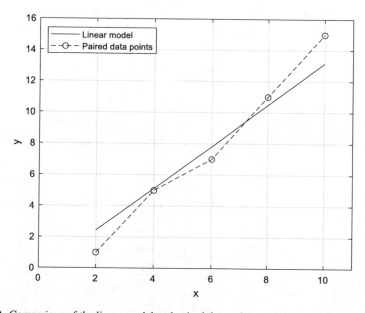

Fig. 4.40 Comparison of the linear model and paired data points

Example 4.10 Multiple Linear Least Square Regression

Consider to design a 6G cellular network and measure channel state information (CSI) in terms of distance from base station and mobility. We obtained the following data set as shown in Table 4.1.

Find $\hat{\beta}$ of a linear model and compare the paired data and the linear model.

Solution

In order to calculate (4.96), we have the matrix $\mathbf{X}$ and $\mathbf{y}$ as follows:

$$\mathbf{X} = \begin{bmatrix} 1 & 9 & 9 \\ 1 & 16 & 14 \\ 1 & 36 & 37 \\ \vdots & \vdots & \vdots \\ 1 & 77 & 70 \\ 1 & 25 & 24 \end{bmatrix}, \ \mathbf{y} = \begin{bmatrix} 8.75 \\ 10.66 \\ 7.22 \\ \vdots \\ 2.95 \\ 9.25 \end{bmatrix}.$$

The first step is to compute the following:

$$\mathbf{X}^T\mathbf{X} = \begin{bmatrix} 1 & 1 & 1 & \cdots & 1 & 1 \\ 9 & 16 & 36 & \cdots & 77 & 25 \\ 9 & 14 & 37 & \cdots & 70 & 24 \end{bmatrix} \begin{bmatrix} 1 & 9 & 9 \\ 1 & 16 & 14 \\ 1 & 36 & 37 \\ \vdots & \vdots & \vdots \\ 1 & 77 & 70 \\ 1 & 25 & 24 \end{bmatrix} = \begin{bmatrix} 20 & 994 & 934 \\ 994 & 64746 & 60606 \\ 934 & 60606 & 57196 \end{bmatrix},$$

Table 4.1 Measurement data of CSI, distance from base station, and mobility

	CSI (y)	Distance from BS (x_1)	Mobility (x_2)		CSI (y)	Distance from BS (x_1)	Mobility (x_2)
1	8.75	9	9	11	5.87	67	52
2	10.66	16	14	12	7.78	32	33
3	7.22	36	37	13	3.98	79	71
4	6.57	55	45	14	3.89	77	71
5	7.55	34	36	15	1.10	98	99
6	9.89	5	8	16	2.01	89	89
7	8.45	27	25	17	5.98	61	51
8	2.53	85	85	18	6.67	54	43
9	8.56	23	26	19	2.95	77	70
10	6.56	45	46	20	9.25	25	24

$$(\mathbf{X}^T\mathbf{X})^{-1} = \begin{bmatrix} 20 & 994 & 934 \\ 994 & 64746 & 60606 \\ 934 & 60606 & 57196 \end{bmatrix}^{-1} = \begin{bmatrix} 0.2122 & -0.0017 & -0.0016 \\ -0.0017 & 0.0019 & -0.002 \\ -0.0016 & -0.002 & 0.0022 \end{bmatrix},$$

$$\mathbf{X}^T\mathbf{y} = \begin{bmatrix} 1 & 1 & 1 & \cdots & 1 & 1 \\ 9 & 16 & 36 & \cdots & 77 & 25 \\ 9 & 14 & 37 & \cdots & 70 & 24 \end{bmatrix} \begin{bmatrix} 8.75 \\ 10.66 \\ 7.22 \\ \vdots \\ 2.95 \\ 9.25 \end{bmatrix} = \begin{bmatrix} 126.2 \\ 4838.3 \\ 4521.1 \end{bmatrix}.$$

The next step is to calculate $\hat{\boldsymbol{\beta}}$ as follows:

$$\hat{\boldsymbol{\beta}} = (\mathbf{X}^T\mathbf{X})^{-1}\mathbf{X}^T\mathbf{y},$$

$$\begin{bmatrix} \hat{\beta}_0 \\ \hat{\beta}_1 \\ \hat{\beta}_2 \end{bmatrix} = \begin{bmatrix} 0.2122 & -0.0017 & -0.0016 \\ -0.0017 & 0.0019 & -0.002 \\ -0.0016 & -0.002 & 0.0022 \end{bmatrix} \begin{bmatrix} 126.2 \\ 4838.3 \\ 4521.1 \end{bmatrix} = \begin{bmatrix} 11.0344 \\ -0.0001 \\ -0.101 \end{bmatrix}.$$

Thus, we obtain the following linear model:

$$\hat{y} = 11.0344 - 0.0001x_1 - 0.101x_2$$

We can compare the pair data and estimation as shown in Table 4.2.

Figure 4.41 illustrates comparison of the linear model and the paired data points. In this figure, the dot (•) and the surface represent the paired data points and the linear model, respectively.

∎

Table 4.2 Comparison of the paired data and estimation

	CSI (y)	Estimation ($\hat{y}$)	Error (ε)		CSI (y)	Estimation ($\hat{y}$)	Error (ε)
1	8.75	10.1242	1.3742	11	5.87	5.7737	−0.0963
2	10.66	9.6183	−1.0417	12	7.78	7.6970	−0.0830
3	7.22	7.2925	0.0725	13	3.98	3.8528	−0.1272
4	6.57	6.4822	−0.0878	14	3.89	3.8530	−0.0370
5	7.55	7.3937	−0.1563	15	1.10	1.0220	−0.0780
6	9.89	10.2256	0.3356	16	2.01	2.0332	0.0232
7	8.45	8.5058	0.0558	17	5.98	5.8754	−0.1046
8	2.53	2.4378	−0.0922	18	6.67	6.6844	0.0144
9	8.56	8.4052	−0.1548	19	2.95	3.9541	1.0041
10	6.56	6.3822	−0.1778	20	9.25	8.6070	−0.6430

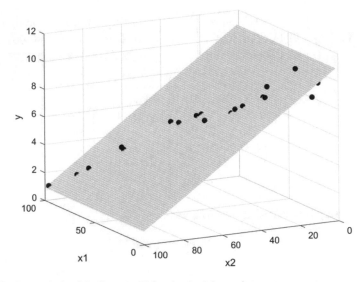

Fig. 4.41 Comparison of the linear model and paired data points

Summary 4.5 Linear regression

1. In linear regression, we assume the underlying behaviour is linear and predict continuous values. If there is only one variable, it is a straight line. Linear regression provides us with good introduction to many regression concepts of machine learning.
2. The least squares approach is to minimize the SSE and find a best fit line. We call this the least squared criterion.
3. We have a linear model $\mathbf{y} = \mathbf{X}\boldsymbol{\beta} + \boldsymbol{\varepsilon}$ and the least squares estimation of $\boldsymbol{\beta}$ are obtained by solving the following optimization problem:

$$\min_{\boldsymbol{\beta}} \left(\sum_{i=1}^{n} \varepsilon_i^2 \right) = \min_{\boldsymbol{\beta}} \left(\sum_{i=1}^{n} (y_i - \mathbf{X}_i \boldsymbol{\beta})^2 \right)$$

where $\mathbf{X}_i$ is a row vector of the matrix $\mathbf{X}$. The least squares estimation of $\boldsymbol{\beta}$ is

$$\hat{\boldsymbol{\beta}} = \left(\mathbf{X}^T \mathbf{X} \right)^{-1} \mathbf{X}^T \mathbf{y}.$$

We can express the predicted value of $\hat{\mathbf{y}}$ as follows:

$$\hat{\mathbf{y}} = \mathbf{X}\hat{\boldsymbol{\beta}} = \mathbf{X} \left(\mathbf{X}^T \mathbf{X} \right)^{-1} \mathbf{X}^T \mathbf{y} = \mathbf{H}\mathbf{y}$$

where $\mathbf{H}$ is a $n \times n$ projection matrix and it projects $\mathbf{y}$ onto the space spanned by the column of $\mathbf{X}$. The matrix $\mathbf{H}$ is often called the hat matrix.

4.3.2 Gradient Descent Algorithms

Gradient descent algorithm can be used to find the coefficient β_i that minimizes some error function. Gradient descent algorithm is an iterative optimization algorithm for finding the minimum of an objective function (or cost function) by moving in the direction of steepest descent. In machine learning, gradient descent algorithm is used to update the parameters of a system model. Consider the unconstrained optimization problem as follows:

$$\min_{\mathbf{x}} f(\mathbf{x}) \tag{4.98}$$

where $\mathbf{x} \in \mathbb{R}^n$ and the objective function $f:\mathbb{R}^n \rightarrow \mathbb{R}$ is convex and differentiable. In this section, the gradient descent algorithm is mainly discussed for unconstrained regression problems. However, it is possible to apply it for constrained optimization problems. The motivation of gradient descent algorithm is based on the observation that the minimum of the objective function $f(\mathbf{x})$ lies at a point $\mathbf{x}^*$ where $\nabla f(\mathbf{x}^*) = 0$ and a negative gradient step decreases the objective function. $\nabla f(\mathbf{x})$ denotes the gradient of $f(\mathbf{x})$ at the point $\mathbf{x}$. Thus, gradient descent algorithm finds the minimum of $f(\mathbf{x})$ by starting from an initial point and iteratively searching the next point in the steepest descent direction until it converges. We start the initial point $\mathbf{x}^{(0)}$. When we have a point $\mathbf{x}^{(k)}$, we can find the next point $\mathbf{x}^{(k+1)}$ using the following equation:

$$\mathbf{x}^{(k+1)} = \mathbf{x}^{(k)} - \alpha^{(k)} \nabla f\left(\mathbf{x}^{(k)}\right) \tag{4.99}$$

where $\alpha^{(k)} > 0$ is the step size (or the learning rate). The step size should be chosen carefully. Too large step size causes overshooting or diverging and too small step size results in very slow converging. Convergence of the gradient descent method depends on the step size and the number of iterations. $\Delta \mathbf{x}^{(k)} = -\nabla f\left(\mathbf{x}^{(k)}\right)$ is a search direction of gradient descent method at the point $\mathbf{x}^{(k)}$. We call this steepest descent direction. The step size can be a small constant or be adapted as the iteration increases. Constant step size is usually not efficient. Adapting the step size is efficient and achieves a faster convergence. The step size $\alpha^{(k)}$ can be chosen to minimize the value of the next point and maximize the decrease of the objective function at each step. Thus, the step size $\alpha^{(k)}$ is defined as follows:

$$\alpha^{(k)} = \arg\min_{\alpha \geq 0} f\left(\mathbf{x}^{(k)} - \alpha \nabla f\left(\mathbf{x}^{(k)}\right)\right) = \arg\min_{\alpha \geq 0} \phi(\alpha). \tag{4.100}$$

We call this step size computation a line search. The stop condition of gradient descent method is ideally when the first-order necessary condition $\left|\nabla f\left(x^{(k)}\right)\right| = 0$. However, this is not practical. Thus, we usually predefine precision (or tolerance) ε and use practical stop conditions as follows: $\left|\nabla f\left(\mathbf{x}^{(k)}\right)\right| < \varepsilon$, $\left|f\left(\mathbf{x}^{(k+1)}\right) - f\left(\mathbf{x}^{(k)}\right)\right| < \varepsilon$, or $\left|\mathbf{x}^{(k-1)} - \mathbf{x}^{(k)}\right| < \varepsilon$. We can summarize the steps of the gradient descent method under the given precision $\varepsilon > 0$ as follows:

Initial conditions: Start with an initial point $\mathbf{x}^{(0)}$, an step size $\alpha^{(k)}$, tolerance $\varepsilon > 0$.

Repeat

– ***Step***: Determine the next point by $\mathbf{x}^{(k+1)} = \mathbf{x}^{(k)} - \alpha^{(k)}\nabla f\left(\mathbf{x}^{(k)}\right)$

Until it converges: $\left|\nabla f\left(\mathbf{x}^{(k)}\right)\right| < \epsilon$.

There are multiple types of the step size as follows:

$$\text{Constant: } \alpha^{(k)} = c, \tag{4.101}$$

$$\text{Constant step length: } \alpha^{(k)} = \frac{c}{\left\|\nabla f\left(\mathbf{x}^{(k)}\right)\right\|}, \ \left|\mathbf{x}^{(k-1)} - \mathbf{x}^{(k)}\right| = c, \tag{4.102}$$

$$\text{Square sum: } \sum_{k=1}^{\infty}\alpha^{(k)^2} < \infty, \sum_{k=1}^{\infty}\alpha^{(k)} = \infty, \tag{4.103}$$

$$\text{Nonsummable diminishing step: } \lim_{k\to\infty}\alpha^{(k)} = 0, \sum_{k=1}^{\infty}\alpha^{(k)} = \infty \tag{4.104}$$

where c is constant. The typical example of square sum is $\alpha^{(k)} = a/(b + k)$ where $a > 0, b \geq 0$. The typical example of nonsummable diminishing step is $\alpha^{(k)} = a/\sqrt{k}$ where $a > 0$. One practical way to choose the learning rate is to take a small amount of the data set initially and then perform a binary search to improve the objective on the subset.

Example 4.11 Gradient Descent   Algorithm for Simple Linear Regression
Consider two-dimensional data set as follows:
    Fit the linear regression parameters to the data set using gradient descent method.

Solution

From Table 4.3, Fig. 4.42 illustrates the scatter plot of the data set.

As we can observe Fig. 4.42, the data set shows us a linear relationship and we find a linear regression line as the form $y = a + bx$. In order to find the best line, we use sum of squared errors (SSE). This metric represents sum difference between actual data and predicted data. The SSE is defined as follows:

$$\text{SSE} = \sum_{i=1}^{n}(y_i - \overline{y})^2$$

Table 4.3 The data set

i	X	Y	i	x	y	i	x	Y
1	1.3167	3.2446	11	5.3893	7.0708	21	24.147	22.203
2	2.2712	4.6107	12	3.1386	6.1891	22	12.592	6.1101
3	13.515	12.231	13	21.767	20.27	23	9.1302	5.5277
4	2.0571	4.634	14	4.263	5.4901	24	13.662	8.5186
5	7.2258	8.4084	15	5.1875	6.3261	25	4.854	7.0032
6	0.71618	5.6407	16	3.0825	5.5649	26	6.8233	5.8598
7	3.5129	5.3794	17	22.638	18.945	27	10.876	7.3828
8	5.3048	6.3654	18	13.501	12.828	28	5.3383	8.4562
9	0.56077	5.1301	19	7.0467	10.957	29	11.23	7.7811
10	3.6518	6.4296	20	14.692	13.176	30	7.4967	7.2761

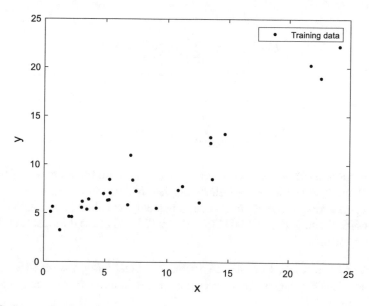

Fig. 4.42 Scatter plot of the data set

where y_i and $\bar{y}$ are dependent variables and average of dependent variables, respectively. We find a linear regression model to make SSE as small as possible. We can use gradient descent method. Gradient descent method is to find a minimum value of the objective function (or cost function). We find minimum errors as difference of predicted values and actual values. Firstly, we represent the hypothesis h_θ as a linear function x as follows:

$$h_\theta(x) = \theta_0 + \theta_1 x$$

where θ_i is the parameter we adjust to minimize the errors. Secondly, we define the objective function (or cost function) as follows:

$$J(\theta_0, \theta_1) = \frac{1}{2n} \sum_{i=1}^{n} \left(h_\theta(x^{(i)}) - y^{(i)} \right)^2$$

where n is the total number of the data set. We can measure the accuracy of the hypothesis by this objective function. From (4.99), we repeatedly perform the following equations:

$$\theta_j = \theta_j - \alpha \frac{\partial}{\partial \theta_j} J(\theta_0, \theta_1)$$

where α is the step size (or learning rate). The above equation can be rewritten as follows:

$$\theta_0 = \theta_0 - \alpha \frac{1}{n} \sum_{i=1}^{n} \left(h_\theta(x^{(i)}) - y^{(i)} \right)$$

and

$$\theta_1 = \theta_1 - \alpha \frac{1}{n} \sum_{i=1}^{n} \left(h_\theta(x^{(i)}) - y^{(i)} \right) x^{(i)}.$$

As an initial condition, we set $\alpha = 0.01$ and maximum iteration $= 100$ and start at $\theta_0 = 0$ and $\theta_1 = 0$. Figure 4.43 illustrates the objective function.

After performing the gradient descent repeatedly, we obtain $\theta_0 = 2.963273$ and $\theta_1 = 0.669276$ and $h_\theta(x) = 2.963273 + 0.669276x$. Figure 4.44 illustrates the best line fit for the data.

Example 4.12 Gradient Descent Algorithm for Linear Regression with Multiple Variables Consider the data set of used cars prices in terms of year and mileage as shown in Table 4.4.

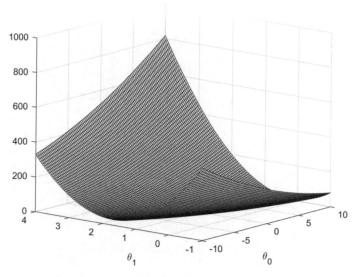

Fig. 4.43 Objective function

Table 4.4 Data set about used car prices

i	Price (Eur)	Year	Miles	i	Price (Eur)	Year	Miles
1	15,493	2015	60,391	11	18,700	2015	33,712
2	17,000	2015	57,165	12	18,800	2015	49,306
3	17,600	2014	39,484	13	19,400	2018	20,257
4	17,800	2014	45,404	14	19,591	2018	18,109
5	17,900	2014	48,949	15	19,600	2018	16,072
6	17,988	2015	72,000	16	20,700	2019	9712
7	17,995	2015	51,259	17	20,800	2019	9306
8	18,400	2015	37,257	18	20,600	2019	6072
9	18,491	2017	28,109	19	21,770	2020	3712
10	18,600	2015	36,072	20	22,800	2020	2306

When we have a used car with year 2020 and 1000 miles, please find the predicted price using gradient descent method.

Solution

In order to sell the used car, we want to know the used car market price. Based on the given data set, we can develop a model of used car prices. As we can observe Table 4.4, each feature has different order of magnitude. We need a normalization of the data set. Normalization process plays important role in gradient descent algorithm. If features are not normalized, features with large scale will be dominated. It takes longer time to reach an optimal point. One simple way of normalization process is

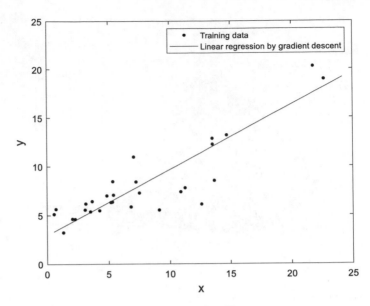

Fig. 4.44 Linear regression by gradient descent method ■

to subtract mean value and divide by standard deviation as follows:

$$x' = \frac{x - \overline{x}}{\sigma}$$

where x', x, $\overline{x}$, and σ are the normalized value, the original value, the mean, and standard deviation, respectively. We can use gradient descent method and find a model. Firstly, we represent the hypothesis h_θ as a linear function x as follows:

$$h_\theta(x_1, x_2) = \theta_0 + \theta_1 x_1 + \theta_2 x_2.$$

Secondly, we define the cost function with multiple variables as follows:

$$J(\theta_0, \theta_1, \ldots, \theta_n) = \frac{1}{2m} \sum_{i=1}^{m} \left(h_\theta(x^{(i)}) - y^{(i)} \right)^2$$

where m is the total number of the data set and θ is n-dimensional. We have 3 parameters and the cost function is

$$J(\theta_0, \theta_1, \theta_2) = \frac{1}{2 \cdot 20} \sum_{i=1}^{20} \left(h_\theta(x^{(i)}) - y^{(i)} \right)^2$$

From (4.99), we repeatedly perform the following equations:

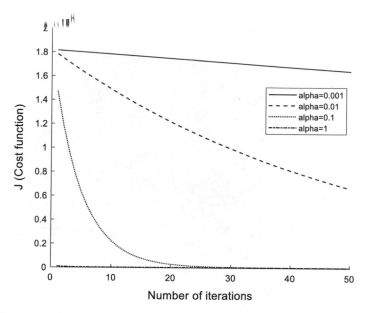

Fig. 4.45 Cost function convergence in terms of different learning rates (α)

$$\theta_j = \theta_j - \alpha \frac{\partial}{\partial \theta_j} J(\theta_0, \theta_1, \theta_2)$$

where α is the step size (or learning rate). As an initial condition, we set $\alpha = 0.01$ and maximum iteration $= 50$ and start at $\theta_0 = 0$, $\theta_1 = 0$ and $\theta_2 = 0$. Figure 4.45 illustrates the cost function.

After performing the gradient descent repeatedly, we obtain $\theta_0 = 19000.895296$, $\theta_1 = -831.655468$, $\theta_2 = 756.924226$ and $h_\theta(x_1, x_2) = 19000.895296 + -831.655468x_1 + 756.924226x_2$. Figure 4.46 illustrates the regression model.

The predicted price of the used car with year 2020 and 1000 miles is 21485.895553 Eur.   ∎

In (4.99), the weight update term $-\alpha^{(k)} \nabla f\left(\mathbf{x}^{(k)}\right)$ is composed of decent $(-)$, learning rate $(\alpha^{(k)})$ and gradient $(\nabla f\left(\mathbf{x}^{(k)}\right))$ as follows:

$$\text{Weight update} = \quad - \quad\quad\quad \alpha^{(k)} \quad\quad\quad \nabla f\left(\mathbf{x}^{(k)}\right).$$
$$\text{(Descent)} \quad \text{(Learning rate)} \quad \text{(Gradient)} \quad\quad\quad (4.105)$$

As we can observe (4.105), we need to calculate the entire data set and then find one weight update. The gradient descent algorithm we discussed is a popular tool in machine learning and accurate to optimize parameters. However, the gradient descent algorithm is not efficient because all gradients are calculated at each point. It takes a long time to compute them and intractable for memory management. In particular, when the training data size is huge, it may be infeasible. Thus, in practical systems, stochastic gradient descent algorithms are widely used. They select random points, and the gradient at the point is calculated. The iterations do not rely on the

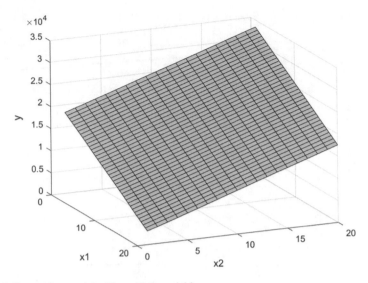

Fig. 4.46 Regression model with multiple variables

total number of the data set. We should select the random points very carefully. If the wrong samples are chosen, it might have in the wrong direction. It typically reaches convergence much faster than the gradient decent algorithm. It performs frequent updates with a high variance and causes fluctuation to find an optimal point heavily. This fluctuation enables us to jump and find a new local optimum. It will typically be overshooting. If we have a small learning rate, stochastic gradient descent algorithms will have similar convergence behaviour to gradient descent algorithms. The objective function of (4.98) can be expressed as sum of a finite number of functions as follows:

$$f(\mathbf{x}) = \frac{1}{m} \sum_{i=1}^{m} f_i(\mathbf{x}) \tag{4.106}$$

where $f_i(\mathbf{x})$ is a loss function at single sample in training data set. The computational cost will linearly scale with the training data size m. Gradient descent algorithm computes the gradient $\nabla f(\mathbf{x})$ but stochastic gradient descent algorithm computes the gradient $\nabla f_i(\mathbf{x})$ at randomly selected samples i. We have the following equation:

$$\nabla f(\mathbf{x}) = \frac{1}{m} \sum_{i=1}^{m} \nabla f_i(\mathbf{x}). \tag{4.107}$$

If we consider a mini-batch B at each iteration, (4.107) can be rewritten as follows:

$$\nabla f_B(\mathbf{x}) = \frac{1}{|B|} \sum_{i \in B} \nabla f_i(\mathbf{x}) \tag{4.108}$$

where $|n|$ represent the number of the mini batch and Eq. (4.99) can be rewritten as follows:

$$\mathbf{x}^{(k+1)} = \mathbf{x}^{(k)} - \alpha^{(k)} \nabla f_B\left(\mathbf{x}^{(k)}\right).   \qquad (4.109)$$

This generalized stochastic gradient descent algorithm is also called a mini-batch stochastic gradient descent algorithm. The stochastic gradient descent algorithm using a mini-batch is widely used at graphic processing. It enables us to increase accuracy without a longer computational time. The min-batch size at graphic processing depends on the memory size but typically sets between 10 and 100 samples. When the training data size is large enough, it enables us to find an optimal solution with small number of iteration and the computational cost will be lower than the gradient descent algorithm. The batch is generally defined as the number of samples to perform before updating the parameters. A training data set is composed of multiple batches. At the end of batch, we can predict the outcomes and errors can be calculated. In gradient descent (also called batch gradient descent), one batch size is same as the training data set. In stochastic gradient descent, one batch size is one data point. In mini-batch gradient descent, one batch size is greater than 1 and less than the training data set. Typically, 32, 64, or 128 samples as one batch size are selected. In addition, an epoch means the number of passes of the entire training data set. At each epoch, we can update the internal parameters. If the batch size is the entire training data set, the number of epochs is same as the number of iterations. Now, we can summarize the steps of the generalized stochastic gradient descent method as follows:

Initial conditions: Start with an initial point $\mathbf{x}^{(0)}$, an step size $\alpha^{(k)}$, tolerance $\varepsilon > 0$.

Repeat

- *Step 1*: Select a mini-batch B randomly
- *Step 2*: Determine the next point by $\mathbf{x}^{(k+1)} = \mathbf{x}^{(k)} - \alpha^{(k)} \nabla f_B\left(\mathbf{x}^{(k)}\right)$

Until it converges: $\left|\nabla f_B\left(\mathbf{x}^{(k)}\right)\right| < \varepsilon$.

In practice, we shuffle the training data points and run stochastic gradient descent algorithm in the shuffled order. If it doesn't converge, we shuffle them again and run. Figure 4.47 illustrates comparison of gradient descent algorithm and stochastic gradient descent algorithm.

Figure 4.48 illustrates comparison example of an optimal point search for gradient descent algorithm and stochastic gradient descent algorithm.

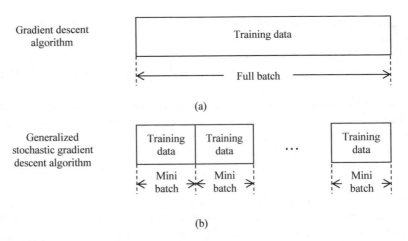

Fig. 4.47 Comparison of gradient descent algorithm (**a**) and generalized stochastic gradient descent algorithm (**b**)

As we can observe Fig. 4.48, stochastic gradient descent algorithm with a mini-batch has fluctuation to find an optimal point x. In Figure 4.29, we can observe the effect of the learning rate. The learning rate affects to implement gradient descent algorithms significantly. In stochastic gradient descent algorithms, we should especially take care of the learning rate at which we decrease it to converge. If the learning rate decreases too quickly, it may not converge. If it decreases too slowly, convergence happens very slowly. Another key parameter is gradient direction. Finding the right direction affects to convergence time. Thus, many extensions on the basic stochastic gradient descent algorithms have been developed in order to improve those two parameters. Stochastic gradient descent algorithm with momentum [20] helps us to accelerate gradient vectors in the right direction and improve the convergence time. The basic concept is to remember the update vector at each iteration and add a momentum term γ of the update vector in the previous step to the current update vector. Like a ball rolling down a hill, it accumulates the momentum. It tends to dampen oscillations and keep going in the same direction as shown in Fig. 4.49.

Equation (4.109) can be modified as follows:

$$\Delta \mathbf{x}^{(k+1)} = \gamma \Delta \mathbf{x}^{(k)} - \alpha^{(k)} \nabla f_B\big(\mathbf{x}^{(k)}\big), \tag{4.110}$$

$$\mathbf{x}^{(k+1)} = \mathbf{x}^{(k+1)} - \Delta \mathbf{x}^{(k+1)}. \tag{4.111}$$

(4.110) and (4.111) can be rewritten as follows:

$$\mathbf{x}^{(k+1)} = \mathbf{x}^{(k)} - \alpha^{(k)} \nabla f_B\big(\mathbf{x}^{(k)}\big) + \gamma \Delta \mathbf{x}^{(k)}. \tag{4.112}$$

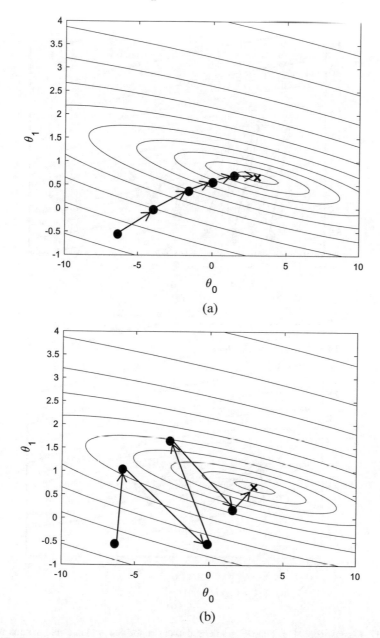

Fig. 4.48 Comparison of gradient descent algorithm (**a**) and generalized stochastic gradient descent algorithm (**b**)

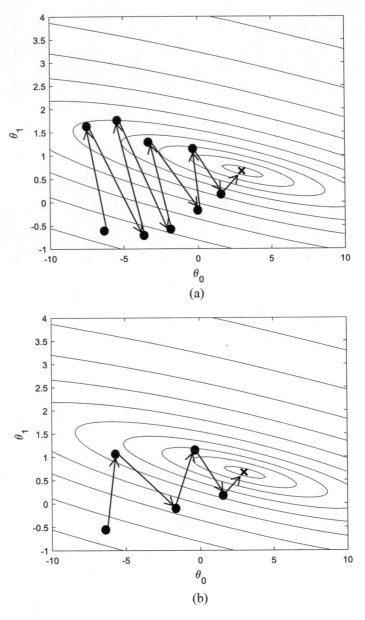

Fig. 4.49 Comparison of stochastic gradient descent algorithm without momentum (**a**) and stochastic gradient descent algorithm with momentum (**b**)

The momentum term γ means the weight to the current gradient and typically sets to 0.9 or above. Adagrad [21] changes the learning rate adaptively. It has a lower learning rate or a higher learning rate for parameters associated with frequent features or infrequent features, respectively. That is to say. Adagrad travels with a large step in the areas we didn't visit and checks them quickly. It travels with a small step in the areas we visit frequently and checks them carefully. This approach works well for sparse data. In (4.109), the same learning rate was used. However, Adagrad uses a different learning rate for parameters x_i at a time step k. (4.109) is modified as follows:

$$x_i^{k+1} = x_i^k - \alpha g_i^k \qquad (4.113)$$

where g_i^k is the gradient (or the partial derivative) of the objective function for parameters x_i at a time step k. The update rule of Adagrad can be expressed as follows:

$$x_i^{k+1} = x_i^k - \frac{\alpha}{\sqrt{G_{ii}^k + \delta}} g_i^k \qquad (4.114)$$

where $G_{ii}^k \in \mathbb{R}^{d \times d}$ is a diagonal matrix whose each diagonal element is the sum of the squares of the gradients with regard to parameters x_i up to a time step k. δ is a smoothing term to avoid division by zero. The main advantage of Adagrad is not to tune the learning rate manually. Typically, we set a default value of 0.01 and leave it. On the other hand, the disadvantage is the accumulation of the squared gradients. It keeps growing, results in the learning rate to shrink, and eventually becomes infinitesimally small. At this point, it doesn't have any additional knowledge. Thus, Adadelta [22] was developed to solve this problem. It restricts the window of accumulated past squared gradients to a fixed size. Instead of storing the previous squared gradients, the sum of the gradients is iteratively defined as a decaying average of all past squared gradients $E\left[g_i^{k^2}\right]$ with regard to parameters x_i up to a time step k. The running averages depend on only both the previous average and the current gradients as follows:

$$E\left[g_i^{k^2}\right] = \gamma E\left[g_i^{k-1^2}\right] + (1-\gamma)g_i^{k^2} \qquad (4.115)$$

where γ is a similar value of the momentum term as 0.9. Thus, the update rule of Adadelta can be expressed as follows:

$$x_i^{k+1} = x_i^k - \frac{\alpha}{\sqrt{E\left[g_i^{k^2}\right] + \delta}} g_i^k = x_i^k - \frac{\alpha}{\text{RMS}\left[g_i^k\right]} g_i^k. \qquad (4.116)$$

Root Mean Square Propagation (RMSProp) [23] is similar to Adadelta and an unpublished method to adapt the learning for the parameters. The basic concept is to divide the learning rate for a weight by a running average of the squared gradients. The running average of RMTProp method is expressed as follows:

$$E\left[g_i^{k^2}\right] = 0.9E\left[g_i^{k-1^2}\right] + 0.1g_i^{k^2}. \tag{4.117}$$

As we can observe (4.117), γ sets to 0.9. A default value for the learning rate was proposed as 0.001 [23]. Adam [24] uses the concept of both RMSProp and momentum. In this method, estimates of the first moment (the mean) and the second moment (the variance) of the gradients are used. We calculate the decaying averages of past gradients m_k and past squared gradients l_k as follows:

$$m_{k+1} = \beta_1 m_k + (1 - \beta_1)g_i^k, \tag{4.118}$$

$$l_{k+1} = \beta_2 l_k + (1 - \beta_2)g_i^{k^2}. \tag{4.119}$$

The update rule of Adam can be expressed as follows:

$$x_i^{k+1} = x_i^k - \frac{\alpha}{\sqrt{\hat{l}_k + \delta}}\hat{m}_k \tag{4.120}$$

where $\hat{m}_k$ and $\hat{l}_k$ are the biased corrected first and second moment estimates as follows:

$$\hat{m}_k = \frac{m_k}{1 - \beta_1^k}, \tag{4.121}$$

$$\hat{l}_k = \frac{l_k}{1 - \beta_2^k}. \tag{4.122}$$

A default value for β_1 and β_2 was proposed as 0.9 and 0.999, respectively [24].δ is set for a small scalar 10^{-8} [24]. It is used to avoid division by zero. Figure 4.50 illustrates summary of extension of gradient descent algorithms.

Summary 4.6 Gradient Descent Algorithms

1. Gradient descent algorithm is an iterative optimization algorithm for finding the minimum of an objective function (or cost function) by moving in the direction of steepest descent. In machine learning, gradient descent algorithm is used to update the parameters of a system model.
2. We can summarize the steps of the gradient descent method under the given precision $\varepsilon > 0$ as follows:

Initial conditions: Start with an initial point $\mathbf{x}^{(0)}$, an step size $\alpha^{(k)}$, tolerance $\varepsilon > 0$.

Repeat.

– *Step*: Determine the next point by $\mathbf{x}^{(k+1)} = \mathbf{x}^{(k)} - \alpha^{(k)} \nabla f\left(\mathbf{x}^{(k)}\right)$

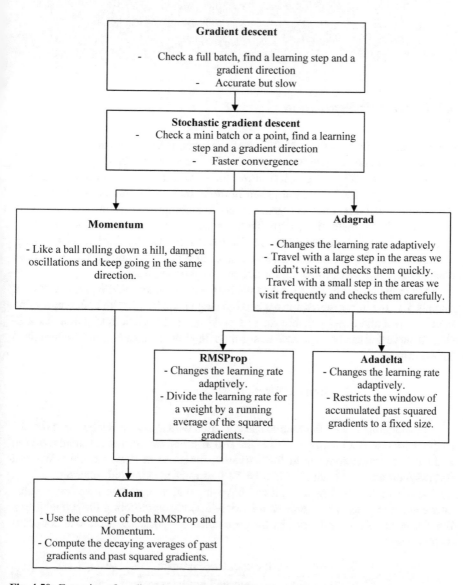

Fig. 4.50 Extension of gradient descent algorithms

Until it converges: $\left|\nabla f\left(\mathbf{x}^{(k)}\right)\right| < \epsilon$.

3. This generalized stochastic gradient descent algorithm is also called a mini-batch stochastic gradient descent algorithm. The stochastic gradient descent algorithm using a mini-batch is widely used at graphic processing. It enables us to increase accuracy without a longer computational time. The min-batch size at graphic processing depends on the memory size but typically sets between 10 and 100 samples.

4. Many extensions (Momentum, Adagrad, Adadelta, RMSProp, Adam) on the basic stochastic gradient descent algorithms have been developed in order to improve gradient direction and learning rate.

4.3.3 Logistic Regression

Logistic regression uses the probability of a class or an event and develops a model to estimate parameters. Logistic regression has been widely used in statistics and machine learning adopted this technique for solving machine learning problems. The name "regression" was established in the statistics community but the logistic regression model refers not to regression but to classification. When the target data is categorical, this method is useful. Thus, we can say that it is a kind of parametric classification in machine learning even if its name is regression. Unlike linear regression models, it doesn't have a straight line to fit a data set. Figure 4.51 illustrates comparison example of linear regression and logistic regression.

As we can observe Fig. 4.51, when we have a binary data set, the predicted target outputs Y in the linear regression exceed the range from 0 to 1. The linear regression is unbounded and is not suitable for this problem. The logistic regression has a S-shaped curve that can be expressed as sigmoid function (also called logistic function) as follows:

$$\text{Sigmoid function: } S(x) = \frac{1}{1 + e^{-x}}. \tag{4.123}$$

This function fits well the sample classification with a probability between 0 and 1 of the observation belonging to one of the two different categories. Sigmoid function and softmax function are used for activation functions of deep learning. We will discuss their usage in Chap. 6. Figure 4.52 illustrates the sigmoid function.

As we can observe Fig. 4.52, it can have any real number and map into a value between 0 and 1. Logistic regression work with odds representing the probability p that the event will occur divided by the probability $1 - p$ that the event will not occur as follows:

$$\text{Odds} = \frac{\text{The probability the event will happen}}{\text{The probability the event will not happen}} = \frac{p}{1 - p}. \tag{4.124}$$

Fig. 4.51 Comparison of linear regression (**a**) and logistic regression (**b**)

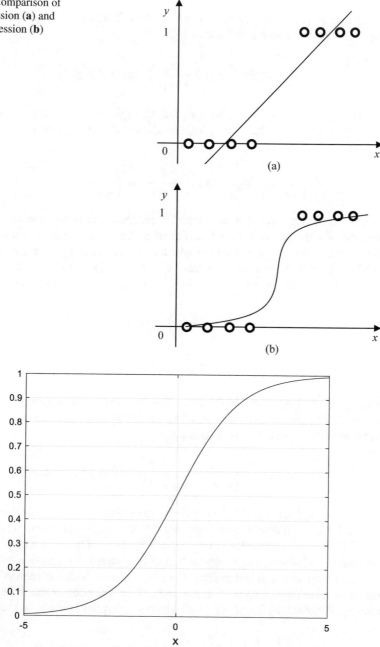

Fig. 4.52 Sigmoid function

Based on this concept, we can model the log of the odds of the probability p as a linear function of the explanatory variable. We call this transformation log odds or logit. The logit function is defined as follows:

$$\text{logit}(p) = \ln\left(\frac{p}{1-p}\right), \quad \text{for } 0 \leq p \leq 1. \tag{4.125}$$

The logistic regression model uses natural logarithms. The logit function takes the range from 0 to 1. The inverse logit function with the range between $-\infty$ and ∞ is

$$\text{logit}^{-1}(x) = \frac{e^x}{1+e^x} = \frac{1}{1+e^{-x}}. \tag{4.126}$$

The simple logistic regression model is related to the proportion of successes in the ratio of one explanatory variable x through the logarithm of the odds of success. We consider a binary logistic model with a predictor (or independent variable) x and one binary response variable Y where $p = P(Y = 1)$. We assume a linear model between two variables. The linear relationship can be expressed as follows:

$$\ln\left(\frac{p}{1-p}\right) = \beta_0 + \beta_1 x, \tag{4.127}$$

$$\frac{p}{1-p} = e^{\beta_0 + \beta_1 x} \tag{4.128}$$

where p and x are a proportion of successes and an explanatory variable, respectively. β_0 and β_1 are regression coefficients of the model. We assume that n success or failure trials are independent. The probability p is

$$p = \frac{e^{\beta_0 + \beta_1 x}}{1 + e^{\beta_0 + \beta_1 x}} = \frac{1}{1 + e^{-(\beta_0 + \beta_1 x)}} = S(\beta_0 + \beta_1 x). \tag{4.129}$$

As we can observe (4.129), we can compute the logit and the probability p that $Y = 1$ for a given observation if parameters β_0 and β_1 are found. The regression coefficients can be estimated from the training data set using maximum likelihood estimation and others. The interpretation of the estimated regression coefficients in logistic regression is different from the interpretation in linear regression because the outcome of logistic regression is a probability. They do not affect the probability linearly. We check an impact of a unit increase of a predictor x, $x + 1$, as follows:

$$\ln\left(\frac{p'}{1-p'}\right) = \beta_0 + \beta_1(x + 1) = \beta_0 + \beta_1 x + \beta_1. \tag{4.130}$$

Now, we compare the ratio of the two predictions and find the slope β_1.

$$\beta_1 = \beta_0 + \beta_1(x + 1) - (\beta_0 + \beta_1 x), \tag{4.131}$$

$$\beta_1 = \ln\left(\frac{p'}{1 - p'}\right) - \ln\left(\frac{p}{1 - p}\right) = \ln\left(\frac{\text{Odds}'}{\text{Odds}}\right), \tag{4.132}$$

$$e^{\beta_1} = \frac{\text{Odds}'}{\text{Odds}}. \tag{4.133}$$

As we can observe (4.132) and (4.133), the slope β_1 is expressed as the log ratio of the odds at x and $x + 1$. If we increase the value of a predictor x by one unit, the estimated odds are changed by a factor of e^{β_1}. We call (4.133) the odds ratio. Another important property of sigmoid function is the derivative as follows:

$$S'(x) = \frac{d}{dx}S(x) = \frac{d}{dx}\frac{1}{1 + e^{-x}} = \frac{1}{(1 + e^{-x})^2}e^{-x}, \tag{4.134}$$

$$= \left(\frac{1}{1 + e^{-x}}\right)\left(1 - \frac{1}{1 + e^{-x}}\right) = S(x)(1 - S(x)). \tag{4.135}$$

As we can observe (4.135), the derivative of sigmoid function is easily computed. Figure 4.53 illustrates sigmoid function and its derivative.

If we have multiple explanatory variable, (4.127) and (4.129) are rewritten as follows:

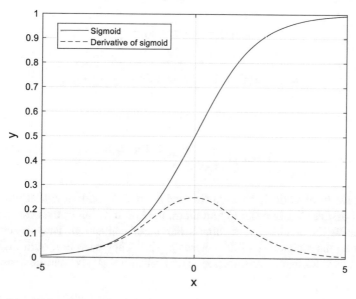

Fig. 4.53 Sigmoid function and its derivative

$$\ln\left(\frac{p}{1-p}\right) = \beta_0 + \beta_1 x_1 + \cdots + \beta_n x_n = \sum_{i=1}^{n} \beta_i x_i, \qquad (4.136)$$

$$p = \frac{1}{1 + e^{-(\beta_0 + \beta_1 x_1 + \cdots + \beta_n x_n)}}. \qquad (4.137)$$

The coefficients β_i of the multiple regression model are estimated using the training data with n independent variables. We consider an input observation $\mathbf{x} = [x_1, x_2, \ldots, x_n]$ and an output classifier y that can be 0 (representing that the observation is in the class) or 1 (representing that the observation is not in the class). In this case, we can't measure the probability directly. Therefore, we observe the occurrence of the event and infer the probability. The data is generated from a noisy target function. Namely, the probability of them can be expressed as $P(y = 0|\mathbf{x})$ or $P(y = 1|\mathbf{x})$. The purpose of the binary classification is to learn how correctly the input observation is classified into one of two classes. Logistic regression introduces a nonlinearity over a linear classifier and solves the task from a training data set. The following model is developed:

$$z = \sum_{i=1}^{n} w_i x_i + b = \mathbf{w}\mathbf{x} + b \qquad (4.138)$$

where w_i and b are the weight associated with the input observation x_i and the bias (also called intercept), respectively. In order to create the probability, sigmoid function is used and we have two cases as follows:

$$P(y = 1) = S(z) = S(\mathbf{w}\mathbf{x} + b) = \frac{1}{1 + e^{-(\mathbf{w}\mathbf{x}+b)}}, \qquad (4.139)$$

$$P(y = 0) = 1 - S(z) = 1 - S(\mathbf{w}\mathbf{x} + b) = \frac{e^{-(\mathbf{w}\mathbf{x}+b)}}{1 + e^{-(\mathbf{w}\mathbf{x}+b)}}. \qquad (4.140)$$

Now, we can have the decision boundary as follows:

$$\hat{y} = \begin{cases} 1, & \text{when } P(y = 1|\mathbf{x}) > 0.5 \\ 0, & \text{otherwise} \end{cases}. \qquad (4.141)$$

What we have to do is to learn the weight $\mathbf{w}$ and the bias b that makes $\hat{y}$ as close as possible to y. In order to find them, the loss function or the cost function is defined using the conditional maximum likelihood estimation. They are chosen by maximizing the probability of the correct $p(y|\mathbf{x})$. They have Bernoulli distribution because this is a binary classification. The probability $p(y|\mathbf{x})$ can be expressed as follows:

$$p(y|\mathbf{x}) = \hat{y}^y (1 - \hat{y})^{1-y}. \qquad (4.142)$$

In order to handle (4.142) easily, we use the log term as follows:

$$\log p(y|\mathbf{x}) = \log\left(\hat{y}^y(1-\hat{y})^{1-y}\right) = y\log\hat{y} + (1-y)\log(1-\hat{y}). \quad (4.143)$$

(4.143) represents the log likelihood. The loss function about how much $\hat{y}$ is close to y is now defined as follows:

$$L(\hat{y}, y) = -\left(y\log\hat{y} + (1-y)\log(1-\hat{y})\right). \quad (4.144)$$

Since the classifier output $\hat{y} = S(\mathbf{wx} + b)$, we have

$$L(\mathbf{w}, b) = -(y\log S(\mathbf{wx} + b) + (1-y)\log(1 - S(\mathbf{wx} + b))). \quad (4.145)$$

As we can observe (4.144) and (4.145), the negative log is convenient and mathematically friendly because it moves from 0 (when we have no loss) to infinity (when we have infinite loss). This loss function is convex. We call this cross-entropy loss function between the estimated distribution $\hat{y}$ and the true distribution y. Now, we minimize the loss function and find the optimal parameters $\mathbf{w}$ and b using gradient descent method. We can define the model as follows:

$$\hat{\boldsymbol{\theta}} = \underset{\boldsymbol{\theta}}{\operatorname{argmin}} \frac{1}{m} \sum_{i=1}^{m} L(y^i, x^i; \boldsymbol{\theta}) \quad (4.146)$$

where $\boldsymbol{\theta}$ denotes $\mathbf{w}$, b and m are the number of the training data. As we discussed gradient descent method in the previous section, we need to find the gradient of the loss function to update $\boldsymbol{\theta}$. The loss function can be rewritten as follows:

$$L(\boldsymbol{\theta}) = -\frac{1}{m} \sum_{i=1}^{m} (y^i \log S(w^i x^i + b) + (1 - y^i)\log(1 - S(w^i x^i + b))). \quad (4.147)$$

The gradient of the loss function means the directional component of the slope. The learning rate or step size means the magnitude of the travelling. The update rule can be expressed as follows:

$$\boldsymbol{\theta}^{(k+1)} = \boldsymbol{\theta}^{(k)} - \alpha^{(k)} \nabla L\left(S(\mathbf{x}; \boldsymbol{\theta}^{(k)}), y\right) \quad (4.148)$$

where $\alpha^{(k)} > 0$ is the learning rate and $S(\mathbf{x}; \boldsymbol{\theta}^{(k)})$ is $\hat{y}$ at the step k. In (4.148), the gradient with respect to a single weight w_k is found by the derivative of the loss function for the observation vector $\mathbf{x}$. Using (4.134) and (4.135), we have as follows:

$$\frac{\partial}{\partial w_k} L(\mathbf{w}, b) = (S(\mathbf{wx} + b) - y)x_k = (\hat{y} - y)x_k. \quad (4.149)$$

Thus, the update rule can be rewritten as follows:

$$\boldsymbol{\theta}^{(k+1)} = \boldsymbol{\theta}^{(k)} + \alpha^{(k)}(y - \hat{y})x_k. \tag{4.150}$$

As we can observe (4.150), the gradient is simply expressed as the difference between the estimated $\hat{y}$ and the true y and then multiplied by the input observation x_k. Now, we can summarize the steps of the stochastic gradient descent method for the logistic regression as follows:

Initial conditions:

– L is the loss function, $\hat{y}$ is the estimated function, $\mathbf{x}$ is the input observation, and y is the output classifier.
– Start with an initial point $\mathbf{x}^{(0)}$, an step size $\alpha^{(k)}$, tolerance $\varepsilon > 0$.

Repeat

– **Step 1**: Select a mini-batch B randomly.
– **Step 2**: Compute $\hat{y}$ and L.
– **Step 3**: Determine the next point by $\boldsymbol{\theta}^{(k+1)} = \boldsymbol{\theta}^{(k)} + \alpha^{(k)}(y - \hat{y})x_k$.

Until it converges: The gradient $< \varepsilon$.

Example 4.13 Logistic Regression Consider a single cell model with 328 mobile stations. We performed the connectivity test in terms of the distance between a base station and mobile stations. The test results are summarized in Table 4.5.

Compute the proportion of the failed connectivity for each distance and then compare with linear regression and logistic regression.

Table 4.5 Connectivity test results

index	Distance between a base station and mobile stations (km)	The number of mobiles stations at the distance	The number of failed connectivity
1	2.0	47	2
2	2.2	43	1
3	2.4	30	0
4	2.6	35	4
5	2.8	31	7
6	3.0	21	8
7	3.2	24	13
8	3.4	22	18
9	3.6	21	19
10	3.8	15	15
11	4.0	18	17
12	4.2	21	21

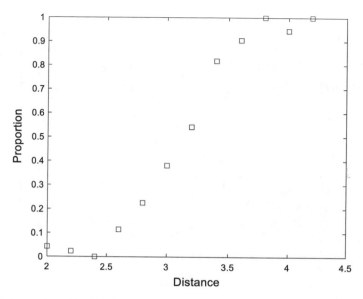

Fig. 4.54 Proportion of the failed connectivity

Solution

From Table 4.5, we can simply calculate the proportion of the failed connectivity for each distance as follows:

$$\text{Proportion} = \frac{\text{The number of failed connectivity}}{\text{The number of mobiles stations at the distance}}.$$

Figure 4.54 illustrates the proportion of the failed connectivity for each distance.

As we discussed the linear regression in Sect. 4.3.2, the linear model can be found as shown in Fig. 4.55.

As we can observe Fig. 4.55, the linear model is not normally distributed and also the linear fit exists less than 0 and greater than 1. In this case, logistic regression is suitable for the data. Figure 4.56 illustrates logistic regression model.

As we can observe Fig. 4.56, the logistic regression is well fitted with the data and also the range of the proportion is between 0 and 1.

■

As we discussed in Example 4.13, the logistic regression is more useful when dealing with the type of binary classification. For example, when we have a data set as shown in Fig. 4.57, the linear model as shown in Fig. 4.57a causes a misclassification but the logistic regression model as shown in Fig. 4.57b fits well with the data points.

Example 4.14 Gradient Descent Method for Logistic Regression Consider a data set with a binary classification as shown in Table 4.6.

Find a linear decision boundary using logistic regression.

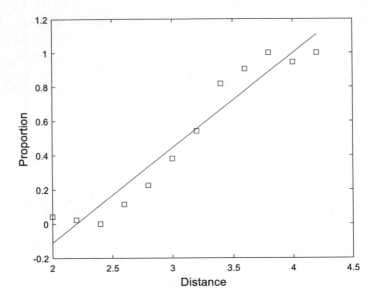

Fig. 4.55 Linear regression

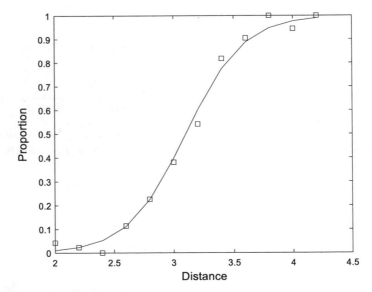

Fig. 4.56 Logistic regression

Solution

The first step is the normalization of the data set. As we did in Example 4.12, we perform the normalization process as follows:

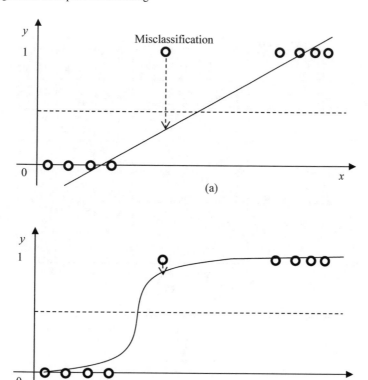

Fig. 4.57 Comparison of linear regression (**a**) and logistic regression (**b**)

Table 4.6 Data set

Index	$x1$	$x2$	y Binary classification
1	5.3	7.0	1
2	3.1	6.1	1
3	1.7	2.2	0
4	4.2	5.4	1
5	7.2	8.4	1
6	3.5	2.8	0
7	7.0	1.9	0
8	4.6	3.1	0
9	0.6	1.1	0
10	6.6	6.1	1
11	4.1	2.2	0
12	2.5	3.1	0
13	7.4	7.2	1

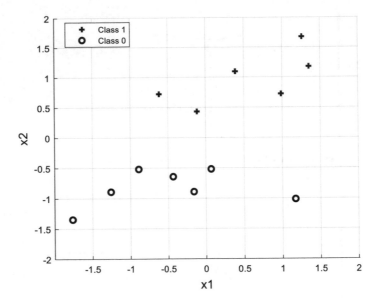

Fig. 4.58 Normalized data set

$$x' = \frac{x - \overline{x}}{\sigma}$$

where x', x, $\overline{x}$, and σ are the normalized value, the original value, the mean, and standard deviation, respectively. Figure 4.58 illustrates the normalized data set.

From (4.139), (4.140), and (4.141), we have the following probability and hypothesis:

$$P(y = 1) = \frac{1}{1 + e^{-(\mathbf{wx}+b)}}, \ P(y = 0) = \frac{e^{-(\mathbf{wx}+b)}}{1 + e^{-(\mathbf{wx}+b)}},$$

$$\hat{y} = \begin{cases} 1, & \text{when } P(y = 1|\mathbf{x}) > 0.5 \\ 0, & \text{otherwise} \end{cases}.$$

In the linear model, we assume that the parameter $\boldsymbol{\theta}$ has only the weight $\mathbf{w}$. The linear decision boundary is defined as follows:

$$h_\theta(\mathbf{x}) = \theta_1 + \theta_2 x_1 + \theta_3 x_2$$

and we find $\boldsymbol{\theta}$ using the loss function (4.147) and the update rule (4.148). We assume that the learning rate is 0.1 and the maximum iteration is 100. Figure 4.59 illustrates the update of $\boldsymbol{\theta}$.

We have the linear decision boundary $h_\theta(\mathbf{x}) = -0.1408 + 0.6069x_1 + 1.9595x_2$ as shown in Fig. 4.60.

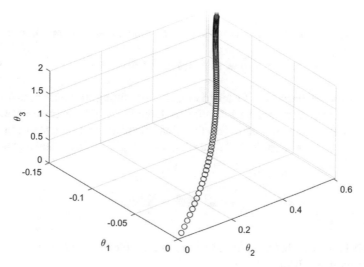

Fig. 4.59 Updating θ

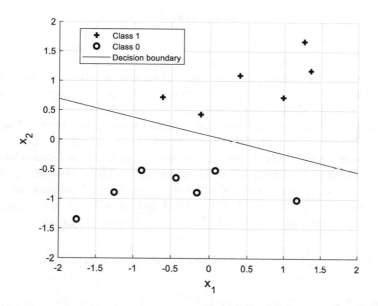

Fig. 4.60 Linear decision boundary

■

Basically, a high order of polynomials provides us with good fit. However, like the linear regression, logistic regression is inclined to overfit if there are large number of features. In order to avoid overfitting, we can reduce the number of features manually, limit the number of training step or learning rate, or apply regularization. In machine learning, Tikhonov regularization using L1 norm and Lasso regularization using L2 norm is widely used. The loss function of the logistic regression can be updated to penalize high values of the parameters. (4.146) can be rewritten as follows:

$$\hat{\theta} = \min L(\theta) + \lambda R(\theta) \tag{4.151}$$

where λ is the regularization factor to control both fitting the training data set and keeping the parameter small. If λ has a very large value, all parameters θ are close to zero and the intercept is left. The decision boundary will be a start line. It can be tuned empirically. $R(\theta)$ is the regularization term. It can be calculated using L2 norm as follows:

$$R(\theta) = \|\theta\|_2^2 = \sum_{i=1}^{n} \theta_i. \tag{4.152}$$

This is Euclidean distance. Another way to compute the regularization term is L1 regularization as follows:

$$R(\theta) = \|\theta\|_1 = \sum_{i=1}^{n} |\theta_i|. \tag{4.153}$$

This is Manhattan distance. If there is no regularization, the logistic regression might cause the overfitting problem. Most of logistic regression models use the regularization to avoid overfitting and dampen complexity of the regression models.

When we need to deal with more than two classes and divide input space into multiple decision regions, multinomial logistic regression (also called softmax regression) is widely used. It generalizes a classification problem of logistic regression in which the class label y can take on any one of k values. Generalization of sigmoid function (also called softmax function) enables us to compute the probability $p(y = c|\mathbf{x})$ where c is one of K classes C_K. The softmax function takes a vector $\mathbf{z} = [z_1, z_2, \ldots, z_K]$ of K real number as input and normalizes it into a probability distribution. The softmax function $\sigma : \mathbb{R}^K \to \mathbb{R}^K$ is defined as follows:

$$\sigma(z_i) = \frac{e^{z_i}}{\sum_{j=1}^{K} e^{z_j}}, \ 1 \leq i \leq K \text{ and } \mathbf{z} = [z_1, z_2, \ldots, z_K] \in \mathbb{R}^K \tag{4.154}$$

and the softmax of the input vector $\mathbf{z}$ is

$$\sigma(\mathbf{z}) = \left[\frac{e^{z_1}}{\sum_{j=1}^{K} e^{z_j}}, \frac{e^{z_2}}{\sum_{j=1}^{K} e^{z_j}}, \ldots, \frac{e^{z_K}}{\sum_{j=1}^{K} e^{z_j}} \right]. \tag{4.155}$$

Now, the probability $p(y = c|\mathbf{x})$ is computed as follows:

$$p(y = c|\mathbf{x}) = \frac{e^{w_c \mathbf{x} + b_c}}{\sum_{j=1}^{K} e^{w_j \mathbf{x} + b_j}}. \tag{4.156}$$

As we can observe (4.153), (4.154), and (4.155), each component will be in the interval [0 1] after applying softmax function. If one component is greater than others, it tends to go its probability to one and reduce the probability of smaller inputs.

Example 4.15 Softmax Function
  Consider the input vector as follows:

$$\mathbf{z} = [0.5\ 1.5\ -1.8\ 2.3\ 2.7\ -1.9\ 4.2\ 5.9\ 1.2\ -3.1]$$

and compute its softmax.

Solution

From (4.155), we compute softmax function as follows:

$$\sigma(\mathbf{z}) = \left[\frac{e^{z_1}}{\sum_{j=1}^{K} e^{z_j}}, \frac{e^{z_2}}{\sum_{j=1}^{K} e^{z_j}}, \ldots, \frac{e^{z_K}}{\sum_{j=1}^{K} e^{z_j}} \right]$$

$$= [0.0035\ 0.0096\ 0.0004\ 0.0214\ 0.0319\ 0.0003\ 0.1430\ 0.7827\ 0.0071\ 0.0001].$$

As we can observe this result, one component with the highest input is increased and others are suppressed.

∎

Summary 4.7 Logistic Regression

1. Logistic regression uses the probability of a class or an event and develops a model to estimate parameters. The name "regression" was established in the statistics community but the logistic regression model refers not to regression but to classification.

2. The logistic regression has a S shaped curve that can be expressed as Sigmoid function:

$$\text{Sigmoid function: } S(x) = \frac{1}{1 + e^{-x}}.$$

3. The steps of the stochastic gradient descent method for the logistic regression are as follows:

Initial conditions:

– L is the loss function, $\hat{y}$ is the estimated function, $\mathbf{x}$ is the input observation, and y is the output classifier.
– Start with an initial point $\mathbf{x}^{(0)}$, an step size $\alpha^{(k)}$, tolerance $\varepsilon > 0$.

Repeat

– *Step 1*: Select a mini batch B randomly.
– *Step 2*: Compute $\hat{y}$ and L.
– *Step 3*: Determine the next point by $\boldsymbol{\theta}^{(k+1)} = \boldsymbol{\theta}^{(k)} + \alpha^{(k)}(y - \hat{y})x_k$.

Until it converges: The gradient $< \varepsilon$.

4.4 Problems

4.1   Compare supervised learning and unsupervised learning.
4.2   Describe the pros and cons of supervised learning.
4.3   Describe the key performance indicators of supervised learning.
4.4   Describe the key applications of supervised learning and how they improve the business.
4.5   The supervised learning process is composed of two phases: training and testing. Describe each step of two phases.
4.6   There are the evaluation metrics of supervised learning algorithms: predictive accuracy, efficiency about model construction, robustness of noise and missing data, scalability, and interpretability. Compare them and describe their limitations.
4.77   One important assumption of supervised learning model is that the training data set should be representative of the test data set. Describe how to overcome in terms of algorithm level if this assumption is violated.
4.8   The performance of supervised learning algorithms depends on size of a data set. The size of training data and test data is highly related to the accuracy of classifiers. Compare the performances of one supervised learning algorithm in terms of different ratio of training data and test data.
4.9   In machine learning, ensemble methods create multiple models and then combine them to produce a better result. Describe different types of ensemble methods and compare them.
4.10   There is a trade-off relationship between interpretability and accuracy. Figure 4.4 illustrates the relationship between interpretability and accuracy for machine learning algorithms. Describe which algorithms are suitable in terms of target applications.
4.11   Supervised learning can be categorized as classification and regression. Describe key algorithms and applications for classification or regression.
4.12   Compare of the lazy learning and the eager learning.

4.13 Decision trees have a hierarchical data structure and are built using recursive partitioning based on a divide-and-conquer strategy. The decision tree is composed of a root node, internal (or test) nodes, leaf (or terminal, decision) nodes, and edges (or branches). Describe the properties of each component.

4.14 The main goal of classification trees is to classify a data set. The attribute with the highest purity (or the lowest impurity) is chosen. Describe types and pros and cons of the impurity measures.

4.15 Describe the recursive procedure of Hunt's algorithm.

4.16 Compare of impurity measures: entropy, Gini, and classification error.

4.17 Describe the relationship between decision tree size and accuracy as well as between prediction error and model complexity.

4.18 Consider the random three classes (class 1, class 2, and class 3) data with target matrix 20 observations with 3 features ($x1$, $x2$ and $x3$) as a training data set as follows:

$$\text{Data} = \begin{bmatrix} 15 & 12 & 8 & 2 & 16 & 12 & 6 & 6 & 6 & 19 & 10 & 11 & 20 & 21 & 23 & 5 & 5 & 15 & 12 & 2 \\ 1 & 0 & 13 & 15 & 12 & 11 & 0 & 0 & 1 & 3 & 4 & 2 & 8 & 10 & 11 & 0 & 0 & 5 & 2 & 4 \\ 6 & 1 & 3 & 0 & 1 & 0 & 11 & 12 & 11 & 11 & 10 & 1 & 0 & 1 & 0 & 12 & 12 & 10 & 1 & 2 \end{bmatrix}$$

and *a* test data set is

$$\text{Data} = \begin{bmatrix} 7 & 0 & 1 & 2 & 6 & 0 \\ 1 & 5 & 0 & 12 & 14 & 7 \\ 6 & 1 & 7 & 0 & 5 & 12 \end{bmatrix}$$

Create the decision tree and then perform classification for the test data set.

4.19 Consider the following labelled data set:

I	x_i	θ_i
1	(14, 13)	Class A
2	(8, 7)	Class A
3	(15, 9)	Class A
4	(3, 6)	Class B
5	(2, 5)	Class B
6	(4, 4)	Class B

When we have a new data point (6, 6) and $K = 3$, find the classification of the data point.

4.20 In the support vector machines, describe how to select support vectors.

4.21 In the support vector machines, explain why kernel function is useful.

Table 4.7 2D data set

I	X	Y	i	X	Y
1	4.38	6.08	11	29.17	26.23
2	2.86	5.11	12	11.52	5.11
3	20.97	21.17	13	10.12	7.77
4	7.63	8.41	14	12.62	9.86
5	4.75	5.61	15	5.54	8.02
6	5.05	7.49	16	6.23	5.98
7	21.68	17.05	17	11.76	8.28
8	14.51	13.88	18	6.83	9.42
9	8.67	11.57	19	12.13	8.78
10	13.92	12.16	20	8.47	8.21

4.22 Consider the following data set with 10 samples:

Class A: (1,4), (2,10), (4,9), (3, 3), (2, 8)
Class B: (6,2), (7,8), (8,3), (9, 10), (7, 3)

Find a classifier using SVM.

4.23 Consider 10 paired data points (1,0), (2, 1), (3,4), (4, 5), (6, 7), (8, 11), (9,9), (10, 15), (12, 17), (15,20), and the paired data satisfies a linear model. Find β_0 and β_1 of the linear model and compare the paired data and the linear model.

4.24 Gradient descent algorithm is an iterative optimization algorithm for finding the minimum of an objective function (or cost function) by moving in the direction of steepest descent. Describe the different types of gradient descent algorithms and compare them in terms of estimation accuracy and complexity.

4.25 Consider two-dimensional data set as in Table 4.7.

Fit the linear regression parameters to the data set using gradient descent method.

4.26 In the logistic regression, the name "regression" was established in the statistics community but the logistic regression model refers not to regression but to classification. Describe the different types of logistic regression algorithms and compare them.

4.27 Consider a data set with a binary classification as in Table 4.8.

Find a linear decision boundary using logistic regression.

Table 4.8 Data set

Index	$x1$	$x2$	y Binary classification
1	4.2	7.1	1
2	1.6	2.1	0
3	5.2	6.4	1
4	8.2	9.4	1
5	3.4	2.7	0
6	6.0	1.3	0
7	4.7	3.3	0
8	0.9	1.8	0
9	7.6	7.2	1
10	4.5	3.8	0

References

1. S.B. Kotsiantis, Supervised Machine Learning: A Review of Classification Techniques. Informatica **31**(3), 249–268 (2007)
2. E. Alpaydin, *Introduction to Machine Learning*, 2nd edn. (The MIT Press, 2010)
3. J. Han, M. Kamber, *Data Mining: Concepts and Techniques* (Morgan Kaufmann, 2006)
4. E.B. Hunt, J. Marin, P.J. Stone, *Experiments in Induction* (Academic Press, 1966)
5. J.R. Quinlan, Induction of decision trees. Mach. Learn. **1**, 81–106 (1986). https://doi.org/10.1007/BF00116251
6. J. R. Quinlan, *C4. 5: Programs for Machine Learning*, vol. 1 (Morgan Kaufmann, 1993)
7. L. Breiman, J. Friedman, C.J. Stone, R.A. Olshen, *Classification and Regression Trees* (CRC Press, 1984)
8. J.N. Morgan, J.A. Sonquist, Problems in the analysis of survey data, and a proposal. J. Am. Stat. Assoc. **58**, 415–434 (1963)
9. P.-N. Tan, M. Steinbach, A. Karpatne, V. Kumar, *Introduction to Data Mining* (Pearson, 2018). ISBN 0133128903
10. J.R. Quinlan, Induction of decision trees. Mach. Learn. **1**(1), 81–106 (1986). https://doi.org/10.1007/BF00116251
11. T.M. Cover, P.E. Hart, Nearest neighbor pattern classification. IEEE Trans. Inform. Theory **IT-13**, 21–27 (1967)
12. T. Bailey, A.K. Jain, A note on Distance weighted k-nearest neighbor rules. IEEE Trans. Syst. Man Cybern. **8**, 311–313 (1978)
13. K. Chidananda, G. Krishna, The condensed nearest neighbor rule using the concept of mutual nearest neighbor. IEEE Trans. Inf.Theory **IT-5**, 488–490 (1979)
14. G.W. Gates, Reduced nearest neighbor rule. IEEE Trans. Inf. Theory **18**(3), 431–433 (1972)
15. S.C. Bagui, S. Bagui, K. Pal, Breast cancer detection using nearest neighbor classification rules. Pattern Recogn. **36**, 25–34 (2003)
16. Z. Yong, An Improved kNN text classification algorithm based on clustering. J. Comput. **4**(3) (2009)
17. V.N. Vapnik, *The Nature of Statistical Learning Theory* (Springer, 1995). ISBN 0-387-98780-0
18. C. Cortes, V. Vapnik, Support vector networks. J. Mach. Learn. **20**(3), 273–297 (1995)
19. C. Burges, A tutorial on support vector machines for pattern recognition. Data Min. Knowl. Disc. **2**(2), 121–167 (1998)
20. N. Qian, On the momentum term in gradient descent learning algorithms. Neural Networks: Official J. Int. Neural Network Soc. **12**(1), 145–151 (1999). https://doi.org/10.1016/S0893-6080(98)00116-6

21. J. Duchi, E. Hazan, Y. Singer, Adaptive subgradient methods for online learning and stochastic optimization. J. Mach. Learn. Res. **12**, 2121–2159 (2011). http://jmlr.org/papers/v12/duchi11a.html

22. M.D. Zeiler, *ADADELTA: An Adaptive Learning Rate Method* (2012). Retrieved from http://arxiv.org/abs/1212.5701

23. G. Hinton, Lecture 6e rmsprop: Divide the gradient by a running average of its recent magnitude (2020). Retrieved from http://www.cs.toronto.edu/~tijmen/csc321/slides/lecture_slides_lec6.pdf

24. D.P. Kingma, J.L. Ba, Adam: a method for stochastic optimization, in *International Conference on Learning Representations*, pp. 1–13 (2015). Retrieved from https://arxiv.org/abs/1412.6980

Chapter 5
Reinforcement Learning

Supervised learning and unsupervised learning enable us to solve classification, regression, dimension reduction, and clustering problems. The decision-making is one of key applications in machine learning. The reinforcement learning (RL) deals with the decision-making problem that is formulated mathematically in a Markov decision process (MDP). Many RL algorithms are based on model-based approaches such as dynamic programming or model-free approaches such as temporal difference learning. The RL methods solve sequential decision-making problem under uncertainty. The goal of sequential decision-making problem is to find the optimal policy to maximize the long-term rewards. Sometimes, the decision-making can be regarded as assigning the label at each step using a training dataset. Choosing a class is making a decision. This is single point decision. Thus, a supervised learning algorithm such as decision tree enables us to make a decision. In addition, it is possible to transform a supervised learning problem to a RL problem. The loss function of supervised learning can be regarded as the reward function. We can map the smaller loss to the larger reward. In more complex environments, the algorithm requires to answer some questions: How much impact will this decision bring to the next step? How to evaluate this decision good or bad? The supervised learning algorithms couldn't satisfy these requirements. Since the RL algorithms interact with an environment, they work in complex environments. Figure 5.1 illustrates the data streaming process of the RL algorithms. As we can observe Fig. 5.1, there is an agent (or decision-maker) in the interaction process. Based on the state of the environment, the agent takes an action. Then, the environment updates the states, and the agent receives a reward or a punishment from the environment. If the agent moves correctly, a reward is given. If not, the agent is punished. The rewards are the numerical values that can be positive for a reward or negative for a punishment. This interaction process is expressed by a finite Markov decision process. The RL algorithm enables the agent to learn behaviours to achieve its goal.

© The Author(s), under exclusive license to Springer Nature Switzerland AG 2022
H. Kim, *Artificial Intelligence for 6G*,
https://doi.org/10.1007/978-3-030-95041-5_5

Fig. 5.1 Interaction process
of the RL

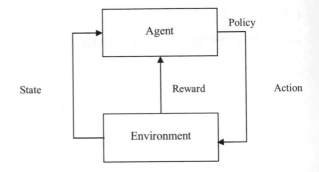

The roles of the RL elements can be summarized as follows: policy (what to do), reward (what is good or bad), value (what is good because it predicts reward), and model (what follows what, Optional element) [1]. The policy defines the behaviours of the agent and maps from states to actions. The reward indicates the intrinsic desirability and defines the goal of the RL problem. The value represents the total amount of rewards an agent expect to accumulate. Unlike the reward, the value function is an estimated value, and there are two forms: state-value function and action-value function. The model is an optional element because there are two forms: model-based and model-free. It predicts a mimic behaviour of the environment. The RL algorithms come from two independent studies. The first study is the experimental psychology of animal learning. When observing the behaviour of animals about the relationship between the occurrence of events and their response, based on trial and error, it tends to have a higher probability that the response happens again in the same situation. That is to say. If an action is followed by a satisfactory state of affairs, the tendency to produce the action is reinforced. This work was published in [2]. Psychologists use the term instrumental conditioning instead of reinforcement learning. The other study is optimal control and dynamic programming. The optimal control problem is to design a controller by minimizing the measure of the system over time. The problem includes system state and value function, and the dynamic programming solves this problem. Many extensions of the dynamic programming have been developed. The RL algorithms affected many AI applications. Table 5.1 summarizes the comparison of unsupervised learning, supervised learning, and RL learning. In this chapter, we discuss the RL algorithms.

5.1 Introduction to Reinforcement Learning and Markov Decision Process

The Markov decision process is a mathematical framework formulating the problem of the RL and describing the environment of the RL, where the environment is observable. Figure 5.1 shows us the interaction between the agent and the environment in the finite MDP. If the state and the action are finite, it is called a finite Markov

Table 5.1   Comparison of unsupervised learning, supervised learning, and reinforcement learning

	Unsupervised learning	Supervised learning	Reinforcement learning
Type of data	Unlabelled data	Labelled data	No predefined data. Learn from interaction with environment
Type of problems	Clustering and association	Classification and regression	Decision-making and control
Strategy and approach	Discovering a pattern and obtaining insight from test data	Mapping a label to test data	Making decision from previous experience
Algorithms	K-means clustering, fuzzy clustering, and so on	Linear regression, K-nearest neighbour, support vector machine, decision tree, and so on	Monte Carlo methods, SARSA, Q learning, and so on

decision process. Most of RL problems can be formulated as the finite MDPs. One important Markov property of the MDP is that an outcome of action depends on only the current state and not the previous history. Namely, the memoryless property implies that a set of possible actions and transitions does not rely on the sequence of events. The successor function cares about only the current state. We can express it mathematically as follows:

$$P(s_{t+1}|s_t) = P(s_{t+1}|s_1, \ldots, s_t) \qquad (5.1)$$

where s_{t+1} and s_t represent the next state and the current state, respectively. As we can observe (5.1), the transition from s_t to s_{t+1} is independent of the past, where the transition means moving from one state to another state and the probability of transition means the probability that an agent moves from one state to another state. We can say that this system has a Markov property. When we have a Markov state s and successor state s', the state transition probability can be expressed as follows:

$$P_{ss'} = P(s_{t+1} = s'|s_t = s). \qquad (5.2)$$

(5.2) means a probability distribution over the next successor states when the current state is given. The transition probability function can be rewritten in a matrix form as follows:

$$\mathbf{P} = \begin{bmatrix} p_{11} & \cdots & p_{1n} \\ \vdots & \ddots & \vdots \\ p_{n1} & \cdots & p_{nn} \end{bmatrix}. \qquad (5.3)$$

In the matrix of (5.3), each row represents the transition probability from the current state to any successor state. For example, p_{ij} means the transition probability from the state i to the state j. Sum of each row should be 1. A Markov process is a memoryless random process, and a sequence of states has a Markov property. It can be defined by a tuple (S, P), where S are P are a finite set of states and transition function (or transition probability matrix), respectively. A Markov reward process (MRP) is a stochastic process with value judgement. We add a reward to each state and extend a Markov chain. It can express how much rewards are accumulated through a sequence of samples. An MRP is defined as a tuple (S, P, R, γ) where S is a finite set of states, P is the transition function, R is a reward function, and $\gamma \in [0, 1]$ is a discount factor. The reward function is defined as follows:

$$R_s = E(R_{t+1}|s_t = s). \tag{5.4}$$

(5.4) represents the expected reward over all possible states transiting from the state s. The RL algorithms find an optimal policy by maximizing not immediate rewards but cumulative rewards. Thus, the return G_t as the total sum of rewards is defined as follows:

$$G_t = R_{t+1} + \gamma R_{t+2} + \gamma^2 R_{t+3} + \cdots = \sum_{k=0}^{\infty} \gamma^k R_{t+k+1} \tag{5.5}$$

where discount factor γ represents how much an agent cares about rewards. This definition represents how much the agent will care more about the immediate reward and future rewards. It is a metric explaining the relationship between immediate reward and future rewards. $\gamma = 0$ means that an agent affects the first reward and the immediate reward is more important. The more immediate rewards have more influence. $\gamma = 1$ means that it affects all future rewards and future rewards are more important. Typical values of the discount factor lie between 0.2 and 0.8. In (5.5), R_{t+1} means the reward received at the current time when transiting from one state to another state. There are two types of tasks: episodic tasks and continuous tasks. The episodic tasks have a terminal state, and the interaction breaks into episodes. We have the finite states. For example, when we play a shooting video game, a player starts the game and is killed. The game is over. We call this an episode. Since a new game starts at an initial state, episodes are independent. In episodic tasks, the return G_t can be expressed as follows:

$$G_t = R_{t+1} + R_{t+2} + \cdots + R_T \tag{5.6}$$

where T is a final step and R_T is the reward at a terminal state when an episode ends. The continuous tasks do not have a terminal. In continuous tasks, the discounted returns are always used as shown in (5.5). In (5.5), the return is sum of an infinite numbers of rewards. If the reward is nonzero constant and the discount factor is less than 1, we have a finite return. For example, when the reward is 1, the return can be expressed as follows:

$$G_t = \sum_{k=0}^{\infty} \gamma^k = \frac{1}{1-\gamma}. \tag{5.7}$$

Markov decision process (MDP) is an extension of Markov reward process (MRP). The MRP is a Markov chain with rewards and the MDP is a MRP with decisions. An MDP is defined as a tuple (S, A, P, R, γ) where A is the set of actions an agent can take actions or make decisions. Since actions are added, it affects to the state and the state transition probability and the reward function are modified as follows:

$$P_{ss'} = P\left(s_{t+1} = s' | s_t = s, a_t = a\right) \tag{5.8}$$

and

$$R_s = E(R_{t+1} | s_t = s, a_t = a). \tag{5.9}$$

As we can observe (5.8) and (5.9), they now depend on the actions. In the MDP, the policy is a mechanism to make decisions or takes actions. It depends on the current state due to Markov property. The policy is a mapping from state space to action space. It is basically stochastic, but, in a special case, we can have deterministic policies. The policy can be evaluated with value functions. The policy π is defined as a probability distribution over a set of actions given states as follows:

$$\pi(a|s) = P(a_t = a | s_t = s). \tag{5.10}$$

A value function represents how good it is for an agent to be in a given state. It is defined as the expected return starting from state s and following policy π for the next state. We define a value function v_π as follows:

$$v_\pi(s) = E_\pi(G_t | s_t = s) = E_\pi\left(\sum_{k=0}^{\infty} \gamma^k R_{t+k+1} | s_t = s\right), \text{for} \forall s \in S. \tag{5.11}$$

This function is called state-value function. This is what we want to optimize. Bellman equation enables us to find an optimal policy and value function. It is a key element of solving the reinforcement learning problems. The value function can be decomposed into two components: immediate reward R_{t+1} and discounted value of successor states $\gamma v(s_{t+1})$. Bellman equation for state-value function is defined as follows:

Fig. 5.2 Bellman equation
backup diagram

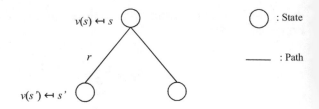

$$v(s) = E(G_t|s_t = s)$$
$$= E\left(R_{t+1} + \gamma R_{t+2} + \gamma^2 R_{t+3} + \dots|s_t = s\right)$$
$$= E(R_{t+1} + \gamma(R_{t+2} + \gamma R_{t+3} + \dots)|s_t = s) \quad (5.12)$$
$$= E(R_{t+1} + \gamma G_{t+1}|s_t = s)$$
$$= E(R_{t+1} + \gamma v(s_{t+1})|s_t = s).$$

Let's consider one simple example in MRP. There is an agent in one state s. It moves from this state s to another state s' as shown in Fig. 5.2.

What we want to know is the value of state s. The state-value function is the expected value of returns we get upon leaving the state s and the discounted average value over the next possible states s'. The expectation operation is distributive, and we can have R_{t+1} and $v(s_{t+1})$ terms separately. Thus, we have the following equation without expectation operation.

$$v(s) = R_s + \gamma \sum_{s' \in S} P_{ss'} v(s'). \quad (5.13)$$

Bellman equation is linear and can be solved directly. (5.13) can be expressed in a matrix form as follows:

$$\mathbf{V} = \mathbf{R} + \gamma \mathbf{PV}, \quad (5.14)$$

$$(\mathbf{I} - \gamma \mathbf{P})\mathbf{V} = \mathbf{R}, \quad (5.15)$$

$$\mathbf{V} = (\mathbf{I} - \gamma \mathbf{P})^{-1}\mathbf{R}. \quad (5.16)$$

This approach works well in a small size of MRP. In a large size of MRP, it is not easy to solve a large size matrix. Thus, we can use an iterative solution such as dynamic programming, Monte Carlo evaluation and temporal difference learning. In MDP, the state-value function $v_\pi(s)$ is the expected return starting from state s and following policy π. Bellman expectation equation for state-value function can be expressed as follows:

$$v_\pi(s) = E_\pi(R_{t+1} + \gamma v_\pi(s_{t+1})|s_t = s). \quad (5.17)$$

(5.17) is similar to the state-value function for MRP. Like (5.12), it represents that the value of a certain state is decided by the immediate reward and the discounted average value over the next possible states when following a policy π. Now, we define action-value function $q_\pi(s, a)$ as the expected return starting state s, taking acting a, and following policy π. The action-value function (or q function) specifies how good it is to take a particular action from a certain state. It is mathematically defined as follows:

$$q_\pi(s, a) = E_\pi(G_t|s_t = s, a_t = a) = E_\pi\left(\sum_{k=0}^{\infty} \gamma^k R_{t+k+1}|s_t = s, a_t = a\right). \quad (5.18)$$

Similar to (5.17), (5.18) can be expressed as follows:

$$q_\pi(s, a) = E_\pi(R_{t+1} + \gamma q_\pi(s_{t+1}, a_{t+1})|s_t = s, a_t = a). \quad (5.19)$$

This is Bellman expectation equation for action-value function (also called state–action value function, state action function, or q function). As we can observe (5.18) and (5.19), it considers actions for state transition and represents the expected return over actions. We need to evaluate the actions along with the states. Both value functions (state-value function and action-value function) are conditional on a certain policy π. The state-value function represents the expected value of achieving a certain state, whereas the action-value function represents the expected value of choosing an action from a certain state and then following a policy π. Figure 5.3 illustrates a backup diagram for state-value function and action-value function.

As we can observe Fig. 5.3, states and actions are expressed as circles and dots, respectively. Figure 5.3a is state-centric view diagram, and Fig. 5.3b is action-centric view diagram. The move from state to action in Fig. 5.3a represents that an agent in state s takes an action a with a policy π. An agent can choose an action, and it is controllable by the agent. The move from action to state in Fig. 5.3b represents that an environment sends an agent to one of states using a transition probability. Since an agent doesn't know how the environment reacts, it is not controllable by the agent. In Fig. 5.3a, there is a state–action valuation function for each action. The state-value function $v_\pi(s)$ can be defined as follows:

$$v_\pi(s) = \sum_{a \in A} \pi(a|s)q_\pi(s, a). \quad (5.20)$$

(5.20) represents the connection between state-value function and action-value function. In Fig. 5.3b, we can average the state values of states and add the immediate reward like Bellman equation. The action-value function $q_\pi(s)$ can be expressed as follows:

$$q_\pi(s, a) = R_s^a + \gamma \sum_{s' \in S} P_{ss'}^a v_\pi(s'). \quad (5.21)$$

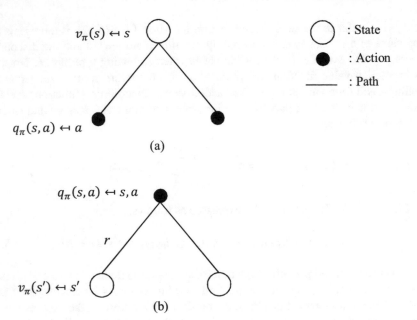

$v_\pi(s) \hookleftarrow s$

$q_\pi(s,a) \hookleftarrow a$

            ○ : State

            ● : Action

           ── : Path

(a)

$q_\pi(s,a) \hookleftarrow s,a$

r

$v_\pi(s') \hookleftarrow s'$

(b)

Fig. 5.3 Backup diagram for state-value function (**a**) and for action-value function (**b**)

where $P_{ss'}^a$ denotes a transition probability from the current state to the next successor states when an action a is taken. (5.21) represents how good it is to take an action a in a state s. Now, we combine Fig. 5.3a with b and create a new backup diagram as shown in Fig. 5.4.

As we can observe Fig. 5.4, an agent takes one of two actions from state s. The transition probability from state s is weighted by a policy π. After taking the action, the transition probability to state s' is weighted by an environment. We can substitute (5.21) in (5.20). We obtain a recursive form of a state-value function as follows:

$$v_\pi(s) = \sum_{a \in A} \pi(a|s)\left(R_s^a + \gamma \sum_{s' \in S} P_{ss'}^a v_\pi(s') \right). \tag{5.22}$$

Fig. 5.4 Combined backup diagram for state-value function

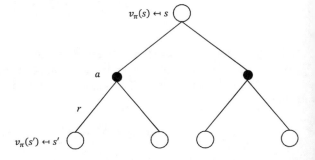

$v_\pi(s) \hookleftarrow s$

a

r

$v_\pi(s') \hookleftarrow s'$

Fig. 5.5 Combined backup diagram for action-value function

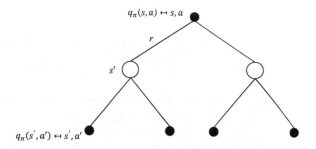

Likewise, we combine them and create a new backup diagram as shown in Fig. 5.5.

This is inverse of state-value function. Similar to (5.22), we obtain a recursive form of an action-value function as follows:

$$q_\pi(s, a) = R_s^a + \gamma \sum_{s' \in S} P_{ss'}^a \sum_{a' \in A} \pi(a'|s') q_\pi(s', a'). \tag{5.23}$$

Both (5.22) and (5.23) are expression of state-value function and action-value function in MDP and satisfy Bellman equation. However, they do not enable us to find the best way in MDP. Now, we find optimal value function and optimal policy function. In the model of MDP, there are many different value functions with regards to different policies. The optimal value function represents what we have the maximum value over all policies comparing with other value functions. Finding the optimal value function means solving the problem of MDP. The optimal state-value function $v^*(s)$ and the optimal action-value function $q^*(s, a)$ are expressed as follows:

$$v^*(s) = \max_\pi v_\pi(s) \tag{5.24}$$

and

$$q^*(s, a) = \max_\pi q_\pi(s, a). \tag{5.25}$$

Now, we define the optimal policy resulting in optimal value function. Basically, when a value function with a policy π for all states is greater than the one with a policy π' for all states, the policy π is better than the policy π'. We can mathematically express them as follows:

$$\pi \geq \pi' \text{ when } v_\pi(s) \geq v_{\pi'}(s) \text{ for } \forall s. \tag{5.26}$$

The value function with the optimal policy π^* is the maximum out of all policies for all states. In order to find an optimal policy, we solve $q^*(s, a)$ and then take the action. Simply, we assign the probability 1 for the action with maximum value of $q^*(s, a)$ and the probability 0 for the rest of actions for all states. We can mathematically express them as follows:

$$\pi^*(a|s) = \begin{cases} 1, & \text{if } a = \underset{a \in A}{\operatorname{argmax}} \, q^*(s, a) \\ 0, & \text{otherwise} \end{cases}. \tag{5.27}$$

Bellman optimality equation has same concept as Bellman equation and represents the relationship between value functions. Basically, it expresses that the value of a state under the optimal policy is equal to the expected return for the best action form the state. We can mathematically express it as follows:

$$v^*(s) = \max_a q^{\pi^*}(s, a)$$

$$= \max_a E_{\pi^*}(G_t | s_t = s, a_t = a)$$

$$= \max_a E_{\pi^*}\left(\sum_{k=0}^{\infty} \gamma^k R_{t+k+1} | s_t = s, a_t = a \right) \tag{5.28}$$

$$= \max_a E_{\pi^*}\left(R_{t+1} + \gamma \sum_{k=0}^{\infty} \gamma^k R_{t+k+2} | s_t = s, a_t = a \right)$$

$$= \max_a E\left(R_{t+1} + \gamma v^*(s_{t+1}) | s_t = s, a_t = a \right) \tag{5.29}$$

$$= \max_a \sum_{s' \in S} P_{ss'}^a \left(R_s^a + \gamma v^*(s') \right) \tag{5.30}$$

where (5.29) and (5.30) are Bellman optimality forms for $v^*(s)$. Figure 5.6 illustrates a backup diagram for state-value function taking the action with greater q value.

Likewise, Bellman optimality equation for $q^*(s, a)$ can be expressed as follows:

$$q^*(s, a) = E\left(R_{t+1} + \gamma \max_{a'} q^*(s_{t+1}, a') | s_t = s, a_t = a \right) \tag{5.31}$$

$$= \sum_{s' \in S} P_{ss'}^a \left(R_s^a + \gamma \max_{a'} q^*(s', a') \right) \tag{5.32}$$

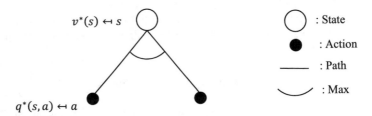

Fig. 5.6 Backup diagram for optimal state-value function

Fig. 5.7 Backup
diagram for optimal
action-value function

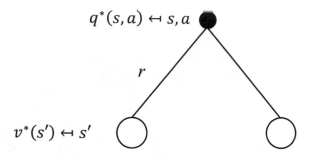

$$q^*(s, a) \leftarrowtail s, a$$

$$v^*(s') \leftarrowtail s'$$

where (5.31) and (5.32) are Bellman optimality forms for $q^*(s, a)$. Figure 5.7 illustrates backup diagram for action-value function. Unlike Bellman equation, we now know the optimal values of each state.

From Fig. 5.7, Bellman optimality equation for action-value function can be expressed as follows:

$$q^*(s, a) = R_s^a + \gamma \sum_{s' \in S} P_{ss'}^a v^*(s'). \tag{5.33}$$

Combining Figs. 5.6 with 5.7, we have Fig. 5.8.

From (5.28) and (5.33), we have the following equation:

$$v^*(s) = \max_a R_s^a + \gamma \sum_{s' \in S} P_{ss'}^a v^*(s') \tag{5.34}$$

where max operation means that the agent takes the action to maximize. (5.34) has same meaning as (5.30). Similar to Fig. 5.8, we can combine and have backup diagram for q^* as shown in Fig. 5.9. An agent in state s takes an action a. The environment enables it to arrive at state s'. From these states, the agent takes an action maximizing

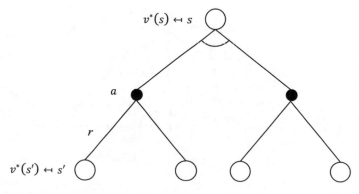

$$v^*(s) \leftarrowtail s$$

$$a$$

$$r$$

$$v^*(s') \leftarrowtail s'$$

Fig. 5.8 Backup diagram for v^*

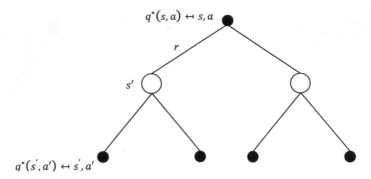

$$q^*(s, a) \leftarrow s, a$$

$$q^*(s', a') \leftarrow s', a'$$

Fig. 5.9 Backup diagram for q^*

q value. We mathematically express it as follows:

$$q^*(s, a) = R_s^a + \gamma \sum_{s' \in S} P_{ss'}^a \max_{a'} q^*(s', a'). \tag{5.35}$$

(5.35) has same meaning as (5.32).

As we can observe (5.34) and (5.35), we can solve the problem iteratively. For a finite MDP, Bellman optimality equation has a unique solution independent of the policy. In order to solve a large size of MDP problems, there are many iterative algorithms to obtain the optimal solution in the MDP.

Example 5.1 MDP Model Consider a simple MDP model as shown in Fig. 5.10.

There are three possible paths (or policies) in the figure: $\pi_1 = s_0 \rightarrow s_1 \rightarrow s_3 \rightarrow s_5$, $\pi_2 = s_0 \rightarrow s_1 \rightarrow s_4 \rightarrow s_5$, and $\pi_3 = s_0 \rightarrow s_2 \rightarrow s_4 \rightarrow s_5$. Find the optimal path (or policy).

Fig. 5.10 Simple MDP model

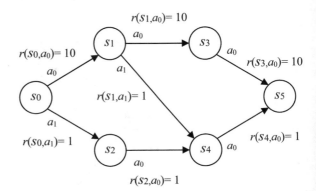

Solution

As we can observe Fig. 5.10, each decision affects to the next decision and each action receives a reward. We should learn a policy by selecting actions to maximize the total rewards and can easily find three possible paths (policies). Each policy receives the following rewards:

$\sum_{\pi_1} r(s, a) = 30$, $\sum_{\pi_2} r(s, a) = 12$, and $\sum_{\pi_3} r(s, a) = 3$.

The value function $v^{\pi}(s)$ can be found for a fixed policy as follows:

$v^{\pi_1}(s) = [v^{\pi_1}(s_0), v^{\pi_1}(s_1), v^{\pi_1}(s_3)] = [30, 20, 10]$,

$v^{\pi_2}(s) = [v^{\pi_2}(s_0), v^{\pi_2}(s_1), v^{\pi_2}(s_4)] = [12, 2, 1]$,

and $v^{\pi_3}(s) = [v^{\pi_3}(s_0), v^{\pi_3}(s_2), v^{\pi_3}(s_4)] = [3, 2, 1]$.

The state action value function $q^{\pi}(s, a)$ can be found without a specific policy as follows:

$q^{\pi}(s_4, a_0) = 1$, $q^{\pi}(s_3, a_0) = 10$, $q^{\pi}(s_1, a_0) = 20$, $q^{\pi}(s_1, a_1) = 2$, $q^{\pi}(s_2, a_0) = 2$, $q^{\pi}(s_0, a_0) = 30$, and $q^{\pi}(s_0, a_1) = 3$.

As we can observe both value functions, both have similar form and we can find the optimal policy by maximizing the value functions. The optimal policy is π_1^*. ∎

The objective of the RL problems in an MDP is to find an optimal policy π^* maximizing the long-term rewards. How to find the optimal policy? There are two approaches: model-based methods and model-free methods. The model-based RL is to learn an approximate model of a transition function and a reward function based on experiences and use them to get an optimal policy as if the model is correct. It can use dynamic programming algorithm on the model and computational cost is high. On the other hands, the model free RL uses a try-and-error method to update its knowledge and is to derive an optimal policy without learning the model. Computational cost and memory usage are low, but it may be slower than the model-based RL. In this section, we formulated the reinforcement learning problem and solved the problem by obtaining Bellman optimality equations. In the next sections, we discuss dynamic programming like policy or value iteration, temporal difference learning, SARSA, and Q learning.

Summary 5.1. Reinforcement learning

1. The reinforcement learning (RL) deals with the decision making problem and solves sequential decision making problem under uncertainty. The goal of sequential decision making problem is to find the optimal policy to maximize the long term rewards.
2. The roles of the RL elements can be summarized as follows: Policy (what to do), Reward (what is good or bad), Value (what is good because it predicts reward), and Model (what follows what, Optional element).
3. Markov decision process (MDP) is defined as a tuple (S, A, P, R, γ) where S is a finite set of states, A is the set of actions an agent can take

actions or make decisions, P is the transition function, R is a reward function, and $\gamma \in [0, 1]$ is a discount factor.

4. The state-value function is the expected value of returns we get upon leaving the state s and the discounted average value over the next possible states s'.

5. The action-value function (or q function) specifies how good it is to take a particular action from a certain state.

6. A recursive form of a state-value function is

$$v_\pi(s) = \sum_{a \in A} \pi(a|s) \left(R_s^a + \gamma \sum_{s' \in S} P_{ss'}^a v_\pi(s') \right).$$

7. A recursive form of an action-value function is

$$q_\pi(s, a) = R_s^a + \gamma \sum_{s' \in S} P_{ss'}^a \sum_{a' \in A} \pi(a'|s') q_\pi(s', a').$$

8. Bellman optimality form for $v^*(s)$ is

$$v^*(s) = \max_a \sum_{s' \in S} P_{ss'}^a (R_s^a + \gamma v^*(s')).$$

9. Bellman optimality equation for $q^*(s, a)$ is

$$q^*(s, a) = \sum_{s' \in S} P_{ss'}^a \left(R_s^a + \gamma \max_{a'} q^*(s', a') \right).$$

5.2 Model-Based Approaches

In order to solve a RL problem, there are two approaches: planning and learning [1]. The main difference between them is that planning is given by a model of the environment and learning solves a problem by interaction with an environment. In addition, an agent in planning computes based on the model and improves its policy. In learning, the environment is initially unknown, but an agent interacts with the model and improve its policy. Using the model, we can predict how the environment responds to actions. There are two types of models: distribution model and sample model [1]. We call it distribution model if a model is stochastic and can produce a description of all probabilities. For example, $P_{ss'}^a$ for all s, s' and $a \in A$. We call it sample model if a model can produce sample experiences, for example, simulation model. Basically, distribution models are more useful than sample model because they can produce samples as well. However, it is easier to produce sample models in many applications. Both types of models are used to simulate the environment and produce simulated experience. If a starting state and action is given, a sample model and a distribution model generate a possible transition and all possible transitions with their probability of occurring, respectively. If a starting state and a policy are given, they produce an entire episode and all possible episodes with their probabilities, respectively. In RL, planning means any computation process that uses a model to produce or improve a policy for interaction with the modelled environment. In machine learning, there are state-space planning and plan-space planning. State-space planning searches the state space to find an optimal policy. Actions results in transitions from one state to another state, and value functions are calculated over states. Plan-space planning searches the plan space. Value functions are defined over the plan space. This is partial-order planning, and it is difficult to apply efficiently to the stochastic sequential decision problems [1]. Thus, the state-space planning is widely used in machine learning area.

Dynamic programming (DP) is a general method for solving complex problems by breaking down into smaller sub-problems. The solutions of multiple sub-problems can be combined to solve the problem. Collection of algorithms can be used to find an optimal policy under a model of the environment. Key idea of dynamic programming in RL is use of value functions to structure the search for good policies. There are two required properties of dynamic programming: optimal substructure and overlapping sub-problems. Optimal substructure means that optimal solution can be decomposed into sub-problems. Overlapping sub-problems means that sub-problems recur many times, and solutions of sub-problems can be cached and reused. The MDP satisfies both properties: (1) Bellman equation enables us to have recursive decomposition and (2) value functions can store and reuse solutions. Basically, dynamic programming assumes full knowledge of the MDP even if it is unrealistic. If we know the structure (such as transitions, rewards, and so on) of the MDP, dynamic programming can be used to solve RL problems. It is used for planning in an MDP to solve prediction problems or control problems. Prediction problem means policy evaluation. When a MDP (S, A, P, R, γ) and a policy π as inputs are given, we find out how good the policy is. Control problem is to find optimal value function v_π and optimal policy π^*. When a MDP (S, A, P, R, γ) as an input is given, we find the optimal policy with the most reward we can receive with the best action. In the previous section, Bellman optimality Eqs. (5.34) and (5.35) have been derived as iterative methods. Using them, we find an optimal policy.

5.2.1 Policy Iteration

Policy iteration algorithm is to find an optimal policy by repeating two steps (policy evaluation and policy improvement) until it converges. Figure 5.11 illustrates the work-flow of policy iteration method.

Based on the current policy, policy evaluation estimates the value function. It selects the actions for some fixed policy and finds the resulting rewards and the

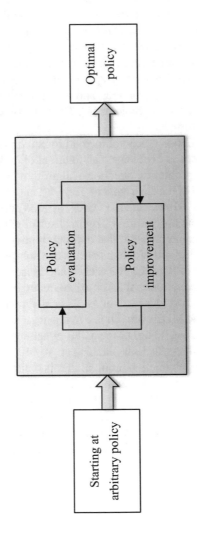

Fig. 5.11 Policy iteration

discounted values of the next states to update. Policy improvement updates the policy, so that it produces the action maximizing the expected one step value. This approach is known as greedy step because it only considers a single step return. Now, we put both policy evaluation and improvement together and obtain policy iteration. The basic concept of policy evaluation is to turn a recursive form of a state-value function (5.22) into update rule as follows:

$$v_{k+1}(s) = \sum_{a \in A} \pi(a|s) \left(R_s^a + \gamma \sum_{s' \in S} P_{ss'}^a v_k(s') \right). \tag{5.36}$$

This is a linear system whose unique solution is v^π. We start from an arbitrary v_0 and then compute (5.36) for all states during every iteration k, which will eventually converge to v^π as $k \to \infty$. At every iteration, each state is back up. The pseudocode of iterative policy evaluation can be summarized as follows:

Procedure Policy evaluation (v)
Load input π; // For example, start off with a random policy
Initialize $v(s) = 0$ for $\forall s \in S^+$, small threshold $\theta > 0$;
Repeat
$\quad \Delta \leftarrow 0$
$\quad$ **for** Each $s \in S$
$\quad\quad$ - $v \leftarrow v(s)$
$\quad\quad$ - $v(s) \leftarrow \sum_{a \in A} \pi(a|s) \left(R_s^a + \gamma \sum_{s' \in S} P_{ss'}^a v_k(s') \right)$
$\quad\quad$ - $\Delta \leftarrow \max(\Delta, |v - v(s)|)$
$\quad$ **end**
until $\Delta < \theta$
Return $v \cong v^\pi$

From the above approach, we evaluated the policy π and associated value function v^π but didn't find the best policy. When should we change the policy? Suppose that we have a deterministic policy π and a new policy π' that is equal to π except for one state s for $\pi'(s) = a$ and $\pi(s) \neq a$. If $q_\pi(s, a) \geq v_\pi(s)$ is satisfied, π' is better than (or as good as)π. We should change the policy. The policy improvement theorem [1] can be summarized as follows: Let π and π' be any pair of deterministic policies such that, $\forall s \in S$,

$$q_\pi(s, \pi'(s)) = \max_a q_\pi(s, a) \geq q_\pi(s, \pi(s)) = v_\pi(s). \tag{5.37}$$

Then, the policy π' must be better than or as good as π, which means the expected return is greater than or equal for all states as follows:

$$v_{\pi'}(s) \geq v_\pi(s). \tag{5.38}$$

If we stop the policy improvement, we have

$$q_\pi\big(s, \pi'(s)\big) = \max_a q_\pi(s, a) = q_\pi(s, \pi(s)) = v_\pi(s) \qquad (5.39)$$

and Bellman optimality equation satisfies

$$v_\pi(s) = \max_a q_\pi(s, a). \qquad (5.40)$$

Therefore, $v_\pi(s) = v^*, \forall s$. π is the optimal policy. Given a policyπ, we evaluate the policy π and improve the policy by acting greedily with respect to v_π: $\pi' = \mathrm{greedy}(v_\pi)$ [1]. More generally, we can update the policy which is greedy in accordance with computing the value v_π as follows:

$$\pi'(s) = \operatorname*{argmax}_a q_\pi(s, a)$$

$$= \operatorname*{argmax}_a \left(R_s^a + \gamma \sum_{s' \in S} P_{ss'}^a v_\pi\big(s'\big) \right) \qquad (5.41)$$

and $v_\pi\big(s'\big) \geq v_\pi(s)$ for $\forall s$. This tell us a local search through the policy space. We can reach the optimal policy when the values of two successive polices are same. The pseudocode of iterative policy improvement can be summarized as follows:

Procedure Policy improvement (π)
Load initial policy π_0; // For example, uniformly and randomly selected
Stable Policy $\leftarrow$ true
   for Each $s \in S$
     - $\Lambda \leftarrow \pi(s)$
     - $\pi(s) \leftarrow \operatorname*{argmax}_a \left(R_s^a + \gamma \sum_{s' \in S} P_{ss'}^a v_\pi\big(s'\big) \right)$
     - If $\Lambda \neq \pi(s)$, then Stable policy $\leftarrow$ false
   end
If Stable policy $\leftarrow$ true,
Stop and Return π; Else go to Policy evaluation.
Policy iteration is to find an optimal policy by combining policy evaluation step (Estimate v^π iteratively) and policy improvement step (Create $\pi' \geq \pi$ by a greedy local search). Thus, we can obtain a sequence of policy and value function improvements. Figure 5.12 illustrates policy iteration.

Since a finite MDP has a finite number of deterministic policies, the policy iteration is a local search in the finite MDP and converges to an optimal policy and value function in a finite number of iterations. One drawback of policy iteration is to include iterative policy evaluation at each iteration. It may require multiple sweeps through the state set and cause a protracted computation. If the state space is large, computational cost is expensive. There are some modified policy iterations including specific stop conditions.

Example 5.2 Policy Iteration Consider a 4×4 gridworld as a simple MDP and it is shown in Fig. 5.13. In the gridworld, we have

Fig. 5.12 Policy iteration

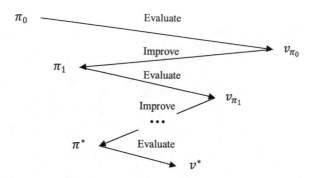

Fig. 5.13 Gridworld example

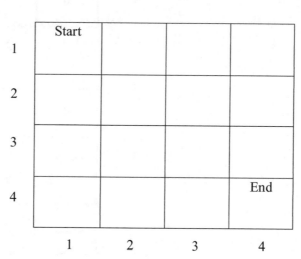

- States $S = \{(i, j)|i$ and $j = 1, 2, 3, 4\}$
- Actions $A = \{$up, down, left and right$\}$
- Terminal states = Start (1,1) and End (4,4)
- Reward $R_s^a = -1$ for nonterminal states
- $\gamma = 1$.

Find an optimal policy using policy iteration algorithm.

Solution

From pseudocode of iterative policy evaluation, we firstly initialize $v(s) = 0$ for $\forall s \in S^+$. All initial values are zero and the initial policy is presented as shown in Fig. 5.14.

As the initial policy, we start off with a random policy. That is to say. The transition probability to the next state in each direction (up, down, left, and right) is same as 1/4. For the terminal states (1,1) and (4,4), we have $v_k((1, 1)), v_k((4, 4)) = 0, \forall k$. From (5.36), we calculate the value of the state (2,2) as follows:

Fig. 5.14 Initial state values
(a) and policy (b) of
gridworld

	1	2	3	4
1	0	0	0	0
2	0	0	0	0
3	0	0	0	0
4	0	0	0	0

(a)

	1	2	3	4
1	0	←↑→↓	←↑→↓	←↑→↓
2	←↑→↓	←↑→↓	←↑→↓	←↑→↓
3	←↑→↓	←↑→↓	←↑→↓	←↑→↓
4	←↑→↓	←↑→↓	←↑→↓	0

(b)

$$v_1((2,2)) = \sum_{a \in A} \pi(a|s) \left(R_s^a + \gamma \sum_{s' \in S} P_{ss'}^a v_0(s') \right).$$

$$= \sum_{a \in \{Up,\ Down,\ Left,\ Right\}} \pi(a|(2,2)) \left(-1 + \sum_{s' \in S} P_{ss'}^a \cdot 0 \right)$$

$$= 0.25 \cdot (-1 - 1 - 1 - 1)$$

$$= -1$$

where $\pi(a|(2,2)) = 0.25$ for $\forall a$, $R_s^a = -1$, $\gamma = 1$, and $v_0(s') = 0$ for $\forall\ s$.
Likewise, we can calculate the values of the remaining states and have the following
figure after the first iteration (Fig. 5.15).

Fig. 5.15 State values after
the first iteration

	1	2	3	4
1	0	-1	-1	-1
2	-1	-1	-1	-1
3	-1	-1	-1	-1
4	-1	-1	-1	0

In the second iteration, we can calculate the value of the state (2,2), (2,1), and (1,2) as follows:

$$v_2((2,2)) = \sum_{a \in \{\text{Up, Down, Left, Right}\}} \pi(a|(2,2)) \left(-1 + \sum_{s' \in S} P^a_{ss'} \cdot v_1(s') \right)$$
$$= 0.25 \cdot (-2 - 2 - 2 - 2) = -2,$$

$$v_2((1,2)) = \sum_{a \in \{\text{Up, Down, Left, Right}\}} \pi(a|(1,2)) \left(-1 + \sum_{s' \in S} P^a_{ss'} \cdot v_1(s') \right)$$
$$= 0.25 \cdot (-2 - 2 - 1 - 2) = -1.75$$

and

$$v_2((2,1)) = \sum_{a \in \{\text{Up, Down, Left, Right}\}} \pi(a|(2,1)) \left(-1 + \sum_{s' \in S} P^a_{ss'} \cdot v_1(s') \right)$$
$$= 0.25 \cdot (-1 - 2 - 2 - 2) = -1.75.$$

Likewise, we can calculate the values of the remaining states and have the following figure after the second iteration (Fig. 5.16).

Likewise, we continuously calculate them, until it converge and have the following figure.

Now, convergence has been achieved. We have a deterministic policy and its value function; we can evaluate a policy change at a single state. We consider selecting action a in state s and checking whether or not we improve the policy π. If we have

Fig. 5.16 State values after
the second iteration

	1	2	3	4
1	0	-1.7	-2	-2
2	-1.7	-2	-2	-2
3	-2	-2	-2	-1.7
4	-2	-2	-1.7	0

$q_\pi\left(s, \pi'(s)\right) \geq v_\pi(s)$, it means that a new policy is better than π. We perform the policy improvement. From pseudocode of policy improvement, we can allocate agent trajectories to state transitions in order to maximize reward. From Fig. 5.17, we have a policy from the converged stated values as shown in Fig. 5.18.

Now, we start calculating the value of the state (1,2) with a new policy as shown in Fig. 5.17 as follows:

$$v_{\pi'}((1, 2)) = \sum_{a \in \{Left\}} \pi'(Left|(1, 2))\left(-1 + \sum_{s' \in S} P_{ss'}^{Left} \cdot v_{\pi'}((1, 1))\right)$$
$$= 1 \cdot (-1 + 0)$$
$$= -1$$

Fig. 5.17 State values when
it converged

	1	2	3	4
1	0	-14	-20	-22
2	-14	-18	-20	-20
3	-20	-20	-18	-14
4	-22	-20	-14	0

Fig. 5.18 Policy from Fig. 5.17

	1	2	3	4
1	0	←	←	←↓
2	↑	←↑	←↓	↓
3	↑	↑→	→↓	↓
4	↑→	→	→	0

where $P_{ss'}^{\text{left}} = \begin{cases} 1, & \text{for } s' = (1, 1) \\ 0, & \text{otherwise} \end{cases}$. As we can observe Fig. 5.17, the value of the state (2,1) is same as the value of the state (1,2). Likewise, we calculate the value of the state (1,3) and (2,2) as follows:

$$v_{\pi'}((1, 3)) = \sum_{a \in \{Left\}} \pi'(Left|(1, 3))\left(-1 + \sum_{s' \in S} P_{ss'}^{Left} \cdot v_{\pi'}((1, 2))\right)$$
$$= 1 \cdot (-1 - 1)$$
$$= -2$$

where $P_{ss'}^{\text{left}} = \begin{cases} 1, & \text{for } s' = (1, 2) \\ 0, & \text{otherwise} \end{cases}$ and

$$v_{\pi'}((2, 2)) = \sum_{a \in \{Up, Left\}} \pi'(a|(2, 2))\left(-1 + \sum_{s' \in S} P_{ss'}^{a} \cdot v_{\pi'}(s')\right)$$
$$= 0.5 \cdot ((-1 + v_{\pi'}((1, 2))) + (-1 + v_{\pi'}((2, 1))))$$
$$= 0.5 \cdot (-2 - 2)$$
$$= -2$$

where $P_{ss'}^{a} = \begin{cases} 1, & \text{for } s' = (1, 2) \text{ and } (2, 1) \\ 0, & \text{otherwise} \end{cases}$. Likewise, we can calculate the values of the remaining states and have the following Fig. 5.19..

Therefore, we can summarize the policy iteration as shown in Fig. 5.20.

As we can observe Fig. 5.20, $v_{\pi_0} \leq 14$ and $v_{\pi_1} \geq -3$ for all non-terminal states. Thus, we have $v_{\pi_1} \geq v_{\pi_0}$ for all states. π_1 is better than π_0. This is small gridworld problem. We found the optimal policy through a couple of iteration. ∎

1	0	-1	-2	-3
2	-1	-2	-3	-2
3	-2	-3	-2	-1
4	-3	-2	-1	0
	1	2	3	4

Fig. 5.19 State values after policy improvement

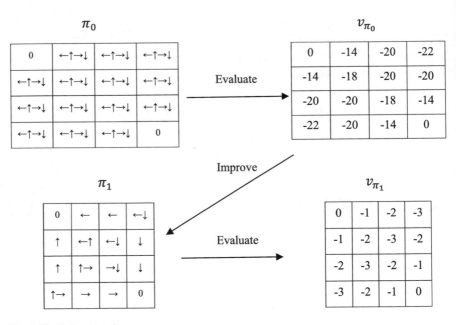

Fig. 5.20 Policy iteration

5.2.2   *Value Iteration*

Value iteration algorithm is similar to policy iteration algorithm, but it has only single backup of each state in the policy evaluation step. Figure 5.21 illustrates the work-flow of policy iteration method.

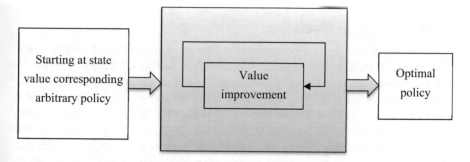

Fig. 5.21 Value iteration

The basic concept is to turn Bellman optimality equation (not Bellman expectation equation) into update rule and combine the truncated policy evaluation and the policy improvement. Bellman optimality equation is about the relationship between optimal value functions. We transform it into an iterative form and have simple update rule as follows:

$$v_{k+1}(s) = \max_a \left(R_s^a + \gamma \sum_{s' \in S} P_{ss'}^a v_k(s') \right). \tag{5.42}$$

If any policy is greedy with respect to v^*, it is an optimal policy. If v^* and the model of an environment are given, the optimal action can be found. We can summarize this as principle of optimality: Any optimal policy can be subdivided into an optimal first action a^* and followed by an optimal policy from successor state s. A policy $\pi(a|s)$ achieves the optimal value from state s, $v_\pi(s) = v^*(s)$, if and only if: for any state s' reachable from s, π achieves the optimal value from state s' [1]. This principle enables us to build value iteration. If we have the solution of sub-problem $v^*(s')$, $v^*(s)$ can be found by (5.40). The basic concept of value iteration comes from the principle of optimality. The pseudocode of value iteration can be summarized as follows:

Procedure Value iteration (π)
Load inputs as state transition function and reward function.
Initialize $v(s)$ for $\forall s \in S^+$, small threshold $\theta > 0$;
Repeat
$\Delta \leftarrow 0$
  for Each $s \in S$
    - $v \leftarrow v(s)$
    - $v(s) \leftarrow \max_a \left(R_s^a + \gamma \sum_{s' \in S} P_{ss'}^a v_k(s') \right)$
    - $\Delta \leftarrow \max(\Delta, |v - v(s)|)$
  end
until $\Delta < \theta$

Return a deterministic policy $\pi \cong \pi^*$ such that $\pi(s) = $ argmax$_a\left(R_s^a + \gamma \sum_{s' \in S} P_{ss'}^a v_k(s')\right)$

As we can observe the above pseudocode, the computational complexity of value iteration algorithm is lower than policy iteration algorithm at each iteration. However, policy iteration algorithm typically converges faster than value iteration algorithm.

Both dynamic programming algorithms we discussed so far updated all states at each iteration, and all states are backed up in parallel. We call this synchronous backups. The major drawback of synchronous backups requires a high computational cost if the state set is large. There are alternatives. If we focus on important states as often visited, we can reduce computational cost significantly. We call this approach asynchronous dynamic programming. It updates value of states individually in any order and applies the appropriate backup. We can select the states based on the magnitude in the Bellman equation error, that is to say, if we have more error, we give it a high priority. It can be applied in a bigger problem and may be suitable for parallel computation. There are three variants of asynchronous dynamic programming: in-place dynamic programming, prioritized sweeping, and real-time dynamic programming. Synchronous value iteration stores old and new of value function for all states as follows:

$$v_{new}(s) = \max_a \left(R_s^a + \gamma \sum_{s' \in S} P_{ss'}^a v_{old}(s') \right),$$
$$v_{old}(s) = v_{new}(s). \tag{5.43}$$

However, in-place dynamic programming stores only one of value function for all states as follows:

$$v(s) = \max_a \left(R_s^a + \gamma \sum_{s' \in S} P_{ss'}^a v(s') \right). \tag{5.44}$$

When we compute $v(s)$, it replaces $v'(s)$ in the next iteration. Prioritized sweeping updates a state with priority. In order to select a state for updating, we use magnitude of Bellman error as follows:

$$\left| \max_a \left(R_s^a + \gamma \sum_{s' \in S} P_{ss'}^a v(s') \right) - v(s) \right|. \tag{5.45}$$

We can backup the state with the largest error and update the state and Bellman error of affected states. It can be implemented by a priority queue. The basic concept of real-time dynamic programming is to use an experience of an agent for selecting states. The biggest problem of dynamic programming is to require a probability model. We assume that the model is correct and use it for dynamic programming. If we can't have accurate model, model-based approaches do not work well. Therefore, we discuss model-free approaches in the next section. Monte Carlo methods do not require both a model and bootstrap. Temporal difference learning methods do not require a model but need to bootstrap. Table 5.2 summarizes the comparison of iterative policy evaluation, policy iteration, and value iteration.

Example 5.3 Value Iteration Consider a 8 × 8 gridworld as a simple MDP, and it is shown in Fig. 5.22. In the gridworld, we have the following conditions:

- States $S = \{(i, j)| i \text{ and } j = 1, 2, 3, 4, 5, 6, 7, 8\}$
- Actions $A = \{up, down, left \text{ and } right\}$, The agent can't move off the grid or into the grey area (Obstacle) at (5.5).

Table 5.2 Comparison of iterative policy evaluation, policy iteration, and value iteration

	Iterative policy evaluation	Policy iteration	Value iteration
Approach	Based on the current policy, estimates the value function. It selects the actions for some fixed policy and finds the resulting rewards and the discounted valued of the next states to update	Find an optimal policy by combining policy evaluation step (Estimate v^π iteratively) and policy improvement step (Create $\pi' \geq \pi$ by a greedy local search)	Turn Bellman optimality equation into update rule and combine the truncated policy evaluation and the policy improvement
Bellman equation	Bellman expectation equation	Bellman expectation equation + Greedy policy improvement	Bellman optimality equation
Problem	Prediction	Control	Control

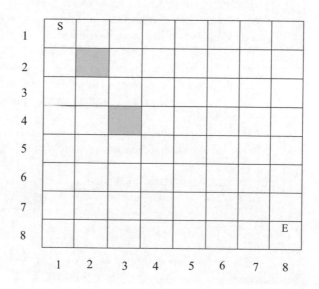

Fig. 5.22 Gridworld example

- Terminal states = Start (1,1) and End (8,8)
- Reward $R_s^a = \begin{cases} -1 & \text{for nonterminal states} \\ 10 & \text{for end terminal state} \\ -50 & \text{for grey area states} \end{cases}$
- $\gamma = 1$.
- The model has different probabilities to move: Probability of moving in same direction $p_s = 0.6$ and otherwise $p_o = 0.4$.
- If an agent hits a wall or boundary, it bounces back to the original position.

Find an optimal policy using value iteration algorithm.

Solution

From pseudocode of value iteration, we can find an optimal policy. The model is deterministic, and we can simplify the value function as follows:

$$v(s) = \max_a \left(R_s^a + \gamma v(s') \right).$$

Using computer simulation, we have state values and policies at each iteration as shown in Fig. 5.23.

As we can observe Fig. 5.23, convergence is achieved at 32 iterations. The state values increase as the number of iteration increases. Therefore, we need a normalization when implementing the algorithm. ∎

Summary 5.2. Model based RL

1. There are planning and learning to solve a RL problem. The main difference between them is that planning is given by a model of the environment and learning solves a problem by interaction with an environment.
2. Dynamic programming (DP) is a general method for solving complex problems by breaking down into smaller sub-problems. The solutions of multiple sub-problems can be combined to solve the problem. Collection of algorithms can be used to find an optimal policy under a model of the environment.
3. Dynamic programming assumes full knowledge of the MDP even if it is unrealistic. It is used for planning in an MDP to solve prediction problems or control problems.
4. Prediction problem means policy evaluation. When a MDP (S, A, P, R, γ) and a policy π as inputs are given, we find out how good the policy is.
5. Control problem is to find optimal value function v_π and optimal policy π^*. When a MDP (S, A, P, R, γ) as an input is given, we find the optimal policy with the most reward we can receive with the best action.

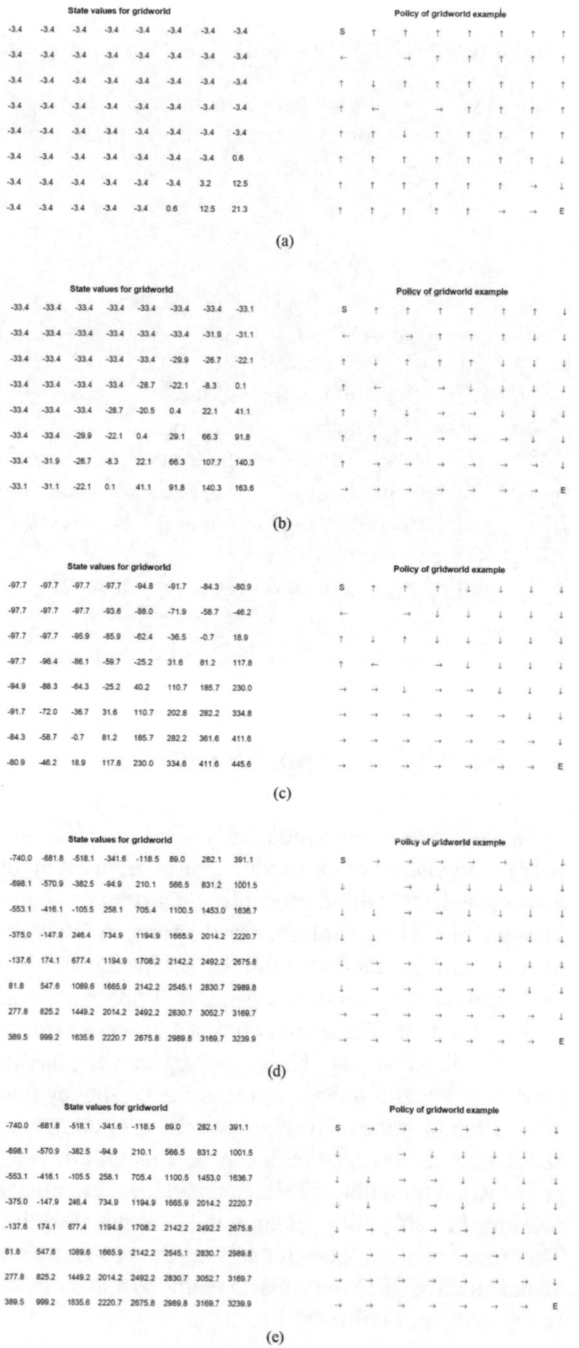

Fig. 5.23 State-value calculation at 5 iteration (**a**), 15 iterations (**b**), 20 iterations (**c**), 32 iterations (**d**), and 33 iterations (**e**)

(a)

State values for gridworld

-3.4	-3.4	-3.4	-3.4	-3.4	-3.4	-3.4	-3.4
-3.4	-3.4	-3.4	-3.4	-3.4	-3.4	-3.4	-3.4
-3.4	-3.4	-3.4	-3.4	-3.4	-3.4	-3.4	-3.4
-3.4	-3.4	-3.4	-3.4	-3.4	-3.4	-3.4	-3.4
-3.4	-3.4	-3.4	-3.4	-3.4	-3.4	-3.4	-3.4
-3.4	-3.4	-3.4	-3.4	-3.4	-3.4	-3.4	0.6
-3.4	-3.4	-3.4	-3.4	-3.4	-3.4	3.2	12.5
-3.4	-3.4	-3.4	-3.4	-3.4	0.8	12.5	21.3

Policy of gridworld example

S	↑	↑	↑	↑	↑	↑	↑
←		→	↑	↑	↑	↑	↑
↑	↓	↑	↑	↑	↑	↑	↑
↑	←		→	↑	↑	↑	↑
↑	↓	↑	↑	↑	↑	↑	↑
↑	↑	↑	↑	↑	↑	↑	↓
↑	↑	↑	↑	↑	↑	→	↓
↑	↑	↑	↑	↑	→	→	E

(b)

State values for gridworld

-33.4	-33.4	-33.4	-33.4	-33.4	-33.4	-33.4	-33.1
-33.4	-33.4	-33.4	-33.4	-33.4	-33.4	-31.9	-31.1
-33.4	-33.4	-33.4	-33.4	-33.4	-29.9	-26.7	-22.1
-33.4	-33.4	-33.4	-33.4	-28.7	-22.1	-8.3	0.1
-33.4	-33.4	-33.4	-28.7	-20.5	0.4	22.1	41.1
-33.4	-33.4	-29.9	-22.1	0.4	29.1	66.3	91.8
-33.4	-31.9	-26.7	-8.3	22.1	66.3	107.7	140.3
-33.1	-31.1	-22.1	0.1	41.1	91.8	140.3	163.6

Policy of gridworld example

S	↑	↑	↑	↑	↑	↑	↓
←		→	↑	↑	↑	↓	↓
↑	↓	↑	↑	↑	↓	↓	↓
↑	←		→	↓	↓	↓	↓
↑	↓	↑	→	→	↓	↓	↓
↑	↑	→	→	→	→	↓	↓
↑	→	→	→	→	→	→	↓
→	→	→	→	→	→	→	E

(c)

State values for gridworld

-97.7	-97.7	-97.7	-97.7	-94.8	-91.7	-84.3	-80.9
-97.7	-97.7	-97.7	-93.6	-88.0	-71.9	-58.7	-46.2
-97.7	-97.7	-95.9	-85.9	-62.4	-36.5	-0.7	18.9
-97.7	-96.4	-86.1	-59.7	-25.2	31.6	81.2	117.8
-94.9	-88.3	-64.3	-25.2	40.2	110.7	185.7	230.0
-91.7	-72.0	-36.7	31.6	110.7	202.8	282.2	334.8
-84.3	-58.7	-0.7	81.2	185.7	282.2	361.6	411.6
-80.9	-46.2	18.9	117.8	230.0	334.8	411.6	445.6

Policy of gridworld example

S	↑	↑	↑	↓	↓	↓	↓
←		→	↓	↓	↓	↓	↓
↑	↓	↑	↓	↓	↓	↓	↓
↑	←		→	↓	↓	↓	↓
→	→	↓	→	→	↓	↓	↓
→	→	→	→	→	→	↓	↓
→	→	→	→	→	→	→	↓
→	→	→	→	→	→	→	E

(d)

State values for gridworld

-740.0	-681.8	-518.1	-341.6	-118.5	89.0	282.1	391.1
-698.1	-570.9	-382.5	-94.9	210.1	566.5	831.2	1001.5
-553.1	-416.1	-105.5	258.1	705.4	1100.5	1453.0	1636.7
-375.0	-147.9	246.4	734.9	1194.9	1665.9	2014.2	2220.7
-137.6	174.1	677.4	1194.9	1708.2	2142.2	2492.2	2675.8
81.8	547.6	1089.6	1665.9	2142.2	2545.1	2830.7	2989.8
277.8	825.2	1449.2	2014.2	2492.2	2830.7	3052.7	3169.7
389.5	999.2	1635.6	2220.7	2675.8	2989.8	3169.7	3239.9

Policy of gridworld example

S	→	→	→	→	→	↓	↓
↓		→	↓	↓	↓	↓	↓
↓	↓	→	→	↓	↓	↓	↓
↓	↓		→	↓	↓	↓	↓
↓	→	→	→	→	↓	↓	↓
↓	→	→	→	→	→	↓	↓
→	→	→	→	→	→	→	↓
→	→	→	→	→	→	→	E

(e)

State values for gridworld

-740.0	-681.8	-518.1	-341.6	-118.5	89.0	282.1	391.1
-698.1	-570.9	-382.5	-94.9	210.1	566.5	831.2	1001.5
-553.1	-416.1	-105.5	258.1	705.4	1100.5	1453.0	1636.7
-375.0	-147.9	246.4	734.9	1194.9	1665.9	2014.2	2220.7
-137.6	174.1	677.4	1194.9	1708.2	2142.2	2492.2	2675.8
81.8	547.6	1089.6	1665.9	2142.2	2545.1	2830.7	2989.8
277.8	825.2	1449.2	2014.2	2492.2	2830.7	3052.7	3169.7
389.5	999.2	1635.6	2220.7	2675.8	2989.8	3169.7	3239.9

Policy of gridworld example

S	→	→	→	→	→	↓	↓
↓		→	↓	↓	↓	↓	↓
↓	↓	→	→	↓	↓	↓	↓
↓	↓		→	↓	↓	↓	↓
↓	→	→	→	→	↓	↓	↓
↓	→	→	→	→	→	↓	↓
→	→	→	→	→	→	→	↓
→	→	→	→	→	→	→	E

Summary 5.3. Policy iteration and value iteration

1. Policy iteration is to find an optimal policy by combining policy evaluation step (Estimate v^π iteratively) and policy improvement step (Create $\pi' \geq \pi$ by a greedy local search).
2. Based on the current policy, policy evaluation estimates the value function. It selects the actions for some fixed policy and finds the resulting rewards and the discounted valued of the next states to update.
3. Policy improvement updates the policy so that it produces the action maximizing the expected one step value.
4. One drawback of policy iteration is to include iterative policy evaluation at each iteration. It may require multiple sweeps through the state set and cause a protracted computation.
5. The basic concept of value iteration is to turn Bellman optimality equation (Not Bellman expectation equation) into update rule and combine the truncated policy evaluation and the policy improvement.
6. The computational complexity of value iteration algorithm is lower than policy iteration algorithm at each iteration. However, policy iteration algorithm typically converges faster than value iteration algorithm.

5.3   Model-Free Approaches

Dynamic programming requires reward functions and state transition probabilities as key components of the model. However, in the model-free approaches, algorithms do not use the transition probability distribution or reward associated with the MDP. They are based on a trial-and-error approach. They focus on finding value functions directly from interactions with the environment. They depend on samples from the environment and do not use generated predictions of the next state and reward to decide an action. There are two types of model-free approaches: on-policy learning and off-policy learning. The on-policy learning methods learn the value based on its current action and policy, whereas the off-policy learning methods learn the value of a different policy. In other words, on-policy learns directly a policy deciding about what action will be taken in a state. Off-policy learning evaluates a target policy while following a behaviour policy. On-policy learning may get stuck in local maxima, but off-policy learning will not get stuck if enough experiences are given. There are few approaches for solving the problem about finding the value function of an unknown MDP: Monte Carlo methods and temporal difference learning methods. In this section, we discuss them.

5.3.1 Monte Carlo Methods

Monte Carlo (MC) methods as a subset of computational algorithms were invented in the development of the Manhattan project. They count on repeated random samples and are widely used when solutions are too difficult to implement. They are applied to various fields and provide us with approximate solutions. The basic steps of MC methods are as follows: (1) Define the statistical properties of inputs. (2) Generate sets of possible inputs with the above properties. (3) Compute the inputs deterministically. (4) Analyse the results statistically. When we have enough number of samples, accuracy of solutions increases. For example, when evaluating the performance around 10^{-5} BER (bit error rates) of channel coding schemes, we need at least 10^6 samples to be computed. Typicallys, the error of the analysis decreases as $1/\sqrt{N}$ where N is the number of random samples. In machine learning, MC methods are learning methods by estimating value functions and finding optimal policies. They learn value functions directly from episodes of experience without any prior knowledge of MDP. Based on averaging sample returns, MC methods update the policy at the end of an episode and solve the problem of reinforcement learning. Basically, the value function of a state means the expected cumulative future discounted reward starting from that state. MC policy evaluation (or prediction) is based on averaging rewards observed after visiting to that state where the visits means each occurrence of the state in an episode. We can say that MC policy evaluation adopts empirical mean return instead of expected return. If we visit a state multiple time in an episode, we need to consider how to compute returns. In order to estimate the value of each state given a policy π, there are two approaches for MC policy evaluation: first-visit MC policy evaluation and every-visit MC policy evaluation. For each state s, we average observed returns after visiting the state s. In the first-visit MC policy evaluation, we average returns only for first time the state s is visited in an episode. In the every-visit MC policy evaluation, we average returns for every times the state s is visited in an episode. Both approaches converge to $v_\pi(s)$ as the number of visits goes to infinity.

Given a policy π, we estimate the value function for all states. The pseudocode of the first-visit MC policy evaluation can be summarized as follows:

Procedure First-visit MC policy evaluation (v_π)
Load inputs as the policy π to be evaluated.
Initialize $v(s) \in \mathbb{R}$ for $\forall s \in S$
Returns(s) $\leftarrow$ an empty list, for $\forall s \in S$
Repeat
Generate an episode $s_0, s_1, \ldots, s_{T-1}$ using the policy π
   for Each $s \in S$, $t = T - 1, \ldots, 0$
      if state s_t is not in the sequence $s_0, s_1, \ldots, s_{T-1}$ **then**
      - Returns(s_t) $\leftarrow$ append returns following the first occurrence of s_t
   end

$v(s) \leftarrow$ average(Returns(s))
   until Convergence

Return $v \cong v^\pi$

Each return is an estimate of v^π and each average is an unbiased estimate. The standard deviation of error falls as $1/\sqrt{n}$ where n is the number of averaged returns. Sequence of averages converges to the expected value of v^π. The pseudocode of the every-visit MC policy evaluation can be summarized as follows:

Procedure Every-visit MC policy evaluation (v_π)

Load inputs as the policy π to be evaluated.

Initialize $v(s) \in \mathbb{R}$ for $\forall s \in S$

Returns(s) $\leftarrow$ an empty list, for $\forall s \in S$

Repeat

Generate an episode $s_0, s_1, \ldots, s_{T-1}$ using the policy π

   for Each $s \in S, t = T - 1, \ldots, 0$

                       - Returns(s_t) $\leftarrow$ append returns following all occurrences of s_t

   end

$$v(s) \leftarrow \text{average}(\text{Returns}(s))$$

until Convergence

Return $v \cong v^\pi$

The every-visit MC policy evaluation is a biased and consistent estimator. The bias value decreases as the number of episodes increases.

Example 5.4 First-Visit And Every-Visit MC Policy Evaluation Consider the following two episodes with two states A and B and rewards (state, reward):

$$\text{Episode 1}: \ (A, +4) \rightarrow (A, +2) \rightarrow (B, -3) \rightarrow (A, +1) \rightarrow (B, -1),$$
$$\text{Episode 2}: \ (B, -4) \rightarrow (A, +3) \rightarrow (B, -2).$$

Compare the state value functions $v(A)$ and $v(B)$ of first-visit and every-visit MC policy evaluation.

Solution

In the first-visit MC policy evaluation, we compute the value function after we visit a state until the end of the episode. If the state appears again, we ignore and don't count again. In the calculation of value function at the state A, $v_f(A)$, the reward sum of the first episode is $+ 4 + 2 - 3 + 1 - 1 = 3$. The reward sum of the second episode is $+ 3 - 2 = 1$. We can find $v_f(A)$ after averaging them as follows:

$$v_f(A) = \frac{3 + 1}{2} = 2.$$

In the calculation of value function at the state B, $v_f(B)$, the reward sum of the first episode is $-3 + 1 - 1 = -3$. The reward sum of the second episode is $-4 + 3 - 2 = -3$. We can find $v_f(B)$ after averaging them as follows:

$$v_f(B) = \frac{-3 - 3}{2} = -3.$$

In the every-visit MC policy evaluation, we compute the value function for every times a state is visited in an episode. In the calculation of value function at the state A, $v_e(A)$, the reward sum of the first episode at the first appearance of the state A is $+4 + 2 - 3 + 1 - 1 = 3$. The reward sum of the first episode at the second appearance of the state A is $+2 - 3 + 1 - 1 = -1$. The reward sum of the first episode at the third appearance of the state A is $+1 -1 = 0$. The reward sum of the second episode at the first appearance of the state A is $+3 - 2 = 1$. We can find $v_e(A)$ after averaging them as follows:

$$v_e(A) = \frac{+3 - 1 + 0 + 1}{4} = 0.75.$$

In the calculation of value function at the state B, $v_e(B)$, the reward sum of the first episode at the first appearance of the state B is $-3 + 1 - 1 = -3$. The reward sum of the first episode at the second appearance of the state B is $-1 = -1$. The reward sum of the second episode at the first appearance of the state B is $-4 + 3 - 2 = -3$. The reward sum of the second episode at the second appearance of the state B is $-2 = -2$. We can find $v_e(B)$ after averaging them as follows:

$$v_e(B) = \frac{-3 - 1 - 3 - 2}{4} = -2.25.$$

∎

We computed the mean and evaluated the value function when all values are given. However, if we don't have all values, how to calculate the mean? Since many problems of reinforcement learning are non-stationary, we should find the means in non-stationary systems. Incremental updates are a useful tool. Given a sequence $x_1, x_2, \ldots, x_k$, the mean μ_k can be computed as follows:

$$\mu_k = \frac{1}{k} \sum_{i=1}^{k} x_i \tag{5.46}$$

and the incremental update rule for μ_k is as follows:

$$\mu_k = \frac{1}{k} \left(x_k + \sum_{i=1}^{k-1} x_i \right) = \frac{1}{k}(x_k + (k-1)\mu_{k-1}) = \mu_{k-1} + \frac{1}{k}(x_k - \mu_{k-1}). \tag{5.47}$$

We can apply this update rule for both first-visit and every-visit MC policy evaluation. For each state s_t with the return G_t, we can update counter $N(s_t)$ and value function $v(s_t)$ incrementally on an episode-by-episode basis as follows:

$$N(s_t) \leftarrow N(s_t) + 1, \tag{5.48}$$

$$v(s_t) \leftarrow v(s_t)\frac{N(s_t) - 1}{N(s_t)} + \frac{G_t}{N(s_t)} = v(s_t) + \frac{1}{N(s_t)}(G_t - v(s_t)). \tag{5.49}$$

In non-stationary problems, we have the following update rule with a running mean:

$$v(s_t) \leftarrow v(s_t) + \alpha(G_t - v(s_t)) \tag{5.50}$$

where G_t is the total return after time t and α is the learning rate or step size. If we have $\alpha = \frac{1}{N(s_t)}$, it is same as the first-visit and every-visit MC policy evaluation. If $\alpha > \frac{1}{N(s_t)}$, it gives more weight to new data and might be suitable for non-stationary problem.

We discussed how to find value functions by both MC policy evaluations. Now, we discuss how to approximate optimal policy. Similar to policy iteration of dynamic programming, MC policy iteration is composed of policy evaluation and policy improvement. In MC methods, we do not assume a probability model. The state values themselves are not sufficient to decide an optimal policy. Thus, we estimate action values. MC methods for state-value estimation are extended to action-value estimation. We can use both first-visit method and every-visit method to estimate them. One problem of MC methods is to only determine the value of a visited state by a policy, that is to say, if states are not visited under a policy, the values associated with the states couldn't be found. Thus, in a model-free reinforcement learning, we need to explore the entire states and interact with the environment to have enough experiences. If we don't have enough experiences, algorithms may get stuck in local maxima. In real world, it is rarely converged to have an optimal policy. Thus, a trade-off between exploitation and exploration is required. If exploitation is emphasized in algorithms, it provides us with the best decision under the current information. If we choose exploration approach, it maximizes long-term rewards. For example, if we have a problem to find a good restaurant in a town, exploitation approach is to go to favourite restaurant, whereas exploration is to try a new restaurant. The exploration approach may include short-term sacrifices, and the exploitation approach may get stuck in a bad policy. Thus, they are a trade-off relationship. In a model-free method, many state–action pairs may not be visited. For example, if a policy is deterministic and we select one action with probability 1 in every state, we can observe returns from one action at each state and, for all other actions, estimates of state–action do not improve no matter what we perform averaging the returns. Thus, in order to maintain the exploration problem, MC estimation of action values assumes that state action pairs have a nonzero probability of being selected when starting an episode. This approach enables all state–action pairs to be visited. Another approach for the exploration problem is to have a stochastic policy with a nonzero policy of all action selections in each state. Similar to policy iteration of dynamic programming, MC estimation in control is to approximate optimal policies based on generalized policy iteration. The policy iteration is composed of policy evaluation and policy improvement. The value function is repeatedly updated to approximate the value function using the current policy and the policy is repeatedly improved in terms of the current value function. In order to evaluate the policy, MC methods used value functions. However, as we reviewed in Sect. 5.2, greedy policy improvement over value functions requires the knowledge of MDP. We can't apply for model-free methods. Thus, we use action-value function. Figure 5.24 illustrates comparison of policy iteration with and without the knowledge of MDP.

As we discussed both first-visit and every-visit MC policy evaluation, given some number of episodes under a policy, we average returns and learn the value functions. In the MC estimation in control, we start with an arbitrary policy, evaluate the state–action function at each step, and then find a new policy. Policy evaluation calculates each q_{π_k} for arbitrary π_k using one of MC policy evaluation. We have two assumptions for exact policy evaluation: (1) we observe an infinite number of episodes and (2) the episodes are created with exploring starts. In order to generate $\pi' \geq \pi$ in policy

Fig. 5.24 Policy iteration without knowledge of MDP (**a**) and with a model of MDP (**b**)

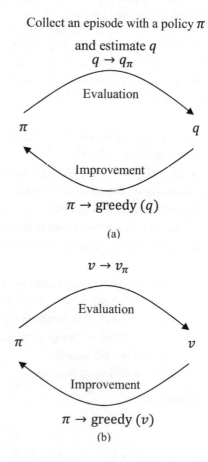

Collect an episode with a policy π

and estimate q

$q \to q_\pi$

Evaluation

π q

Improvement

$\pi \to \text{greedy}(q)$

(a)

$v \to v_\pi$

Evaluation

π v

Improvement

$\pi \to \text{greedy}(v)$

(b)

improvement, we construct π_{k+1} as the greedy policy in terms of the current action-value function q_{π_k} as follows:

$$\pi_{k+1}(s) = \operatorname*{argmax}_a q_{\pi_k}(s, a). \tag{5.51}$$

Convergence of MC policy iteration is guaranteed by the policy improvement theorem [1]. Suppose that we can observe an infinite number of episodes with exploring starts, we have for all π_k, π_{k+1} and $s \in S$ [1],

$$
\begin{aligned}
q_{\pi_k}(s, \pi_{k+1}(s)) &= q_{\pi_k}\left(s, \operatorname*{argmax}_a q_{\pi_k}(s, a)\right) \\
&= \max_a q_{\pi_k}(s, a) \geq q_{\pi_k}(s, \pi_k(s)) \\
&= v_{\pi_k}(s).
\end{aligned}
\tag{5.52}
$$

Thus, we have $\pi_{k+1} \geq \pi_k$ and $q_{\pi_{k+1}} \geq q_{\pi_k}$ by the policy improvement theorem. In order to remove two unrealistic assumptions: infinite number of episodes and access to exploring starts, we give up complete policy evaluation and collect enough episodes until the margin of error is small enough. In addition, we can have an episode by episode basis approach. After each episode, we use the observed return for policy evaluation and update the policy at all visited states in an episode. The pseudocode of MC control assuming exploring starts can be summarized as follows:

Procedure MC control with exploring starts (π^*)

Load inputs to be evaluated.

Initialize $\pi(s) \in A(s), q(s, a) \in \mathbb{R}$ arbitrarily, for $\forall s \in S, a \in A(s)$,

Returns$(s, a) \leftarrow$ an empty list, for $\forall s \in S$

Repeat forever for each episode

Choose state and action randomly such that all pairs have probability > 0 $\big\rbrace$— Exploring starts

Generate an episode using exploring starts and π

$G \leftarrow 0$

    for each pair s, a appearing in the episode, $t = T\text{-}1, \ldots, 0$

        - $G \leftarrow$ return following the first occurrence of s, a — Evaluation

        - Append G to Returns(s, a)

        - $q(s, a) \leftarrow$ average (Returns(s, a))

    for each s in the episode   — Improvement

        - $\pi(s) \leftarrow \operatorname*{argmax}_a q(s, a)$

Return $\pi \cong \pi^*$

In the above MC control algorithm with exploring starts, all returns for each state and action pair are accumulated and averaged irrespective of specific policy. We can converge an optimal fixed point as the variation of action-value function is getting smaller over time. However, proving convergence to optimal fixed point is still open problem [1]. In real world, it is unrealistic to select all actions infinitely. Thus, in order to avoid exploring starts assumption, we have two approaches: on-policy methods and off-policy methods. On-policy methods evaluate and improve a policy while using for control. Off-policy methods maintain two policy for exploring the one and optimizing another one. On-policy methods are generally simpler. There are many variations of on-policy methods. In on-policy control methods, we select the policy

$$\pi(a|s) > 0, \quad \text{for } \forall s \in S, a \in A(s). \tag{5.53}$$

We call this a soft policy. An ϵ-soft policy is

$$\pi(a|s) > \epsilon/|A(s)|, \quad \text{for } \forall s \in S, a \in A(s) \tag{5.54}$$

and an ϵ-greedy policy is

$$\pi(a|s) = \begin{cases} 1 - \varepsilon + \frac{\epsilon}{|A(s)|}, & \text{for the greedy action} \\ \frac{\varepsilon}{|A(s)|}, & \text{for the non} - \text{greedy action} \end{cases} \tag{5.55}$$

The ϵ-greedy policy as a stochastic policy is an example of ϵ-soft policy. All actions in all states have nonzero probability. At each step, we choose the action determined by a policy with probability 1-ϵ, and we choose a random action with probability ϵ. Among ϵ-soft policy methods, the ϵ-greedy policy is closest to greedy policy. The pseudocode of on-policy MC control with ϵ-soft policy can be summarized as follows:

Procedure On-policy MC control with ϵ-soft policy (π^*)

Load inputs to be evaluated.

 Small $\epsilon > 0$

Initialize $\pi(a|s) \leftarrow$ arbitrarily ϵ − soft policy

$q(s,a) \in \mathbb{R}$ arbitrarily, for $\forall s \in S, a \in A(s)$,

Returns(s,a) $\leftarrow$ an empty list, for $\forall s \in S$

Repeat forever for each episode

Generate an episode using exploring starts and π

$G \leftarrow 0$

    for each pair s, a appearing in the episode, $t = T$-1, ...,0

        - $G \leftarrow$ return following the first occurrence of s, a

        - Append G to Returns(s, a)

        - $q(s, a) \leftarrow$ average (Returns(s, a))

    for each s in the episode

        - $A^* \leftarrow \text{argmax}_a\, q(s, a)$

        for $\forall a \in A(s)$

$$\pi(a|s) \leftarrow \begin{cases} 1 - \epsilon + \frac{\epsilon}{|A(s)|}, & \text{if } a = A^* \\ \frac{\epsilon}{|A(s)|}, & \text{if } a \neq A^* \end{cases} \left.\right\} \quad \epsilon\text{-greedy exploration}$$

Return $\pi \cong \pi^*$

If we need to estimate state-value functions or action-value functions of a policy π but can't test them directly, how do we obtain an experience? In off-policy methods, we use another policy μ to generate data. Thus, we have two different policies: target policy π and behaviour policy μ. The target policy is the policy we want to evaluate and improve. The behaviour policy is the policy to generate experiences. Off-policy methods have greater variance and slower to converge. However, they are more useful in general because they enable us to learn many policy in parallel. When an episode is generated by a different policy, can we learn the value function for a policy given only off-policy experience? In order to learn a target policy π while following a behaviour policy μ, the behaviour policy should satisfy the following condition:

$$\pi(a, s) > 0 \text{ implies } \mu(a, s) > 0, \quad \forall a \in A(s). \tag{5.56}$$

We call this assumption of convergence. The behaviour policy μ must have a nonzero probability of all action selections that might be selected by the target policy π. The behaviour policy μ needs to explore sufficiently. The behaviour policy μ should be soft in states where it is not identical to the target policy π. Let $p_i(s)$ and $p'_i(s)$ represent the probability of a complete episode occurring under the target policy π and the behaviour policy μ, respectively. Let $R_i(s)$ represent the corresponding observed reward from the state s. After observing n numbers of episodes from the state s under the target policy π, the value function can be estimated as follows:

$$v^\pi(s) = \frac{\sum_{i=1}^n \frac{p_i(s)}{p'_i(s)} R_i(s)}{\sum_{i=1}^n \frac{p_i(s)}{p'_i(s)}}.$$

(5.57)

As we can observe (5.57), it depends on the environmental probabilities $p_i(s)$ and $p'_i(s)$. The relative probability $\frac{p_i(s)}{p'_i(s)}$ is weighted. It is up-weighted or down-weighted if the reward is more or less likely to occur under the target policy π, respectively. We don't need to compute the individual probability but their ratio. When $T_i(s)$ is the time of termination of the ith episode involving the state s, we have

$$p_i(s_t) = \prod_{k=t}^{T_i(s)-1} \pi(s_k, a_k) P^{a_k}_{s_k s_{k+1}}$$

(5.58)

where $P^{a_k}_{s_k s_{k+1}}$ is state transition probability and

$$\frac{p_i(s)}{p'_i(s)} = \frac{\prod_{k=t}^{T_i(s)-1} \pi(s_k, a_k) P^{a_k}_{s_k s_{k+1}}}{\prod_{k=t}^{T_i(s)-1} \mu(s_k, a_k) P^{a_k}_{s_k s_{k+1}}} = \prod_{k=t}^{T_i(s)-1} \frac{\pi(s_k, a_k)}{\mu(s_k, a_k)}.$$

(5.59)

As we can observe (5.59), the weights depends on only two policies. We call this estimation tool importance sampling. It is widely used for estimating expected values under one distribution given samples from another. Most of off-policy MC methods use importance sampling. In off-policy MC method, we generate episodes in terms of a behaviour policy μ and estimate the value function v^π as expected returns. The pseudocode of off-policy MC control can be summarized as follows:

Procedure Off-policy MC control (π^*)

Load inputs to be evaluated.

Initialize $\pi(a|s) \leftarrow$ arbitrarily determinstic policy,

$q(s,a) \in \mathbb{R}$ arbitrarily, for $\forall s \in S, a \in A(s)$,

$N(s,a) \leftarrow 0, \ D(s,a) \leftarrow 0$, for $\forall s \in S, a \in A(s)$, // Numerator and denominator of

$q(s,a)$

Repeat forever for each episode

Select a soft policy μ and generate an episode,

$\quad \tau \leftarrow$ latest time at $a_\tau \neq \pi(s_\tau)$

$\qquad$ **for** each pair s, a appearing in the episode at time τ or later

$\qquad\qquad$ - $t \leftarrow$ the time of the first occurrence of $s, a, \ t \geq \tau$

$\qquad\qquad$ - $\omega \leftarrow \prod_{k=t+1}^{T-1} \frac{1}{\mu(s_k,a_k)}$

$\qquad\qquad$ - $N(s,a) \leftarrow N(s,a) + \omega R_t$

$\qquad\qquad$ - $D(s,a) \leftarrow D(s,a) + \omega$

$\qquad\qquad$ - $q(s,a) \leftarrow N(s,a)/D(s,a)$

$\qquad$ **for** each $s \in S$

$\qquad\qquad$ - $\pi(s) \leftarrow \text{argmax}_a \, q(s,a)$

Return $\pi \cong \pi^*$

MC methods directly learn from interaction with environment and do not require a full model. They simply average returns for each state–action pair. However, the limitations of MC methods are summarized as follows: (1) only working for episodic environments, (2) requiring complete episodes and (2) waiting until the end of the episode. Temporal difference learning methods can overcome some limitations of MC methods. For example, it enables us to work in non-terminating environments and learn from incomplete sequences. We discuss it in the next section.

Example 5.5 On-Policy MC Method Consider a 10×10 gridworld as a simple MDP, and it is shown in Fig. 5.25. In the gridworld, we have

- States $S = \{(i, j) | i \text{ and } j = 1, 2, 3, 4, 5, 6, 7, 8, 9, 10\}$
- Actions $A = \{up, \ down, \ left \text{ and } right\}$, The agent can't move off the grid or into the grey area (Obstacle) at (5.5).
- Terminal states = Obstacle (5,5) and Goal (10,10)
- Reward $R_s^a = \begin{cases} -1, & \text{for obstacle } (5, 5) \\ 1, & \text{for goal } (10, 10) \\ 0, & \text{for non} - \text{terminal} \end{cases}$
- Discount factor $\gamma = 0.9$.
- $\epsilon = 0.1$ to 0.8.
- It is an episodic task.

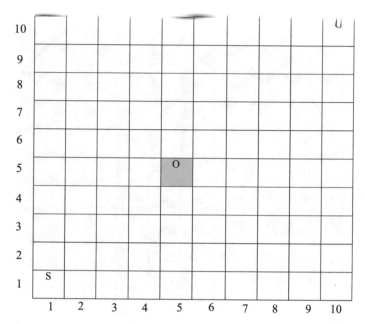

Fig. 5.25 Gridworld example

– The model has different probabilities to move: Probability of moving in same
  direction $p_s = 0.8$ and otherwise $p_o = 0.2$.

Find an optimal policy using on-policy MC method.

Solution

The on-policy MC method doesn't need a dynamic of the gridworld environment. The
optimal policy can be found by the interaction with the gridworld environment. The
MC method finds an action-value function by averaging the sample returns. When we
have enough number of samples, it converges to the expected value. According to the
pseudocode of on-policy MC control with ϵ-soft policy, we can run the algorithm. It
enables us to have the probabilities for all actions in a state. Using the probabilities,
one action can be selected. If one episode is complete and it arrives at the goal with
the final return, the action-value function is updated as follows:

$$q(s_k, a_k) = \frac{1}{n} \sum_{i=1}^{n} \text{Returns}(s_k, a_k)$$

where a_k is chosen in s_k in the episode and n is the number of episodes. The action
can be found by

$$A^* \leftarrow \underset{a}{\text{argmax}}\, q(s, a).$$

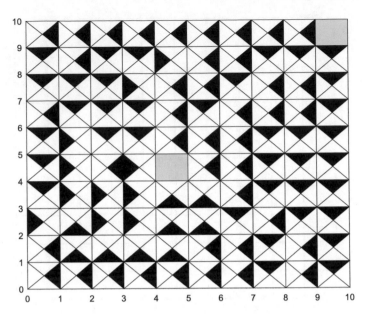

Fig. 5.26 Policy calculation

The value of ϵ starts at 0.8 and decreases to 0.1. The on-policy MC methods keep exploring with the probability ϵ. Using computer simulation, we perform 100 iterations and have the optimal policy as shown in Fig. 5.26.

In each cell of Fig. 5.26, ▲, ▼, ◄, and ► represent down, up, right, and left, respectively. Figure 5.27 illustrates an agent's move to search. In MC methods, we update after the agent arrived at the goal.

∎

Example 5.6 Off-Policy MC Method Consider a 10×10 gridworld as a simple MDP, and it is shown in Fig. 5.25. In the gridworld, we have same conditions as Example 5.5. Find an optimal policy using off-policy MC method.

Solution

According to the pseudocode of off-policy MC control, we can run the algorithm including the incremental update rule. Using (5.55), the general form of the update rule is

$$q_n(s_k, a_k) \leftarrow q_{n-1}(s_k, a_k) + \alpha(G_{n-1} - q_{n-1}(s_k, a_k)).$$

Using computer simulation, we perform 100 iterations and have the optimal policy as shown in Fig. 5.28.

Figure 5.29 illustrates an agent's move to search. In MC methods, we update after the agent arrived at the goal.

∎

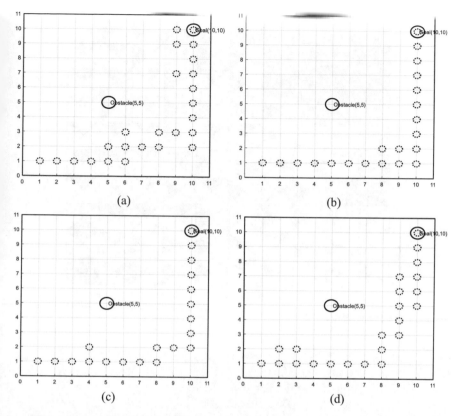

Fig. 5.27 First iteration with 32 actions (**a**), the second iteration with 29 actions (**b**), the third iteration with 25 actions (**c**), and the forth iteration with 24 actions (**d**)

Summary 5.4. Monte-Carlo Methods

1. In machine learning, MC methods are learning methods by estimating value functions and finding optimal policies. Based on averaging sample returns, MC methods update the policy at the end of an episode and solve the problem of reinforcement learning.

2. MC policy evaluation (or prediction) is based on averaging rewards observed after visiting to that state where the visits means each occurrence of the state in an episode.

3. There are two approaches for MC policy evaluation: First-visit MC policy evaluation and Every-visit MC policy evaluation. For each state s, we average observed returns after visiting the state s. In the first-visit MC policy evaluation, we average returns only for first time the state s is visited in an episode. In the every-visit MC policy evaluation, we average

Fig. 5.28 Policy calculation

returns for every times the state s is visited in an episode. Both approaches converges to $v_\pi(s)$ as the number of visits goes to infinity.

4.   There is a trade-off between exploitation and exploration. If exploitation is emphasized in algorithms, it provides us with the best decision under the current information. If we choose exploration approach, it maximizes long-term rewards.

5.   In order to avoid exploring starts assumption, we have two approaches: On-policy methods and off-policy methods. On-policy methods evaluate and improve a policy while using for control. Off-policy methods maintain two policy for exploring the one and optimizing another one.

6.   The limitations of MC methods are summarized as follows: (1) Only working for episodic environments, (2) requiring complete episodes and (2) waiting until the end of the episode. Temporal-difference learning methods can overcome some limitations of MC methods.

5.3.2   *Temporal difference learning methods*

Temporal difference (TD) learning methods can be regarded as the combination of MC methods and dynamic programming. TD learning methods use the value function and update an estimate of the value function by learning from every experience. TD

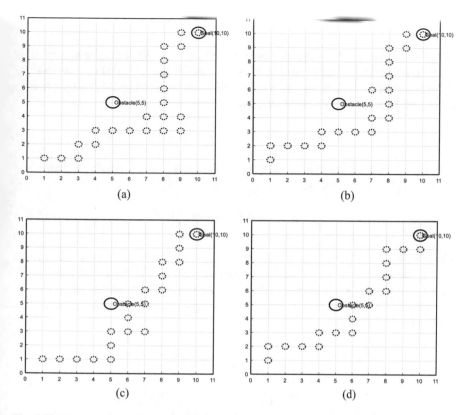

Fig. 5.29 First iteration with 25 actions (**a**), the second iteration with 22 actions (**b**), the third iteration with 20 actions (**c**), and the forth iteration with 18 actions (**d**)

learning methods learn from environment directly like MC methods and update state-value functions from the current estimates like dynamic programming. Thus, it is combined with sampling of MC method and bootstrapping of dynamic programming method. TD learning methods can overcome some limitation of MC methods. MC methods wait until the end of an episode and update the value function. However, TD learning methods only waits until the next time step. Thus, they require less memory and peak computation and learn from incomplete sequences. MC methods have high variance and zero bias, while TD learning methods have low variance and some bias. MC methods are not much sensitive to initial value, while TD learning methods are very sensitive to initial value. As we discussed in the previous section, there are prediction task and control task in reinforcement learning. Assuming that the policy is fixed, the goal of prediction is to measure how good the policy is. When the policy is not fixed, the control task is to find an optimal policy by maximizing the expected total reward from any given state. TD learning methods have prediction such as TD prediction and control such SARSA and Q learning.

5.3.2.1   TD Prediction

The prediction problem is regarded as policy evaluation. We recall the update rules of MC methods and dynamic programming. As we discussed in the previous section, MC methods work well in non-stationary environment. From (5.50), the update rule of MC methods can be described as follows:

$$v(s_t) \leftarrow v(s_t) + \alpha(G_t - v(s_t)) = (1 - \alpha)v(s_t) + \alpha G_t \qquad (5.60)$$

and the target is actual return after time t. The update rule of dynamic programming can be simply described as follows:

$$v(s_t) \leftarrow E\big[R + \gamma v(s_t)\big] \qquad (5.61)$$

where $R + \gamma v(s_t)$ is an estimate of the return and γ is discount factor. If we set $\gamma \sim 1$, the future rewards have a big impact as much as the immediate rewards. If we set $\gamma = 0$, there is no impact from the future rewards. Assuming that s_{t+1} is the only possible next state, we can express the simplest TD methods from (5.60) and (5.61) as follows:

$$
\begin{aligned}
v(s_t) \leftarrow \quad &\underbrace{v(s_t)}_{\text{Existing state value}} \quad +\underbrace{\alpha(R_{t+1} + \gamma v(s_{t+1}) - v(s_t))}_{\text{TD error}} \\
= \;&\underbrace{(1 - \alpha)v(s_t)}_{\text{Existing state value}} + \underbrace{\alpha(R_{t+1} + \gamma v(s_{t+1}))}_{\text{TD target}}.
\end{aligned}
\qquad (5.62)
$$

As we can observe (5.60) and (5.62), MC methods need the total return from state s_t to the end of an episode, but TD learning methods wait until the next time step to update the value function. The target of TD learning methods is an estimate of the return $R + \gamma v(s_t)$. We update it using the observed reward R_{t+1} and the current estimate $v(s_{t+1})$. We call the term $R_{t+1} + \gamma v(s_{t+1}) - v(s_t)$ TD error. The TD error is a temporal difference estimate error. TD error > 0 means that we have more rewards than expected. TD error < 0 means that we have less rewards than expected. When we have a backup diagram as shown in Fig. 5.30,

We can update an estimate of $v(s)$ as follows:

$$\Delta v_t(s_t) = \alpha\big(R_{t+1} + \gamma v_{t+1}(s_{t+1}) - v_t(s_t)\big). \qquad (5.63)$$

$$v_{t+1}(s_t) - v_t(s_t) \qquad\qquad \text{Initial estimate of future reward}$$

$$\text{Discounted estimate of future reward}$$

$$\text{Actual reward for one step}$$

$$(5.63)$$

Fig. 5.30 Backup diagram
of TD prediction

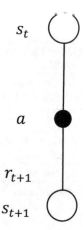

Based on A. Samuel's temporal difference learning [3], TD(λ) algorithm was invented by R.S. Sutton [1], where the parameter λ represents the trace decay parameter with $0 \leq \lambda \leq 1$. It is used to combine estimates obtained from various distances. The parameter λ is defined as an average of all n step returns in order to update the estimate of the value function. If λ is close to 1, we trace longer. TD(λ) algorithm tunes the decay rate. Thus, TD(1) is roughly identical to every-visit MC method. In TD(0), we look ahead one step. It updates only current state based on the estimate of the next state. TD(0) algorithm uses 1 step returns to update the state-value function, but the update rule of TD(λ) algorithm is given as multi-step returns. The pseudocode of TD prediction algorithm can be summarized as follows:

Initial conditions: $v(s)$ for $\forall\, s$
Repeat for each episode

– Initialize s

Repeat for each step in episode

– *Step 1*: Sample action in the current state by a policy π: $a \leftarrow p_\pi(a|s)$
– *Step 2*: Transition to the next state: Take action and observe the reward and the next state.
– *Step 3*: Update $v(s_t) \leftarrow v(s_t) + \alpha(R_{t+1} + \gamma v(s_{t+1}) - v(s_t))$
– *Step 4*: Reset the current state

Until: s is terminal.

TD(0) algorithm is converged for a fixed policy if the learning rate α is sufficiently small. The learning rate should decrease as the state visit increases. In order to find the optimality of TD(0) algorithm, we need to understand the batch updating. For example, when an agent interests with an environment with 5 episodes (episode A, episode B, ..., episode E), we consider all experiences in each episode as one batch and compute iteratively until the algorithm is converged. We learn from batch (episode A, ..., episode E) and then play episode F. We call this batch updating. The

Table 5.3 Comparison of DP, MC, and TD methods

	DP	MC	TD
Model based vs model free	Model based	Model free	Model free
Bootstrapping (Update involves an estimate)	Y	N	Y
Sampling (Update involves an actual return)	N	Y	Y
Update period	Time step	Episode-by-episode	Time step
Pros and cons	Efficient but a full model is required	Simple but slower than TD	Faster than MC and lower variance. However, it may not be less stable than MC
Remarks		Work in non-Markov environments	Exploit the Markov property

updates are made only after each batch is complete. The value function is changed only by the sum of increments. For any finite Markov prediction task, both TD(0) and MC methods converge under batch updating but a different answer. Table 5.3 summarizes the comparison of dynamic programming (DP), Monte Carlo (MC) methods, and temporal difference (TD) learning methods.

Example 5.7 TD Prediction 1 Consider a small Markov reward process shown in Fig. 5.31.

All episodes start in the centre state, S_3, and proceed either left or right by one state on each step, with equal probability. Episodes terminate either on the extreme left or the extreme right. This behaviour is presumably due to the combined effect of a fixed policy and an environment's state transition probabilities, but we do not care which we are concerned only with predicting returns. Episodes terminate either on the extreme left or the extreme right. When an episode terminates on the right a reward of 1 occurs; all other rewards are zero. Because this task is undiscounted and episodic, the true value of each state is the probability of terminating on the right if starting from that state. The true values of all the states, S_1 through S_5, are 1/6, 2/6, 3/6, 4/6, and 5/6.

Compare the effects of the learning rate α and the trace decay parameter λ.

Note: This example is same as random walk example in [1]. *In this example, we observe the different aspect.*

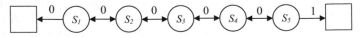

Fig. 5.31 Small Markov reward process [1]

Solution

Using computer simulation with the configuration: $\alpha = 0.1, 0.2, 0.3$ and 0.4, $\lambda = 0, 0.1, \ldots, 0.9, 1$, 400 training sets, and 40 episodes, we compare the root mean-squared (RMS) error between the value function learned and the true value function in terms of the learning rate α and the race decay parameter λ. Figure 5.32 illustrates the comparison of the effects of the learning rate α and the trace decay parameter λ.

As we can observe Fig. 5.32, the learning rate of TD(0) prediction should be small enough. As the learning rate increases, the RMS error increases. At $\alpha \geq 0.4$, it may not be converged. As the race decay parameter λ varies from 0 (single-step TD learning) to 1 (MC learning), the RMS error increases when $\alpha = 0.1$ and TD(0) provides us with a better performance. In this example, the value between 0.2 and 0.8 enables us to have good performance regardless of the learning rate. However, the best value depends on the problem. ∎

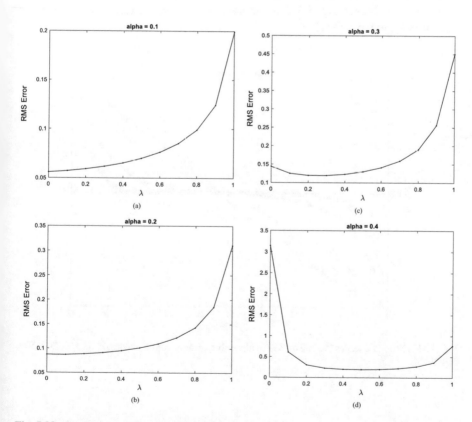

Fig. 5.32 Comparison of the effect of the learning rate α and the race decay parameter λ: $\alpha = 0.1$ (**a**), $\alpha = 0.2$ (**b**), $\alpha = 0.3$ (**c**) and $\alpha = 0.4$ (**d**)

Fig. 5.33 Small Markov reward process

Example 5.8 TD Prediction 2 Consider a small Markov reward process shown in Fig. 5.33.

All episodes start in the centre state, S_{50}, and proceed either left or right by one state on each step, with equal probability. Episodes terminate either on the extreme left or the extreme right. The total nodes are 100 from S_1 to S_{100}. The other conditions are same as Example 5.7.

Compare the estimated value with various numbers of episodes.

Solution

Using computer simulation with the configuration: $\alpha = 0.1$, $\lambda = 0$, $\gamma = 1$, 100 nodes, and 1 10 100 200 500 1000 episodes, we compare estimated values learned by TD(0) algorithm as shown in Fig. 5.34.

As we can observe Fig. 5.34, as the number of episodes increase, the estimate value is close to the true value. ∎

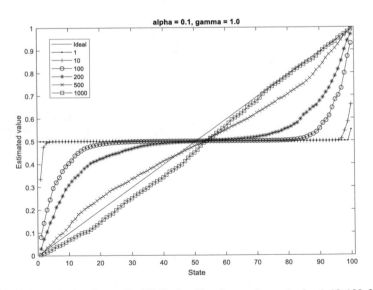

Fig. 5.34 Estimated values learned by TD(0) algorithm from various episodes: 1, 10, 100, 200, 500, and 1000

5.3.2.2 SARSA

TD methods for control problems find an optimal policy. The exploration can be achieved by on-policy or off-policy methods. The on-policy methods and the off-policy methods update an action value function (Q value) from experience following the current policy and another policy, respectively. State–action–reward–state–action (SARSA) as the on-policy method learns an action value function and estimate $q_\pi(s, a)$ following the current policy π. When we have a backup diagram as shown in Fig. 5.35,

The update rule of SARSA can be described as follows:

$$q(s_t, a_t) \leftarrow \underset{\text{Existing action value}}{q(s_t, a_t)} + \alpha \left(\underset{\text{Estimate of future action value}}{R_{t+1} + \gamma q(s_{t+1}, a_{t+1})} - \underset{\text{Existing action value}}{q(s_t, a_t)} \right). \quad (5.64)$$

As we can observe (5.64), the update is performed after every transition from a state s_t. If s_{t+1} is a terminal state, $q(s_t, a_t)$ is zero. This update rules use the quintuple $(s_t, a_t, R_{t+1}, s_{t+1}, a_{t+1})$. The name SARSA comes from this quintuple elements. (5.64) is rewritten as follows:

$$
\begin{aligned}
q(s_t, a_t) &\leftarrow q(s_t, a_t) + \alpha \underbrace{(R_{t+1} + \gamma q(s_{t+1}, a_{t+1}) - q(s_t, a_t))}_{\text{Action value SARSA error}} \\
&= (1 - \alpha)q(s_t, a_t) + \alpha \underbrace{(R_{t+1} + \gamma q(s_{t+1}, a_{t+1}))}_{\text{Action value SARSA target}}.
\end{aligned} \quad (5.65)
$$

SARSA uses the current estimate of the optimal policy to generate the behaviour while learning the optimal policy. It converges to an optimal policy as the number of visiting to state–action pairs increases. We can use ϵ-greedy policy for policy improvement. The pseudocode of SARSA algorithm can be summarized as follows:

Fig. 5.35 Backup diagram of SARSA

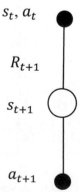

s_t, a_t

R_{t+1}

s_{t+1}

a_{t+1}

Initial conditions: $q(s, a)$ for $\forall s$ and a

Repeat for each episode

-   Initialise s
-   Choose a in s using a policy derived from q

Repeat for each step of episode

-   **Step 1**: Take a_t , observe R_{t+1} and s_{t+1}
-   **Step 2**: Choose a_{t+1} in s_{t+1} from q $\qquad\qquad\qquad$ $\}$ On-policy, ϵ-greedy
-   **Step 3**: Update $q(s_t, a_t) \leftarrow (s_t, a_t) + \alpha\big(R_{t+1} + \gamma q(s_{t+1}, a_{t+1}) - q(s_t, a_t)\big)$.
-   **Step 4**: Reset current state $s_t \leftarrow s_{t+1}$, $a_t \leftarrow a_{t+1}$, $\}$ behaviour policy = target policy

Until: s is terminal

Example 5.9 SARSA

Consider a gridworld as a simple MDP as shown in Fig. 5.36. In the grid world, we have

-   States $S = \{(i, j)| i$ and $j = 1, 2, \ldots, 8\}$
-   Actions $A = \{$up, down, left, right, Diagonally down-right, Diagonally down-left, Diagonally up-right and Diagonally up-left$\}$. The agent can't move off the grid.
-   Reward $R(s, a) = \begin{cases} 1 & \text{for nonterminal states} \\ 10 & \text{for goal state } (8, 8) \\ 0 & \text{for start state } (1, 1) \\ -50 & \text{for black area states (Obstacles)} \end{cases}$
-   Maximum iteration $= 100$

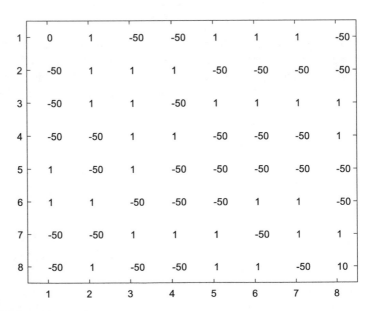

Fig. 5.36 Rewards at each state

- $\alpha = 0.1$ and $\gamma = 0.5$.

Find an optimal path using SARSA learning.

Solution

From the pseudocode of SARSA algorithm, we initialize the state and update Q values using the following update rule:

$q(s_t, a_t) \leftarrow (s_t, a_t) + 0.1(R_{t+1} + 0.5q(s_{t+1}, a_{t+1}) - q(s_t, a_t))$.

The matrix size is 8×8, and it is not simple calculation without a computer. Using computer simulation, we obtain the grid map as shown in Fig. 5.37a and the optimal path as shown in Fig. 5.37b.

As we can observe Fig. 5.37b, the colour represents the Q values, and we obtain the optimal path: $(1,1) \rightarrow (2,2) \rightarrow (3,2) \rightarrow (4,3) \rightarrow (5,3) \rightarrow (6,2) \rightarrow (7,3) \rightarrow (7,4)$ $\rightarrow (7,5) \rightarrow (8,6) \rightarrow (7,7) \rightarrow (8,8)$. ∎

5.3.2.3 Q Learning

Q learning is an off-policy TD control algorithm seeking to find the optimal policy using a Q function. It does not specify what the agent should do. The agent could exploit the knowledge one of the actions maximizes the Q function or explore to build a better estimate of the optimal Q function. Thus, both exploration and exploitation might be achieved at the same time. We evaluate a target policy π while following a behaviour policy μ. In off-policy learning, the behaviour policy μ is not same as the target policy π. The next action is selected using a behaviour policy μ. We consider alternative successor action using a target policy π and update the action value function towards value of alternative action. The target policy π is greedy as follows:

$$\pi(s_{t+1}) = \operatorname*{argmax}_a q(s_{t+1}, a) \tag{5.66}$$

and the behaviour policy μ can be ϵ-greedy. The Q learning target is

$$R_{t+1} + \gamma q(s_{t+1}, A)$$
$$= R_{t+1} + \gamma q\left(s_{t+1}, \operatorname*{argmax}_a q(s_{t+1}, a)\right) \tag{5.67}$$
$$= R_{t+1} + \gamma \max_a q(s_{t+1}, a).$$

When we have a backup diagram as shown in Fig. 5.38,
The update rule of Q learning can be described as follows:

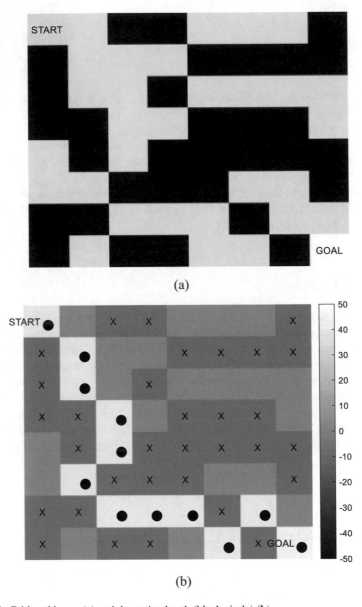

Fig. 5.37 Gridworld map (**a**) and the optimal path (black circle) (**b**)

$$q(s_t, a_t) \leftarrow \underset{\substack{\downarrow \\ \text{Current action as per } \mu}}{q(s_t, a_t)} + \alpha \left(R_{t+1} + \gamma \underset{a}{\max} \underset{\substack{\downarrow \\ \text{Next action as per } \pi}}{q(s_{t+1}, a)} - q(s_t, a_t) \right). \qquad (5.68)$$

Fig. 5.38 Backup diagram
of Q learning

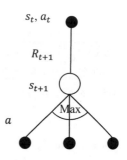

As we can observe (5.68), we always update it using maximum value of Q value available from the next state. Q learning target $R_{t+1} + \gamma \max_a q(s_{t+1}, a)$ is composed of immediate reward (R_{t+1}) and the estimated value of the best available next action discounted by γ $(\gamma \max_a q(s_{t+1}, a))$. When we have $R_{t+1} + \gamma \max_a q(s_{t+1}, a) \geq q(s_t, a_t)$ or $R_{t+1} + \gamma \max_a q(s_{t+1}, a) \leq q(s_t, a_t)$, the action value is increased or decreased, respectively. Comparing with the update rule of SARSA, the different term is only "$\max_a q(s_{t+1}, a)$", but it makes a big difference. The Q learning learns quickly values for the optimal policy based on the maximum rewards, whereas the SARSA learns from the Q values consistent with the policy and chooses a longer and safer path. We can say that SARSA is conservative. The Q learning brings great impact on the machine learning area. It allows us to learn many policies at same time while following a single behaviour policy. In addition, we can reuse experience generated by old policies. The pseudocode of Q learning algorithm can be summarized as follows:

Initial conditions: $q(s, a)$ for $\forall s$ and a

Repeat for each episode
- Initialise s

Repeat for each step of episode
- **Step 1**: Choose a_t in s_t using a policy derived from q $\qquad$ $\}$ ϵ-greedy
- **Step 2**: Take action a_t, observe R_{t+1} and s_{t+1}
- **Step 3**: Update $q(s_t, a_t) \leftarrow q(s_t, a_t) + \alpha\big(R_{t+1} + \gamma \max_a q(s_{t+1}, a) - q(s_t, a_t)\big)$
- **Step 4**: Reset current state $s_t \leftarrow s_{t+1}$ $\qquad$ off-policy

Until: s is terminal

Table 5.4 summarizes comparison of TD(0), SARSA, and Q learning.
Table 5.5 summarizes RL algorithms.

Table 5.4 Comparison of TD learning, Q learning, and SARSA

	Update rule	Features
TD(0)	$v(s_t) \leftarrow v(s_t) + \alpha(R_{t+1} + \gamma v(s_{t+1}) - v(s_t))$	Prediction
SARSA	$q(s_t, a_t) \leftarrow q(s_t, a_t) + \alpha(R_{t+1} + \gamma q(s_{t+1}, a_{t+1}) - q(s_t, a_t))$	On-policy control
Q learning	$q(s_t, a_t) \leftarrow q(s_t, a_t) + \alpha\left(R_{t+1} + \gamma \max_a q(s_{t+1}, a) - q(s_t, a_t)\right).$	Off-policy control

Table 5.5 RL algorithms

	Model-based approaches	Model-free approaches
Prediction problem	Dynamic programming policy evaluation	MC prediction, TD(0) method, TD(λ) method
Control problem	Dynamic programming policy/value iteration	MC control, SARSA, Q learning

Example 5.10 Q Learning I Consider a gridworld as a simple MDP as shown in Fig. 5.39. In the grid world, we have

- States $S = \{(i, j) | i$ and $j = 1, 2, 3, 4, 5\}$
- Actions $A = \{$up, down, left and right$\}$, The agent can't move off the grid. If the agent move into the grey area, it should move back to the start.
- Reward $R(s, a) = \begin{cases} 0 \text{ for nonterminal states} \\ 10 \text{ for terminal state} \\ -100 \text{ for grey area states} \end{cases}$
- Deterministic transitions
- $\alpha = 0.5$ and $\gamma = 1$.

Find an optimal path using Q learning.

Fig. 5.39 Gridworld example

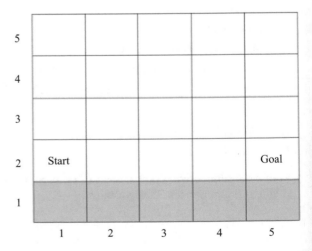

	1	2	3	4	5
5	$q(s,u)=0$ $q(s,d)=0$ $q(s,l)=0$ $q(s,r)=0$	$q(s,u)=0$ $q(s,d)=0$ $q(s,l)=0$ $q(s,r)=0$	$q(s,u)=0$ $q(s,d)=0$ $q(s,l)=0$ $q(s,r)=0$	$q(s,u)=0$ $q(s,d)=0$ $q(s,l)=0$ $q(s,r)=0$	$q(s,u)=0$ $q(s,d)=0$ $q(s,l)=0$ $q(s,r)=0$
4	$q(s,u)=0$ $q(s,d)=0$ $q(s,l)=0$ $q(s,r)=0$	$q(s,u)=0$ $q(s,d)=0$ $q(s,l)=0$ $q(s,r)=0$	$q(s,u)=0$ $q(s,d)=0$ $q(s,l)=0$ $q(s,r)=0$	$q(s,u)=0$ $q(s,d)=0$ $q(s,l)=0$ $q(s,r)=0$	$q(s,u)=0$ $q(s,d)=0$ $q(s,l)=0$ $q(s,r)=0$
3	$q(s,u)=0$ $q(s,d)=0$ $q(s,l)=0$ $q(s,r)=0$	$q(s,u)=0$ $q(s,d)=0$ $q(s,l)=0$ $q(s,r)=0$	$q(s,u)=0$ $q(s,d)=0$ $q(s,l)=0$ $q(s,r)=0$	$q(s,u)=0$ $q(s,d)=0$ $q(s,l)=0$ $q(s,r)=0$	$q(s,u)=0$ $q(s,d)=0$ $q(s,l)=0$ $q(s,r)=0$
2	$q(s,u)=0$ $q(s,d)=0$ $q(s,l)=0$ $q(s,r)=0$	$q(s,u)=0$ $q(s,d)=0$ $q(s,l)=0$ $q(s,r)=0$	$q(s,u)=0$ $q(s,d)=0$ $q(s,l)=0$ $q(s,r)=0$	$q(s,u)=0$ $q(s,d)=0$ $q(s,l)=0$ $q(s,r)=0$	$q(s,a)=0$
1	$q(s,a)=0$	$q(s,a)=0$	$q(s,a)=0$	$q(s,a)=0$	$q(s,a)=0$

Fig. 5.40 Initial condition of $q(s_t, a_t)$

Solution

From $\alpha = 0.5$, $\gamma = 1$ and Q learning update rule, we have

$$q(s_t, a_t) \leftarrow q(s_t, a_t) + 0.5\left(R_{t+1} + \max_a q(s_{t+1}, a) - q(s_t, a_t)\right)$$

As initial condition, $q(s_t, a_t)$ for all s and a is set as zero as shown in Fig. 5.40.

In the Fig. 5.40, the actions u, d, l, and r represent up, down, left, and right, respectively. When the agent moves right until the end point, we have the following state action values:

For the first iteration,

$$q((5, 2), a) \leftarrow 0 + 0.5(10 + 0 - 0) = 5,$$

For the second iteration,

$$q((4, 2), r) \leftarrow 0 + 0.5(0 + 5 - 0) = 2.5,$$

$$q((5, 2), r) \leftarrow 5 + 0.5(10 + 0 - 5) = 7.5,$$

For the third iteration,

$$q((3, 2), r) \leftarrow 0 + 0.5(0 + 2.5 - 0) = 1.25$$

$$q((4, 2), r) \leftarrow 2.5 + 0.5(0 + 7.5 - 2.5) = 5$$

$$q((5, 2), r) \leftarrow 7.5 + 0.5(10 + 0 - 7.5) = 8.75$$

For the third iteration,

$$q((2, 2), r) \leftarrow 0 + 0.5(0 + 1.25 - 0) = 0.625$$

$$q((3, 2), r) \leftarrow 1.25 + 0.5(0 + 5 - 1.25) = 3.125$$

$$q((4, 2), r) \leftarrow 5 + 0.5(0 + 8.75 - 5) = 6.875$$

$$q((5, 2), r) \leftarrow 8.75 + 0.5(10 + 0 - 8.75) = 9.375$$

For the fourth iteration,

$$q((1, 2), r) \leftarrow 0 + 0.5(0 + 0.625 - 0) = 0.3125$$

$$q((2, 2), r) \leftarrow 0.625 + 0.5(0 + 3.125 - 0.625) = 1.875$$

$$q((3, 2), r) \leftarrow 3.125 + 0.5(0 + 6.875 - 3.125) = 5$$

$$q((4, 2), r) \leftarrow 6.875 + 0.5(0 + 9.375 - 6.875) = 8.125$$

$$q((5, 2), r) \leftarrow 9.375 + 0.5(10 + 0 - 9.375) = 9.6875$$

Figure 5.41 illustrates the action–state values when moving right to the end point.
Likewise, we keep updating the Q values. When they converges to specific values, we can find the optimal policy. Q learning finds the shortest path as an optimal path. ∎

Example 5.11 Q Learning II Consider a gridworld as a simple MDP as shown in Fig. 5.42. In the grid world, we have

- States $S = \{(i, j)|i \text{ and } j = 1, 2, \ldots, 16\}$
- Actions $A =\{$up, down, left, right, Diagonally down-right, Diagonally down-left, Diagonally up-right and Diagonally up-left$\}$. The agent can't move off the grid.
- Reward $R(s, a) = \begin{cases} 0 & \text{for nonterminal states} \\ 10 & \text{for goal state } (16, 16) \\ 1 & \text{for start state } (1, 1) \\ -100 & \text{for black area state (Obstracles)} \end{cases}$

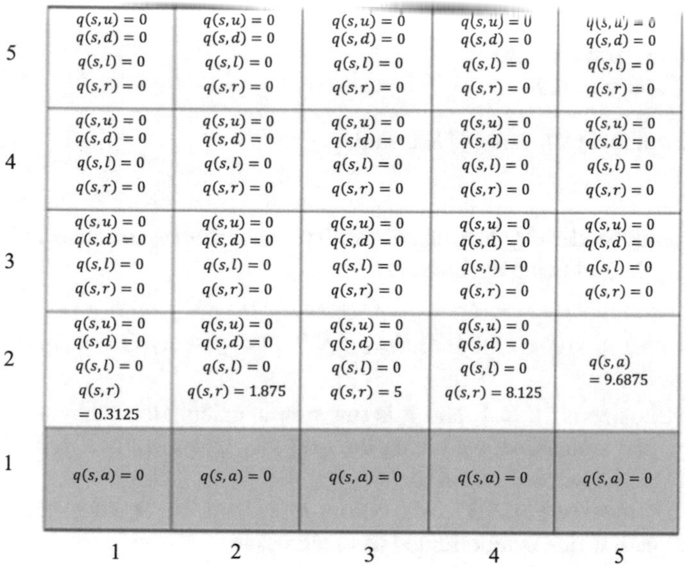

	1	2	3	4	5
5	$q(s,u)=0$ $q(s,d)=0$ $q(s,l)=0$ $q(s,r)=0$	$q(s,u)=0$ $q(s,d)=0$ $q(s,l)=0$ $q(s,r)=0$	$q(s,u)=0$ $q(s,d)=0$ $q(s,l)=0$ $q(s,r)=0$	$q(s,u)=0$ $q(s,d)=0$ $q(s,l)=0$ $q(s,r)=0$	$q(s,u)=0$ $q(s,d)=0$ $q(s,l)=0$ $q(s,r)=0$
4	$q(s,u)=0$ $q(s,d)=0$ $q(s,l)=0$ $q(s,r)=0$	$q(s,u)=0$ $q(s,d)=0$ $q(s,l)=0$ $q(s,r)=0$	$q(s,u)=0$ $q(s,d)=0$ $q(s,l)=0$ $q(s,r)=0$	$q(s,u)=0$ $q(s,d)=0$ $q(s,l)=0$ $q(s,r)=0$	$q(s,u)=0$ $q(s,d)=0$ $q(s,l)=0$ $q(s,r)=0$
3	$q(s,u)=0$ $q(s,d)=0$ $q(s,l)=0$ $q(s,r)=0$	$q(s,u)=0$ $q(s,d)=0$ $q(s,l)=0$ $q(s,r)=0$	$q(s,u)=0$ $q(s,d)=0$ $q(s,l)=0$ $q(s,r)=0$	$q(s,u)=0$ $q(s,d)=0$ $q(s,l)=0$ $q(s,r)=0$	$q(s,u)=0$ $q(s,d)=0$ $q(s,l)=0$ $q(s,r)=0$
2	$q(s,u)=0$ $q(s,d)=0$ $q(s,l)=0$ $q(s,r)=0.3125$	$q(s,u)=0$ $q(s,d)=0$ $q(s,l)=0$ $q(s,r)=1.875$	$q(s,u)=0$ $q(s,d)=0$ $q(s,l)=0$ $q(s,r)=5$	$q(s,u)=0$ $q(s,d)=0$ $q(s,l)=0$ $q(s,r)=8.125$	$q(s,a)=9.6875$
1	$q(s,a)=0$	$q(s,a)=0$	$q(s,a)=0$	$q(s,a)=0$	$q(s,a)=0$

Fig. 5.41 Q value updates when moving right

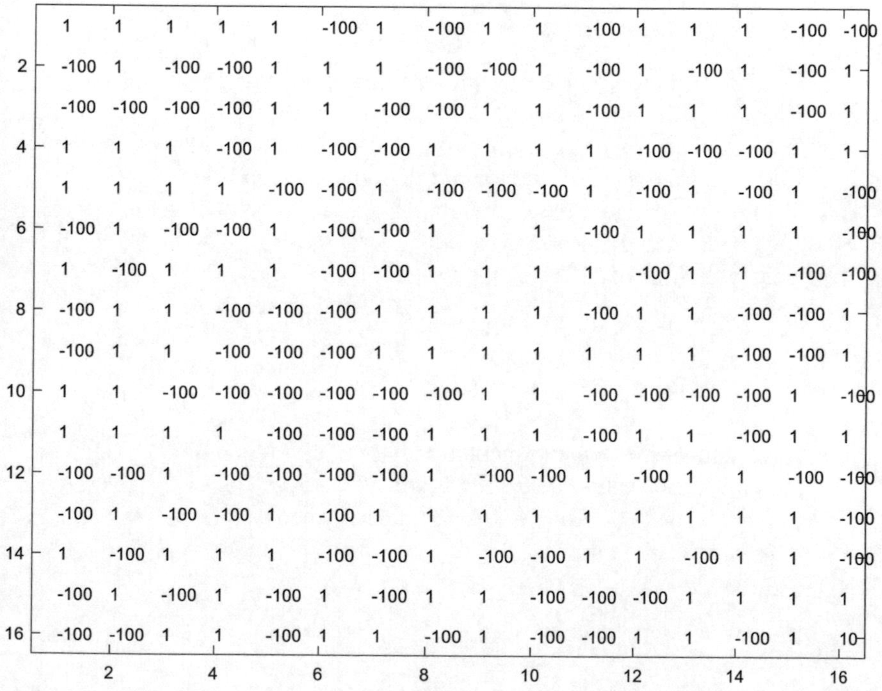

1	1	1	1	1	-100	1	-100	1	1	-100	1	1	1	-100	-100
-100	1	-100	-100	1	1	1	-100	-100	1	-100	1	-100	1	-100	1
-100	-100	-100	-100	1	1	-100	-100	1	1	-100	1	1	1	-100	1
1	1	1	-100	1	-100	-100	1	1	1	1	-100	-100	-100	1	1
1	1	1	1	-100	-100	1	-100	-100	-100	1	-100	1	-100	1	-100
-100	1	-100	-100	1	-100	-100	1	1	1	-100	1	1	1	1	-100
1	-100	1	1	1	-100	-100	1	1	1	1	-100	1	1	-100	-100
-100	1	1	-100	-100	-100	1	1	1	1	-100	1	1	-100	-100	1
-100	1	1	-100	-100	-100	1	1	1	1	1	1	1	-100	-100	1
1	1	-100	-100	-100	-100	-100	-100	1	1	-100	-100	-100	-100	1	-100
1	1	1	1	-100	-100	-100	1	1	1	-100	1	1	-100	1	1
-100	-100	1	-100	-100	-100	-100	1	-100	-100	1	-100	1	1	-100	-100
-100	1	-100	-100	1	-100	1	1	1	1	1	1	1	1	1	-100
1	-100	1	1	1	-100	-100	1	-100	-100	1	1	-100	1	1	-100
-100	1	-100	1	-100	1	-100	1	1	-100	-100	-100	1	1	1	1
-100	-100	1	1	-100	1	1	-100	1	-100	-100	1	1	-100	1	10

Fig. 5.42 Rewards at each state

- Deterministic transitions
- Maximum iteration $= 30$
- $\alpha = 0.2$ and $\gamma = 0.9$.

Find an optimal path using Q learning.

Solution

From the pseudocode of Q learning algorithm, we initialize the state and update Q values using the following update rule:

$$q(s_t, a_t) \leftarrow q(s_t, a_t) + 0.2\left(R_{t+1} + 0.9 \max_a q(s_{t+1}, a) - q(s_t, a_t)\right).$$

The matrix size is 16×1, and it is not simple calculation without a computer. Using computer simulation, we obtain the grid map as shown in Fig. 5.43a and the optimal path as shown in Fig. 5.43b.

As we can observe Fig. 5.37b, the colour represents the Q values and we obtain the optimal path which is represented as black circles.   ■

Summary 5.5. TD learning

1.   Temporal difference (TD) learning methods can be regards as the combination of MC methods and dynamic Programming. TD learning methods learn from environment directly like MC methods and update state value functions from the current estimates like dynamic programming.
2.   TD learning methods have prediction such as TD prediction and control such SARSA and Q learning.
3.   The update rule of TD(0) method is as follows:

$$v(s_t) \leftarrow v(s_t) + \alpha(R_{t+1} + \gamma v(s_{t+1}) - v(s_t)).$$

4.   State–action–reward–state–action (SARSA) as the on-policy method learns an action value function and estimate $q_\pi(s, a)$ following the current policy π. The update rule of SARSA is described as follows:

$$q(s_t, a_t) \leftarrow q(s_t, a_t) + \alpha(R_{t+1} + \gamma q(s_{t+1}, a_{t+1}) - q(s_t, a_t)).$$

5. Q learning methods evaluate a target policy π while following a behaviour policy μ. The next action is selected using a behaviour policy μ. The update rule of Q-learning is described as follows:

$$q(s_t, a_t) \leftarrow q(s_t, a_t) + \alpha \left(R_{t+1} + \gamma \max_a q(s_{t+1}, a) - q(s_t, a_t) \right).$$

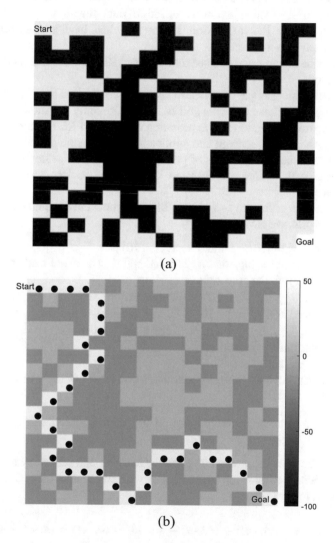

Fig. 5.43 Gridworld map (**a**) and the optimal path (black circle) (**b**)

5.4 Problems

5.1.   Compare unsupervised learning, supervised learning, and reinforcement learning in terms of applications.

5.2.   Most of RL problems can be formulated as the finite MDPs. Explain why the property of MDP is important in RL problems.

5.3.   Describe elements and their role of Markov decision process.

5.4.   Compare state-value function and action-value function.

5.5.   Compare the RL interaction processes when having deterministic and stochastic environment.

5.6.   Describe the pros and cons of model-based RL methods and model-free RL methods.

5.7.   Describe an example of episodic tasks and continuous tasks in real life.

5.8.   Compare Bellman equation and Bellman optimality equation.

5.9.   Compare exploitation and exploration approach of RL methods.

5.10. Describe the key concept and applications of dynamic programming.

5.11. Describe the pros and cons of policy iteration and value iteration.

5.12. Describe the difference between synchronous backup and asynchronous backup.

5.13. Consider a gridworld as a simple MDP as shown in the below figure. In the grid world, we have

  – States $S = \{(i, j)|i \text{ and } j = 1, \ldots, 10\}$
  – Actions $A = \{\text{up, down, left and right}\}$, The agent can't move off the grid or into the grey area.
  – Reward $r(s, a) = \begin{cases} 0 & \text{for nonterminal states} \\ 10 & \text{for terminal state} \\ -100 & \text{for black area states} \end{cases}$
  – Deterministic transitions
  – $\alpha = 0.1$ and $\gamma = 0.8$.
  – If an agent hits a wall or boundary, it bounces back to the original position (Fig. 5.44).

   Find optimal paths using policy iteration and value iteration. Then, compare the results.

5.14. In model-free RL, compare on-policy learning and off-policy learning.

5.15. Consider the following two episodes with two states A and B and rewards (state, reward):

$$\text{Episode 1 : } (A, +3)) \rightarrow (B, -2) \rightarrow (A, +5) \rightarrow (B, -1) \rightarrow (A, +2) \rightarrow (B, -2),$$
$$\text{Episode 2 : } (B, -2) \rightarrow (A, +1) \rightarrow (A, +3) \rightarrow (B, -3).$$

   Compare the state-value functions $v(A)$ and $v(B)$ of first-visit and every-visit MC policy evaluation.

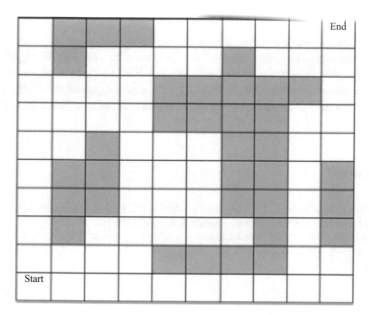

Fig. 5.44 Gridworld example

Fig. 5.45 Markov reward process

5.15. Temporal difference (TD) learning methods can be regarded as the combination of MC methods and dynamic programming. Find the lines about MC methods and dynamic programming in the pseudocode.

5.17. Consider a small Markov reward process shown in the below Fig. 5.45.

All episodes start in the centre state, S_{15}, and proceed either left or right by one state on each step, with equal probability. Episodes terminate either on the extreme left or the extreme right. The total nodes are 30 from S_1 to S_{30}. The other conditions are same as Example 5.7. Compare the estimated value with various numbers of episodes and also the effects of α and λ.

5.18. Compare the update rule of TD(0) method, SARSA, and Q learning, and find the best approach in terms of convergence time, accuracy, and complexity.

5.19. Reinforcement learning is well established in machine learning area. It is widely used in many applications. However, there are still key research challenges such as sample efficiency problem, exploration problem, catastrophic interference problem, and others. Find the recent development to solve these problems and discuss their pros and cons.

5.20. The RL methods solve sequential decision-making problem under uncertainty. There are many sequential decision making problem in wireless communications and networks. For example, the Viterbi algorithm is dynamic programming to find the most likely sequence of hidden states in hidden Markov models. Find the recent development about RL methods in wireless communications and networks and discuss their 1 and cons.

References

1. R.S. Sutton, A.G. Barto, R. Learning, *An Introduction* (MIT Press, Cambridge, MA, 2018)
2. E.L. Thorndike, *Animal Intelligence* (Hafner, Darien, Conn, 1911)
3. A.L. Samuel, Some studies in machine learning using the game of checkers. IBM J. Res. Dev. **3**(3), 211–229 (1959)

Chapter 6
Deep Learning

Deep learning is a sub-field of machine learning inspired by artificial neural networks (ANNs). Deep learning can be regarded as ANNs, and we often use the terms interchangeably. ANNs use artificial neurons (nodes) and synapses (connections), and they are built on same idea as how our brains work. The input, output, node, and interconnection of the artificial neuron are matched with dendrites, axon, cell nucleus, and synapse of the biological neuron, respectively. The nodes receive data, calculate them, and pass the new data to other nodes via the connections. The connections include weights or biases affecting the calculation of the next nodes. Deep learning uses a neural network with multiple layers of nodes between input and output. We can make the neural network deeper as the number of hidden layers increases. In deep learning, the term "learning" means finding optimal weights from data. The term "deep" means the multiple hidden layers in the networks in order to extract high-level features from inputs. Figure 6.1 illustrates an example of simple neural network and deep learning.

The process of deep learning can be summarized as follows: (1) training data, (2) forward to neural network for prediction, (3) finding errors (loss, cost, or empirical risks) between true values and predictions, and (4) updating the weight of connections. In many applications, deep learning may include thousands of nodes and connections. The layer is composed of multiple nodes. There are input layers, output layers, and intermediate (or hidden) layers between input layers and output layers. Basically, as the number of nodes, connections, and layers increases, we will have more accurate outcomes. Deep learning requires enough number of training samples. Untrained machine learning doesn't work properly as we discuss in the previous chapters. Backpropagation algorithms are widely used for training samples. They observe an error in the output layers and update the weights to reduce the error. The adjustment of weights is performed using stochastic gradient descent. Backpropagation of deep neural networks requires a substantial amount of data. Gradient descent method with random initialization may lead to poor local minima. Depending on layers and nodes, the complexity will increase. For example, we have a problem about handwriting recognition with 10 digit classes and 28×28 input image scale.

H. Kim, *Artificial Intelligence for 6G*,
https://doi.org/10.1007/978-3-030-95041-5_6

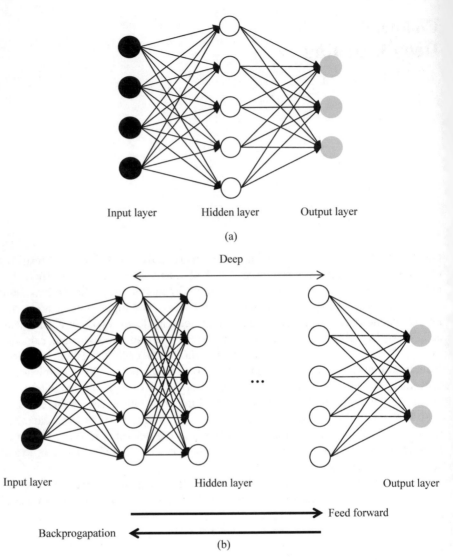

Fig. 6.1. Simple neural network (**a**) and deep learning (**b**)

There is one hidden layer with 500 nodes. The input layer requires 784 (=28 × 28) nodes. The output layer has 10 nodes corresponding to the number 1, 2, …, 10. We need 0.4 Million (=784 × 500 + 500 × 10) weights. Thus, we need to solve an optimization problem with 0.4 million dimensional space. It is very difficult to find the global optimum by gradient descent method with random initialization. The key design parameters of deep learning algorithms are to find an optimal point between the complexity and the performance. This process is performed repeatedly until the deep learning algorithm produces consistent outcomes with the minimum error. If we have stable structure of networks, the deep learning algorithm is ready to use for classification, prediction, regression, or others.

6.1 Introduction to Deep Learning

The basic concept of a neural network was invented in 1943 [1]. McCulloch and Pitts [1] published the first model of a neural network in the paper "A Logical Calculus of the Ideas Immanent in Nervous Activity". In 1957, Rosenblatt [2] published the concept of the perceptron as a binary classifier in the paper "The perceptron: A probabilistic model for information storage and organization in the brain". The perceptron is a mathematical model of a biological neuron. It is regarded as an artificial neuron. Neural networks were a key research area in both neuroscience and computer science until 1969. The research of neural networks had risen and declined repeatedly. In 1986, Rumelhart et al. [3] published a backpropagation learning method in the paper "Learning representations by back-propagating errors". This simple method for training deep neural networks enables us to implement deep learning efficiently. It opened the second generation of neural networks. The term deep learning was coined by Dechter in 1986 [4]. Deep learning requires a huge data for training. Due to significant improvement of cellular networks and VLSI technology, the creation and availability of a big data have grown exponentially. We have enough computing power. The resurgence of deep learning happens from the mid-2000s. Nowadays, deep learning is in a boom. Deep learning mimics human behaviour and performs human-like actions. As evolving communications, networks, and VLSI technologies, the ability of deep learning at specific area has been reached at same or even better level as the human. We can find a lot of similarity and difference between an artificial neural network and a human brain. They can be summarized as follows: (1) When same inputs are given in both, ANNs produce same outputs but a human brain may not give the same output. (2) ANNs learn from training while a human brain can learn from recalling information. (3) ANNs never forget if it is fully trained but a human brain can forget. (4) The data processing of ANNs can be monitored by a control unit while a human brain does not have a separate control unit. In [5], a human brain and AI computer are compared. The microprocessors have 106 time faster than the biological neuron. The instructions of AI computer are performed sequentially while a human brain has a massive parallel interconnection and simpler and slower processing units. Thus, due to these structural and operational differences, a human brain works more efficiently at the moment. Table 6.1 is simple comparison of a human brain and AI computer.

When comparing of machine learning and deep learning, machine learning steps are composed of manual feature extraction and classification. The manual feature extraction step is tedious and costly. However, deep learning performs end-to-end learning including feature extraction, representations, and tasks from the inputs. It

Table. 6.1 Comparison of a human brain and AI computer

	Human brain	AI computer
A basic unit	Biological neuron	Transistor
Size of a basic unit	10^{-6} m	10^{-9} m
Volume	200 billion biological neurons and 32 trillion synapses	1 billion byte RAM and multiple trillion bytes hard disk
Energy consumption	6 ~ 10 J operation/sec	10 ~ 16 J operation/sec
Transmission time	Slow	Fast
Fault tolerance	Yes	No
Storage	Storage in synapses	Contiguous memory location such as RAM or Hard disk

provides us with much better results than machine learning. For example, we have thousands of photographs about apples and oranges. We should identify the images of them using machine learning and deep learning. Machine learning requires structured data to determine the differences between apples and oranges and performs the classification. Thus, in order to classify them, we should mark the images of apples and oranges manually. Based on the characteristics of both apples and oranges, machine learning is trained. Then, the trained machine learning determines thousands of the images. On the other hands, deep learning does not require structured data to classify them. In deep learning, the input data are sent through different levels of neural network layers and each network layer hierarchically determines the specific features. They depend on different outputs at each neural network layer, and the final output is combined to form a single way of classifying images. After performing at different levels of neural network layers, deep learning finds classifiers for both apples and oranges. Table 6.2 is simple comparison of machine learning and deep learning.

There are multiple types of deep learning algorithms. They are not universal, and they all have pros and cons. Some algorithm is well-matched with a specific environment and condition. Therefore, it is important to understand how they work and choose the correct one in a specific environment. We focus on 4 important types of neural networks: artificial neural networks (ANN), deep neural networks (DNN), convolution neural networks (CNN), recurrent neural networks (RNN). ANN is a multi-layer fully connected neural network as shown in Fig. 6.1a. ANN makes up the backbone of deep learning algorithms, and many deep learning algorithms have been derived from ANN. It is a group of multiple perceptrons at each layer. It typically has a couple of hidden layers. It is also known as feedforward neural network because data is processed in the forward direction. A multi-layer perceptron enables us to learn nonlinear functions between the data. ANN works well on tasks such as classification and regression but the disadvantages are overfitting problem, slow learning rate, difficulty of optimal parameters, complexity, and others. Some of them are already solved. For example, overfitting can be solved by pre-training. The earlier version of ANN was shallow, and only single hidden layer was placed between the

Table. 6.2 Comparison of machine learning and deep learning

	Machine learning	Deep learning
Volume of data and data dependencies	Medium size of data is enough. Typically, thousands of data for optimal point	Big data is required. Typically, millions of data for optimal point
HW dependencies	Require a less powerful machine than deep learning	Require a powerful machine for a big size of matrix calculation
Output type	Mainly numerical data	Numerical data, image, text, and others
Feature extraction	Yes	No
Training time	Low	High
Interpretability	Easy or difficult	Difficult or impossible
Basic approach	Learn through data to solve the problem about classification, prediction, and others	Build neural networks and automatically discover patterns, classification, and others

input layer and the output layer. In order to improve accuracy and solve some disadvantages of ANN, DNN has a feedback loop and multiple hidden layers. Typically, more than three hidden layers qualify as deep learning. The earlier version of DNN is a feed-forward neural network without a feedback loop. Backpropagation was added to adjust the weights. Based on DNN structure, many other deep learning structures were developed. One drawback of DNN is to connect all artificial nodes between layers. It brings high complexity, easy to overfitting, large memory consumption, and so on. CNN solves some of them by weight sharing, local connectivity, and pooling. The basic concept of CNN is to connect artificial nodes with only some of them. CNN has convolution layers, rectified linear unit layers, and pooling layers. They extract the features from the input data. These layers are acting like filters to help extracting the relevant features from the input data. CNN captures the spatial invariance information from the input data, and the spatial information refers to the arrangement and relationship between the input data elements. It enables us to identify the data accurately and find the relationship with them. In particular, CNN is widely used in the image processing and video recognition. After the feature extraction, fully connected hidden layers perform classification and regression with the flattened matrix from the pooling layer. CNN is based on spatial invariance information but RNN models temporal information. It applies recurrence formula at every time step and processes a sequence of vectors. RNN is used in time series forecasting. RNN has a directed cycle to pass the information back to the neural network. It allows RNN to remember past data and use it for prediction. The parameters in different time steps are shared in RNN. We call this parameter sharing. It enables us to have fewer parameters to train the data and reduce the complexity. Many deep learning algorithms including RNN suffer from the vanishing and exploding gradient problem. They are related to the backpropagation method. When updating the weights by the backpropagation method in a neural network with a large number of hidden layers, the gradient may vanish or explode. There are several solutions for them including residual neural networks with skip connections or residual connections. From the next sections, we discuss DNN, CNN, and RNN.

Summary 6.1. Deep Learning

1. Deep learning is built on same idea as how our brains work. The input, output, node and interconnection of the artificial neuron is matched with dendrites, axon, cell nucleus, and synapse of the biological neuron, respectively.
2. The process of deep learning can be summarized as follows: (1) Training data, (2) Forward to neural network for prediction, (3) Finding errors (loss, cost, or empirical risks) between true values and predictions, and (4) Updating the weight of connections.
3. Similarity and difference between an artificial neural network and a human brain can be summarized as follows: (1) When same inputs are given in both, ANNs produce same outputs but a human brain may not give the same output. (2) ANNs learn from training while a human brain can learn from recalling information. (3) ANNs never forget if it is fully trained but a human brain can forget. (4) The data processing of ANNs can be monitored by a control unit while a human brain does not have a separate control unit.
4. Machine learning steps are composed of manual feature extraction and classification. The manual feature extraction step is tedious and costly. However, deep learning performs end-to-end learning including feature extraction

6.2   Deep Neural Network

As we discussed in the previous section, a perceptron is a linear classifier in supervised learning. The perceptron is a single layer neural network. A multi-layer perceptron (MLP) is a feedforward artificial neural network composed of multiple layers of perceptrons. A multi-layer perceptron (MLP) is a subset of deep neural networks (DNN). MLPs are feedforward, and DNN can have feedback loops. However, they are often used interchangeably. DNNs are artificial neural networks with multiple hidden layers between the input layer and the output layer. We build DNN structure from the perceptron. Figure 6.2 illustrates a perceptron.

In Fig. 6.2, the signal flows from left to right. The inputs are weighted. The bias term is constant and does not depend on the inputs. The weighted sum of the inputs is passed through a nonlinear activation function. The objective of activation function is

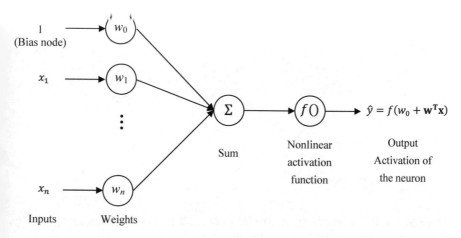

Fig. 6.2 Perceptron

to add nonlinearities in the neural networks. The activation function can be sigmoid function, tanh function, or rectified linear unit function as follows:

$$\text{Sigmoid function:} f(x) = \frac{1}{1 + e^{-x}}, \tag{6.1}$$

$$\text{Tanh function:} f(x) = \tanh(x), \tag{6.2}$$

$$\text{Rectified linear unit function:} f(x) = \begin{cases} 0 \text{ for } x < 0 \\ x \text{ for } x \geq 0 \end{cases}. \tag{6.3}$$

The perceptron is regarded as a single artificial neuron. The objective of the perceptron is to adjust weights and classify the inputs into one of the classes correctly. For example, when we have training samples $[x_1, x_2, \ldots, x_n]$, we calculate the output as follows:

$$\hat{y} = f\left(w_0 + \mathbf{w}^{\mathsf{T}}\mathbf{x}\right) \tag{6.4}$$

where $\mathbf{x} = [x_1, x_2, \ldots, x_n]$ and $\mathbf{w} = [w_1, w_2, \ldots, w_n]$ and update the weights as follows:

$$\mathbf{w}_i \leftarrow \mathbf{w}_i + \Delta\mathbf{w}_i \tag{6.5}$$

where

$$\Delta\mathbf{w}_i = \alpha(d - \hat{y})\mathbf{x}_i \tag{6.6}$$

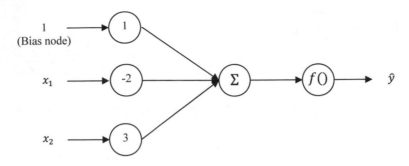

Fig. 6.3 Simple perceptron model

where α, d, and $\hat{y}$ are learning rate, desired value, and perceptron output, respectively. The index i is time or batch of training data set.

Example 6.1 Perceptron Consider a simple perceptron model as shown in Fig. 6.3.

Compare a linear activation function, tanh activation function, and sigmoid activation function.

Solution

As we can observe Fig. 6.3 we have $w_0 = 1$, $w_1 = -2$, and $w_2 = 3$. The output function is

$$\hat{y} = f\left(w_0 + \mathbf{w}^{\mathsf{T}}\mathbf{x}\right) = f(1 - 2x_1 + 3x_2)$$

Figure 6.4 illustrates the output functions with a linear activation function, tanh activation function, and sigmoid activation function.

Single perceptron is a linear classifier. The activation function of the perceptron is used to determine the output as binary decision. The output is mapped in between 0 and 1 or -1 and 1. Thus, the linear activation function $f(x) = x$ as shown in Fig. 6.4a doesn't help for this. As we discussed sigmoid function $f(x) = \frac{1}{1+e^{-x}}$ in the previous chapter, the sigmoid function is a S-shaped curve. The output of sigmoid function as shown in Fig. 6.4c is placed between 0 and 1. It could be good choice. In addition, the tanh function is a S-shaped curve as well and the output range of the tanh function is from -1 to 1 as shown in Fig. 6.4b. As we can observe Fig. 6.4b, the negative inputs are mapped to negative outputs and the zero input is mapped to zero output. The function is differentiable and monotonic. The tanh function is useful for classification with two classes.   ∎

Single perceptron doesn't directly apply to the neural network with hidden layers. In order to build a neural network, single perceptron can be simplified as shown in Fig. 6.5a and multiple output perceptrons can be expressed as shown in Fig. 6.5b.

Using single perceptron as shown in Fig. 6.5a, we can implement any form of logic gate. Three basic operations NOT, AND, and OR can be constructed by finding a proper connection, weight, and threshold. When we have inputs x_1, x_2 and output

Fig. 6.4 Output function with a linear activation function (**a**), tanh activation function (**b**), and sigmoid activation function (**c**)

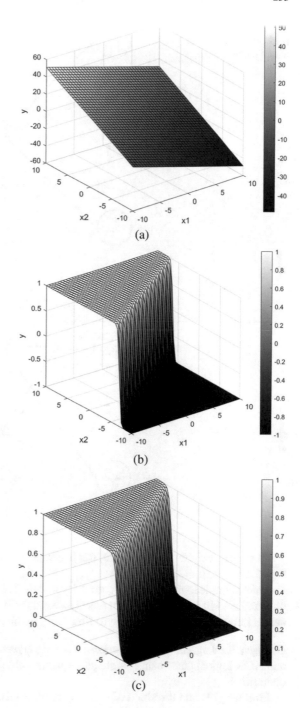

(a)

(b)

(c)

Fig. 6.5 Simplified perceptron (**a**) and multiple output perceptron (**b**)

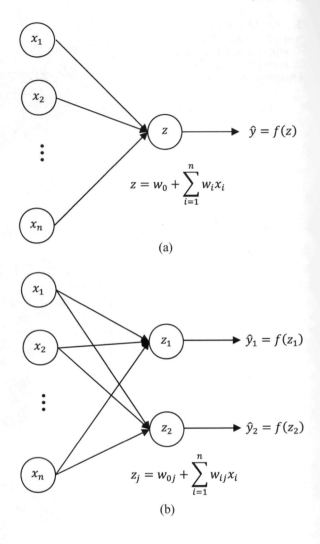

$$\hat{y} = f(z)$$

$$z = w_0 + \sum_{i=1}^{n} w_i x_i$$

(a)

$$\hat{y}_1 = f(z_1)$$

$$\hat{y}_2 = f(z_2)$$

$$z_j = w_{0j} + \sum_{i=1}^{n} w_{ij} x_i$$

(b)

$\hat{y}$ and proper weights and threshold, the logic gates can be implemented as shown in Fig. 6.6

This is simple logical gate. We can easily find the connections, weights, and thresholds and find the decision boundaries. However, when dealing with more complex neural networks, it is not possible to find them by hand.

Example 6.2 Logical Gate Operation Using Perceptron We want to construct AND and XOR logical gates using single perceptron. Table 6.3 describes AND and XOR operations.

Find weights and threshold of single perceptron for AND and XOR operations.

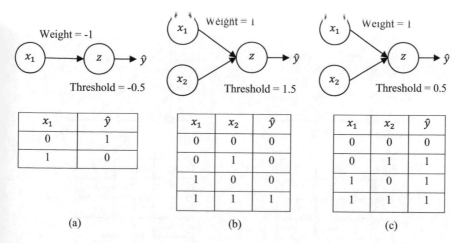

Fig. 6.6 Implementation of a logic gate NOT (a), AND (b), and OR (c)

Table. 6.3 AND (left) and XOR (right) operations

x_1	x_2	$\hat{y}$		x_1	x_2	$\hat{y}$
0	0	0		0	0	0
0	1	0		0	1	1
1	0	0		1	0	1
1	1	1		1	1	0

Solution

From Table 6.3a, we have four inequalities for AND operation as follows:

$$w_1 0 + w_2 0 - \theta < 0 \rightarrow \theta > 0$$

$$w_1 0 + w_2 1 - \theta < 0 \rightarrow w_2 < \theta$$

$$w_1 1 + w_2 0 - \theta < 0 \rightarrow w_1 < \theta$$

$$w_1 1 + w_2 1 - \theta \geq 0 \rightarrow w_1 + w_2 \geq \theta$$

where w_1 and w_2 are weights and θ is threshold. Likewise, we have four inequalities for XOR operation as follows:

$$w_1 0 + w_2 0 - \theta < 0 \rightarrow \theta > 0$$

$$w_1 0 + w_2 1 - \theta \geq 0 \rightarrow w_2 \geq \theta$$

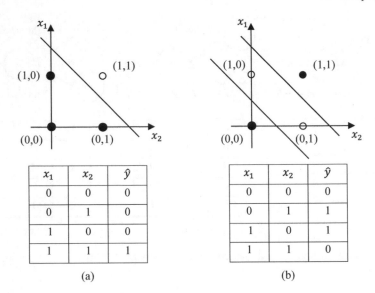

Fig. 6.7 Decision boundaries for AND (**a**) and XOR (**b**)

$$w_1 1 + w_2 0 - \theta \geq 0 \rightarrow w_1 \geq \theta$$

$$w_1 1 + w_2 1 - \theta < 0 \rightarrow w_1 + w_2 < \theta.$$

Now, we can plot the decision boundaries for them as shown in Fig. 6.7.

As we can observe Fig. 6.7a, the threshold θ determines the intersection of decision boundary and the decision boundary clearly classifies them satisfying four inequalities for AND operation. There are many solutions for them. One of them is w_1 and $w_2 = 1$ and $\theta = 1.5$. As we can observe both the inequalities for XOR operation and Fig. 6.7b, there is no simple solution for decision boundary for XOR operation. We need two separate lines for them and define three different areas. In order to find the solution, we should use more complex neural network to create more complex decision boundaries. Therefore, we can summarize the limitations of single perceptron as follows: (1) Too simple model and a linear classifier, (2) If the input is not linearly separable, it doesn't work. ∎

The neural networks can be expressed as weighted directed graphs. The common structures of neural networks are single-layer feedforward neural networks, multi-layer feedforward neural networks, and recurrent neural networks as shown in Fig. 6.8.

As we can observe Fig. 6.8, single-layer feedforward neural networks have input and output layers only. Multi-layer feedforward neural networks have input, output, and hidden layers but there is not feedback connection. Recurrent neural networks have at least one feedback connection and may or may not include hidden layers.

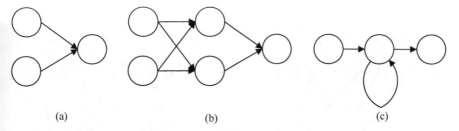

Fig. 6.8 Single-layer feedforward neural networks (**a**), multi-layer feedforward neural networks (**b**), and recurrent neural networks (**c**)

As we can observe Fig. 6.5b, multiple output perceptron as single layer has two layers: input layer and output layer and doesn't have a hidden layer. The input nodes are fully connected to multiple output nodes with a weighted sum of all inputs. A MLP is a neural network of perceptrons and consists of input layer, output layer, and hidden layers. The hidden layers enable us to have specific transformations of the input data and help producing the accurate outputs. We introduce a hidden layer and have multi-layer neural network with a hidden layer as shown in Fig. 6.9

From Fig. 6.9, each node of the hidden layer is an output of a perceptron. The nodes of the hidden layer can be expressed as follows:

$$h_j = w_{0j}^1 + \sum_{i=1}^{n} w_{ij}^1 x_i \tag{6.7}$$

the output nodes are

Fig. 6.9 Multi-layer feed-forward neural network with one hidden layer

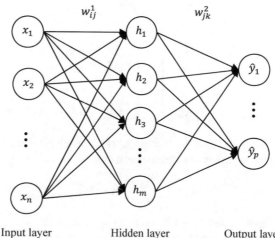

Input layer   Hidden layer   Output layer

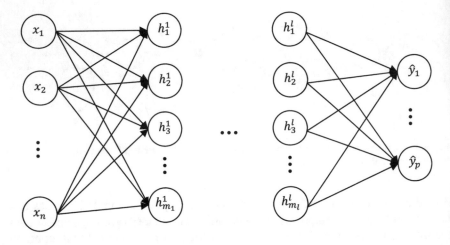

Input layer   Hidden layers   Output layer

Fig. 6.10 Multi-layer feed-forward neural network with multiple hidden layer

$$\hat{y}_k = f\left(w_{0k}^2 + \sum_{j=1}^{m} w_{jk}^2 h_j \right). \tag{6.8}$$

where $1 \le k \le p$. Figure 6.10 illustrates a multi-layer feed-forward neural network with multiple hidden layers.

When we have zero bias, the nodes of the first hidden layer can be expressed in the matrix form as follows:

$$\mathbf{h} = \mathbf{W}^T \mathbf{x}, \tag{6.9}$$

$$
\begin{bmatrix} h_1^1 \\ h_2^1 \\ \vdots \\ h_{m_1}^1 \end{bmatrix}
=
\begin{bmatrix}
w_{11}^1 & w_{12}^1 & \cdots & w_{1n}^1 \\
w_{21}^1 & w_{22}^1 & \cdots & w_{2n}^1 \\
\vdots & \vdots & \ddots & \vdots \\
w_{m_1 1}^1 & w_{m_1 2}^1 & \cdots & w_{m_1 n}^1
\end{bmatrix}
\begin{bmatrix} x_1 \\ x_2 \\ \vdots \\ x_n \end{bmatrix},
\tag{6.10}
$$

When we have m_l hidden layers (namely, it is a deep learning), the nodes of the hidden layers in the matrix form are

$$\mathbf{h} = \mathbf{W}_{m_l}^T \cdots \mathbf{W}_{m_1}^T \mathbf{x}. \tag{6.11}$$

Each node of the hidden layer represents certain features and the neural network identifies the whole feature of the input data. In order to learn the weight efficiently, we should minimize the loss function

Table 6.4 XOR operation

x_1	x_2	$\hat{y}$
0	0	0
0	1	1
1	0	1
1	1	0

$$J\left(\mathbf{W}_{m_{l+1}} \cdots \mathbf{W}_{m_1}\right) = \sum_{i=1}^{N} \left\| \mathbf{W}_{m_{l+1}}^{T} \cdots \mathbf{W}_{m_1}^{T} \mathbf{x}_i - \mathbf{y}_i \right\|^2 \tag{6.12}$$

and we can find them using gradient descent methods and backpropagation method. In practical problems, a MLP has often high-dimensional inputs and many hidden layers. If we have a large number of inputs N, we need 2^N nodes in hidden layers. The complexity increases exponentially with N. Thus, it is important to find the efficient numbers of hidden layers. The key design problems are overfitting and the slow computation if the neural network has many nodes and weights. What we calculated in the neural network is forward propagation. The input is fed in the forward direction from input to output. Each hidden layer receives the input, sequentially calculates it, and passes to the next layer in the forward direction. The forward propagation computes the output $\hat{y}$ of the neural network.

Example 6.3 Multi-layer Feed-Forward Network for XOR Operation We want to construct XOR logical gates using multi-layer feed-forward neural network. Table 6.4 describes XOR operation.

Find a neural network with one hidden layer for XOR operation.

Solution

The XOR operation can be constructed by OR operation, NAND operation, and AND operation. Table 6.5 illustrates OR, NAND, and operations with inputs A and B.

As the algebraic expression, the XOR gate with inputs A and B can be $A \cdot \overline{B} + \overline{A} \cdot B$, $(A + B) \cdot \overline{(A + B)}$ or $(A + B) \cdot \overline{(A \cdot B)}$ where $+$, $\cdot$, and $[\overline{}]$ are AND, OR, and NOT operations. Figure 6.11 illustrates XOR gate circuit.

Based on Fig. 6.11, we can construct a neural network with one hidden layer as shown in Fig. 6.12.

From Fig. 6.12, we have Tables 6.6 and 6.7.

Table 6.5 OR (Left), NAND (Center), and (Right) AND operations

A	B	Out		A	B	Out		A	B	Out
0	0	0		0	0	1		0	0	0
0	1	1		0	1	1		0	1	0
1	0	1		1	0	1		1	0	0
1	1	1		1	1	0		1	1	1

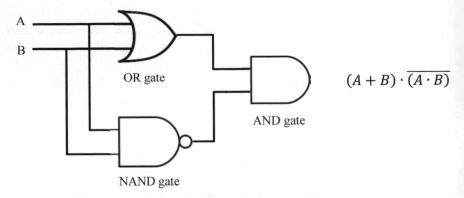

Fig. 6.11 XOR gate circuit using OR, NAND, and AND operations

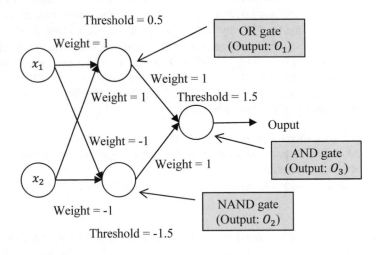

Fig. 6.12 Multi-layer feedforward neural networks for XOR operation

Table 6.6 OR (left) and NAND (right)

x_1	x_2	O_1		x_1	x_2	O_2
0	0	0		0	0	1
0	1	1		0	1	1
1	0	1		1	0	1
1	1	1		1	1	0

As we can observe Table 6.7, the output O_3 of the multi-layer neural network with one hidden layer expresses the XOR operation. ∎

Table 6.7 Output as the multiplayer neural network

x_1	x_2	O_1	O_2	O_3
0	0	0	1	0
0	1	1	1	1
1	0	1	1	1
1	1	1	0	0

Fig. 6.13 Neural network example with one hidden layer

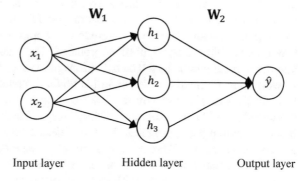

Input layer   Hidden layer   Output layer

Example 6.4 Forward Propogation Consider a multi-layer feed-forward neural network as shown in Fig. 6.13.

The neural network has sigmoid activation function ($f(x) = \frac{1}{1+e^{-x}}$) and the following input and weight matrices:

$$\mathbf{x} = \begin{bmatrix} 2 \\ -1 \end{bmatrix}, \mathbf{W}_1 = \begin{bmatrix} 2.5 & 4.2 & 1.5 \\ -0.6 & -1.5 & 5.6 \end{bmatrix}, \mathbf{W}_2 = \begin{bmatrix} -0.4 \\ 1 \\ 2.5 \end{bmatrix}.$$

Compute the output $\hat{y}$ of the neural network.

Solution

From (6.9), we calculate the nodes of the first hidden layer as follows:

$$\mathbf{W}_1^T \mathbf{x} = \begin{bmatrix} 2.5 & -0.6 \\ 4.2 & -1.5 \\ 1.5 & 5.6 \end{bmatrix} \begin{bmatrix} 2 \\ -1 \end{bmatrix} = \begin{bmatrix} 5.6 \\ 9.9 \\ -2.6 \end{bmatrix},$$

$$\mathbf{h} = f(\mathbf{W}_1^T \mathbf{x}) = \begin{bmatrix} f(5.6) \\ f(9.9) \\ f(-2.6) \end{bmatrix} = \begin{bmatrix} 0.996 \\ 0.999 \\ 0.069 \end{bmatrix}.$$

Now, we calculate the output $\hat{y}$ as follows:

$$\mathbf{W}_2^T \mathbf{h} = \begin{bmatrix} -0.4 & 1 & 2.5 \end{bmatrix} \begin{bmatrix} 0.996 \\ 0.999 \\ 0.069 \end{bmatrix} = 0.774,$$

$$\hat{y} = f\left(\mathbf{W}_2^T \mathbf{h}\right) = f(0.774) = 0.684.$$

■

A neural network is more widely used for supervised learning problems. The training data sets enable the neural network to learn the weights and biases. A multi-layer feedforward neural network is just mapping from input values to output values by simple calculation. It doesn't include a learning mechanism. The weights and biases of the hidden layers must be updated. The backpropagation enables us to find the optimal parameters. The backpropagation algorithm for multi-layer neural networks was invented by Werbos in his Ph.D. dissertation in 1974 [6]. After that, Rumelhart et al. [3], rediscovered the backpropagation training algorithm with a framework in 1986 and was more widely known. Their algorithm includes a set of input and output pairs, produces the actual neural network output from the inputs, and compares them. If there is a difference between them, the weights of the neural network are updated to reduce the difference. The basic idea of the backpropagation training is same as the one of DNN. By using large training data sets, we can discover intricate structures of the neural network. The optimal values of the weights and biases can be found by backpropagation algorithms and gradient descent algorithms. The basic concept of the backpropagation is to compare the actual outputs with the desired outputs to obtain an error signal. In terms of the different parameters, the partial derivatives of the loss (or cost, error) function are calculated in the backward direction of the neural network. The chain rule of the backpropagation is similar to the forward propagation. The weights can be iteratively updated using gradient descent algorithms until they converge. The update of the weights is realized by minimizing the loss function (6.12). When we have an activation function, the loss function can be rewritten as follows:

$$J\left(\mathbf{W}_{m_{l+1}} \dots \mathbf{W}_{m_1}\right) = \sum_{i=1}^{N} \left\| f\left(\mathbf{W}_{m_{l+1}}^T \dots f\left(\mathbf{W}_{m_1}^T \mathbf{x}_i\right)\right) - \mathbf{y}_i \right\|^2 \qquad (6.13)$$

In order to optimize the loss function, gradient descent algorithm is used. For example, for the weights of the first hidden layer $\mathbf{W}_1$, the update rule is

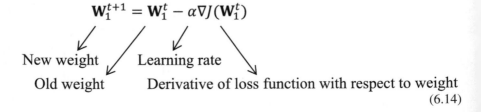

$$\mathbf{W}_1^{t+1} = \mathbf{W}_1^t - \alpha \nabla J(\mathbf{W}_1^t)$$

New weight   Learning rate

Old weight   Derivative of loss function with respect to weight

$$(6.14)$$

where t and α are the parameters for the gradient direction and the learning rate, respectively. In order to perform this for all weights, we calculate the partial derivatives of the loss function with respect to each weight. Roughly speaking, the backpropagation starts updating the weight of the last layer, propagates the error signal back to the previous layer, and updates the weights of the previous layer. For example, we consider a loss function with two weight matrices as follows:

$$J(\mathbf{W}_1, \mathbf{W}_2) = \left\| f\left(\mathbf{W}_2^T f\left(\mathbf{W}_1^T \mathbf{x}\right)\right) - \mathbf{y} \right\|^2 = \|\delta_2 - \mathbf{y}\|^2 \tag{6.15}$$

where

$$\delta_2 = f\left(\mathbf{W}_2^T f\left(\mathbf{W}_1^T \mathbf{x}\right)\right) = f(\varepsilon_2). \tag{6.16}$$

The variable ε_i in the hidden layer can be expressed as $\varepsilon_i = \mathbf{W}_i^T \mathbf{h}_{i-1}$. If we have $\mathbf{h}_0$, this is an input layer. The variable δ_i is $\delta_i = f(\varepsilon_i)$. Now, we perform the partial derivatives of the loss function with respect to $\mathbf{W}_2$ as follows:

$$\frac{\partial J}{\mathbf{W}_2} = \frac{\partial J}{\partial \delta_2} \frac{\partial \delta_2}{\mathbf{W}_2} = \frac{\partial J}{\partial \delta_2} \frac{\partial \delta_2}{\partial \varepsilon_2} \frac{\partial \varepsilon_2}{\mathbf{W}_2} \tag{6.17}$$

(6.15) is rewritten as follows:

$$J(\mathbf{W}_1, \mathbf{W}_2) = \left\| f\left(\mathbf{W}_2^T f\left(\mathbf{W}_1^T \mathbf{x}\right)\right) - \mathbf{y} \right\|^2 = \left\| f\left(\mathbf{W}_2^T \delta_1\right) - \mathbf{y} \right\|^2 \tag{6.18}$$

where

$$\delta_1 = f\left(\mathbf{W}_1^T \mathbf{x}\right) = f(\varepsilon_1). \tag{6.19}$$

δ_1 is same as $\mathbf{h} = f\left(\mathbf{W}_1^T \mathbf{x}\right)$. Likewise, we perform the partial derivatives of the loss function with respect to $\mathbf{W}_1$ as follows:

$$\frac{\partial J}{\partial \mathbf{W}_1} = \frac{\partial J}{\partial \delta_2} \frac{\partial \delta_2}{\partial \mathbf{W}_1} = \frac{\partial J}{\partial \delta_2} \frac{\partial \delta_2}{\partial \delta_1} \frac{\partial \delta_1}{\partial \mathbf{W}_1} = \frac{\partial J}{\partial \delta_2} \frac{\partial \delta_2}{\partial \delta_1} \frac{\partial \delta_1}{\partial \varepsilon_1} \frac{\partial \varepsilon_1}{\partial \mathbf{W}_1}. \tag{6.20}$$

As we can observe (6.17) and (6.20), we have a sequence of chain rule. We can regard the backpropagation as the process to adjust the neural network's weight to reduce the overall loss function as follows:

$$\Delta \mathbf{W} \propto -\frac{\partial J}{\partial \mathbf{W}} \tag{6.21}$$

The computation begins at the output layer with a particular weight using chain rule as follows:

$$\Delta w \propto -\frac{\partial J}{\partial w}$$

$$= -\frac{\partial(\text{Loss function})}{\partial(\text{Activation function})} \frac{\partial(\text{Activation function})}{\partial(\text{Network input})} \frac{\partial(\text{Network input})}{\partial(\text{Weight})}. \quad (6.22)$$

Example 6.5 Backpropagation Consider a neural network as shown in Fig. 6.13. The neural network has sigmoid activation function ($f(x) = \frac{1}{1+e^{-x}}$), and the input and weight matrices are as follows:

$$\mathbf{x} = \begin{bmatrix} 2 \\ -1 \end{bmatrix}, y = 1, \mathbf{W}_1 = \begin{bmatrix} 2.5 & 4.2 & 1.5 \\ -0.6 & -1.5 & 5.6 \end{bmatrix}, \mathbf{W}_2 = \begin{bmatrix} -0.4 \\ 1 \\ 2.5 \end{bmatrix}, \text{ and } \alpha = 1.$$

Compute the back propagation of the neural network and update the weights.

Solution

From (6.13), we can compute the loss function as follows:

$$J\left(\mathbf{W}_{m_{l+1}} \cdots \mathbf{W}_{m_1}\right) = \sum_{i=1}^{N} \left\| f\left(\mathbf{W}_{m_{l+1}}^T \cdots f\left(\mathbf{W}_{m_1}^T \mathbf{x}_i\right)\right) - \mathbf{y}_i \right\|^2$$

However, we need two assumptions to compute the loss (or cost, error) function. Firstly, the loss function J can be expressed as an average over loss functions for individual training sample as follows:

$$J = \frac{1}{n} \sum_{x} J_x$$

where J, n, and J_x are average, the total number of training sample, and the loss function for individual training sample, respectively. J_x can be expressed as follows:

$$J_x = \frac{1}{2} \left\| \mathbf{y}_i - \hat{\mathbf{y}}_i \right\|^2$$

This assumption allows us to compute the partial derivatives for the individual training sample and the recover them by averaging. Secondly, the loss function can be described as the output function from the neural network. Thus, the loss function for a single training sample can be rewritten as follows:

$$J = \frac{1}{2}(y - \hat{y})^2.$$

Gradient descent algorithm is used to update the weights. From (6.14), we should calculate the partial derivative (or gradient) of the loss function. In order to compute the gradient of $\mathbf{W}_2$, we have the following chain rule:

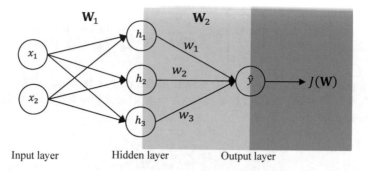

Fig. 6.14 Gradient calculation of $\mathbf{W}_2$

$$\frac{\partial J}{\partial \mathbf{W}_2} = \frac{\partial J}{\partial \hat{y}} \frac{\partial \hat{y}}{\partial \mathbf{W}_2}$$

and Fig. 6.14 illustrates the part of computation.

In order to compute the gradient of $\mathbf{W}_1$, we have the following chain rule:

$$\frac{\partial J}{\partial \mathbf{W}_1} = \frac{\partial J}{\partial \hat{y}} \frac{\partial \hat{y}}{\partial \mathbf{h}} \frac{\partial \mathbf{h}}{\partial \mathbf{W}_1}$$

and Fig. 6.15 illustrates the part of computation.

As we can observe the gradients of $\mathbf{W}_1$ and $\mathbf{W}_2$, the loss function J is a scalar and the weights $\mathbf{W}_1$ and $\mathbf{W}_2$ are a matrix. The gradients $\frac{\partial J}{\partial \mathbf{W}_1}$ and $\frac{\partial J}{\partial \mathbf{W}_2}$ will be a matrix as follows:

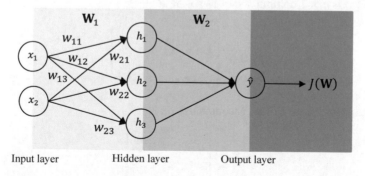

Fig. 6.15 Gradient calculation of $\mathbf{W}_1$

$$\frac{\partial J}{\partial \mathbf{W}_2} = \begin{bmatrix} \frac{\partial J}{\partial w_1} \\ \frac{\partial J}{\partial w_2} \\ \frac{\partial J}{\partial w_3} \end{bmatrix} \text{ and } \frac{\partial J}{\partial \mathbf{W}_1} = \begin{bmatrix} \frac{\partial J}{\partial w_{11}} & \frac{\partial J}{\partial w_{12}} & \frac{\partial J}{\partial w_{13}} \\ \frac{\partial J}{\partial w_{21}} & \frac{\partial J}{\partial w_{22}} & \frac{\partial J}{\partial w_{23}} \end{bmatrix}$$

We repeat this process for each weight in the neural network. When we have $\mathbf{x} \in \mathbb{R}^m$, $\mathbf{h} \in \mathbb{R}^n$ and $y \in \mathbb{R}$ and functions $\mathbf{h} = g(\mathbf{x})$ and $y = l(\mathbf{h})$, the derivative of y can be expressed as follows:

$$\frac{\partial y}{\partial x_i} = \sum_j \frac{\partial y}{\partial h_i} \frac{\partial h_i}{\partial x_i}.$$

In matrix form, we have

$$\begin{pmatrix} \frac{\partial y}{\partial x_1} \\ \vdots \\ \frac{\partial y}{\partial x_m} \end{pmatrix} = \begin{pmatrix} \sum_j \frac{\partial y}{\partial h_i} \frac{\partial h_i}{\partial x_1} \\ \vdots \\ \sum_j \frac{\partial y}{\partial h_i} \frac{\partial h_i}{\partial x_m} \end{pmatrix} = \nabla_{\mathbf{x}} y = \left(\frac{\partial \mathbf{h}}{\partial \mathbf{x}} \right)^T \nabla_{\mathbf{h}} y$$

where $\frac{\partial \mathbf{h}}{\partial \mathbf{x}}$ is the $n \times m$ Jacobian matrix of g. As we can see from the above equation, the gradient of $\mathbf{x}$ is a multiplication of Jacobian matrix with a vector. Jacobian matrix of $f(\mathbf{x})$ with respect to $\mathbf{x}$ (column vector $m \times 1$) is defined as follows:

$$\frac{\partial f}{\partial \mathbf{x}} = \begin{bmatrix} \frac{\partial f_1}{\partial x_{11}} & \cdots & \frac{\partial f_1}{\partial x_{m1}} \\ \vdots & \ddots & \vdots \\ \frac{\partial f_n}{\partial x_{11}} & \cdots & \frac{\partial f_n}{\partial x_{m1}} \end{bmatrix}$$

where $f(\mathbf{x})$ is $n \times 1$ column vector as follows:

$$f(\mathbf{x}) = \begin{bmatrix} f_1(x_{11}, \ldots, x_{m1}) \\ \vdots \\ f_n(x_{11}, \ldots, x_{m1}) \end{bmatrix}.$$

If $\mathbf{x}$ is a row vector $1 \times m$, Jacobian matrix is

$$\frac{\partial f}{\partial \mathbf{x}} = \begin{bmatrix} \frac{\partial f_1}{\partial x_{11}} & \cdots & \frac{\partial f_1}{\partial x_{1m}} \\ \vdots & \ddots & \vdots \\ \frac{\partial f_n}{\partial x_{11}} & \cdots & \frac{\partial f_n}{\partial x_{1m}} \end{bmatrix}.$$

The backpropagation algorithm performs Jacobian gradient produces to each operation in the neural network. We can rewrite this in terms of tensors operation. Let the gradient of y with respect to a tensor $\mathbf{X}$ be $\nabla_{\mathbf{X}} y$. When we have $\mathbf{H} = g(\mathbf{X})$ and

$y = l(\mathbf{H})$, $\nabla_{\mathbf{X}} y$ can be expressed as follows:

$$\nabla_{\mathbf{X}} y = \sum_j (\nabla_{\mathbf{X}} H_j) \frac{\partial y}{\partial H_j}.$$

Now, we compute the partial derivative of J with respect to $\mathbf{W}_2$ as follows:

$$\frac{\partial J}{\partial \mathbf{W}_2} = \begin{bmatrix} \frac{\partial J}{\partial w_1} \\ \frac{\partial J}{\partial w_2} \\ \frac{\partial J}{\partial w_3} \end{bmatrix}.$$

We firstly consider $\frac{\partial J}{\partial w_1}$ as follows:

$$\frac{\partial J}{\partial w_1} = \frac{\partial J}{\partial \hat{y}} \frac{\partial \hat{y}}{\partial w_1} = \frac{\partial J}{\partial \hat{y}} \frac{\partial \hat{y}}{\partial \varepsilon_2} \frac{\partial \varepsilon_2}{\partial w_1} = \frac{\partial J}{\partial \hat{y}} \frac{\partial \hat{y}}{\partial \varepsilon_2} \frac{\partial}{\partial w_1} (\mathbf{W}_2^T \mathbf{h}) = (\hat{y} - y) f'(\varepsilon_2) h_1.$$

The loss function is calculated as follows:

$$J = \frac{1}{2} (0.684 - 1)^2 = 0.0498.$$

Next, we compute the derivative of J with respect to $\hat{y}$ as follows:

$$\frac{\partial J}{\partial \hat{y}} = \frac{\partial}{\partial \hat{y}} \left(\frac{1}{2} (y - \hat{y})^2 \right) = -(y - \hat{y}) = -0.3156.$$

For example, the meaning of the last derivative $\frac{\partial \varepsilon_2}{\partial w_1}$ is how much weight and the previous neuron node influences the next neuron node. The derivative of the sigmoid function is

$$f'(x) = f(x)(1 - f(x))$$

and the variable ε_i in the hidden layer can be described as follows:

$$\varepsilon_i = \mathbf{W}_i^T \mathbf{h}_{i-1}.$$

The variable ε_2 is

$$\varepsilon_2 = \mathbf{W}_2^T \mathbf{h} = 0.774.$$

We can calculate $f'(\varepsilon_2)$ as follows:

$$\begin{aligned} f'(\varepsilon_2) &= f(\varepsilon_2)(1 - f(\varepsilon_2)) \\ &= f(0.774)(1 - f(0.774)) \end{aligned}$$

$$= 0.6844(1 - 0.6844) = 0.216.$$

Thus, we have

$$\frac{\partial J}{\partial w_1} = (\hat{y} - y) f'(\varepsilon_2) h_1$$

$$= (-0.3156)(0.216)(0.996) = -0.0679.$$

Likewise, we find $\frac{\partial J}{\partial w_2}$ and $\frac{\partial J}{\partial w_3}$ and obtain

$$\frac{\partial J}{\partial \mathbf{W}_2} = \begin{bmatrix} -0.0679 \\ -0.0682 \\ -0.0047 \end{bmatrix}.$$

From (6.14), we can update $\mathbf{W}_2$ as follows:

$$\mathbf{W}_2^{t+1} = \mathbf{W}_2^t - \alpha \frac{\partial J}{\partial \mathbf{W}_2} = \begin{bmatrix} -0.4 \\ 1 \\ 2.5 \end{bmatrix} - 1 \cdot \begin{bmatrix} -0.0679 \\ -0.0682 \\ -0.0047 \end{bmatrix} = \begin{bmatrix} -0.3321 \\ 1.0682 \\ 2.5047 \end{bmatrix}$$

and Fig. 6.16 illustrates the update of $\mathbf{W}_2$.

In the next layer, we have 6 weights to update for the loss function. There are multiple paths. We consider chain rule. When implementing the backpropagation, the chain rule is important because we don't want to recompute the same equation. The chain rule is very useful to speed up the computation. The chain rule in this example can be summarized as follows:

Fig. 6.16 $\mathbf{W}_2$ update

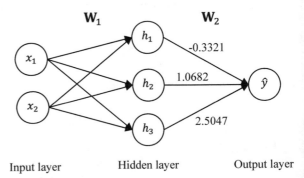

Input layer   Hidden layer   Output layer

$$\frac{\partial J}{\partial \mathbf{W}_2} = \frac{\partial J}{\partial \hat{y}} \frac{\partial \hat{y}}{\partial \mathbf{W}_2} = \frac{\partial J}{\partial \hat{y}} \frac{\partial \hat{y}}{\partial \varepsilon_2} \frac{\partial \varepsilon_2}{\partial \mathbf{W}_2},$$

$$\frac{\partial J}{\partial \mathbf{W}_1} = \frac{\partial J}{\partial \hat{y}} \frac{\partial \hat{y}}{\partial \mathbf{h}} \frac{\partial \mathbf{h}}{\partial \mathbf{W}_1} = \frac{\partial J}{\partial \hat{y}} \frac{\partial \hat{y}}{\partial \varepsilon_2} \frac{\partial \varepsilon_2}{\partial \mathbf{h}} \frac{\partial \mathbf{h}}{\partial \varepsilon_1} \frac{\partial \varepsilon_1}{\partial \mathbf{W}_1}.$$

Now, we compute the partial derivative of J with respect to $\mathbf{W}_1$ using a chain rule as follows:

$$\frac{\partial J}{\partial \mathbf{W}_1} = \begin{bmatrix} \frac{\partial J}{\partial w_{11}} & \frac{\partial J}{\partial w_{12}} & \frac{\partial J}{\partial w_{13}} \\ \frac{\partial J}{\partial w_{21}} & \frac{\partial J}{\partial w_{22}} & \frac{\partial J}{\partial w_{23}} \end{bmatrix}.$$

We firstly consider $\frac{\partial J}{\partial w_{11}}$ as follows:

$$\frac{\partial J}{\partial w_{11}} = \frac{\partial J}{\partial \hat{y}} \frac{\partial \hat{y}}{\partial \varepsilon_2} \frac{\partial \varepsilon_2}{\partial \mathbf{h}} \frac{\partial \mathbf{h}}{\partial \varepsilon_1} \frac{\partial \varepsilon_1}{\partial w_{11}}$$

where $\frac{\partial J}{\partial \hat{y}}$ and $\frac{\partial \hat{y}}{\partial \varepsilon_2}$ are found previously. We compute $\frac{\partial \varepsilon_2}{\partial \mathbf{h}}$ as follows:

$$\frac{\partial \varepsilon_2}{\partial \mathbf{h}} = \mathbf{W}_2^{\mathsf{T}} = [-0.4 \; 1 \; 2.5], \frac{\partial \varepsilon_2}{\partial h_1} = -0.4$$

where

$$\varepsilon_2 = \mathbf{W}_2^{\mathsf{T}} \mathbf{h}.$$

Since w_{11} is connected to h_1, $\frac{\partial \varepsilon_2}{\partial \mathbf{h}} \approx \frac{\partial \varepsilon_2}{\partial h_1} = -0.4$. The next term $\frac{\partial \mathbf{h}}{\partial \varepsilon_1}$ is computed as follows:

$$\frac{\partial \mathbf{h}}{\partial \varepsilon_1} = f'(\varepsilon_1) = f(\varepsilon_1)(1 - f(\varepsilon_1)) = \begin{bmatrix} 0.0037 \\ 0.0001 \\ 0.0644 \end{bmatrix}$$

where $\mathbf{h} = f(\mathbf{W}_1^{\mathsf{T}} \mathbf{x}) = f(\varepsilon_1)$ and $\varepsilon_1 = \mathbf{W}_1^{\mathsf{T}} \mathbf{x}$. We can compute

$$\varepsilon_1 = \mathbf{W}_1^T \mathbf{x} = \begin{bmatrix} 2.5 & -0.6 \\ 4.2 & -1.5 \\ 1.5 & 5.6 \end{bmatrix} \begin{bmatrix} 2 \\ -1 \end{bmatrix} = \begin{bmatrix} 5.6 \\ 9.9 \\ -2.6 \end{bmatrix},$$

$$\mathbf{h} = f(\varepsilon_1) = \begin{bmatrix} f(5.6) \\ f(9.9) \\ f(-2.6) \end{bmatrix} = \begin{bmatrix} 0.996 \\ 0.999 \\ 0.069 \end{bmatrix} \text{ and}$$

$$f'(\boldsymbol{\varepsilon}_1) = f(\boldsymbol{\varepsilon}_1)(1 - f(\boldsymbol{\varepsilon}_1)) = \begin{bmatrix} 0.0037 \\ 0.0001 \\ 0.0644 \end{bmatrix}.$$

Since w_{11} is connected to h_1, $\frac{\partial \mathbf{h}}{\partial \boldsymbol{\varepsilon}_1} \approx \frac{\partial h_1}{\partial \boldsymbol{\varepsilon}_1} = 0.0037$. The next term $\frac{\partial \boldsymbol{\varepsilon}_1}{\partial w_{11}}$ is computed as follows:

$$\frac{\partial \boldsymbol{\varepsilon}_1}{\partial w_{11}} = 2$$

where $\boldsymbol{\varepsilon}_1 = \mathbf{W}_1^T \mathbf{x}$. Now, we have

$$\frac{\partial C}{\partial w_{11}} = (\hat{y} - y) f'(\boldsymbol{\varepsilon}_2) \mathbf{W}_2^T f'(\boldsymbol{\varepsilon}_1) x_1$$
$$= (-0.3156)(0.216)(-0.4)(0.0037)(2) = 0.0002.$$

Likewise, we find the other elements of $\frac{\partial J}{\partial \mathbf{W}_1}$ and obtain

$$\frac{\partial J}{\partial \mathbf{W}_1} = \begin{bmatrix} 0.0002 & -0.00001 & -0.022 \\ 0.0001 & -0.000007 & -0.011 \end{bmatrix}.$$

From (6.14), we can update $\mathbf{W}_1$ as follows:

$$\mathbf{W}_1^{t+1} = \mathbf{W}_1^t - \alpha \frac{\partial J}{\partial \mathbf{W}_1} = \begin{bmatrix} 2.5 & 4.2 & 1.5 \\ -0.6 & -1.5 & 5.6 \end{bmatrix}$$
$$- 1. \begin{bmatrix} 0.0002 & -0.00001 & -0.022 \\ 0.0001 & -0.000007 & -0.011 \end{bmatrix}$$
$$= \begin{bmatrix} 2.4998 & 4.2 & 1.522 \\ -0.6001 & -1.5 & 5.611 \end{bmatrix}$$

and Fig. 6.17 illustrates the update of $\mathbf{W}_1$.   ■

As we discussed in Example 6.5, a key calculation of backpropagation is the partial derivative of the loss (or cost, error) function with respect to weights in the neural network. This equation indicates how the loss function affects to the change of weights and biases. Now, we generalize the backpropagation algorithm and redefine the notation. The weight w_{ji}^l represents from the ith node in the l-1 layer to the jth node in the l layer. The activation a_j^l represents the jth node activation in the l layer. The bias b_j^l is the jth node bias in the l layer. Thus, similar to (6.4), we have

$$a_j^l = f\left(\sum_i w_{ji}^l a_i^{l-1} + b_j^l \right) = f(z_j^l) \tag{6.23}$$

Fig. 6.17 W_1 update

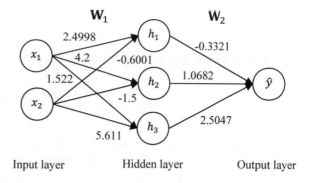

Input layer   Hidden layer   Output layer

where

$$z_j^l = \sum_i w_{ji}^l a_i^{l-1} + b_j^l. \tag{6.24}$$

Figure 6.18 illustrates (6.23) and (6.24) in the perceptron when we have 3 nodes in l-1 layer.

As we discussed in Example 6.5, the quadratic loss function can be expressed as follows:

$$J = \frac{1}{2} \sum_j (y_j - a_j^L)^2 \tag{6.25}$$

where a_j^L is the output activations and L is the last layer. For example, we have two nodes, the loss function is $J = \frac{1}{2}\left((y_1 - a_1^L)^2 + (y_2 - a_2^L)^2\right)$. The error in the jth node in the l layer is defined as follows:

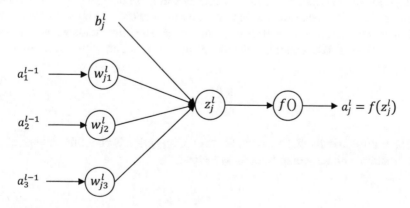

Fig. 6.18 Perceptron with 3 nodes in l-1 layer

$$\delta_j^l = \frac{\partial J}{\partial z_j^l} \tag{6.26}$$

and we have this error at the output layer as follows:

$$\delta_j^L = \frac{\partial J}{\partial a_j^L} \frac{\partial a_j^L}{\partial z_j^L} = \frac{\partial J}{\partial a_j^L} f'\left(z_j^L\right). \tag{6.27}$$

Equation (6.27) is the first fundamental equation for backpropagation. The first term $\frac{\partial J}{\partial a_j^L}$ represents how quickly the loss function is changing with respect to a_j^L. The second term $f'\left(z_j^L\right)$ means how quickly the activation function is changing with respect to z_j^L. From (6.25) and (6.27), we have

$$\frac{\partial J}{\partial a_j^L} = \left(a_j^L - y_j\right). \tag{6.28}$$

Equation (6.27) is a component wise computation. Thus, in matrix form, we should use the elementwise product. (6.27) can be rewritten in the matrix form as follows:

$$\delta^L = \nabla_{\mathbf{a}} J \odot f'\left(z^L\right) \tag{6.29}$$

where $\odot$ is the elementwise multiplication (or Hadamard product) and $\nabla_{\mathbf{a}} J$ is a vector whose components are the partial derivatives $\frac{\partial J}{\partial a_j^L}$. The error δ^l in terms of the error in the next layer is expressed as follows:

$$\delta^l = \left(\left(w^{l+1}\right)^T \delta^{l+1}\right) \odot f'\left(z^l\right). \tag{6.30}$$

Equation (6.30) is the second fundamental equation for backpropagation. It means to move the error backward through the neural network. Using both (6.27) and (6.30), we can determine the error δ^l at any layer. The third fundamental equation for backpropagation is about the change of the loss function with respect to biases as follows:

$$\frac{\partial J}{\partial b_j^l} = \delta_j^l. \tag{6.31}$$

The fourth fundamental equation for backpropagation is about the change of the loss function with respect to weights as follows:

$$\frac{\partial J}{\partial w_{ji}^l} = a_i^{l-1} \delta_j^l. \tag{6.32}$$

As we can observe (6.32), we can find the change of the loss function with respect to weights by local error δ^l_j and local input a^{l-1}_i. Both (6.31) and (6.32) are easy to compute. We can summarize backpropagation algorithm as follows:

- Initialize weights and bias in the neural network.
- For each training sample

    - Feedforward: Compute $a^l_j = f\left(z^l_j\right)$ and $z^l_j = \sum_i w^l_{ji} a^{l-1}_i + b^l_j$

    - Output error: Compute the error of the neural network $\delta^L = \nabla_a J \odot f'\left(z^L\right)$

    - Back propagation: Compute $\delta^l = \left(\left(w^{l+1}\right)^T \delta^{l+1}\right) \odot f'\left(z^l\right)$ from $l = L$ to $l = 1$ recursively.

- Update the weights and biases using the gradient of the loss function $\frac{\partial J}{\partial b^l_j} = \delta^l_j$ and $\frac{\partial J}{\partial w^l_{ji}} = a^{l-1}_i \delta^l_j$.

In order to avoid overfitting problems, regularization techniques are widely used in deep learning. They make slight modifications to deep learning algorithms and improve the performance of deep learning algorithm. As the regularization techniques were discussed in Chap. 4, Tikhonov regularization using L1 norm and Lasso regularization using L2 norm are widely used. The regularization factor (λ) controls both fitting the training data set and keeping the parameter small.

Example 6.6 Deep Neural Network We develop handwritten digit (0 ~ 9) recognition using deep learning. The handwritten digits images are represented as a 20×20 matrix containing grayscale pixel value at each cell. We have 5000 training data images. Consider a fully connected neural network with 2 hidden layers, sigmoid activation function, randomly initialized weights, regularization factor $\lambda = 0.1$, and learning rate $\alpha = 1$. Compare the performances of the neural networks according to the different number of epoch and neuron nodes.

Solution

Firstly, the input 20×20 digits images cannot be fed directly into the neural network. Thus, we process the data and reshape the matrix as one-dimensional data for the input layer. The neural nodes of the input layer are 400. The neural nodes of the output layer are 10 (0 ~ 9). Secondly, we develop the neural network model with 2 hidden layers in terms of different training data set (10, 20, and 50) and neural nodes (first hidden layer with 20 or 40 neural nodes and second hidden layer with 10 or 20 neural nodes). Evaluating deep learning algorithm is an essential part of developments. There are many different types of evaluation metrics such as classification accuracy, logarithmic loss, confusion matrix, F1 score, mean absolute error, mean-squared error, and so on. We use the confusion matrix. The confusion matrix is similar to contingency table we discussed in Chap. 3. The contingency table is used to describe data and represent frequency distribution of the variables. The confusion matrix is used to evaluate classifiers. Using computer simulation, we perform classification according to the number of epochs and neural nodes. One epoch is one iteration (one feed-forward + one back-propagation) through training data set. In the first simulation, we perform the digit image recognition with 10, 20, and 50 epochs and 20 and 10 neural nodes at each hidden layer. Figures 6.19, 6.20 and 6.21 illustrates

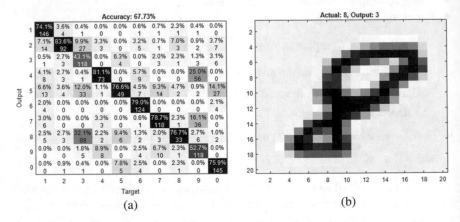

Fig. 6.19 Confusion matrix (**a**) and an example of handwritten digit (**b**) for 10 epochs and the first hidden layer with 20 neural nodes and the second hidden layer with 10 neural nodes

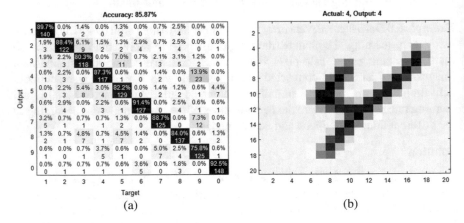

Fig. 6.20 Confusion matrix (**a**) and an example of handwritten digit (**b**) for 20 epochs and the first hidden layer with 20 neural nodes and the second hidden layer with 10 neural nodes

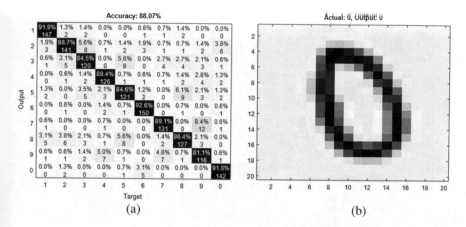

Fig. 6.21 Confusion matrix (**a**) and an example of handwritten digit (**b**) for 50 epochs and the first hidden layer with 20 neural nodes and the second hidden layer with 10 neural nodes

the output of the handwritten digit recognition.

In the second simulation, we perform the digit image recognition with 10, 20, and 50 epochs and 40 and 20 neural nodes at each hidden layer. Figures 6.22, 6.23 and 6.24 illustrates the output of the handwritten digit recognition.

As we can observe the above results, accuracy increases with more epochs and neural nodes at each hidden layer. However, in general, the accuracy depends on not only them but also the number of layer, model and training data quality, and so on. ∎

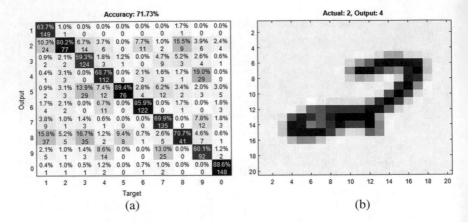

Fig. 6.22 Confusion matrix (**a**) and an example of handwritten digit (**b**) for 10 epochs and the first hidden layer with 40 neural nodes and the second hidden layer with 20 neural nodes

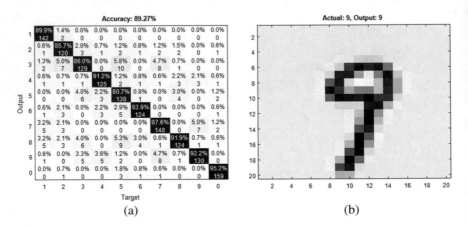

Fig. 6.23 Confusion matrix (**a**) and an example of handwritten digit (**b**) for 20 epochs and the first hidden layer with 40 neural nodes and the second hidden layer with 20 neural nodes

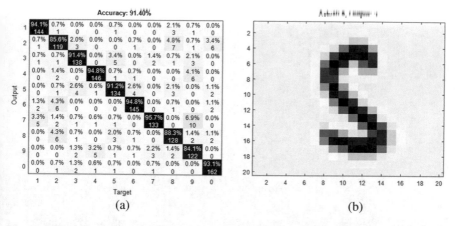

Fig. 6.24 Confusion matrix (**a**) and an example of handwritten digit (**b**) for 50 epochs and the first hidden layer with 40 neural nodes and the second hidden layer with 20 neural nodes

Summary 6.2. Deep Neural Network

1. The perceptron is a single layer neural network. A multilayer perceptron (MLP) is a feedforward artificial neural network composed of multiple layers of perceptrons.
2. The perceptron is regarded as a single artificial neuron. The objective of the perceptron is to adjust weights and classify the inputs into one of classes correctly.
3. The basic concept of the back propagation is to compare the actual outputs with the desired outputs to obtain an error signal. The update rule of the weights is

$$\mathbf{W}_1^{t+1} = \mathbf{W}_1^t - \alpha \nabla J(\mathbf{W}_1^t)$$

where t and α are the parameters for the gradient direction and the learning rate, respectively.
4. The computation begins at the output layer with a particular weight using chain rule as follows:

$$\Delta w \propto -\frac{\partial J}{\partial w}$$
$$= -\frac{\partial (\text{Loss function})}{\partial (\text{Activation function})} \frac{\partial (\text{Activation function})}{\partial (\text{Network input})} \frac{\partial (\text{Network input})}{\partial (\text{Weight})}.$$

5. A key calculation of back propagation is the partial derivative of the loss (or cost, error) function with respect to weights in the neural network. This

equation indicates how the loss function affects to the change of weights and biases.

6. We can summarize back propagation algorithm as follows:

  – Initialize weights and bias in the neural network.
  – For each training sample
  – Feedforward: Compute $a_j^l = f\left(z_j^l\right)$ and $z_j^l = \sum_i w_{ji}^l a_i^{l-1} + b_j^l$.
  – Output error: Compute the error of the neural network $\delta^L = \nabla_a J \odot f'\left(z^L\right)$
  – Back propagation : Compute $\delta^l = \left(\left(w^{l+1}\right)^T \delta^{l+1}\right) \odot f'\left(z^l\right)$ from $l = = L$ to $l = 1$ recursively.
  – Update the weights and biases using the gradient of the loss function $\frac{\partial J}{\partial b_j^l} = \delta_j^l$ and $\frac{\partial J}{\partial w_{ji}^l} = a_i^{l-1} \delta_j^l$.

6.3 Convolutional Neural Network

Conventional neural networks are not suitable for images because it doesn't consider the characteristic of the pixel image. We should take into account the spatial organization of the image and apply a local transformation to a set of pixel. The convolutional technique is useful for this approach. The basic concept of convolution was proposed by Fukushima in 1980. He called it neocognitron [7]. The modern concept of convolutional neural networks (CNNs) was developed for handwriting recognition by LeCun [8] in 1998. He called it LeNet. A. Krizhevsky developed AlexNet to make faster training of CNN [9] in 2017. Now, CNNs are widely used for image recognition such as object detection for automated vehicles, face recognition, medical image analysis, and so on. CNNs take into account spatial structure of inputs by using convolution operation and enable us to have much fewer parameters. This approach allows us to train a huge data more efficiently even if the performance is slightly degraded. When we have two continuous functions $f(t)$ and $g(t)$, the convolution operation is defined as follows:

$$(f * g)(t) = \int_{-\infty}^{\infty} f(t - \tau)g(\tau)d\tau = \int_{-\infty}^{\infty} f(\tau)g(t - \tau)d\tau. \tag{6.33}$$

As we can observe (6.33), they are commutative. When they are discrete functions $f[n]$ and $g[n]$, the convolution operation is defined as follows:

$$(f * g)[n] = \sum_{m=-\infty}^{\infty} f[n - m]g[m] = \sum_{m=-\infty}^{\infty} f[m]g[n - m] \tag{6.34}$$

and if the discrete function $g[n]$ has a finite set $\{-M, \ldots, M\}$, (6.34) is rewritten as follows:

$$(f * g)[n] = \sum_{m=-M}^{M} f[n-m]g[m].$$ (6.35)

The discrete convolution operation can be regarded as matrix multiplication. We call this discrete function $g[n]$ a kernel function. The function $f[n]$ can be an input. The kernel will be learned in CNN. If we have concatenation of multiple kernels, we call it a filter. Basically, the dimension of a filter is one more than the dimension of a kernel. For example, if a kernel has height x width dimension and an input has depth dimension, the filter has height x width x depth dimension. When the functions are multidimensional, the convolution operation can be defined in a similar way. For example, when $f[x, y]$ and $g[x, y]$ are two-dimensional functions with finite sets $\{-N, \ldots, N\}$ and $\{-M, \ldots, M\}$, the convolution operation is defined as follows:

$$(f * g)(x, y) = \sum_{m=-M}^{M} \sum_{n=-N}^{N} f[x-n, y-m]g[n, m].$$ (6.36)

For example, if we have an input $\mathbf{X}$ and kernel function $\mathbf{K}$ as follows:

$$\mathbf{X} = \begin{bmatrix} x_1 & x_2 & x_3 & x_4 \\ x_5 & x_6 & x_7 & x_8 \\ x_9 & x_{10} & x_{11} & x_{12} \end{bmatrix} \text{ and } \mathbf{K} = \begin{bmatrix} k_1 & k_2 \\ k_3 & k_4 \end{bmatrix},$$ (6.37)

we perform the convolution operation and the output $\mathbf{Y}$ is as follows:

$$\mathbf{Y} = \begin{bmatrix} x_1 k_1 + x_2 k_2 & x_2 k_1 + x_3 k_2 & x_3 k_1 + x_4 k_2 \\ + x_5 k_3 + x_6 k_4 & + x_6 k_3 + x_7 k_4 & + x_7 k_3 + x_8 k_4 \\ x_5 k_1 + x_6 k_2 & x_6 k_1 + x_7 k_2 & x_7 k_1 + x_8 k_2 \\ + x_9 k_3 + x_{10} k_4 & + x_{10} k_3 + x_{11} k_4 & + x_{11} k_3 + x_{12} k_4 \end{bmatrix}$$ (6.38)

Figure 6.25 illustrates convolution operation.

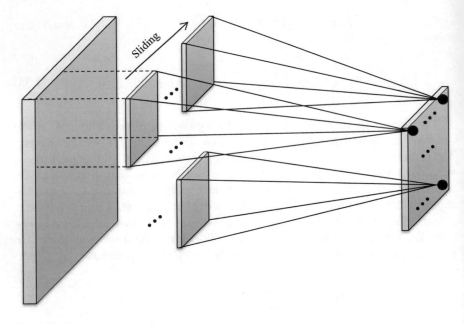

Fig. 6.25 Convolution operation

Example 6.7 Convolution Operation Consider the following input and kernel function:

$$\mathbf{X} = \begin{bmatrix} 1\ 1\ \ 1\ 0 \\ 0\ 1\ \ 1\ 1 \\ 0\ 0\ \ 1\ 1 \end{bmatrix} \text{ and } \mathbf{K} = \begin{bmatrix} 1\ 0 \\ 1\ 1 \end{bmatrix}.$$

Perform convolution operation and find the output **Y**.

Solution

From (6.38), we can obtain the first element y_1 of output **Y** as follows:

$$y_1 = x_1 k_1 + x_2 k_2 + x_5 k_3 + x_6 k_4 = 1 \cdot 1 + 1 \cdot 0 + 0 \cdot 1 + 1 \cdot 1 = 2$$

Likewise, we obtain the output **Y** as follows:

$$\mathbf{Y} = \begin{bmatrix} 2\ 3\ 3 \\ 0\ 2\ 3 \end{bmatrix}.$$

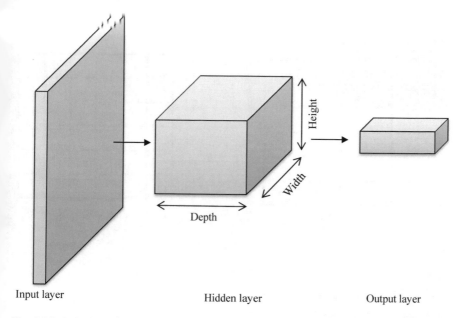

Input layer   Hidden layer   Output layer

Fig. 6.26 CNN architecture

CNN architecture is similar to conventional neural network. However, since it is specialized in image recognition, it takes advantage of the fact that the inputs are images. The CNNs arrange neurons in three dimensions: width, height, and depth (Sometimes, depth is called as channel.) where the depth is not the hidden layer depth of neural networks. As we can observe Fig. 6.26, each layer transforms the three-dimensional input volume to the three-dimensional output volume of neuron activations. The width and height would be the two-dimensional image, and the depth would be the colour channels such as red, green, and blue (RGB). This is very helpful for reducing complexity. For example, when we have $1024 \times 1024 \times 1$ image (which means that the image as a black and white image contains 1024×1024 two-dimensional image and 1 black and white channel), the total number of neurons in the input layer is about 1 million ($=1024 \times 1024$). The huge number of neurons is required for each operation. This is computationally inefficient. If we use CNN, it enables us to extract the feature of the image and transform to lower dimension with a slight degradation. In convolutional layer, there are 4 key parameters (kernel size, depth (or channel), stride, and zero-padding) to determine the output volume size. Among them, the stride represents how much we move the filter horizontally or vertically. Figure 6.27 illustrates an example of the strides when the input is 8×8 and the kernel is 3×3.

The zero-padding represents how many zeros we will add at the boarders. It is a process of adding zeros to the input volume symmetrically and is used to maintain a constant size. This is very useful for controlling the dimensionality of the

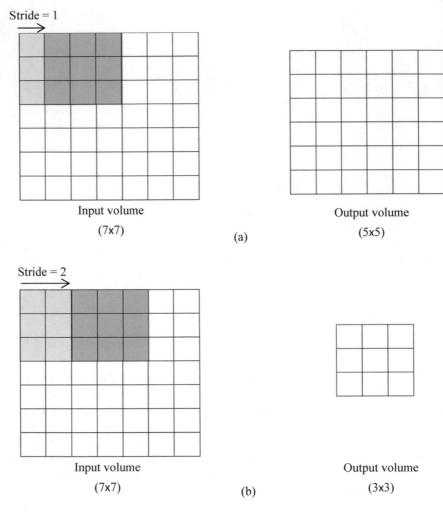

Fig. 6.27 Example of stride size $= 1$ (**a**) and 2 (**b**)

output volume. The depth represents the number of filters to search for a different characteristic. The output volume size is determined by the following equation:

$$O = \frac{2P + I - K}{S} + 1 \tag{6.39}$$

where O, P, I, K, and S represent the output volume, the zero-padding size, the input volume, the filter size, and the stride, respectively.

Example 6.8 Spatial Arrangement of Convolution Layer Consider the following input **I**, kernel **K**, zero padding P, and stride S:

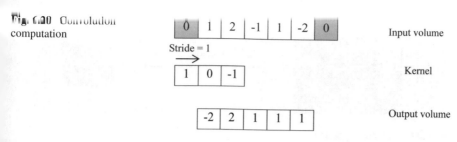

Fig. 6.20 Convolution computation

Input volume

Stride = 1

Kernel

Output volume

$$\mathbf{I} = [1\ 2\ -1\ 1\ -2], \mathbf{K} = [1\ 0\ -1], P = 1, S = 1.$$

Compute the output volume size and find the output.

Solution

From (6.39), we can compute the output volume as follows:

$$O = \frac{2P + I - K}{S} + 1 = \frac{2 \cdot 1 + 5 - 3}{1} + 1 = 5.$$

Figure 6.28 illustrates the process of convolution process. ∎

A CNN is composed of four different layers: convolution, activation (ReLU), pooling, and fully connected. Convolutional layer plays important role in CNNs. It is composed of a set of learnable filter sliding over the input. They are applied to local regions. Each filter is spatially small but spread along with the full depth of the input volume where the depth will learn different features. Each filter always extends the full depth of the input volume. For example, if we have 6 filters, 6 separate activation maps are created. As we can observe Figs. 6.25, 6.26, and 6.27, the kernel slides over the input and we can obtain a specific output at a given spatial position. The kernels produce the corresponding activation maps. They will be stacked along the depth to determine the output volume. The weight (or parameter) in the kernels may start with a random value and be learned. When a convolutional layer $l-1$ is connected to a convolution layer l with the depth D_{l-1} and D_l, respectively, and we have a filter size K, the associated weights W to each neuron for a convolutional layer can be calculated as follows:

$$W = K \cdot K \cdot D_{l-1} \cdot D_l + D_l \tag{6.40}$$

The weights can be shared to reduce the number of the weights. The weights sharing is based on one assumption: If one region feature is useful to compute at some spatial region, it would be useful to compute at a different region as well. Therefore, we can constrain an individual activation map within the output to same weights and reduce the number of parameters in the convolutional layer. This approach enables us to improve memory efficiency and computation speed. Typically, we have millions of input pixel values. If we use convolution techniques, we can extract small size of meaningful features with kernels. The convolutional layer creates nonlinearity in

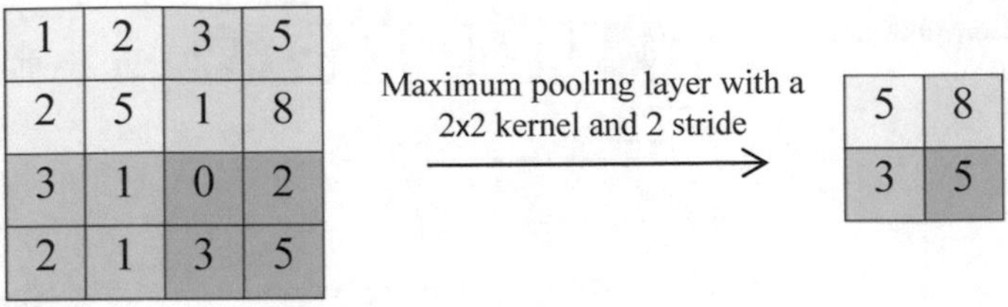

1	2	3	5
2	5	1	8
3	1	0	2
2	1	3	5

Maximum pooling layer with a
2x2 kernel and 2 stride
$\longrightarrow$

5	8
3	5

Fig. 6.29 Example of pooling

the neural network. The activation layer (also known as rectified linear units (ReLU) layer) allows the neural network to learn harder decision functions and reduce the overfitting. The activation makes all negative values to zero using the following function

$$f(x) = \begin{cases} 0, & \text{when } x < 0 \\ x, & \text{when } x \geq 0 \end{cases} \tag{6.41}$$

The pooling layer is also known as a sub-sampling layer. Its aim is to downsample the output of a convolutional layer and gradually reduce the spatial size of the representation. In the pooling layer, we perform the average or maximum function on each depth slice of the input and reduce the dimensionality. The maximum pooling layer with a 2×2 kernel and 2 stride is widely used. It enables us to scale down to 25% of the input size and maintains the depth. Figure 6.29 illustrates an example of maximum pooling.

In the layers of convolution, activation (ReLU), and pooling, the beat features are extracted from the input volume. In the fully connected layer, we learn how to classify the features. The fully connected layer is analogous to conventional neural networks. The neurons in the fully connected layer are fully connected to all activations in the previous layer. In addition to them, softmax (or logistic) layer, normalization layers, and dropout layers can be added. Softmax layer is located in the end of fully connected layer. Softmax and logistic layers are used for multi-classification and binary classification, respectively. The softmax function turns a vector of K real values into a vector of K real values whose summation is 1. The softmax function transforms the input values into the value between 0 and 1. The softmax function can be expressed as follows:

$$\text{softmax}(x_i) = \frac{e^{x_i}}{\sum_j e^{x_j}}. \tag{6.42}$$

There are many types of normalization layers. One approach is to compute the average over all images for each red, green, and blue channels and then subtract the value to each channel of the image. The dropout layer is used to reduce overfitting and drops out units in a neural network. It can be added in any hidden layers or input layers. The dropout layer randomly sets input units to zero during training. Thus, the

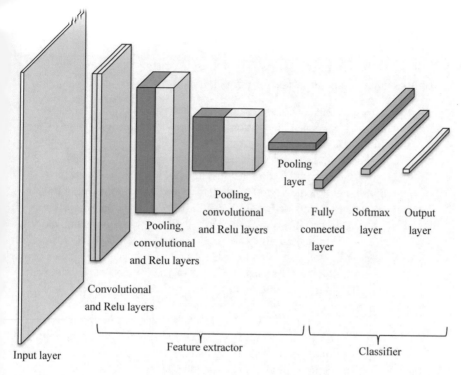

Fig. 6.30 Common CNN architecture example

common process of CNN is summarized as follows: input layer → Convolution + ReLU layer → Pooling layer → .... → Convolution + ReLU layer → Pooling layer → Fully connected layer → Softmax layer → Output layer. Figure 6.30 illustrates the common CNN architecture example.

As we can observe Fig. 6.30, the spatial resolution is reduced and the number of map is increased as going deeper layer. We extract high-level features from the input and classify them in fully connected layers. Based on common CNN architecture, there are many variations and improvements of CNNs. In [10], ZFNet as the improved version of AlexNet [9] was proposed to find out what feature maps search. It expands the size of the middle convolutional layers and reduces the stride and filter size in the first layer. Thus, it obtains less parse connection between layers than AlexNet. In [11], GoogLeNet was proposed to approximate more complex functions while it is robust to overfitting. GoogLeNet developed an inception module and dramatically reduced the number of parameters in the neural network. The basic concept of the inception module is to run multiple convolutional and pooling operations with multiple filter sizes in parallel. Parallel convolutions of different sizes are used to learn objects of different sizes and positions. In [12], different types of CNNs are summarized. CNNs can be categorized into seven types: spatial exploitation-based CNNS, depth-based CNNS, multi-path-based CNNs, width-based multi-connection CNNs, feature map exploitation-based CNNs, channel exploitation-based CNNs, and attention-based CNNs [12].

Summary 6.3. Convolutional Neural Network

1. The convolutional neural networks (CNNs) take into account spatial structure of inputs by using convolution operation and enable us to have much fewer parameters.
2. The CNNs arranges neurons in 3 dimensions: width, height and depth. each layer transforms the 3 dimensional input volume to the 3 dimensional output volume of neuron activations. The width and height would be the 2 dimensional image and the depth would be the colour channels such as red, green and blue (RGB). This is very helpful for reducing complexity.
3. A CNN is composed of four different layers: convolution, activation (Relu), pooling and fully connected. Convolutional layer plays important role in CNNs. It is composed of a set of learnable filter sliding over the input. The activation layer allows the neural network to learn harder decision functions and reduce the overfitting. The pooling layer is also known as a sub-sampling layer. Its aim is to downsample the output of a convolutional layer and gradually reduce the spatial size of the representation. The fully connected layer is analogous to conventional neural networks. The neurons in the fully connected layer are fully connected to all activations in the previous layer.
4. The common process of CNN is summarized as follows: input layer $\rightarrow$ $\rightarrow$ Convolution + Relu layer $\rightarrow$ $\rightarrow$ Pooling layer $\rightarrow$ $\rightarrow$ .... $\rightarrow$ $\rightarrow$ Convolution + Relu layer $\rightarrow$ $\rightarrow$ Pooling layer $\rightarrow$ $\rightarrow$ Fully connected layer $\rightarrow$ $\rightarrow$ Softmax layer $\rightarrow$ $\rightarrow$ Output layer.

6.4 Recurrent Neural Network

Recurrent neural networks (RNNs) are neural networks with feedbacks allowing information to persist based on Rumelhart's work in 1986 [3]. RNNs are developed to recognize sequential characteristics of data and use the sequential data or time series data to solve common temporal problems in speech recognition, natural language processing, and others. Feedforward neural networks can't take time dependencies into account and can't deal with variable length inputs. For example, a feedforward neural network classifies three word sentences such as "I love you". If we receive four word sentences such as "I love you too.", it is not easy to classify them even if it is just the slight modification. One approach is to drop one word from the sentence and match the input size. However, it might lose the meaning. It is not suitable for this application. RNNs allow us to process arbitrary sequences of inputs. They have dynamic hidden states that can store a long-term information. The internal state of the RNNs enables us to have internal memory and exploit the dynamic behaviours. RNNs are very powerful in a sequential pattern because they employ key properties such as temporal and accumulative, distributed hidden states, and nonlinear dynamics. The input and output of RNNs could have different structures depending on various input and output pairs such as one-to-many, many-to-one, and many-to-many. One-to-many RNNs are used for generative models such as generating text or music, drawing an image, and so on. Many-to-one RNNs are used for binary classification such as yes or no grammatical correction of a sentence and so on. Many-to-many RNNs can be used for image or text translations. If we want to translate one English sentence to another language sentence, the length of words between two languages may not be equal. When we have an input vector $\mathbf{x} = [x_1, x_2, \ldots, x_n]$ and we should receive n times at each time instead of receiving n inputs at a time, RNNs are very useful because it reads the input sequentially and the input is processed in sequential order. The basic idea of RNNs is similar to CNNs. They both use parameter sharing. CNNs share the parameters across spatial dimension on pixel image data while RNNs share the parameters across temporal dimension on speech or text data. We can obtain gradients by concatenation because the parameters among hidden states are shared. However, as we have a long-time sequence and the depth grows, RNNs are in trouble about long-term dependencies. We call this gradient exploring and vanishing problem. In order to solve this problem, long short-term memory (LSTM) is widely adopted. LSTMs are an extension of RNNs. In addition, the computation of CNNs is fixed because the input size is fixed while RNNs have a variable number of computations and the hidden layers and the outputs rely on the previous states of the hidden layers. The main advantages of RNNs can be summarized as follows: (1) variable input length, (2) model complexity not depending on the input size, and (3) shared computation across time. The disadvantages of RNNs are (1) slow computation and (2) gradient vanishing problem. Figure 6.31 illustrates the comparison of CNNs and RNNs.

One simple structure of RNNs is an MLP including the previous data set of hidden states feeding back into the neural network along with the inputs. Thus, RNNs introduce cycles and time step in their structure as shown in Fig. 6.32.

In Fig. 6.32a, a RNN has a recurrence structure to operate over a sequence of vectors. It processes the input sequence $\mathbf{x}_t$ at time step t and produces the output sequence $\mathbf{y}_t$ at time step t. $\mathbf{h}_t$ is internal representation (hidden states or memories) at time step t. They all are real-valued vectors. The time step depends on the operation of neurons. RNNs can be unrolled across multiple time steps and converted into a feedforward neural network as shown in Fig. 6.32b. RNNs are regarding as multiple copies of the same neural network passing information to a successor. The length of the unrolling RNN depends on the input vector size. RNNs are recurrent because the same functions and parameters are used at each time step and the output of the current input relies on the past computation. We can represent discrete, time-independent difference equations of the RNN state $\mathbf{h}_t$ and output $\hat{\mathbf{y}}_t$ as follows:

$$\mathbf{h}_t = f(\mathbf{U}\mathbf{x}_t + \mathbf{W}\mathbf{h}_{t-1} + \mathbf{b}), \tag{6.43}$$

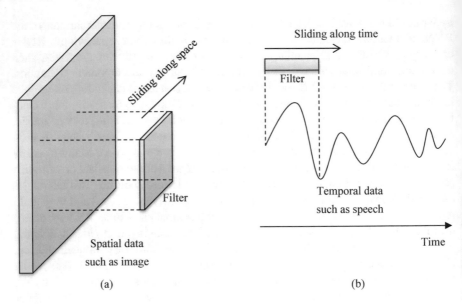

Fig. 6.31 Comparison of CNNs (**a**) and RNNs (**b**)

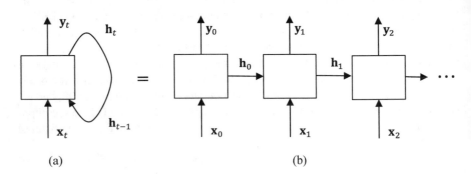

Fig. 6.32 RNN structure (**a**) and unrolled RNN (**b**)

and

$$\hat{\mathbf{y}}_t = g(\mathbf{V}\mathbf{h}_t + \mathbf{c}) \qquad (6.44)$$

where **U**, **W**, and **V** represent input hidden weight matrix transforming the current input, hidden-hidden weight matrix transforming the previous hidden state, and hidden-output weight matrix transforming the new activated hidden state, respectively. **b** and **c** are bias vectors. $f()$ is the activation function (e.g. tanh() or sigmoid()) implemented by each neuron. Basically, it is same nonlinear function for all neurons. $g()$ is output transformation function. It can be any function such as softmax and sigmoid and provided by the internal processing. We consider one simple example.

In the example, the model is trained. The initial hidden state $\mathbf{h}_0$ is randomly given. When the input $\mathbf{x}_1$ is given, we make a prediction. Figure 6.33 illustrates an example of RNN computation.

As we can observe Fig. 6.33, we compute the hidden state $\mathbf{h}_1$ and make a prediction $\hat{\mathbf{y}}_1$. At time step 2, the prediction $\hat{\mathbf{y}}_1$ from the previous time step is considered as the input $\mathbf{x}_2$. In the same way, we make a prediction at each time step. The weight matrices $\mathbf{U}$, $\mathbf{W}$, and $\mathbf{V}$ are shared in the computation. There are no probabilistic components. In the deep neural networks, we typically assume that each input is independent. However, RNNs deal with sequential data. Thus, the input in the current time step depends on the previous time step. As we performed training in DNN and CNN, the parameters of RNNs can be trained with gradient descent. A loss function can be defined as the error and regularization term of the training data set as follows:

$$L_t = E_t(\mathbf{y}_t, \hat{\mathbf{y}}_t) + \lambda R \tag{6.45}$$

where $E_t()$, $\mathbf{y}_t$ and $\hat{\mathbf{y}}_t$ are the error function, true target, and predicted output, respectively. λ and R are the regularization parameter and the regularization term, respectively. Multiple loss functions can be used to train a neural network. The selection of loss function directly affects to the activation functions in the output layer. Typically, when solving a regression problem, we deal with a real value quantity. Mean-squared error (MSE) is used for the loss function, and output layer includes a linear activation unit. When we have a binary classification problem, we predict the likelihood 0 or 1 of the training data. Binary cross-entropy is used for the loss function, and sigmoid activation is used at output layer. If we solve multi-class classification problem, we deal with training data set belonging to multiple classes. Cross-entropy is used for the loss function, and softmax activation is used at output layer. The loss function is simply the sum of losses over all time steps for a training data set. The loss function as cross-entropy without regularization can be expressed as follows:

$$E_t(\mathbf{y}_t, \hat{\mathbf{y}}_t) = -\mathbf{y}_t \log \hat{\mathbf{y}}_t, \tag{6.46}$$

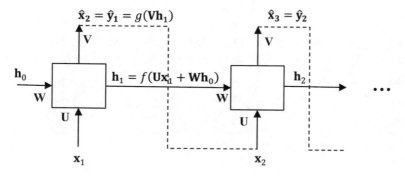

Fig. 6.33 RNN computation

$$E = \sum_t E_t\left(\mathbf{y}_t, \hat{\mathbf{y}}_t\right) = -\sum_t \mathbf{y}_t \log \hat{\mathbf{y}}_t. \tag{6.47}$$

Training the RNNs is straightforward. We don't need a new algorithm. We can apply the backpropagation algorithm to the unrolled RNN. We call it backpropagation through time (BPTT). The training data for RNNs is an ordered sequence. We redraw the unrolled RNN including the loss as shown in Fig. 6.34

The RNN is regarded as a layered feed-forward neural network with shared weights. We can train the layered feed-forward neural network with weight constraints. As we can observe Fig. 6.34, each neural network in the unrolled RNN is relative to a different time interval and the weight parameters $\mathbf{U}$, $\mathbf{V}$, and $\mathbf{W}$ are constrained to be same. The feedforward depth means the longest path from the inputs to the outputs at the same time step in the RNN. The recurrent depth means the longest path between same hidden state in successive time step. In order to train the neural network, as we obtained the gradients using chain rule and matrix derivatives in deep neural network, we compute gradients of the loss with respect to the weight parameters $\mathbf{U}$, $\mathbf{V}$, and $\mathbf{W}$ and learn them using stochastic gradient descent. For each node, we need to compute the gradient iteratively. Suppose that we have the following vanilla RNN with

$$\mathbf{h}_t = \tanh(\mathbf{U}\mathbf{x}_t + \mathbf{W}\mathbf{h}_{t-1}) = \tanh(\mathbf{z}_t) \tag{6.48}$$

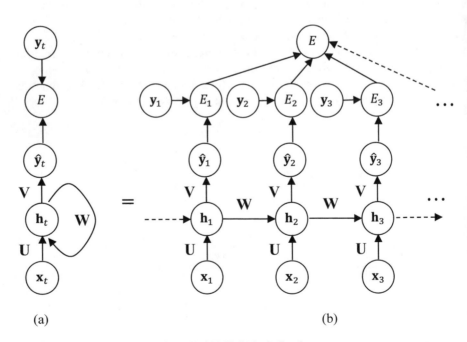

(a)   (b)

Fig. 6.34 RNN structure (**a**) and unrolled RNN (**b**) including loss

where

$$\mathbf{z}_t = \mathbf{U}\mathbf{x}_t + \mathbf{W}\mathbf{h}_{t-1} \tag{6.49}$$

and

$$\tanh(x) = \frac{\sinh(x)}{\cosh(x)} = \frac{e^{2x} - 1}{e^{2x} + 1} \text{ and } \frac{d \tanh(x)}{dx} = 1 - \tanh^2(x). \tag{6.50}$$

The term "Vanilla" represents a simple or regular neural network. The prediction $\hat{\mathbf{y}}_t$ at the time step t is

$$\hat{\mathbf{y}}_t = \text{softmax}(\mathbf{V}\mathbf{h}_t) = \text{softmax}(\mathbf{q}_t) \tag{6.51}$$

where

$$\mathbf{q}_t = \mathbf{V}\mathbf{h}_t. \tag{6.52}$$

We have $\mathbf{x}_t \in \mathbb{R}^N, \hat{\mathbf{y}}_t \in \mathbb{R}^N, \mathbf{h}_t \in \mathbb{R}^M, \mathbf{U} \in \mathbb{R}^{M \times N}, \mathbf{V} \in \mathbb{R}^{N \times M}, \text{and } \mathbf{W} \in \mathbb{R}^{M \times M}$. We assume that the varying length for a sequential training data, the weight parameters at each time step are shared in the computation, and the loss function is defined in (6.46) and (6.47). The loss function uses the dot products between $\mathbf{y}_t$ and elementwise logarithm of $\hat{\mathbf{y}}_t$. By taking the derivative with respect to $\mathbf{V}$, we have

$$\frac{\partial E_t}{\partial V_{ij}} = \frac{\partial E_t}{\partial \hat{y}_{t_k}} \frac{\partial \hat{y}_{t_k}}{\partial q_{t_l}} \frac{\partial q_{t_l}}{\partial V_{ij}}. \tag{6.53}$$

The first term in the right-hand side of (6.53) is calculated as follows:

$$\frac{\partial E_t}{\partial \hat{y}_{t_k}} = \frac{\partial}{\partial \hat{y}_{t_k}} \left(-y_{t_k} \log \hat{y}_{t_k} \right) = -\frac{y_{t_k}}{\hat{y}_{t_k}}. \tag{6.54}$$

The second term in the right-hand side of (6.53) is calculated as follows:

$$\frac{\partial \hat{y}_{t_k}}{\partial q_{t_l}} = \frac{\partial}{\partial q_{t_l}} \left(\text{softmax}(\hat{y}_{t_k}) \right) = \begin{cases} -\hat{y}_{t_k} \hat{y}_{t_l} & \text{when } k \neq l \\ \hat{y}_{t_k}(1 - \hat{y}_{t_k}) & \text{when } k = l \end{cases}. \tag{6.55}$$

From (6.54) and (6.55), we obtain

$$\begin{aligned}
\frac{\partial E_t}{\partial \hat{y}_{t_k}} \frac{\partial \hat{y}_{t_k}}{\partial q_{t_l}} &= \sum_{k \neq l} \left(-\frac{y_{t_k}}{\hat{y}_{t_k}} \right)(-\hat{y}_{t_k} \hat{y}_{t_l}) + \left(-\frac{y_{t_l}}{\hat{y}_{t_l}} \right)(\hat{y}_{t_l}(1 - \hat{y}_{t_l})) \\
&= \sum_{k \neq l} (y_{t_k} \hat{y}_{t_l}) - y_{t_l} + y_{t_l} \hat{y}_{t_l} = -y_{t_l} + \hat{y}_{t_l} \sum_k (y_{t_k}).
\end{aligned} \tag{6.56}$$

The prediction is based on softmax function. The sum is equal to 1. Therefore, we have

$$\frac{\partial E_t}{\partial \hat{y}_{t_k}} \frac{\partial \hat{y}_{t_k}}{\partial q_{t_l}} = -y_{t_l} + \hat{y}_{t_l}.$$ (6.57)

The third term in the right-hand side of (6.53) is calculated as

$$\frac{\partial q_{t_l}}{\partial V_{ij}} = \frac{\partial}{\partial V_{ij}} \left(V_{lm} h_{t_m} \right) = \delta_{il} h_{t_j}$$ (6.58)

where δ is the delta function. From (6.57) and (6.58), we have

$$\frac{\partial E_t}{\partial V_{ij}} = \left(\hat{y}_{t_i} - y_{t_i} \right) h_{t_j}$$ (6.59)

and

$$\frac{\partial E_t}{\partial \mathbf{V}} = \left(\hat{\mathbf{y}}_t - \mathbf{y}_t \right) \otimes \mathbf{h}_t$$ (6.60)

where $\otimes$ is the outer product. By taking the derivative with respect to $\mathbf{W}$, we have the following chain rule:

$$\frac{\partial E_t}{\partial W_{ij}} = \frac{\partial E_t}{\partial \hat{y}_{t_k}} \frac{\partial \hat{y}_{t_k}}{\partial q_{t_l}} \frac{\partial q_{t_l}}{\partial h_{t_m}} \frac{\partial h_{t_m}}{\partial W_{ij}}.$$ (6.61)

The first two terms in the right-hand side of (6.61) are already computed. The third term in the right-hand side of (6.61) is

$$\frac{\partial q_{t_l}}{\partial h_{t_m}} = V_{lm}.$$ (6.62)

In order to compute the fourth term in the right-hand side of (6.61), we should consider the hidden state h_t partially depending on h_{t-1}. We have

$$\frac{\partial h_{t_m}}{\partial w_{ij}} = \sum_{k=0}^{t} \frac{\partial h_{t_m}}{\partial h_{k_d}} \frac{\partial h_{k_d}}{\partial W_{ij}}$$ (6.63)

where k is less than t and d is dummy index. From (6.57), (6.62), and (6.3), we have

$$\frac{\partial E_t}{\partial W_{ij}} = \left(\hat{y}_{t_l} - y_{t_l} \right) V_{lm} \sum_{k=0}^{t} \frac{\partial h_{t_m}}{\partial h_{k_d}} \frac{\partial h_{k_d}}{\partial W_{ij}}.$$ (6.64)

In similar way to the gradient of $\mathbf{W}$, we take the derivative with respect to $\mathbf{U}$ and have

$$\frac{\partial E_t}{\partial U_{ij}} = \frac{\partial E_t}{\partial \hat{y}_{t_k}} \frac{\partial \hat{y}_{t_k}}{\partial q_{t_l}} \frac{\partial q_{t_l}}{\partial h_{t_m}} \frac{\partial h_{t_m}}{\partial U_{ij}} = \left(\hat{y}_{t_l} - y_{t_l}\right) V_{lm} \sum_{k=0}^{t} \frac{\partial h_{t_m}}{\partial h_{k_d}} \frac{\partial h_{k_d}}{\partial U_{ij}}. \tag{6.65}$$

As we sum up the loss, we sum up the gradients for one training data, which are computed over all relevant time steps. Finally, we can update the weight matrices as follows:

$$\mathbf{V}_{\text{new}} = \mathbf{V}_{\text{old}} - \alpha \frac{\partial E}{\partial \mathbf{V}}, \tag{6.66}$$

$$\mathbf{W}_{\text{new}} = \mathbf{W}_{\text{old}} - \alpha \frac{\partial E}{\partial \mathbf{W}}, \tag{6.67}$$

and

$$\mathbf{U}_{\text{new}} = \mathbf{U}_{\text{old}} - \alpha \frac{\partial E}{\partial \mathbf{U}} \tag{6.68}$$

where α is the learning rate. The optimal value of the learning rate can be found by cross-validation. There are several methods to improve learning rates, such as annealing of the learning rate, second-order methods, adaptive learning rates, and so on. The weights occur at each time step. Each path from the weight to the loss includes both an implicit dependence and a direct dependence on the weights. Thus, we should find all paths from each weight to the loss. For example, if we find paths from $\mathbf{W}$ to E through E_3 in Fig. 6.35, there are two paths: one from $\mathbf{h}_1$ and the other one from $\mathbf{h}_2$.

As we can observe the gradients of the loss function with respect to $\mathbf{V}$, $\mathbf{W}$, and $\mathbf{U}$, the computation cost is expensive. We can't reduce the run time by parallelization. The memory cost is expensive as well because the parameters must be stored until we reuse them. Thus, RNNs are powerful technique in sequential data but training cost is expensive. In addition, as we briefly discussed, RNNs have gradient exploring and vanishing problem. For example, we have many hidden layers and backpropagate through many layers. If the weights are big, the gradients will grow exponentially as we propagate and we call this exploding gradient problem. Due to the exploding gradient problem, the RNN will have a big change in the loss at each update and can't learn much on the training. The RNN becomes unstable. On the other hands, if the weights are small, the gradient will decrease exponentially as we propagate and we call this vanishing gradient problem. Due to the vanishing gradient problem, the RNN improves very slowly in the training. The weights nearby the output layer affect to the change significantly while the weights close to the input layer couldn't change much. Even the weights may become 0 in the training. In the RNN, we basically deal with a long data sequence and the gradients are easily exploding or vanishing. In order to avoid them, we should carefully initialize the weights. Although we have good initial

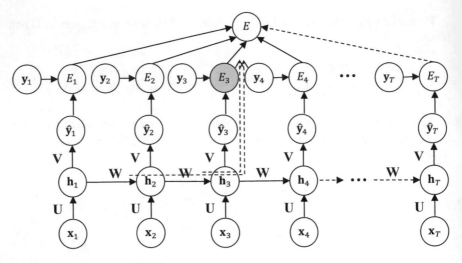

Fig. 6.35 Possible paths of RNN

weights, it is very difficult to manage the long-range dependencies. Long short-term memory networks (LSTMs) were proposed by Hochreiter and Schmidhuber [13]. If we have an extremely small gradient value, it doesn't contribute to the learning process of the RNN. Thus, we can stop updating if a layer detects the small gradient value and determine not to affect the learning process. The basic concept of LSTMs is to regulate the information flow using some internal gate mechanisms. LSTMs suppress vanishing gradients through a gating mechanism and provide us with great performance. They introduce a memory unit (or cell) and handle long sequential well while considering the gradient vanishing problem. In order to capture short-term and long-term memories, LSTMs contain a cell with an input gate, an output gate, and a forget gate. The contents of cells are regulated by the gates. The gates are implemented by a sigmoid activation function and a pointwise multiplication. Figure 6.36 illustrates the LSTM unit.

As we can observe Fig. 6.36, the cell keeps values over arbitrary time intervals so that we can control the information flow along with the cells. The information can be written or removed to the cell regulated by gates. The state of a cell at time step t is $\mathbf{c}_t$. The gates are implemented by sigmoid function and pointwise operation. The output of sigmoid function generates 0 (remove the information) and 1 (keep the information). We can determine how much of each component should pass through. The pointwise multiplication applies the decision. The forget gate determines which old data we remove. As we can observe forget gate in Fig. 6.36, the forget gate concatenates the previous hidden states and the input and then applies sigmoid function $\sigma()$ as follows:

$$\boldsymbol{f}_t = \sigma\left(\mathbf{W}_f\left[\mathbf{h}_{t-1}, \mathbf{x}_t\right] + \mathbf{b}_f\right) = \begin{cases} 0, & \text{Remove the content in } \mathbf{c}_t \\ 1, & \text{Keep the content in } \mathbf{c}_t \end{cases}. \tag{6.69}$$

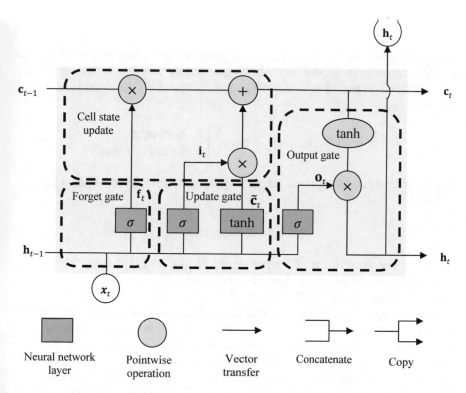

Fig. 6.36 LSTM unit

The forget gate is controlled by the current input $\mathbf{x}_t$ and the previous cell output $\mathbf{h}_{t-1}$. In the update gate, we determine how much we update the old state $\mathbf{c}_{t-1}$ with a new candidate state $\tilde{\mathbf{c}}_t$. The information enters in the cell whenever the input gate is on. The input acts like a gatekeeper to determine which information enter into the cell. There are two types of input gates: input gate and input modulation gate. The input gate determines the extent of information to be written in the cell state. The input modulation gate modulates the information by adding nonlinearity and making the information zero-mean via the activation function. It enables us to reduce the run time as zero-mean input converges quickly. The input gate $\mathbf{i}_t$ is expressed as follows:

$$\mathbf{i}_t = \sigma\left(\mathbf{W}_i\left[\mathbf{h}_{t-1}, \mathbf{x}_t\right] + \mathbf{b}_i\right) = \begin{cases} 0, & \text{No update} \\ 1, & \text{Update} \end{cases} \tag{6.70}$$

and the input modulation gate is expressed as follows:

$$\tilde{\mathbf{c}}_t = \tanh\left(\mathbf{W}_c\left[\mathbf{h}_{t-1}, \mathbf{x}_t\right] + \mathbf{b}_c\right) \tag{6.71}$$

where $\tilde{\mathbf{c}}_t$ is a new candidate cell state. Now, we can update the cell state as follows:

$$\mathbf{c}_t = \mathbf{f}_t * \mathbf{c}_{t-1} + \mathbf{i}_t * \tilde{\mathbf{c}}_t \tag{6.72}$$

where the operation * is the elementwise product. As we can observe (6.72), the previous cell state is updated by the input gate and the forget gate. The output gate determines which part of cell state will be in the output from the current cell state as follows:

$$\mathbf{o}_t = \sigma\left(\mathbf{W}_o\left[\mathbf{h}_{t-1}, \mathbf{x}_t\right] + \mathbf{b}_o\right) = \begin{cases} 0, & \text{No output} \\ 1, & \text{Return cell state} \end{cases} \tag{6.73}$$

and the hidden state as the final output is generated from $\tanh(\mathbf{c}_t)$ filtered by $\mathbf{o}_t$ as follows:

$$\mathbf{h}_t = \mathbf{o}_t * \tanh(\mathbf{c}_t). \tag{6.74}$$

As we can observe Fig. 6.36 and (6.69)–(6.74), LSTM decides whether the information should be kept or forgotten and controls the gradients values at each time step. It allows to neural network to have desired behaviour from the error gradient by updating the learning process at each time step. RNN has emerged in many applications such as speech recognition and language translation with great prospects. LSTMs are widely used because they solve the vanishing gradients problem. However, LSTMs have their disadvantages. For example, the data should move in the cell for evaluation. The additional process increases the complexity. The data transmission to the cells results in more bandwidth and memory usages. LSTMs are easy to overfit. It is not simple to apply the dropout algorithm to reduce the overfitting problem. Thus, many variants of LSTMs have been developed to improve the performance and simplify the cell structure. RNN with peephole connection was proposed by Gers and Schmidhuber [14]. This structure allows the gates to know about the cell state and inspect the current cell state. Gated recurrent unit (GRU) [15] improved the complexity by reducing the number of gates. The basic approach is to combine the forget gate and the input gate into an update gate and merge the cell and the hidden state.

Example 6.9 LSTM Consider that a city reports COVID-19 cases for 450 days. Fig. 6.37 illustrates COVID-19 daily cases.
    Predict the next 50 days cases using the LSTM network.

Solution

Firstly, we need to train the LSTM network using the 450 days cases training data. The training data includes the new cases and a single time series. In order to avoid from diverging, we perform data standardization using the following equation:

$$x_{\text{stand}} = \frac{x - \mu}{\sigma}$$

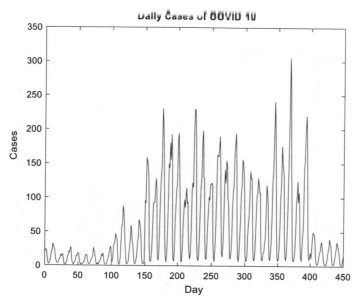

Fig. 6.37 COVID-19 daily cases

where x, μ, and σ are the input data, the mean of the inputs, and standard deviation, respectively. The absolute value of x_{stand} represents the distance between the input x and the population mean μ in units of the standard deviation σ. Secondly, we define the LSTM network structure with 200 hidden layers. We consider maximum 100 epochs, 1 gradient threshold, and the initial learning rate 0.005. Using computer simulation, we perform LSTM network and obtain the predicted data about 50 days new cases. Figure 6.38 illustrates prediction for 50 days new cases.

After obtaining the ground true data for 50 days, we compare the prediction with the ground true data using the metric the root mean square error (RMSE). Figure 6.39 illustrates the comparison of the observed data and prediction data for 50 days and RMSE values. ∎

Summary 6.4. Recurrent Neural Network

1. Recurrent neural networks (RNNs) are developed to recognize sequential characteristics of data and use the sequential data or time series data to solve common temporal problems in speech recognition or natural language processing.
2. The input and output of RNNs could have different structures depending on various input and output pairs such as one-to-many, many-to-one or and many-to-many. One-to-many RNNs are used for generative models

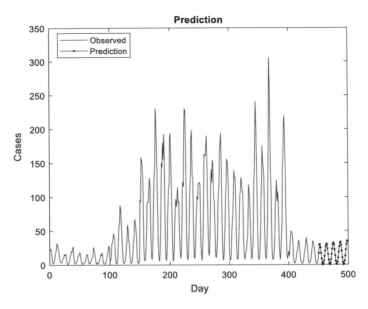

Fig. 6.38 Prediction for 50 days new cases

such as generating text or music, drawing an image and so on. Many-to-one RNNs are used for binary classification such as yes or no grammatical correction of a sentence and so on. Many-to-many RNNs can be used for image or text translations. If we want to translate one English sentence to another language sentence, the length of words between two languages may not be equal.

3. Training the RNNs is straightforward. We can apply the back propagation algorithm to the unrolled RNN. We call it backpropagation through time (BPTT). The training data for RNNs is an ordered sequence.

4. The basic concept of LSTMs is to regulate the information flow using some internal gate mechanisms. LSTMs suppress vanishing gradients through a gating mechanism and provide us with great performance.

5. LSTM decides whether the information should be kept or forgotten and controls the gradients values at each time step. It allows to neural network to have desired behaviour from the error gradient by updating the learning process at each time step.

6.5 Problems

6.1. Describe the process of DNN, CNN, and RNN.

6.2. There are multiple design partners of deep learning algorithms. Discuss how to find an optimal point among them.

6.3 Compare key characteristics of human brain and deep learning.

6.4 Describe the pros and cons of reinforcement learning and deep learning.

6.5 Construct neural networks from single perception and explain the role of each component.

6.6 Describe the characteristics of the activation functions: sigmoid, tanh, and rectified linear unit functions.

6.7 Construct a neural network with one hidden layer for XOR operation in different ways of example 6.8 Consider a multi-layer feed-forward neural network as shown in the below (Fig. 6.40)

The neural network has Tanh activation function and the following input and weight matrices:

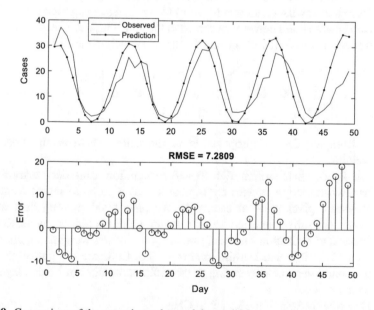

Fig. 6.39 Comparison of the ground true data and the prediction

Fig. 6.40 Neural network
with one hidden layer

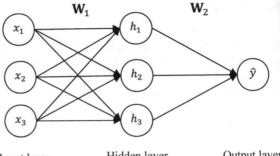

Input layer   Hidden layer   Output layer

$$\mathbf{x} = \begin{bmatrix} 1 \\ 2 \\ -3 \end{bmatrix}, \mathbf{W}_1 = \begin{bmatrix} 2.1 & 3 & 2.3 \\ -2.7 & 3.6 & -2 \\ -0.6 & 2 & -1.5 \end{bmatrix}, \mathbf{W}_2 = \begin{bmatrix} 1.5 \\ 4 \\ -1.5 \end{bmatrix}.$$

Compute the output $\hat{y}$ of the neural network.

6.9   Describe why the chain rule is useful for the backpropagation calculation.

6.10   Consider a neural network as shown in the above figure. The neural network
has tanh activation function, and the input and weight matrices are as follows:

$$\mathbf{x} = \begin{bmatrix} 1 \\ 2 \\ -3 \end{bmatrix}, \mathbf{W}_1 = \begin{bmatrix} 2.1 & 3 & 2.3 \\ -2.7 & 3.6 & -2 \\ -0.6 & 2 & -1.5 \end{bmatrix}, \mathbf{W}_2 = \begin{bmatrix} 1.5 \\ 4 \\ -1.5 \end{bmatrix}, \text{ and } \alpha = 0.5.$$

Compute the backpropagation of the neural network and update the
weights.

6.11   We develop handwritten digit (0 ~ 9) recognition using deep learning. The
handwritten digits images are represented as a 25×25 matrix containing
grayscale pixel value at each cell. We have 5000 training data images.
Consider a fully connected neural network with 1 ~ 4 hidden layers, Rectified
linear unit activation function, randomly initialized weights, regularization
factor $\lambda = 0.1$, and learning rate $\alpha = 0.5$. Compare the performances of
the neural networks according to the different number of hidden layers and
neural nodes.

6.12   Describe the role of the kernel in CNN.

6.13   Consider the following input $\mathbf{I}$, kernel $\mathbf{K}$, zero padding P, and stride S:

$$\mathbf{I} = [2 \ -1 \ 2 \ 3 \ -1 \ 3 \ 1 \ -5], \mathbf{K} = [-1 \ 0 \ 1], P = 1, S = 1.$$

Compute the output volume size and find the output.

6.14   With problem 6.11 configuration, design DNN and CNN structure and
compare them in terms of performance and complexity.

6.15   Describe the pros and cons of conventional RNN, LSTM, and GRU.

6.16 Discuss how to combine RNN and other neural networks.

6.17 Reinforcement learning uses sequential trial and error to find the best action to take in every situation and deep learning evaluates complex inputs and selects the best response. Deep reinforcement learning combines reinforcement learning and deep learning. Describe the pros and cons of deep reinforcement learning and identify how to improve the disadvantages.

References

1. W.S. McCulloch, W. Pitts, A logical calculus of the ideas immanent in nervous activity. Bull. Math. Biophys. **5**, 115–133 (1943). https://doi.org/10.1007/BF02478259
2. F. Rosenblatt, The perceptron: a probabilistic model for information storage and organization in the brain. Psychol. Rev. **65**(6), 386–408 (1958). https://doi.org/10.1037/h0042519
3. D. Rumelhart, G. Hinton, R. Williams, Learning representations by back-propagating errors. Nature **323**, 533–536 (1986). https://doi.org/10.1038/323533a0
4. B.G. Buchanan, A (Very) brief history of artificial intelligence. AI Mag. 53–60. Retrieved 30 Aug 2007
5. B. Yegnanarayana, *Artificial Neural Networks*, 1st edn. (Prentice-Hall, August 30, 2004). ISBN-10: 8120312538
6. P. Werbos, Beyond regression: new tools for prediction and analysis in the behavioral sciences. Ph.D. Dissertation, Committee on Applied Mathematics, Harvard University, Cambridge, MA, (1974)
7. K. Fukushima, Neocognitron: a self-organizing neural network model for a mechanism of pattern recognition unaffected by shift in position. Biol. Cybern. **36**, 193–202 (1980). https://doi.org/10.1007/BF00344251
8. Y. LeCun, L. Bottou, Y. Bengio, P. Haffner, Gradient-based learning applied to document recognition. Proc. IEEE **86**(11), 2278–2324 (1998)
9. A. Krizhevsky, I. Sutskever, G.E. Hinton, ImageNet classification with deep convolutional neural networks. Commun. ACM **60**(6), (2017). https://doi.org/10.1145/3065386
10. M.D. Zeiler, R. Fergus, Visualizing and understanding convolutional networks. in *Computer Vision—ECCV 2014. ECCV 2014. Lecture Notes in Computer Science*, vol. 8689 (Springer, Cham, 2014), eds. By D. Fleet, B. Schiele, T. Tuytelaars. https://doi.org/10.1007/978-3-319-10590-1_53
11. C. Szegedy, et al., Going deeper with convolutions. In *2015 IEEE Conference on Computer Vision and Pattern Recognition (CVPR)* (2015), pp. 1–9. https://doi.org/10.1109/CVPR.2015.7298594
12. A. Khan, A. Sohail, U. Zahoora, et al., A survey of the recent architectures of deep convolutional neural networks. Artif. Intell. Rev. **53**, 5455–5516 (2020). https://doi.org/10.1007/s10462-020-09825-6
13. S. Hochreiter, J. Schmidhuber, Long short-term memory. Neural Comput. **9**(8), 1735–1780 (1997). https://doi.org/10.1162/neco.1997.9.8.1735
14. F.A. Gers, J. Schmidhuber, Recurrent nets that time and count. Neural Netw. IJCNN 2000 **3**, 189–194 (2000)
15. K. Cho, B. Merriënboer, C. Gulcehre, D. Bahdanau, F. Bougares, H. Schwenk, Y. Bengio, Learning phrase representations using RNN encoder-decoder for statistical machine translation. in *Proceedings of the 2014 Conference on Empirical Methods in Natural Language Processing (EMNLP)* (Doha, Qatar, 2014). https://doi.org/10.3115/v1/D14-1179

Part II
AI-Enabled Communications and Networks Techniques for 6G

Chapter 7
6G Wireless Communications and Networks Systems

6G systems are now actively under discussion in academy, standardization body, and industry. It is always difficult to predict the future. However, we can learn from the past and explore what 6G will be. The weakness of the current cellular system will be the main part to improve in the future system. For example, privacy and security are one weakness in the cellular system in order to expand the private service. They will be the main part of technology development. In 3G, the high data rate service was one of key features and the targeted main application was web browsing. However, it was not successful in 3G systems but 4G achieved the high data rate service including video transmission and web browsing by cost-effective solutions. All features of 6G systems will not be satisfied at the same time. It will take some time to meet all requirements. Typically, one generation of cellular networks is 10 years. The 6G systems will be gradually evolved and deployed while including new features. They will operate about 10 years and gradually fade out. In the stage of 6G technology development, use cases of the technologies are actively discussed but actual killer applications and business models of 6G systems may not be in-line with the use cases. In addition, there is a big gap between theoretical work and practical implementation. Although some technologies provide us with good performance theoretically and their implementation in laboratory scale are successful, commercial production and deployment may not be possible. In real life and environment, they do not work well as much as we expected. For example, there are many successful demonstrations of antenna array techniques in laboratory scale but they are still not ready to deploy in practical cellular systems.

As of 2022, 5G systems are now deploying in many cities but their functions are not fully operated. Most of 5G deployment is enhanced mobile broadband (eMBB) applications and non-standalone. The phase II of 5G systems covering ultra-reliable low latency communications (URLLC) and standalone system of 5G systems will deploy gradually until 2030. Like the previous generation, 6G system will expand the 5G capabilities and provide a market with new applications and services. As cross-industry collaboration is key aspect of 5G systems, and 5G systems affect to telecommunication industry as well as other industries such as automotive, health care, food, factory automation, and others, 6G will be in-line with 5G approaches including cross-industry collaboration and new applications creation. The initial discussion on 6G developments has been started in International Telecommunication Union (ITU) Focus Group on technologies for network 2030 (FG NET 2030). They studied the future network architectures, requirements, use cases, and capabilities for the year 2030 and beyond, and published the initial results including technology, applications, and market drivers for beyond 5G [1, 2]. They drop a hint about 6G systems. The initial 6G deployments may start from 2030 and will be overlapped with 5G systems as the previous cellular systems did. In academy and telecommunication industry, many research groups have been conceptualizing 6G systems and new features and applications have been proposed. Among them, AI-enabled components of communications and networks may play an important role in 6G systems. AI and ML algorithms will be helpful for optimizing wireless systems, improving the network performance, and creating new services and applications. In this chapter, we review the progress of 6G developments and the role of AI and ML algorithms in the 6G systems.

7.1 6G Wireless Communications and Networks

In cellular communications and networks, the 6th generation (6G) will be a successor to the 5th generation (5G). It will support more applications and services than 5G. As of 2022, it is still under development. How do we develop the 6G? Who are the main players? It will be similar to the previous developments in 4G and 5G. Firstly, the international regulator such as the ITU defines the 6G spectrum allocation for different services and applications. Basically, wider bandwidths enable us to support higher data rates and larger amount of traffics. Thus, 6G spectrum allocation will play an important role of 6G system design. 5G frequency bands are categorized as (1) Sub-1 GHz for wide range coverage, (2) prime 5G mid bands around 3.5 GHz for dense urban high date rates, (3) 26, 28, and 39 GHz for hotspot, and (4) WRC-19 (ITU World Radiocommunication Conference in 2019, above 24 GHz) bands for future mmWAVE options. In 6G, the spectrum allocation is still under discussion but terahertz frequency may be a part of 6G spectrums. The regional regulatory bodies such as European Conference of Postal and Telecommunications Administrations (CEPT) in Europe and Federal Communications Commission (FCC) in the USA play quite important role in the decision-making process in the ITU. They build a regional consensus even if there are national representatives in ITU meetings. After spectrum allocation for a certain service and application, the ITU may label certain allocations for a certain technology and determine the requirements for the technology. In addition, ITU will define the 6G requirements and potential applications and services. Secondly, in order to meet the requirements, some standardization bodies such as the 3rd Generation Partnership Project (3GPP) and the Institute of Electrical and Electronics Engineers (IEEE) make their contributions to the ITU, where they guarantee that their technology will meet the requirements. The actual technology is specified

in the internal standardization such as 3GPP and IEEE. Some standards are established in the regional level. The regulators are responsible for spectrum licensing and
protection. Actual 6G specifications can be found from 3GPP or IEEE works. Thirdly,
in order to develop technical specification of 6G systems, a 6G wireless channel is
defined in 3GPP or IEEE. Its mathematical model is developed and the test scenario
is planned for the evaluation of the proposed technical solutions or algorithms. For
example, in cellular systems, we basically use empirical channel models such as ITU
channel models. Their test scenarios are different as well. Fourthly, each block of
wireless communication and network systems is selected to satisfy the requirements
of 6G systems. This step is discussed and defined in wireless communication standards such as 3GPP and IEEE. Lastly, vendors or operators of cellular networks will
develop networks components and services. In this chapter, we review 6G use cases
and requirements, timeline to develop and 6G key enabling techniques.

7.1.1 6G Use Cases and Requirements

In 6G systems, many new applications and services will be introduced and more
interactions between a human and a machine are expected. The new applications and
services will require higher requirements than 5G systems. In [3], five key use cases
of 6G systems are described as follows: huge scientific data applications, application-
aware data burst forwarding, emergency and disaster rescue, socialized Internet of
things, and connectivity and sharing of pervasively distributed AI data, models and
knowledge. As the name said, huge scientific data applications cover large-scale
scientific applications such as astronomical telescopes and so on. The data traffics
and network throughputs have been significantly evolved from 1G to 5G. The 6G
throughput may reach around 1 Tbps, and a massive amount of data may be handled
in 6G networks. Many scientific data applications require a huge data traffic. For
example, Very Long Baseline Interferometry [4] is composed of multiple distributed
telescope in order to observe the sky. Each telescope creates a huge amount of data and
transfer to a gateway for the integrated data analysis. Therefore, a high throughput
is one of key requirements. The current demand for the throughput is about 100
Gbps, and in near future, they will require about 1 Tbps. Typically, the distributed
networks are based on real-time analysis. The low latency and high reliability are
other important requirements. In order to analysis the data, the data collection from
all nodes is needed. If data collection from one node is delay and we can't synchronize
them, the integrated analysis will be delayed as well. The scientific data should be
accurate. The low packet loss and high reliability are also crucial.

In order to minimize the end-to-end transmission delay and improve the network
performance, application-aware data burst forwarding will be useful. In particular, it
can reduce the data transmission time. A burst is the basic data packet processed by the
application. Depending on applications, the burst will be managed in the network.
The burst is transmitted in sequence and received at the destination nodes with
low latency. For example, when streaming a video, the burst is the video including

multiple pictures and sounds. The network needs to create the virtual channel to avoid any congestion. The destination nodes process the bursts immediately in sequence. This approach enables us to accelerate the end-to-end transmission and also optimizes the network resources. In order to support application-aware data burst forwarding in the network, low latency transmission, virtual channel supporting burst forwarding and resource management and scheduling for burst transmission are required.

Due to the coronavirus pandemic, many governments paid attention to the needs for reliable and robust digital infrastructure. The use case "Mass service of individualized control for the population rescue" is now emphasized in cellular systems. Emergency and disaster rescue will be helpful for managing natural or man-made emergency situations such as blizzards, chemical spills, droughts, earthquake, fire, floods, and so on. In [3], two emergency situations are described according to the response time. In the first case $T > 10$ min, we can organize an individualized control of the rescue of the people from the emergency location. The system notifies people and encourages the self-evacuation before the time T. In the first case $T = 0$ min such as earthquake or explosion, the warning signals should be notified at least 10 min before the emergency situation. Thus, massive IoT systems should be designed to increase sensitivity and detect the emergency as early as possible. The network requirements should be considered in two types of data: (1) well-structured and small volume control information and (2) large volume information from expert object. The traffic volume in the networks should be small, and the data packets should be prioritized. Thus, we can reduce the latency and respond emergency situation at anytime and anywhere. When transmitting the large volume of the data such as a high-quality video, enough bandwidth should be reserved while it doesn't harm to a priority transmission. Key requirements are different from non-emergency data transmission. Reliability and accuracy of transmission are one of key requirements in conventional cellular systems. In emergency and disaster rescue use cases, they are especially important because some warning notices should be broadcasted to the emergency and disaster rescue areas with high precision and low latency. In addition, the remote interventional supports including remote surgery and remote ultrasound examination require ultra-low latency and small jitter. Using 5G networks, network infrastructures have been implemented and system requirements of this use case are partially satisfied. Higher performance of networks results in greater percentage of saved people. Thus, 6G systems expect to improve the emergency and disaster rescue use cases targeting to anywhere, anytime, and for everyone.

The number of IoT devices is now close to 10 billion and increases rapidly [5]. These IoT devices generate a huge amount of data, and the data is managed in one centralized system. The network traffics might be overloaded. Unlike this approach, socialized Internet of things (SIoT) is decentralized to make interconnection among objects. The SIoT enables us to have simple establishment of object connections with different levels of security and also makes service discovery across different platforms possible. For example, logistics companies consider this use case to minimize costs and improve sustainability. The network establishes functions and APIs to implement virtual entities and enables them to interact with other virtual entities.

Thus, it is possible to monitor logistics assets, collect the data from the environment, and optimize the process. SIoT will play an important role in the context of the network infrastructure. The key network requirements of SIoT are similar to conventional IoT requirements such as energy efficiency, security, computing power and storage at the edge networks, and so on. In addition to this, they require open network service interfaces and the object virtualization to support social relationship among SIoT devices. If we consider more intelligent and moving objects such as UAV, low latency and high mobility should be supported.

The use case connectivity and sharing of pervasively distributed AI data are about a new IoT solution. 5G IoT systems are based on dummy sensors and actuators and low rate connectivity links. In 6G IoT systems, we expect intelligent objects to make decision autonomously and interactively with humans using AI algorithms. One of key applications can be found in healthcare domain. For example, a wearable device equipped sensors and AI algorithms can monitor health data, predict anomalies, and trigger alarms to prevent critical situation. The pervasive distribution of AI algorithms would be the radical shift of the IoT system. The workloads and data of AI will be distributed over a pervasive AI system in terms of specific applications and system requirements. Many AI components inside IoT devices can be pooled. This decentralized and cooperative system will be enabled to share AI resources with other systems. In order to implement, the networks should support device-to-device connectivity, self-organization, flexible authentication and network extension, and so on. Key network requirements can be summarized as follows: (1) Virtualization. AI solutions would fit with virtualization concept. It will be easy to reduce the deployment cost. (2) Network orchestration. In order to bring synergy of computing, caching, and communication resource, the network orchestration concept will be helpful for operating AI systems. (3) Network capacity optimization. Massive intelligent objects will generate large amounts of data and AI system will use them for training. Thus, large bandwidth is required and network capacity should be optimized. (4) Low latency. If an application requires a real-time solution, the data transmission among objects should be as fast as possible and support real-time decision-making. (5) Interoperability. In order to support multiple AI solutions, unified network interfaces are required to serve them and reuse the AI and network resources. (6) Network programmability. In order to exchange the variable size data and AI components in dynamic manner, network programmability is required. Thus, we can recognize the objects to be reached and forwarded to different networks. (7) Security. Most of data in the network are associated with private or secured devices. The different levels of security should be provided. Table 7.1 represents the relevant network requirements in terms of five use cases.

In Table 7.1, five dimensions (BW, Time, Security, AI, and MN) are identified. The score scales from 1 to 10. Bigger value represents more important. Bandwidth (BW) includes multiple aspects such as bandwidth, QoS, flexibility, and adaptable transport. The bandwidth is the most important dimension for HSD, ABF, and CSAI use cases. The huge amount of transmitted data will be handled in those use case. Definitely, it is a key metric of 6G systems. Time covers latency, synchronization, jitter, accuracy, scheduling, and geolocation accuracy. The time is important for HDS, SIoT, and

Table 7.1 6G system abstract requirement scores of five use cases [3]

Use cases	BW	Time	Security	AI	MN
Huge Scientific Data Applications (HSD)	10	9	6	6	9
Application-aware Data Burst Forwarding (ABF)	8	5	2	2	2
Emergency and Disaster Rescue (EDR)	5	6	9	8	5
Socialized Internet of Things (SIoT)	7	9	9	7	8
Connectivity and Sharing of pervasively distributed AI data, models and knowledge (CSAI)	8	9	8	8	8

CSAI use cases. In particular, latency plays a key role in these use cases. Security includes security, privacy, reliability, and trustworthiness. The security is related to most of the use cases. The different levels of security should be provided in terms of specific use case scenarios. In 6G and beyond systems, the importance of security and privacy protection will be increased. AI means data computation, storage, modelling, collection and analytics, and programmability. The use case CSAI is highly related to the AI dimension. The importance of the AI dimension will be increased in 6G and beyond networks. ManyNet (MN) includes addressing, mobility, network interface, and heterogeneous network convergence. The ManyNetwork is the heterogeneity of physical and logical patterns. It implies seamless internetworking of heterogeneous networks and devices.

Another view of 6G landscape by NTT Docomo [6] can be summarized by (1) solving social problems, (2) communication between humans and things, (3) expansion of communication environment, and (4) sophistication of cyber-physical fusion. In 6G era, many applications such as telework, telemedicine, distance learning, and autonomous operation will be implemented by high-speed and low latency communications and networks. This application will be helpful for solving the social problems such as regional creation, low birth rate, again and labour shortage. The advance of the 6G technologies enables us to reduce the gap between rural and urban areas and achieves the well-being life. In 6G systems, more communications between humans and things are required to proliferate wearable devices and innovative entertainment services. As implementing IoT services, the demand of the communications between them is getting larger. The IoT devices with higher requirements such as 4K or 8K image processing and low latency control will be adopted in 6G systems. In addition, communication environment will be expanded in many places including ground, sky, and sea. High-rise buildings, drones, airplanes, and flying taxi will definitely require good connectivity solution. 6G will cover these area. The needs of autonomous factories and unmanned vehicles are growing. The connectivity solutions without human being will be provided in any place. Lastly, many things such as vehicles, machines, cameras, and sensors will be connected in cyberspace. Many more services using cyber-physical fusion will be implemented in 6G systems. A large amount of data processing between cyberspace and physical space is one of key requirements. 6G systems will meet the requirement and achieve enough safety and security for a rich life. In order to support four 6G use cases, new combinations of requirements should

be satisfied, For example, massive machine type communication xxxxxx (mMTC) application requires a low data rate and massive connectivity in 5G systems. However, the
IoT devices are getting smarter and need more computing power. Higher data rate is
required in 6G IoT services. Thus, both high data rates and massive connectivity are
a new combination of requirements in 6G.

7.1.2 6G Timeline, Technical Requirements, and Technical Challenges

A new generation of cellular networks appears every 10 years. Now, 6G developments are ongoing. Many standard groups, major vendors, and mobile operators
announce their timeline of 6G developments. Table 7.2 summarizes the timeline of
6G developments.

3GPP release 15 defined 5G phase I including new radio (NR) air interface, next-
generation radio access network (NG-RAN), 5th generation core (5GC), network
slicing, and edge computing. 5G phase I focused on eMBB technologies and applications. 3GPP release 16 defined 5G phase II including enhanced vehicle-to-everything
(eV2X), URLLC, industrial IoT, integrated access and backhaul (IAB), and service
enabler architecture layer (SEAL) for verticals. 5G phase II has been extended to
URLLC and V2X technologies and applications. IAB techniques enable us to have
cost-effective deployment by simplifying radio core connectivity and reducing the
complexity and time for network deployments. The IAB allows us to share the spectrum between access networks and backhaul networks. The key features of IAB can
be summarized as follows: (1) In-band (overlapping between access link and backhaul link is allowed.) and out-of-band (the overlapping is not allowed) backhaul, (2)
support non-standalone (NSA) and standalone (SA) NR mode, (3) autonomously
reconfiguration of the backhaul networks. The NR enhancement for URLLC covers
high reliability, latency reduction, and tight synchronization. In link layer, compact
downlink control information (DCI) for faster processing is supported in physical

Table 7.2 Timeline of 6G developments

	2021	2022	2023	2024	2025	2026	2027	2028	2029	2030
3GPP	Release 17 and 18 (5G Advanced)			Release 19 and 20 (5G Advanced Pro)			Release 21, 22 and 23 (6G)			
ITU-R	IMT2030 Vision and Technology Trends			IMT2030 Requirements			Towards IMT2030 (Evaluation and methodology)			
ITU-T	Future FG-ML5G and Net2030									
EU projects	Smart Networks and Services (SNS): Exploratory research			Smart Networks and Services (SNS): Pre-standard and proof-of-concepts			Smart Networks and Services (SNS): Trials			
IEEE	WiFi7, THz			WiFi8 and 9, Evolution of THz						

downlink control channel (PDCCH). Physical uplink shared channel (PUSCH) is repeated in a slot. Uplink control information (UCI) supports more than on physical uplink control channel (PUCCH) for HARQ. The NR enhancement includes UL cancellation, prioritization, enhanced power control, and grant-free transmissions. In addition, SEAL supports various verticals and provides us with group management, configuration management, location management, identity management, network resource managements, and so on. The application layer support of SEAL has on-network and off-network models. The on-network model enables the UE to connect to the radio network via the Uu interface. The off-network model is used when the UE needs the PC5 interface. The Uu interface is defined to implement V2X communication, and the PC5 interface is defined to use for direct communication among vehicles. As we can observe Table 7.2, 3GPP release 17 is now being developed. The potential features will be non-terrestrial networks, new frequency bands, NR sidelink, and NR light. 5G NR light (previously, 5G NR RedCap (new radio reduced capability device)) is one of key features in release 17. The NR devices have trade-off capability between the conventional eMBB services and NB-IoT (or LTE-M). They are targeting to new use cases such as industrial IoT. New features are continuously included in a new release. 3GPP releases covering 6G will be developed until 2030, and new features will be continuously discussed and adopted from release 17 to 23. ITU-R started IMT2030 discussion in 2021. The requirement, evaluation, and methodology for 6G will be developed by 2030. ITU-T started Focus Group on machine learning for 5G (FG-ML5G) in study group 13. FG-ML5G worked from 2018 to 2020 and published 10 technical specifications including interfaces, network architecture, protocols, algorithms, and data formats. Their outcomes will affect to 6G developments. In Europe, a strategic partnership on smart networks and services (SNS) towards 6G was adopted by European Commission (EC). EC has a plan to invest 900 million euro for 6G development from 2021 to 2027. There are two objectives of SNS: (1) EC fosters 6G technology sovereignty, implement research programme and prepare for early market adoption of 6G systems. (2) EC boosts 5G deployment in Europe and enables the digital and green transition of economy. The research projects of SNS will be composed of three phases (exploratory research, proof-of-concept, and large-scale trials), which are similar to 5G developments. IEEE 802.11 (WiFi) was firstly released in 1997. WiFi has been evolving from a few Mbps to a few Gbps data rate. The speed and coverage are improved, and new features are added. The revisions of WiFi can be summarized as follows: 802.11b, 802.11a, 802.11g, 802.11n (WiFi4), 802.11ac (WiFi5), 802.11ax (WiFi6), 802.11be (WiFi7). The original WiFi is 802.11b. It supports a maximum 11 Mbps data rate and operates in 2.4 GHz frequency. 802.11a operates in 5 GHz frequency, adopts OFDM technique, and supports a maximum 54Mbps data rate. It is the most popular version of WiFi in the market. 802.11g is backward compatible with 802.11b and supports a maximum 54 Mbps data rate in 2.4 GHz frequency. There are needs for faster transmission in small cell. 802.11n (WiFi4) is released in 2009 and supports up to 450 Mbps with a MIMO technique. It operates in both 2.4 and 5 GHz frequency bands. 802.11ac (WiFi5) adopts beamforming and multi-user MIMO. It can create multiple streams to multiple devices and improve overall throughput of the network.

802.11ax (WiFi6) adopts many new features: (1) Up to 8 × 8 MIMO, (2) Up to 9.6
Gbps, and (3) traffic congestion avoidance in public area. 802.11be (WiFi7) is now
under discussion. We expect to have 320 MHz bandwidth, multiband multichannel
aggregation, 16 streams of MIMO, 4096 QAM, and so on. WiFi will keep evolving
and may take a meaningful role in 6G systems. In addition, 3GPP defines various
methods of WiFi and LTE internetworking in order to support WiFi offloading technologies. LTE-WLAN aggregation (LWA) is defined by the 3GPP. Most of mobile
phones are equipped with both LTE and WiFi. LWA enables us to use both links for a
single traffic by coordination at lower protocol layers. A mobile user can experience
seamless data services with a higher network throughput, and a mobile operator can
improve system utilization and reduce OPEX cost. We expect synergy between WiFi
and 6G systems and improve the throughput in small area.

In academy, industry, and standard, 6G technical requirements are actively
discussed. They all have different perspectives to 6G technical requirements.
Academy expects 6G systems in theoretical point of view. Mobile vendors and
telecom industry anticipate them in terms of implementation and realization.
Tables 7.3 and 7.4 summarize the 6G requirements in terms of physical layer and
networks.

As we can observe Tables 7.3 and 7.4, the 6G technical requirements have been
set very high. In order to achieve the requirements, many technical problems should
be solved. Cross-industry collaboration may not be fully satisfied in 5G systems.
In 6G, a telecom industry will focus on the improvement of the technologies to
meet the requirements from other industries. Wireless communications and networks
designers encounter many technical and non-technical challenges when creating new
services and improving the performances. As the requirements of 6G systems have
been set higher, they face more challenges. Main vision and technical trend of 6G

Table 7.3 Requirements of physical layer [7, 8]

	Carrier frequency	Bandwidth	Data rate	Latency	Connection density	Mobility
IMT2020	Up to 100 GHz	Up to 1 GHz	Peak (DL/UL): 20/10 Gbps User (DL/UL): 100/50 Mbps	User plane: 1 ms Control plane: 20 ms	1 device/m^2	Up to 500 km/h
IMT2030	Up to 300 GHz	Up to 10 GHz	Peak (DL/UL): 200/100 Gbps User (DL/UL): 1 Gbps/500 Mbps	User plane: 0.5 ms Control plane: 5 ms	10 device/m^2	Up to 1000 km/h

Table 7.4 Requirements of network layer [9]

	Slicing	Service deployment by slicing	Network bandwidth	Data-driven analysis	Energy consumption	Coverage
NET 2020	Limited slicing service	Few hours	100 Gbps and few billion devices	Centralized in cloud	Moderate	Separate terrestrial and satellite
NET 2030	End-to-end slicing service	Less than hours	1 Tbps and trillion devices	Distributed and AI-based	Low	Integrated terrestrial and satellite

can be summarized as follows: (1) Support of verticals, (2) Support of multiple types of network architectures, (3) Widespread of virtualization, (4) Support of millions connectivity, and (5) Data-driven communication and network with AI and ML. Verticals (also called as industrial applications) support would be one of key growth areas for wireless communications and networks. The collaboration between different industries would be a main driver in the evolution of wireless systems. We expect new value chains to be emerged. The new value chains will serve different verticals. They enable us to save public expenditure and create a sustainable society. Heterogeneous networks have been discussed in 4G and 5G systems but their deployments have been limited. In 6G, in order to meet variable requirements, multiple types of networks architectures will be considered and deployed. For example, distributed MIMO support, outdoor small cell support, indoor and outdoor distributed antenna systems (DAS), and so on. 5G systems have been built using virtualization concept and allow us to provide customers with tailored networks services using slicing techniques. 6G systems will maintain this momentum and expand the virtualization concepts such as O-RAN. The connectivity of things will be expanded in 6G since IoT techniques are applied to many industries such as factory automation. In addition, AI and ML have been already used in many applications because AI algorithms are matured and big data are available in cellular networks. Thus, we can't stop the main trend to use AL and ML in wireless communications and networks. Wireless system architecture based on AI and ML solutions will change the design of the communication and network components. New features of base station, edge computing, and mobile switch centre will be included. In order to resolve the technical problems and realize the 6G systems, the first step is to identify the challenges. The technical challenges can be summarized as follows: (1) High throughput at ultra-dense networks (UDNs), (2) Low latency of E2E connectivity, (3) Diverse network deployments, (4) Flexibility using distributed systems across the network from edge to core, (5) Efficient AL- or ML-powered communications and networks, (6) Cost-efficient network solutions for sustainable society, (7) Seamless connectivity of terrestrial, satellite, and UAV-based networks, (8) Mobility management to edge networks, (9) Network virtualization and cloud, (10) New spectrum strategy between unlicensed bands and licensed bands, (11) New physical layer techniques such as terahertz communications, distributed massive MIMO, and so on.

Summary 7.1. 6G Wireless Communications and Networks

1. 6G systems are now under discussion in academy, standardization body
    and industry. The weakness of the 5G cellular systems will be the main
    part to improve in the 6G systems.
2. 5 key use cases of 6G systems are (1) Huge Scientific Data Applications,
    (2) Application-aware Data Burst Forwarding, (3) Emergency and disaster
    rescue, (4) Socialized Internet of Things, and (5) Connectivity and sharing
    of pervasively distributed AI data, models and knowledge.
3. The potential features of 3GPP release 17 will be non-terrestrial networks,
    new frequency bands, NR side-link and NR light. 3GPP releases will be
    developed until 2030 and new features will be continuously discussed and
    adopted from release 17 to 23.
4. Main trends to 6G can be summarized as follows: (1) Support of verticals,
    (2) Support of multiple types of network architectures, (3) Widespread of
    virtualization, (4) Support of millions connectivity, and (5) Data-driven
    communication and network with AI and ML.
5. The collaboration between different industries would be a main driver in
    the evolution of wireless systems.
6. The 6G technical challenges can be summarized as follows: (1) High
    throughput at ultra dense networks (UDNs), (2) Low latency of E2E
    connectivity, (3) Diverse network deployments, (4) Flexibility using
    distributed systems across the network from edge to core, (5) Efficient
    AL or ML powered communications and networks, (6) Cost efficient
    network solutions for sustainable society, (7) Seamless connectivity of
    terrestrial, satellite and UAV based networks, (8) Mobility management
    to edge networks, (9) Network virtualization and cloud, (10) New spec-
    trum strategy between unlicensed bands and licensed bands, and (11)
    New physical layer techniques such as THz communications, distributed
    massive MIMO.

7.1.3 6G Key Enabling Techniques

In this section, we shed light on key 6G enabling techniques. The key technologies of 6G systems may be in-line with the key technologies of 5G systems. The key requirements of 5G systems have not been fulfilled yet. Most of 5G key technologies should be matured further, and they may be adopted to 6G systems. In addition, 6G needs an infusion of new blood. It will improve the fundamental cellular networks by adding new features such as ultra-wide spectrum, virtualized concept, distributed systems, integrated services, AI-empowered networks components, and so on. Based on these new features, we expect a paradigm shift to cost-efficient and service-oriented networks.

7.1.3.1 6G Frequency Bands and Terahertz Communications

One of key requirements in 6G systems is significantly faster transmission than 5G systems. Basically, wider bandwidths enable us to support higher data rates and larger amount of network traffics. Wider bandwidths than 5G is required. 6G frequency bands may work in the wavelength ranges above 95 GHz. Some experiment works between 95 GHz and 3THz are now being tested in many research groups. Figure 7.1 illustrates the spectrums of cellular systems.

The Federal Communications Commission (FCC) in the USA has opened up experimental 6G spectrum licenses in March 2019, and the spectrum is placed in the 95 GHz to 3 THz. This spectrum license enables 6G developers to perform their experiments for up to 10 years. Many mobile vendors take into account terahertz bands as the starting point for 6G experiments. The terahertz bands provide us with huge bandwidths but there are many research challenges. The terahertz bands like mmWAVE spectrum depend on environmental conditions and weather, and the coverage of the terahertz bands will be about 100 m. In addition, the interference and noise level are high. Thus, the receiver should be coordinated and maintain the light of sight. Due to these challenges, the terahertz bands would be useful for a fixed service or backhaul connection. Similar to 5G spectrum, 6G spectrum may have a mid-band around 3 or 4 GHz bands.

Terahertz communication (also known as sub-millimetre communication) systems will be a key part of 6G systems because 6G use cases including virtual reality and 8K video streaming require an ultra-wide bandwidth. Their frequency bands are placed between 0.1 and 10 THz, and the corresponding wavelengths are between 3 and 0.03 mm. The main characteristics of terahertz can be summarized as follows:

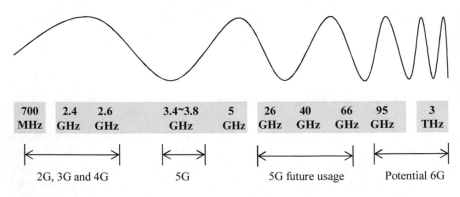

Fig. 7.1 6G spectrum

absorption loss, attenuation with distance, ultra-wideband, and extremely directional beam. Most importantly, in terms of implementation, high processing powers are required to deal with an ultra-wide bandwidth. We need to develop a very high-speed processing chip with high power consumption because power consumption is highly related to sampling rate and broadband terahertz analogue to digital conversion. In addition, terahertz power amplifiers are another big challenge because the efficiency of the current power amplifiers is very low at the high frequency. Since terahertz waves are extremely high-frequency waves with extremely short wavelength, we could have big benefits and research challenges at the same time. The main advantages are (1) huge bandwidth, (2) miniaturized antennas, (3) penetrating non-conducting materials such as clothes, woods, plastics, ceramics, and papers, and (4) minimum effects on human body. On the other hand, key research challenges are (1) short-range communication due to scattering and absorption by dust and rain, (2) the limited propagation length (it does not penetrate metal or water), (3) difficult to detect, (4) expensive devices of detectors, generators, and modulators.

7.1.3.2 6G Open RAN

The mobile networks are composed of radio access network (RAN) and core network (CN). The RAN connects the mobile devices to the CN. The main component of the RAN is a base station. The base station covers a specific area in terms of frequency bands and cell planning. The RANs of cellular systems have been evolved such as GERAN (GSM EDGE radio access network), UTRAN (universal mobile telephone system RAN), E-UTRAN (evolved universal terrestrial RAN), and so on. The CN provides us with access controls, handover, routing, switching, billing, and so on. It allows a mobile user to connect to the Internet or establish a phone call. Typically, mobile operators have one vendor to deploy their core network but multiple vendors for the RAN deployment. The interoperability is one important issue, and it causes an additional cost. The mobile operators seek a flexible use of mobile equipment. Due to significant improvements of VLSI technologies such as central processing units (CPUs) and graphic processing units (GPUs), it is possible to implement virtual network components and the networks have more flexibility and scalability. This approach makes open-source software become more attractive option to implement radio access networks. Due to many benefits of open-source software such as interoperability and development cycles reduction, many researchers are actively working for open and intelligent radio access networks. Open RAN (O-RAN) enables them to use mixed components with more flexibility. The main benefits of Open RAN include lower network equipment cost, wider adaptation, higher network performance, higher interoperability, better security, and so on. More importantly, vendors may have a wider network access with more flexibility and mobile operators may select the network equipment and solutions for the best fit. The basic concept of Open RAN is to open the protocol and build interfaces between many different blocks. The main building blocks of O-RAN consist of the Radio Unit (RU), the Distributed Unit (DU), and the Centralized Unit (CU). The role of the RU is to transmit and receive the radio signal. In the DU and CU, the computation is performed and the radio signal is transmitted to the network. The DU and CU are located nearby the RU and CN, respectively. The AI or ML may take an important role of RAN intelligent controller in Open RAN to optimize the network. The Open RAN architecture includes AI-enabled RAN intelligent controller (RIC) for near real-time and non-real-time services. The O-RAN alliance [10] defined the multiple interfaces including Fronthaul between the RU and the DU, Midhaul between the DU and the CU, and Backhaul connecting the RAN to the Core. In 6G, Open RAN may be adopted. However, there are still many challenges. For example, Open RAN may have higher network complexity and more difficulty for system integration and maintenance due to multi-vendors support. Figure 7.2 illustrates comparison of traditional RAN, 5G vRAN, and Open RAN.

As we can observe 5G vRAN and Open RAN in Fig. 7.2, the basic idea of them is to decouple software from hardware and operate the RAN function on Commercial

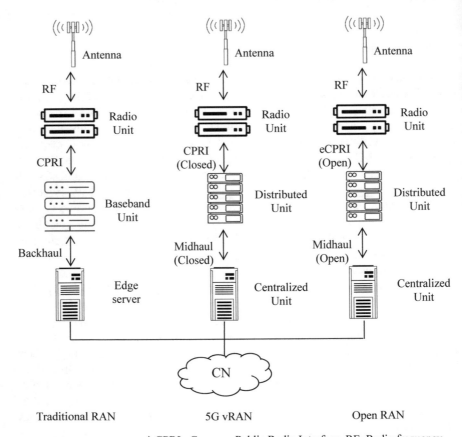

* CPRI : Common Public Radio Interface, RF: Radio frequency

Fig. 7.2 Comparison of traditional RAN, 5G vRAN, and Open RAN

Off-the-Shelf (COTS) servers. This approach allows mobile operators to modularize network functions and reduce OPEX. Both vRAN and Open RAN introduce virtualization concept and improve the efficiency of network equipment. However, the key characteristic of Open RAN is to pursue openness that any software vendor can work on COST-based hardware via open interface.

7.1.3.3 Diverse Network Deployments

As the network traffics grow exponentially and the numbers of the macro-, pico-, and femtocells increase significantly in 6G, multiple types of cells are getting more packed and eventually ultra-dense networks (UDNs) are deployed in urban area. The UDN encounters new challenges such as different types of network deployments and high intercell interferences. In particular, a higher cell density causes a higher intercell interference. It would be the major bottleneck of the UDNs. As long as cell-centric networks are deployed, the intercell interference is inevitable. In the traditional cellular networks, mobile devices (user equipment (UE) or mobile station (MS)) are connected to one cell among multiple cells. Access points (APs) or base stations (BSs) have multiple numbers of mobile devices, and they affect to neighbouring access points or mobile devices. From 4G systems, the concept of signal co-processing has been adopted and cellular networks have improved spectral efficiency. The basic idea is to perform data processing at multiple APs and transmit to one mobile device via multiple APs. Coordinated multipoint transmission/reception (CoMP) in LTE, coordinated multipoint with joint transmission (CoMP-JT) [11] and multi-cell MIMO cooperative networks [12] are based on this co-processing concept. These techniques divide the APs into disjoint clusters and are implemented in a network-centric way. It enables multiple APs in a cluster to transmit one data set to one mobile device jointly. This is similar to the distributed antenna system. However, those techniques are not widely used in LTE systems because of gap between theoretical gain and practical gain. Another approach as a user-centric way has been studied for a joint transmission from multiple APs. The user-centric way considers interference and transmits the signal to specific user. A mobile device is jointly served by multiple APs in a cluster. It enables us to have much lower interference by removing the cell boundaries. Thus, cooperative MIMO [13], cooperative small cells [14], and Cloud/Centralized Radio Access Network (C-RAN) [15] adopted this concept. In order to implement a user-centric multiple APs, synchronization and reference signals among APs are an important part. A mobile device should find and select the best cell by acquiring time and frequency synchronization and detecting the corresponding cell ID and reference signals. Figure 7.3 illustrates the comparison of traditional cellular networks and user-centric networks.

The user-centric networks enable users to move from one network to another seamlessly. The networks that can provide the best service with the user are automatically selected using the available communication links. Basically, frequent handover in dense networks may result in data losses, handover delays, ping-pong effect, and so on. It allows us to overcome the limits of cell-based wireless communications

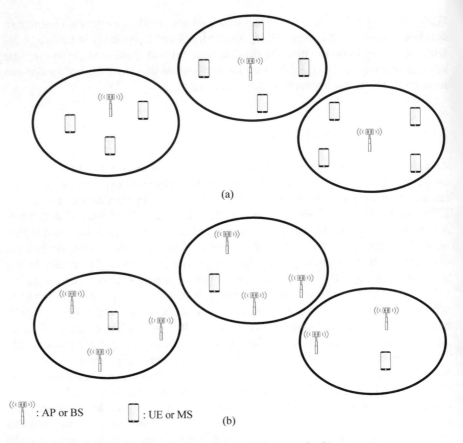

: AP or BS   : UE or MS   (b)

Fig. 7.3 Traditional cellular networks (a) and user-centric networks (b)

and provide us with a better quality of network services. In addition, it provides us with good environments many new techniques such as multi-connectivity, multi-tier hybrid techniques, and multiple radios on one device can be adopted.

7.1.3.4   Integrated Networks

Non-terrestrial networks such as unmanned aerial vehicles (UAV) communications, satellite communication (Satcom), and maritime communications have been widely investigated. They are more suitable for some use cases such as artic area, a high mountain, and so on. In these locations, it is difficult to implement cellular networks due to unstable environments and high network deployment costs. Thus, Satcom covers these areas as a reasonable solution. The utilization of geostationary satellites (GEO), low earth orbit satellites (LEO), and high altitude pseudosatellites (HAPS) is considered in 6G systems. As we briefly discussed the integrated network between

WiFi and cellular networks, there are many benefits including offloading. The 3GPP is continuously studying the integrated networks between WiFi and cellular networks. In addition, internetworking between cellular systems and Satcom has been discussed in 5G standards. One ground gateway receives data packets from satellites, acts like packet routers, and connects to the Internet. The Satcom provides us with many benefits such as wider coverage, better adaptability to disaster events, and flexibility. Thus, the integrated networks between Satcom and cellular communications will improve cell edge coverage. However, there are multiple research challenges. For example, co-channel interference between satellites and cellular networks would be one of key challenges because millimetre wave bands are now used in satellite, and also, these bands have been adopted in cellular small cell networks. The 24–29 GHz bands are allocated in 5G millimetre wave bands. The Ka-band (26.5–40 GHz) is a part of satellite communications. They are overlapping. The main use cases of the internetworking between Satcom and cellular networks are (1) offloading cellular network traffic for non-time sensitive data, (2) support moving platforms such as ships, trains, cars, and so on, (3) IoT services in rural area. For example, smart agriculture, smart factory, and wind farm may be located in rural area and investing terrestrial networks may not be economically efficient. Thus, Satcom network may be feasible for this use cases. (4) Backbone connection support. For example, if a 5G base station in rural area, high mountain, or artic area cannot connect to the 5G core network due to the unstable environments or the lack of backhaul facilities, Satcom will be helpful for the implementation of backhaul. An UAV is a flying vehicle that can hover at a low altitude and provide connectivity for a short time (typically up to 1 h) in a limited area. Multi-access edge computing (MEC) allows mobile stations to offload their computation tasks to the edge of the network and improve computing capabilities of the networks. The communication link using the UAV equipped with MEC will be useful for low latency applications. The use cases of UAV communications can be 360 live broadcasting, temporary relay node in disaster situations, private cellular networks, and so on. Key challenge of UAV communications is to find an optimal positioning in a deployment scenario. If the UAV communication links are used for aerial relays or backhauls of cellular networks, it will provide us with many benefits such as high mobility and quick deployment. In addition, we can avoid blockages and path losses and enhance the next reliability and capacity. However, due to the limited operation time, we should find optimal positions of UAVs to provide enhanced connectivity. The integrated network between non-terrestrial networks and cellular networks increases the value of 6G networks. Figure 7.4 illustrates an example of integrated networks between non-terrestrial network and cellular networks.

7.1.3.5 Channel Coding, Modulation, and Waveform

In 6G systems, key components of physical layer should be more efficient for a better radio link to optimize channel capacity. Channel coding schemes in cellular systems have been evolved from convolutional coding to turbo coding and LDPC. In cellular systems, the main target of channel coding is to reach the Shannon limit of single user

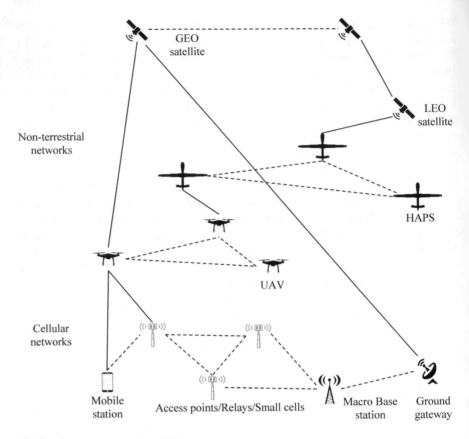

Fig. 7.4 Integrated networks of 6G

channel. Turbo coding and LDPC coding provide us with the performance nearby
the Shannon limit. In 5G and 6G, multi-user channels are considered to improve
the performance. However, the channel coding study in the multi-user channels is
very limited because the complexity in a receiver is much higher than single user
channel. Another research challenge in channel coding is to cover non-binary codes.
Basically, non-binary LDPC and turbo coding provide us with a better performance
but the decoding complexity is much higher than binary LDPC and turbo coding.
Another big research challenge in channel coding is to have a robust channel coding
scheme with short length codeword. Basically, both turbo codes and LDPC codes
require a big packet length to achieve good performance. However, in order to have
a short latency, the data packet size should be small and we can't use turbo codes
or LDPC codes for URLLC applications. A new coding scheme is required to reach
the near Shannon limit performance when the codeword length is short. The quadrature amplitude modulation (QAM) is widely used in cellular systems because the
modulation and demodulation of QAM are simpler than others. However, its constellation points are not arranged in equally spaced positions and the distribution is far

|||||| transmission. On the other hand, amplitude phase shift keying (ASK) is robust to the nonlinearity of the PA and also is tolerant to phase noise compared to others. Thus, this is used in broadcasting networks and satellite communications. It might be used in terahertz communications of 6G systems. In 5G, many new waveforms are considered such as CP-OFDM, Filter Bank Multi-Carrier (FBMC), Universal Filtered Multi-Carrier (UFMC), and Generalized Frequency Division Multiplexing (GFDM). The waveforms should satisfy the following requirements: high spectrum efficiency, numerology, scalable bandwidth extension, high-order modulation support, efficient structure for MIMO, and so on. The 6G waveforms design should be in-line with the 5G waveform requirements. In addition to this, terahertz supports, more complexity receivers, or analogue processing might be new challenges.

7.1.3.6   Backscatter Communications

The backscatter communications (BackCom) use the reflected or backscattered signals to transmit data. They rely on passive reflection and modulation of an incident signal and do not require any active RF components on a tag. Since a BackCom transceiver consumes very small power compared to conventional transceivers and has very low complexity architecture, it is suitable for IoT communications. The battery-less IoT devices can be used in 6G massive connectivity. The BackCom system is composed of a tag and a reader. The tag includes an RF energy harvester, a battery, a modulation block, and an information decoder. It is a passive node harvesting energy from an incident signal. An intentional mismatch between antenna and load impedance causes the reflection of the signal. The load impedance change enables the reflection coefficient to change a random sequence modulating the reflected signal with the information of the tag. We call this backscatter modulation. The tag is operated by RF energy harvesting, and we don't need active RF components. On the other hand, the reader needs power supply and conventional RF components for transmission. There are three types of BackCom as shown in Fig. 7.5.

As we can observe Fig. 7.5a, the reader in the traditional backscatter communication generates RF signals to activate the tag. The tag modulates and reflects the RF signals to transmit the data. If the distance between the tag and the reader is far away, a higher energy outage problem occurs and a lower modulated backscatter signal is generated due to the signal loss between them. Thus, the traditional backscatter communication link can be used for short-range RFID applications. In bi-static backscatter communications as shown in Fig. 7.5b, there is a carrier emitter. If we can allocate the carrier emitters around tags, the quality of the communication link can be significantly improved. The carrier emitter can send RF signals to tags so that they can harvest energy. Thus, the coverage can be expanded. Ambient backscatter communication as shown in Fig. 7.5c is more generalized version of bi-static backscatter communications. It doesn't use the dedicated carrier emitters and allocates RF sources like TV towers, base stations, access points, relay, and so on. Since they use the existing RF signals, there are no additional costs and we don't

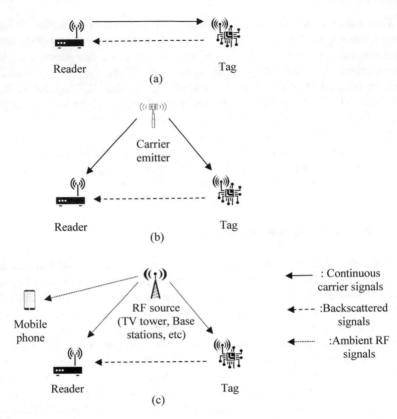

Fig. 7.5 Three types of BackCom: Traditional backscatter (**a**), bi-static backscatter (**b**), ambient backscatter (**c**)

need to assign separate frequency bands for this communications. However, ambient RF signals are not dedicated signal. Thus, it is not easy to predict and the communication link may not be stable. In addition, the operation of ambient backscatter communication may be tricky to control. It would be difficult to achieve the optimal performance. The BackCom has many benefits and potential for many applications but there are many research challenges. For example, in order to implement BackCom in a cellular system, it requires the phase and channel state information of the cellular system. It causes a higher complexity. Thus, non-coherent communication is considered and it enables us to have a better resource utilization and service. In 6G systems, the conventional RFID may be replaced by BackCom or they may coexist as one important part of 6G systems. Table 7.5 summarizes the comparison of conventional RFID communication and BackCom.

Table 7.5 Comparison of
RFID communication and
BackCom

	RFID	BackCom
Distance	Less than 1 m	Less than 1 km
Data rate	Less than 640 Kbps	Less than 10 Mbps
Modulation	BPSK	ASK, FSK, PSK, QAM High-order modulation is possible
Network structure	Point-to-point	Multiple access
Energy source	Dedicated source	Dedicated or ambient source

7.1.3.7 Programmable Metasurfaces

Recently, many research groups pay attention to programmable metasurface [16]. It may change the fundamental hardware structure of wireless communication systems. According to the recent study [17], it is useful for improving the channel environment for wireless communications. Basically, metasurfaces are artificial conductive structure and two-dimensional materials with subwavelength structural inclusions. They enable wireless communication engineers to manipulate the propagation of electromagnetic waves. The conventional metasurfaces have a fixed structure and can't manipulate electromagnetic waves because the reflection and their transmission coefficient are constant. Thus, the initial version of metasurfaces is called as analogue metasurfaces. However, in order to use them in communications systems, we should dynamically adjust them and control electromagnetic waves. Programmable metasurfaces with reconfigurable electromagnetic parameters overcome the analogue metasurfaces and manipulate the phase, amplitude, polarization, and orbital angular momentum of electromagnetic waves over their surfaces [18]. The electromagnetic waves can be tuned by some control signal. There are many types of tuning to implement programmable metasurfaces. For example, the electric permittivity of the liquid crystal can be adjusted under different gate voltages [19]. It can shift the resonance frequency. In addition, we can control the on and off states of a diode and switch the metasurface performance from perfect absorption to reflection [20]. In [21], wireless transceivers using metasurfaces are investigated. Figure 7.6 illustrates the transceiver architecture using programmable metasurface.

As we can observe Fig. 7.6a, the transmitter is composed of coding and modulation, control signal, and programmable metasurface using reflection coefficient. The basic concept of programmable metasurface in wireless communications is to control the reflection coefficient $\Gamma(t)$ in terms of the modulated or source data. The incident wave is a carrier signal. The reflection coefficient is defined as the ratio of the complex amplitude of the reflected wave to that of the incident wave [22]. The reflected wave includes the information at the frequency of the carrier signal as follows:

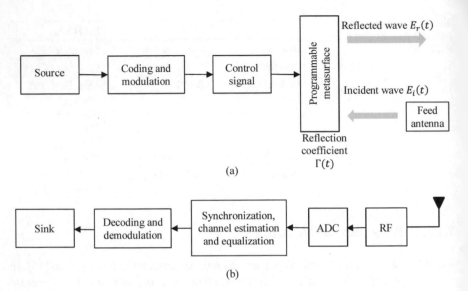

Fig. 7.6 Transmitter (a) and receiver (b) using programmable metasurfaces

$$E_r(t) = E_i(t)\Gamma(t) \tag{7.1}$$

where $E_r(t)$, $E_i(t)$, and $\Gamma(t)$ are the reflected wave, incident wave, and reflection coefficient, respectively. In terms of the reflection coefficient, we can construct the modulation of the wireless system. The information data is encoded and then modulated. Multiple types of channel encoding and modulation schemes can be selected. They are mapped into the reflection coefficient control signal. For example, if QPSK modulation is considered, the reflection coefficient can be expressed as follows:

$$\Gamma(t) = \sum_{n=1}^{N} P_n h(t - nT) \tag{7.2}$$

where $P_n \in [P_1, P_2, P_3, P_4]$ is the complex reflection coefficient at the nth symbol and can have the four different reflection coefficients. T denotes the symbol duration. $h(t)$ is the sampling function. As we can observe Fig. 7.6b, the conventional receiver can be used. It includes synchronization, channel estimation, equalization, demodulation, and channel decoding.

7.1.3.8   Advanced Dynamic Spectrum

6G systems will face the problem of spectrum shortage as the previous generation experienced this problem. The mobile operators buy a spectrum license from the government and operate their cellular networks with exclusive right. Although same spectrums are not used frequently, other operators or third parties cannot use them. Spectrum sharing techniques have been investigated to use the limited spectrums efficiently. They are helpful for improving the spectrum utilization rate. However, in practical cellular systems, the spectrum sharing techniques have never been deployed because they are not yet matured techniques to deploy. Most of mobile operators expect not to harm their subscribers when spectrum sharing techniques work. However, the existing spectrum sharing techniques do not guarantee to interfere with the customers of the mobile operators. In 6G systems, the matured spectrum sharing techniques are expected to improve the resource utilization and flexibility of the spectrum. In addition, unlicensed bands may be considered in 6G spectrums. Thus, they may be shared in many operators, third-party operators, or local operators using spectrum sharing techniques. We can obtain a hint from a licensed-assisted access (LAA) of 3GPP. LAA uses carrier aggregation in licensed and unlicensed spectrum to transmit in parallel. When interference is below a threshold, it uses a large amount of unlicensed spectrum. Based on this approach, 3GPP release 16 adopts the concept of LAA as NR-Unlicensed and supports network deployment using unlicensed spectrum. The initial version of NR-Unlicensed includes below 7 GHz spectrum, and, in the next version, high-frequency spectrum is expected.

More practical problem in terms of the spectrum shortage is the imbalance between uplink and downlink spectrums as well as between different networks. The imbalance results in the low utilization rate of the cellular systems. The spectrum sharing techniques cover this problem but advanced duplex systems are paid attention to solve this imbalance problem. Full duplexing techniques that have been discussed in 5G systems enable mobile operators to allocate flexible spectrum and increase the utilization rate of spectrums. However, they have not been adopted in 5G systems because of not enough theoretical and trial results. Instead, the 5G systems adopted flexible spectrum operation such as dynamic TDD. Depending on the use cases or spectrum configurations, uplink and downlink spectrums can be adjusted. In 6G, more flexible spectrum allocations are expected by sharing not only spectrum but also time and other network resources.

7.1.3.9 Intelligent Radio

5G systems included network virtualization. This new design approach enables us to decouple software from hardware. It allows us to deliver the network component to edge networks or others when a user needs specific function. Due to this virtualization, mobile operators can obtain many benefits such as including capital expenditures (CAPEX) and operating expenses (OPEX) reduction, time reduction to deploy a network service, and network scalability improvement. The approach is helpful for optimizing network resource provision. Network slicing in 5G systems is based on this network virtualization. It can establish virtual end-to-end networks tailored to different 5G applications such as eMBB, URLLC, and mMTC. It enables 5G systems to be more scalable and faster to adapt a new service. The network slicing became one of key function of 5G system. 6G systems will make a more deliberate effort on virtualization. In order to support the fast upgrade of networks equipment and end-user devices, software and hardware separated architecture will be investigated further in 6G systems. In terms of self-upgrade, the device can estimate the capability of hardware over which software is suitable or protocol runs and installs the most suitable software to optimize. Traditionally, the hardware and software co-design is common and their functions and capabilities such as the number of antenna, resolution of ADC, and error correction coding schemes are not possible to upgrade. Software-defined radio is in-line with this approach. However, it was not successful because the computational power was not enough previously, and also, it was not efficient to implement software-based radio system. The recent advance of VLSI and antenna system technologies make it possible to improve the hardware capability and allow us to have agile adaptation to fast upgrade at the hardware system. In addition, AI and ML algorithms will support for optimal configuration of transceiver algorithms on the hardware platform and construction of intelligent physical layer.

Summary 7.2. 6G Key Enabling Techniques

1. Many 5G key technologies may be adopted to 6G systems. In addition, 6G needs an infusion of new blood.
2. Wider bandwidths than 5G is required in 6G. 6G frequency bands will cover the wavelength ranges above 95 GHz. The main advantages of Terahertz communications are (1) huge bandwidth, (2) miniaturized antennas, (3) penetrating non-conducting materials such as clothes, woods, plastics, ceramics and papers, and (4) minimum effects on human body.
3. The mobile operators seek a flexible use of mobile equipment. Open RAN (O-RAN) enables them to use mixed components with more flexibility. The main benefits of O-RAN are lower network equipment cost, wider adaptation, higher network performance, higher interoperability, better security, and so on.
4. As long as cell centric networks are deployed, the inter-cell interference is inevitable. The user centric networks enable mobile users to move from one network to another seamlessly. The networks that can provide the best service with the user are automatically selected using the available communication links.
5. The integrated network between non-terrestrial networks and cellular networks increases the value of 6G networks.
6. In 6G channel coding, multi-user channels are considered to improve the performance. Non-binary LDPC and turbo coding and short length codeword transmission are key research topics. In addition, new modulation and waveform are investigated to meet a high throughput and low latency.

7. Since a BackCom transceiver consumes very small power compared to conventional transceivers and has very low complexity architecture, it is suitable for IoT communications.
8. Programmable metasurface is useful for improving the channel environment for wireless communication and can manipulate the propagation of electromagnetic waves.
9. In 6G, more flexible spectrum allocations are expected by sharing not only spectrum but also time and other network resources.
10. The network slicing became one of key function of 5G system. 6G systems will make a more deliberate effort on virtualization.

7.2 AI-Enabled 6G Wireless Communications and Networks

The traditional cellular communications and networks are designed and deployed with pre-defined systems configuration by iterative trial-and-error manner for each scenario. Their values or thresholds of cellular systems can be adjusted by networks operators. This design approach has worked well so far. However, 6G systems consider more complicated communication and network scenarios and support many different use cases such as automated vehicles, factory automation, telemedicine, and so on. The traditional approach may not work well for supporting multiple 6G requirements. Thus, 5G NR systems contain many new features such as network slicing in order to support multiple services flexibly and scale-up for feature applications. In 6G systems, data-driven approach of AI algorithms will be equipped with heuristic parameter setting and thresholds. AI-enabled communications and networks will be helpful for containing new features. As AI algorithms significantly affect many research fields such as image recognition by new achievement like deep learning, adopting AI algorithms to wireless systems became a key driver of 6G systems. Many research groups in wireless communications and networks have studied the use of AI algorithms in various areas such as handover mechanism, network traffic classification, elephant flow detection, network traffic prediction, network fault detection, and so on. Practical implementation of AI algorithms is considered in 6G systems. In particular, it will improve the 6G system design and optimization. The optimization of wireless communications and networks systems is basically entangled with many parameters such as complexity, cost, energy, latency, throughput, and so on. Thus, the optimization problem is non-deterministic polynomial (NP) hard and also is defined as a multi-objective optimization problem subject to complex constraints. The AI and ML algorithms can be applied to solve the optimization problems and contribute to improve 6G performance as well as develop a new service. AI and ML algorithms are useful for classification, clustering, regression, dimension reduction, and decision-making. Many components of 6G systems are highly related to them. For example, a resource allocation and scheduling problem is a sort of classification and clustering problem. Channel estimation is a regression problem. Viterbi decoding is based on dynamic programming. Network traffic management is highly related to a sequential decision-making problem. Thus, the AI and ML algorithms will be 6G key enablers. In this section, we discuss in which domains AI and ML algorithms will helpful for 6G systems and also how they face research challenges. In addition, the key technical questions to apply AI algorithms to wireless communications are networks can be summarized as follows: (1) How to collect and handle the data of wireless systems for AI algorithms? (2) Which applications and functions of AI algorithms are well-matched with the components of wireless systems? (3) How do AI algorithms improve the wireless systems performance in terms of throughput, energy efficiency, latency, and others? (4) Are AI algorithms capable of replacing some functions in physical, MAC, and network layers? (5) How to overcome the limitation of the requirements such as huge training data set and high computational power? (6) How AI algorithms will be helpful for designing wireless systems? (7) Is it possible to redesign Shannon-based communication architecture? (8) How to harmonize AI algorithms and communication theories and find common theoretical background? (9) How AI algorithms will impact to design and implement 6G systems? In part II, we will discuss these questions.

7.2.1 AI and ML Contributions to Physical Layers

In cellular networks, the main purpose of physical layer is to achieve a reliable transmission over a wireless channel. The transmitter includes a baseband processing such as a channel coding, modulation, OFDM, and MIMO, and a RF processing such as a low-noise amplifier (LNA), mixer, power amplifier (PA), and so on. The receiver recovers the original data from the received data including channel impairments such as noise, fading, and interference. AI and ML algorithms will play an important role in physical layer optimization. One of key design parameters in physical layer is to overcome impairments such as channel noise, fading, interference, IQ imbalance, phase noise, thermal noise, and distortion of RF devices. It is not easy to find an optimal design in physical layer blockchain. AI and ML algorithms may give us an opportunity to find an optimal co-design between hardware and software and also have new capabilities such as self-optimization and self-learning. We call this an intelligent radio as we discuss in Sect. 7.1.3.9. The intelligent radio is one study field combining AI and ML algorithms, signal processing, communication theories, and VLSI technologies. Figure 7.7 illustrates comparisons of the conventional physical layer components based on OFDM, physical layer components including ML, and ML-based physical layer processing.

As we can observe Fig. 7.7, the ML algorithms may contribute to the transceiver blockchain as component-wise in the first stage. In the final stage, the physical layer processing may be replaced by ML-based physical layer processing. One approach

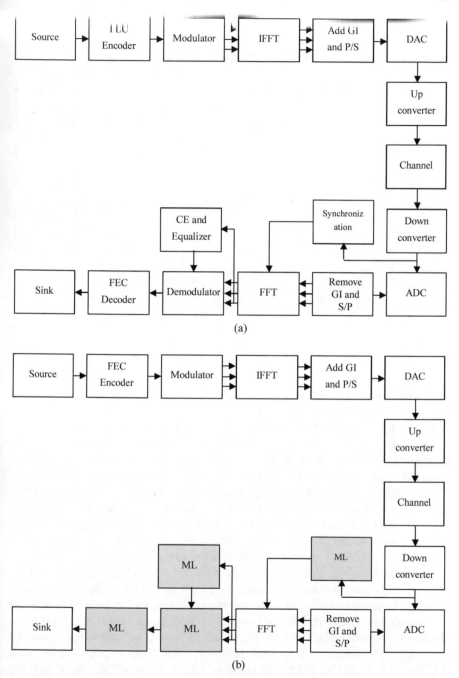

Fig. 7.7 OFDM-based physical layer blocks (**a**), physical layer blocks including ML algorithm in the receiver, and ML-based physical layer processing (**c**)

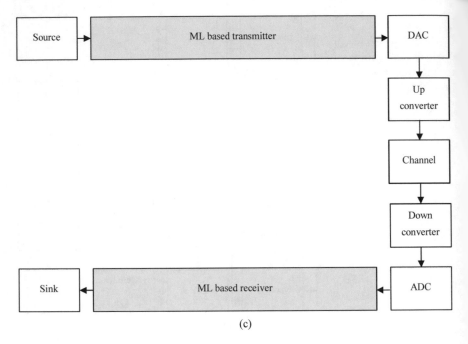

(c)

Fig. 7.7 (continued)

about a ML-based physical layer processing is an end-to-end autoencoder for physical layer. We will discuss this in Chap. 8.

Signal detection and channel estimation are the appropriate components AI and ML can improve. Deep learning can improve the performance of signal detection in physical layer. We can apply deep learning for equalization with nonlinear symbol distortion. It enables us to map the received signal to transmitted signal without classical detection threshold. This approach provides us with a better performance than classical MMSE detection [23]. In the modern wireless communications, it is important to acquire accurate channel state information (CSI) because the CSI is directly related to the performance of MIMO and others. The channel estimation using pilots requires a long pilot length in order to increase accuracy but the long pilot reduces the spectral efficiency. Thus, it is a key research challenge to have the optimal length and pilot design between the pilot length and the channel estimation accuracy. In addition, a blind channel estimation finds statistical properties from the received signals. Although it doesn't require the pilot symbols, it is more difficult to obtain accurate channel estimation. ML techniques can contribute to those research challenges. Recurrent neural network (RNN) is suitable for anticipating new data set. The RNN will be helpful for predicting in the blind channel estimation. In many different environments of 6G systems, we need to build AI-based channel estimation. There are many obstacles to implement. In fast fading channel, the predicted period should be short and the limited training data will be given. Thus, the key research challenge in this topic would have the generalized channel estimation with the limited training data. Most importantly, AI and ML techniques assume that the training data distribution is identical to the test data distribution. However, in real world, it is very difficult to have same distribution between them. Thus, it is a key research challenge to implement AI and ML techniques to physical layers. In Chap. 8, we will discuss how AI and ML techniques can be adopted in physical layer.

7.2.2 AI and ML Contribution to Data Link and Network Layers and Open Research Challenges

The data link layer includes two key sublayers: the media access control (MAC) and the logical link control (LLC). The data link layer is responsible for media access, error detection, and multiplexing data streams and enables us to have a reliable connectivity link. The main purpose of network layer is to interconnect different networks, determine the best routing paths for transferring data packet, and manage the network traffic flows. The data-driven approach of AI algorithms is suited to the tasks of MAC and network layer. We can collect a huge amount of data associated with massively connected devices and use them for 6G systems. They have a big potential to improve the overall performance in MAC and network layer, reduce the overall network operation cost, and create new services in 6G systems. For example, they can improve handover operating while observing geographical environments. They can help the network planners to optimize the network deployment. In terms of network operation, they can help the network fault tolerance by predicting the network traffics and detecting the potential faults. The purpose of MAC layer is to encapsulate the data from the network and upper layer and to control access to the media. The main functions of MAC layer are to determine the channel access methods for data transmission and perform multiple access. In addition, security service is one of key functions. If the received data packet is not secure, it can initiate the retransmission. AI and ML algorithms can improve the performance of MAC layer. The most appropriate functions to apply for AI and ML algorithms are resource allocation, scheduling, carrier aggregation, handover, and so on. In addition, it can reduce overhead of MAC data packet and improve the reliability of transmission by AI empowered a hybrid automatic repeat request (HARQ). The key role of network layer is to exchange data packets from one source to one destination across multiple nodes. The key functions are to route data packet by finding the best path across the nodes and assemble or reassemble the packet. AI and ML techniques will be helpful for discovering an optimal path, choosing serving cells, providing good quality of service, and so on. In addition, it is suitable to balance the traffic load of the networks and improve the overall services of the networks by finding an optimal load balancing. Table 7.6 summarizes the potential AI and ML contributions to wireless communications and networks.

As we discussed in Sect. 7.1, 6G systems will require much higher throughput, lower latency, better energy efficiency, and so on. They face new challenges and

Table 7.6 Example of potential AI and ML contributions to wireless systems tasks

Wireless system tasks / AI tasks	Physical layer	Data link layer	Network and upper layers
Clustering	Signal detection, Symbol mapping	User clustering and association	Critical network establishment, Controlling cluster size in dynamic heterogeneous networks
Classification	Positioning, localization, MIMO user paring	Dynamic scheduling, Carrier aggregation, Optimal multi-connectivity	Slicing admission, scheduling, Load balancing
Regression	Channel estimation, Equalization	Adaptive power control	Dynamic slicing management
Prediction	Blind channel estimation	Adaptive power control, Beam prediction, Radio resource allocation	Network monitoring, Orchestration, Traffic prediction, Mobility prediction
Decision-making	Error correction coding	Interference mitigation, Handover	Mobility management, Optimal routing path

build smart and intelligent communications and networks with high level of flexibility, scalability, sustainability, security, efficiency, and distributed intelligence. AI and ML technologies will be helpful for satisfying the requirements and creating new services. However, AI and ML tasks are computationally heavy and have multiple bottlenecks to apply for wireless systems. For example, in order to have accurate outcomes, AI and ML tasks need a lot of training data. Sometimes the personal data may be shared with others if the data processing is performed on device. The data protection and privacy will be one open question. If AI and ML tasks are carried out in IoT networks and the sensor nodes do not have enough computing power, the frequent communications between the sensor nodes and cloud networks are required and it causes additional usages of network resources and high energy consumption. In particular, in the distributed networks, computation units, distribution units, and storage units may be separated and it may be the main problem to overcome. In addition, if we deploy AI and ML algorithms in the heterogeneous environments, it makes networks more complicated and the complicated network deployments degrade network performances including spectral efficiency, resource efficiency, latency, and energy efficiency. We need the unified interface to support a dispersed computing environment for both training and inference of deep neural networks. In Chaps. 8, 9, and 10, we will discuss how AI and ML algorithms can improve wireless systems.

Summary 7.3. AI-Enabled 6G Wireless Communications and Networks

1. Many research groups in wireless communications and networks have studied the use of AI algorithms in various areas. Practical implementation of AI algorithms is considered in 6G systems.
2. The AI and ML algorithms can be applied to solve the optimization problems and improve 6G performances as well as develop new services.
3. AI and ML algorithms give us an opportunity to find an optimal co-design between hardware and software and also have new capabilities such as self-optimization and self-learning.
4. AI and ML algorithms will contribute to the transceiver block chain as component wise in the first stage. In the last stage, the physical layer processing may be replaced by ML based physical layer processing.
5. In data link layer, the most appropriate functions to apply for AI and ML algorithms are resource allocation, scheduling, carrier aggregation, handover, and so on.
6. In network layer, AI and ML techniques will be helpful for discovering an optimal path, choosing serving cells, predicting the network traffic, finding network faults, classifying the network traffics, providing good quality of service and so on.

7.3 Problems

7.1. Describe the pros and cons of 1G to 5G cellular systems and then explain how the weakness of the previous generation cellular systems is overcome.
7.2. Describe the key requirements of 6G use cases: (1) Huge Scientific Data Applications, (2) Application-aware Data Burst Forwarding, (3) Emergency and disaster rescue, (4) Socialized Internet of Things, and (5) Connectivity and sharing of pervasively distributed AI data, models, and knowledge.
7.3. Explain the key outcomes of 6G developments in the timeline (2021–2030) of 3GPP.
7.4. Discuss the 6G requirements of physical layer and network layers in terms of academy and industry. How much gap do they have? What requirements are the biggest gap?
7.5. In Sect. 7.1.2, the 6G technical challenges can be summarized. Select one and discuss how to overcome the challenges in terms of technology and business development.
7.6. The terahertz bands provide us with huge bandwidths but there are many research challenges. Discuss the research challenges and solutions.
7.7. The basic concept of Open RAN is to open the protocol and build interfaces between many different blocks. The main building blocks of O-RAN consist

of the Radio Unit (RU), the Distributed Unit (DU), and the Centralized Unit (CU). Discuss how AI and ML algorithms can contribute to the O-RAN.

7.8. The user-centric networks allow us to overcome the limits of cell-based wireless communications and provide us with a better quality of network services. Discuss the key research challenges.

7.9. Non-terrestrial networks such as unmanned aerial vehicles (UAV) communications, satellite communication (Satcom), and maritime communications have been widely investigated. They are very useful at specific use cases. Discuss the most suitable use case of each network.

7.10. Compare binary and non-binary LDPC and turbo coding.

7.11. Discuss why new modulation and waveform are required in 6G. What are the pain points of the 5G modulation and waveform?

7.12. Describe the pros and cons of backscatter communications. What is the key challenge in terms of technology and business model?

7.13. Describe the communication mechanism of programmable metasurface.

7.14. Describe the weakness of cognitive radio and benefits of spectrum and network component sharing.

7.15. Discuss the pros and cons of network virtualization.

7.16. Discuss which tasks of communications and networks are related to AI and ML algorithms.

References

1. R. Li, Network 2030: Market drivers and prospects. in *Proceedings of the First International Telecommunications Union (ITU-T) Workshop on Network 2030* (October 2018)
2. ITU-T, *Network 2030: A Blueprint of Technology, Applications, and Market Drivers toward the Year 2030* (November 2019)
3. ITU-T, *FG NET-2030 Technical Report on Network 2030. Additional Representative Use Cases and Key Network Requirements for Network 2030* (June 2020)
4. http://www.jive.nl/e-vlbi
5. K.L. Lueth, State of the IoT 2018: number of IoT devices now at 7B—market accelerating. *IoT Analytics* (August 2018)
6. T. Nakamura, 5G evolution and 6G. in *International Symposium on VLSI Design, Automation and Test (VLSI-DAT)* (2020), pp. 1–1. https://doi.org/10.1109/VLSI-DAT49148.2020.9196309
7. ITU-R Doc 5/40-E, *Minimum Requirements Related to Technical Performance for IMT-2020 Radio Interface(s)* (Feb 2017)
8. ITU-R M.2370–0, *IMT Traffic Estimates for the Years 2020 to 2030* (July 2015)
9. Strategic Research and Innovation Agenda 2021–2027, European Technology Platform NetWorld2020, Smart Networks in the context of NGI (September 2020). https://5g-ia.eu/sns-horizon-europe/. Download from https://bscw.5g-ppp.eu/pub/bscw.cgi/d367342/Networld2020%20SRIA%202020%20Final%20Version%202.2%20.pdf
10. https://www.o-ran.org/
11. R. Irmer, H. Droste, P. Marsch, M. Grieger, G. Fettweis, S. Brueck, H.P. Mayer, L. Thiele, V. Jungnickel, Coordinated multipoint: concepts, performance, and field trial results. IEEE Commun. Mag. **49**(2), 102–111 (2011)

12. D. Gesbert, S. Hanly, H. Huang, S. Shamai O Simeone, W. Yu, Multi cell MIMO cooperative networks: a new look at interference. IEEE J. Sel. Areas Commun. **28**(9), 1380–1408 (2010)

13. S. Kaviani, O. Simeone, W. Krzymien, S. Shamai, Linear precoding and equalization for network MIMO with partial cooperation. IEEE Trans. Veh. Technol. **61**(5), 2083–2095 (2012)

14. V. Jungnickel, K. Manolakis, W. Zirwas, B. Panzner, V. Braun, M. Lossow, M. Sternad, R. Apelfrojd, T. Svensson, The role of small cells, coordinated multipoint, and massive MIMO in 5G. IEEE Commun. Mag. **52**(5), 44–51 (2014)

15. J. Yuan, S. Jin, W. Xu, W. Tan, M. Matthaiou, K. Wong, User-centric networking for dense C-RANs: high-SNR capacity analysis and antenna selection. IEEE Trans. Commun. **65**(11), 5067–5080 (2017)

16. S.B. Glybovski, S.A. Tretyakov, P.A. Belov et al., Metasurfaces: from microwaves to visible. Phys. Rep. **634**, 1–72 (2016)

17. C. Liaskos, S. Nie, A. Tsioliaridou, A. Pitsillides, S. Ioannidis, L. Akyildiz, A new wireless communication paradigm through software-controlled metasurfaces. IEEE Commun. Mag. **56**(9), 162–169 (2018)

18. T.J. Cui, S. Liu, L. Zhang, Information metamaterials and metasurfaces. J. Mater. Chem. C, 3644–3668 (2017)

19. F. Zhang, Q. Zhao, W. Zhang et al., Voltage tunable short wire-pair type of metamaterial infiltrated by nematic liquid crystal. Appl. Phys. Lett. **97**(13) (2010)

20. B. Zhu, Y. Feng, J. Zhao et al., Switchable metamaterial reflector/absorber for different polarized electromagnetic waves. Appl. Phys.Lett. **97**(5) (2010)

21. W. Tang, X. Li, J.Y. Dai, S. Jin, Y. Zeng, Q. Cheng, T.J. Cui, Wireless communications with programmable metasurface: transceiver design and experimental results. J. China Commun. **16**(5), 46–61 (2019)

22. K. Zhang, D. Li, *Electromagnetic Theory for Microwaves and Optoelectronics* (Springer-Verlag, Berlin, Germany, 1998)

23. H. Ye, G.Y. Li, B. Juang, Power of deep learning for channel estimation and signal detection in OFDM systems. IEEE Wirel. Commun. Lett. **7**(1), 114–117 (2018)

Chapter 8
AI-Enabled Physical Layer

In wireless communications, the main purpose of the physical layer is to provide us with the reliable and efficient data transmission links by modelling wireless channels, designing the signal to transmit, detecting the impaired signal, and mitigating channel impairments. Many research works on the physical layer of cellular systems have been done to improve the reliability such as bit error rates and the performances such as throughput. The physical layer architecture has been optimized in terms of throughput and reliability. In this matured research field, it is a big challenge to meet the 6G requirements and improve the performance further by new techniques. In particular, the physical layer is very sensitive to complexity, latency, energy efficiency, and computational power. There are many obstacles to overcome physical layer research problems. Cellular systems such as 5G and 6G will set a high bar and deal with huge data in their systems. This environment enables us to apply AI algorithms to cellular systems. Many AI experts and wireless system designers are forecasting that conventional wireless system models will be complemented by AI-based data-driven approaches. Conventional wireless systems have been designed with approximation of the end-to-end simple mathematical model. It is basically difficult to find accurate mathematical model for wireless system and obtain optimal values of parameters of wireless communication systems. For example, pre-distortion techniques in a baseband enable us to linearize the model with nonlinear distortion. Communication theory provides us with simple mathematical analysis. However, the nonlinearity problems in real world are still tricky parts to analysis by modelling the system. Even it doesn't fit well especially at a high frequency. The learning capability of AI algorithms can be a powerful tool to adapt the time-varying channel environments and linearize the model. The data-driven approach of AI algorithms will improve their capability by learning the relationships between input and output. This approach makes wireless systems simplify and gives us a new opportunity to design physical layer systems in terms of throughput, latency, energy efficiency, and connection density. Key features of AI contributions to physical layer can be summarized as follows: data-driven approach, high parallelism, hardware friendly computation, and flexibility. AI algorithms will improve the efficiency of the transmission link by optimizing components of physical layers such as modulation, channel coding, power level, channel estimation, interference mitigation, and so on. In 6G physical layer, AI algorithms will play key roles. In this chapter, we discuss how AI algorithms will improve physical layer performance and what obstacles we will face in order to adapt AI algorithms in physical layer.

8.1 Design Approaches of AI-Enabled Physical Layer

The key concepts of AI algorithms are adaptation and iteration. As we discussed in Part I, the back propagation of deep learning provides us with the optimal weight values at hidden layers using adaptation and iteration. While training the neural networks with huge data, the weight values are converging to the optimal values. These key concepts adaptation and iteration are already adopted and widely used in physical layer of wireless communications. For example, belief propagation of low-density parity check (LDPC) decoding is based on iteration concept. Adaptive modulation and coding technique improve throughput by channel adaptation. From 1G to 5G, the system performance has been improved while adapting these concepts to each component of physical layer. The physical layer architecture of cellular systems is stable and close to the optimal architecture. Thus, AI algorithms have been initially used to solve higher layer problems such as resource allocation problem, network traffic classification, fault detection, and so on. Mobile operators have many complex tasks to manage network traffics, share network resources, and optimize the network performance. AI and VLSI techniques enable us to process large amounts of data, learn characteristics of dynamic network traffics and resources, obtain key knowledge about network operation and management, and find optimal network solution. Mobile operators are expecting to optimize their networks, increase the network capacity, reduce operational costs, and create new services by integrating AI algorithms into mobile networks. The intelligent network operation using AI techniques allows us to have high network performance and efficient network resource management as well as create a new service to mobile users. Adopting AI algorithms to networks requires a new challenge such as efficient network data management including huge data collection, structure, and analysis. Adopting them to physical layer was not so much active so far. Some people believe that it would be difficult to expect a tremendous leap and achieve a breakthrough in physical layer. However, due to high requirements of 6G physical layer and success models in many different research fields, adopting AI algorithms to physical layer is now actively studied in academy and industry. People believe that AI techniques will be a promising tool to solve the complex problems of wireless communication systems. When developing AI-enabled physical layer, the key challenges can be identified as (1) good balance between wireless system models and data-driven approach of AI algorithms, (2) global optimization in modular physical layer blocks, (3) trade-off between training efficiency and performance, and (4) new metrics or learning capability in terms of different communication system metrics such as throughput, latency, energy efficiency, and so on. The physical layer

*ʳ cellular system depends on the mathematical model for each component. Each
component is optimized individually. In real world, physical layer block chains are
complex systems and there are many unknown impairments and imperfection. If the
mathematical model can contain them all, the mathematical model is too complex
to analyse. In addition, it may suffer from inaccurate prior knowledge between the
model and real world. This is the limitation of the model-based approach. On the other
hand, the data-driven approach of AI algorithms uses the relationship between input
and output. It has less presumption. The physical layer system can be optimized over
a huge amount of training data set. We don't need a tractable mathematical model.
The trained AI algorithms can run faster and have lower power consumption than the
model-based conventional physical layer algorithms. AI algorithms can compensate
for the limitation of traditional wireless systems. For example, AI algorithms can
exploit the physical layer model and use a prior knowledge like channel information from the model. It can reduce the training time and improve the performance
more efficiently. If we can find good balance of two approaches, the physical layer
systems can be efficiently developed. Modular physical layer blocks are composed
of signal detection, synchronization channel coding, modulation, and so on. Each
block is optimized locally and independently. Thus, global optimization cannot be
guaranteed. Since wireless channel is varying and mobile devices face different environment, the physical layer should adapt them efficiently. The learning capability of
AI algorithms can help wireless systems adapt the different environments. As we
reviewed in Part I, it is important to train AI algorithm with enough data set. It
may take a long time to train each physical layer block. Thus, it is important to find
good trade-off point between training efficiency and overall performance. From 5G
systems, wireless system designers focus on not only throughput but also latency,
energy efficiency and others. In 6G systems, we definitely take into consideration
of multiple communication and network metrics. They should be optimized in end-
to-end physical layer. In addition, one metric will be dealt with different priority at
each different block or system layers. For example, bit error rate is important metric
to measure the performance of channel coding and modulation schemes in physical
layer. However, in higher layers, it is not good metric to evaluate the performance
because they have application specific functions. They consider packet error rates or
retransmission rates more importantly. Thus, we may need to redefine metrics to evaluate and optimize the whole system. AI algorithms need to be trained accordingly.
The key physical layer design approaches using AI techniques can be summarized
as follows:

Redesign an End-to-End Physical Layer Chain The conventional physical layer
systems are designed in a block-by-block manner. The block-by-block design concept
provides us with efficient design process, and each block is close to the optimal
structure individually. However, the optimization of each block typically does not
guarantee the optimization of a whole physical layer system. AI techniques may help
us to achieve end-to-end performance optimization in physical layer. Combination
of AI techniques and communication techniques will bring us new benefits. It will
give opportunity to rethink the physical layer architecture in the different views. The

recent advances of VLSI and AI techniques allow us to have a powerful tool for solving many intractable physical layer problems and improve the performance in terms of throughput as well as latency, security, and energy efficiency. For example, the secure wireless communications play an important role in cellular networks. In particular, cellular networks are widely deployed in dense area. Multiple devices in heterogeneous networks are widely used. Traditionally, security and privacy issues were studied in an independent research area with little relation to wireless communications physical layer. Encryption techniques didn't consider wireless channel characteristics. However, one important research trend is to exploit wireless channel characteristics and combine encryption techniques and information theory. The basic concept of physical layer security is to design secure transmission link by exploiting the wireless channel characteristics and keep confidential from eavesdroppers. AI algorithms will be helpful for analysing the wireless channel characteristics and making decision about thresholds. Adopting AI algorithms to physical layer security will enhance security paradigm compared with conventional approaches and increase secrecy capacity of wireless links. There are few research results to achieve end-to-end optimization and performance improvement by combination between AI and communications techniques [1–3]. They give us a sneak peek of redesigning physical layer. In addition, a new end-to-end physical layer design using an autoencoder is proposed in [1, 2]. It will be discussed in Sect. 8.2. In [4], reinforcement learning (RL) was proposed to avoid the missing gradient problem from channels and optimize the transmitter. As we reviewed reinforcement learning in Chap. 5, the RL algorithm deals with the decision-making problem that is formulated mathematically in a Markov decision process (MDP). There is an agent in the interaction process. This interaction process is expressed by a finite Markov decision process (MDP). The RL algorithm enables the agent to learn behaviours to achieve its goal. In the RL-based end-to-end physical layer, the transmitter is regarded as an agent. The channel and the receiver are regarded as the environment. The source data and the transmit signals are state and action, respectively. The reward is the end-to-end loss. Like conventional RL algorithm, the transmitter learns to take an action to maximize the cumulative rewards that optimize the end-to-end loss. However, it is not clear whether or not this approach can outperform the conventional physical layer architecture. The theoretical background should be developed further, and their performance should be compared under the different channel environments. Figure 8.1 illustrates the RL-based end-to-end physical layer.

Implementation of AI algorithms on Lightweight Platforms and HW-SW Co-design When adopting AI algorithms to physical layer, we consider key characteristics of physical layer design and implementation. Typically, AI algorithms require training and intensive computation with large volumes of data. For example, the computation of AI algorithms should be fast enough to less than channel coherence time or system parameter change time. In particular, when applying AI algorithms for ultra-reliable low latency communication (URLLC), the latency is key performance indicator. For example, in 6G URLLC, the end-to-end latency requires less than 1 ms.

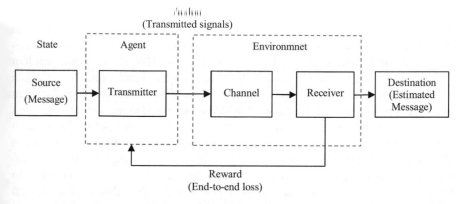

Fig. 8.1 RL-based end-to-end physical layer structure

The latency is approximately composed of device processing time 0.3 ms, transmission time 0.1 ms, network processing time 0.5 ms, and other margin 0.1 ms. Thus, AI algorithms should run with latency less than 0.1 ms. It causes higher sampling rate of the 6G waveforms and more computational powers to analyse time-varying wireless channels. In radio access networks, the base band processing in physical layer runs on an user equipment and a base station. Although the computational power of the user equipment increases in every generation, the workload of the base band processing is already heavy at the user equipment and it would not be efficient to run AI algorithms on user equipment. Thus, the base band processing at base stations would be suitable for adopting AI algorithms to physical layer. In order to run AI algorithms, the intelligent application with high computations will be deployed at the edge of radio networks. The wireless links with high reliability, low latency, and high capacity between the user equipment and the edge networks will be essential in 6G systems. The distributed computation and storage with enough data protection should be supported in 6G systems. In addition, AI algorithms run on dedicated graphics processing units (GPU) or central processing unit (CPU). The processing power of the base band is limited so that AI algorithms should run with limited computing power, storage, and energy sources. Thus, lightweight platforms with respect to computing power, energy efficiency, and reliability should be developed to accelerate training speed and operate AI algorithms efficiently. Another view point of AI algorithms implementation is hardware (HW)–software (SW) co-design (or AI algorithm-hardware co-design). The HW-SW co-design enables us to develop hardware and software concurrently and save the development time. We can optimize a system performance more efficiently by the iterative process of HW-SW co-design. The first step of HW-SW co-design is system specification and architecture definition of physical layer. The specification is to define the physical layer tasks and describe the behaviour including concurrency, hierarchy, state transition, timing, and so on. The unified representation such as data flow graphs and Petri Nets can describe specific tasks of physical layer that can be implemented in HW or SW. The second step is HW-SW partition. This step is the key part of HW-SW co-design because the partition affects to overall performance, development time, and cost. The main purpose of HW-SW partition is to design a physical layer system to meet the requirements (throughput, latency, energy efficiency, and so on) under the given constraints (cost, power, complexity, and so on). The HW-SW partition problem is typically known as a multivariate optimization problem and a non-deterministic polynomial-time hard (NP-hard) problem. In order to solve this problem, there are two main approaches: constructive algorithms and iterative algorithms. The next steps are HW/SW interface design, co-synthesis, and co-verification. Many electronic design automation (EDA) tools are supporting this step. HW-SW co-design is efficient to implement the end-to-end physical layer chain. It improves good system performance and provides us with lower costs and smaller development cycle. In terms of AI algorithms implementation, HW-aware AI algorithm implementation would be helpful for overcoming the physical layer limitations. Task partitions in HW and SW platform would be key functions to implement AI algorithm efficiently.

New AI Algorithms for Physical Layer Functions New AI algorithms should be developed to satisfy both low complexity and reasonable accuracy. Traditionally, many signal processing techniques (such as signal detection, channel estimation, coding and modulation, interference mitigation, and so on) in physical layer are based on maximum likelihood (ML), maximum a posteriori (MAP), minimum mean squared error (MMSE), zero forcing (ZF), and so on. These algorithms provide us with an optimal solution under the given environments but typically involve complex computation such as high-dimensional matrix operation and a large number of iterations. Thus, wireless system designers should find good trade-off between performance and computational complexity by heuristic parameter tuning. In practice, the heuristic parameter tuning is performed by simulation in various environments. One selection works well under one channel condition, but it doesn't provide us with good trade-off if channel is varying. Thus, heuristic approaches are instable. We expect that AI algorithms improve the parameter tuning problem by training and learning ability of deep learning. Most of AI algorithms rely on the training data set. Using the big data analysis, many new features can be adopted in 6G systems. For example, future states of 6G systems can be predicted from the historical data. It also can improve 6G system design, fault detection, optimal configuration, and so on. In particular, it is possible to refine the end-to-end communication chain and achieve a global optimal construction by well trained physical layer models. However, if there is no availability of huge data set, accuracy of AI algorithms are very low. The data collection is directly related to the performance of AI algorithms, and we should carefully select data sets from cellular systems. The data set generated by physical layer contains information about channel impairments in time, frequency, and space domains, for example, channel noises, fading, and interferences over different frequency and time. We need to extract characteristics of radio data and channel impairments data from those data sets and adjust the parameters such as transmit power, beamforming directions, spectrum, modulation, and so on. In addition, the data-driven approach of AI algorithms will provide us with a new opportunity to design physical layer. The AI algorithms for physical layer should be developed while considering specific physical layer environment.

Selection of AI Algorithms to Improve and Combination Between AI Algorithms and Physical Layer Algorithms As we discussed in Part I, key applications of AI algorithms are classification, clustering, regression, and decision making. AI algorithms are categorized as unsupervised learning, supervised learning, reinforcement learning, and deep learning. Unsupervised learning does not need labelled data and extracts the structures and patterns of the data from the trained model. Supervised learning has labelled data and builds a model from training data set. The trained model enables us to predict the future data and find patterns. Reinforcement learning makes an agent to interact with an environment while maximizing the rewards. The agent learns how to take an action and achieve the goal in the environment. We can make a sequence of decisions optimally. Deep learning mimics human brain. It allows us to learn from large amounts of data and predict the data patterns. Many physical layer algorithms are in line with these key applications and the basic mechanism of AI techniques. We can reformulate the physical layer problems to apply AI algorithms. The learning methods of AI algorithms would improve the performance of physical layer algorithms and models. It is possible to simply adapt AI algorithms to physical layer. For example, K-mean clustering will be helpful for spectrum sensing, power allocation, and relay node selection. In physical layer, the block chain in a transmitter and a receiver can be decomposed into multiple independent components. Each component of the physical layer block chain performs specific functions, and it can be designed independently. For example, channel modelling is one of key research topics in physical layer. Conventional channel modelling makes assumptions to reduce the complexity in various environment and simplify signal propagation model. Channel characteristics are extracted from huge measurement data, and wireless channel models are constructed from the key chancel characteristics. The important point of the conventional channel modelling is to find good trade-off between model accuracy and assumptions. We can predict pass loss, channel impulse response, and signal propagation model in 6G heterogeneous networks using AI algorithms. The classification of waveforms is one of key selections. The legacy low-dimensional techniques can distinguish up to 20 or 30 devices, but AI algorithms enables us to distinguish hundreds of devices by learning key features in noisy channels [5]. Physical layer algorithms are based on model-driven optimization. When the model assumptions do not match with the actually wireless systems or channel conditions, the physical layer algorithms do not provide us with the optimal performance. For example, one physical layer algorithm is designed under the low mobility channel environment. The physical layer algorithm couldn't show us the best performance at the high-mobility channel environment. However, AI algorithms using data-driven optimization can be adapted their parameters and may perform fine tuning in real time through gradient descent methods.

As we discussed in Chap. 7, AI algorithms may contribute to the physical layer block chain as component wise at first. In the final stage, the physical layer processing may be replaced by AI-based physical layer processing with learning mechanism.

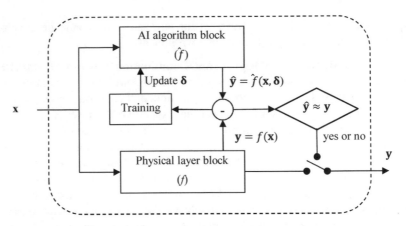

Fig. 8.2 Architecture of AI-supporting physical layer component

There are three main design approaches to improve physical layer blocks by AI algorithms. The first approach is to support each physical layer block. Many physical layer blocks find an optimal solution iteratively and adaptively. It takes a long time to converge it and requires a high complexity. AI algorithms can help to find the optimal point and reduce the computational time and complexity. Figure 8.2 illustrates the architecture of AI-enabled physical layer component based on the first approach.

As we can observe Fig. 8.2, the input $\mathbf{x}$ is fed into both physical layer block and AI algorithm block. Both function blocks can be expressed as physical layer block: $\mathbf{y} = f(\mathbf{x})$ and AI algorithm block: $\hat{\mathbf{y}} = \hat{f}(\mathbf{x}, \boldsymbol{\delta})$ where $\boldsymbol{\delta}$, $\hat{\mathbf{y}}$, and $\mathbf{y}$ are AI algorithm parameters, AI algorithm block output, and physical layer block output, respectively. Firstly, we train AI algorithm block at certain time with input training data set by comparing the output of both blocks. Secondly, we update the AI algorithm block parameters $\boldsymbol{\delta}$. If we have the trained AI algorithm block, we can improve the physical layer block by comparing the outputs between them. When we have convex or non-convex optimization problems of physical layer, the optimization problem is formulated and solved by an iterative process. It requires huge computational complexity. The AI-enabled physical layer component can learn the relationship between the input and the output and extract key properties. The approximation allows us to train the AI algorithm block. Using this knowledge, we can reach the optimal point quickly and also find good balance among convergence time, accuracy, and complexity. The second approach is to include AI algorithm block as supplementary block in the physical layer component as shown in Fig. 8.3.

In physical layer, there are many nonlinear and unpredicted distortions such as nonlinear power amplifier. They are typically modelled by an unknown function. We have the distorted output $\mathbf{x} = g(\mathbf{y})$ from the unknown function. The distortion is estimated and compensated by the inversion function. As we can observe Fig. 8.3, AI algorithm block can act like the inversion function. In the AI algorithm block, the training is performed, the AI algorithm parameter $\boldsymbol{\delta}$ is updated, and the model of the inversion function is created. The trained AI algorithm block reduces the

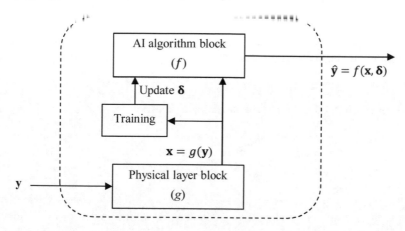

Fig. 8.3 Architecture of physical layer component including AI algorithm block

approximation errors and achieve the optimal form of the inversion function. The third approach is to combine both a physical layer component and an AI algorithm block or replace a physical layer component by an AI algorithm block as shown in Fig. 8.4.

Some sub-functions of physical layer block already have key concept of iteration and adaptation. The sub-functions can be improved by new AI algorithms.

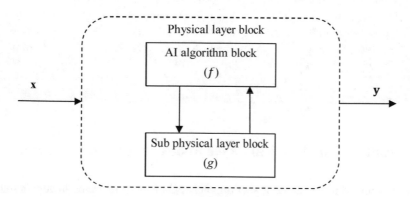

Fig. 8.4 Architecture of combining physical layer and AI algorithm block

Many parts of them can be optimized. We can reduce complexity and iteration time of the conventional physical layer blocks and compensate many distortions of the transmitter and receiver more efficiently.

Summary 8.1. AI-Enabled Physical Layer Design

1. The key concepts of AI algorithms are adaptation and iteration. These concepts were already adopted to each component of physical layer. The physical layer architecture of cellular systems is stable and close to the optimal architecture.
2. Due to high requirements of 6G physical layer and success models in many different research fields, adopting AI algorithms to physical layer is now actively studied in academy and industry.
3. The key challenges of AI-enabled physical layer can be summarised as (1) good balance between wireless system models and data driven approach of AI algorithms, (2) global optimization in modular physical layer blocks, (3) trade-off between training efficiency and performance and (4) new metrics or learning capability in terms of different communication system metrics.
4. AI algorithms can compensate for the limitation of traditional wireless systems by data-driven approach.
5. AI-enabled physical layer design approaches can be summarised as (1) Re-design an end-to-end physical layer chain, (2) Implementation of AI algorithms on lightweight platforms and HW-SW co-design, (3) New AI algorithms for physical layer functions, and (4) Selection of AI algorithms to improve and combination between AI algorithms and physical layer algorithms.

8.2   End-To-End Physical Layer Redesign with Autoencoder

As we discussed in the previous section, physical layer system architecture is stable to establish connectivity. Nevertheless, there are many practical physical layer problems such as lack of optimal solutions, models, and algorithms. A new physical layer view in terms of AI systems might be helpful for finding optimal physical layer solutions. An autoencoder is an unsupervised neural network to reconstruct unlabelled data by training the neural network to reduce their dimension and then learn how efficiently we compress them [1, 6, 7]. This idea has been very popular in the area of deep learning. It was used for image processing in order to have the compressed representation. The simplest form is similar to multilayer perceptrons (MLP). An autoencoder consists of the encoder and the decoder as shown in Fig. 8.5.

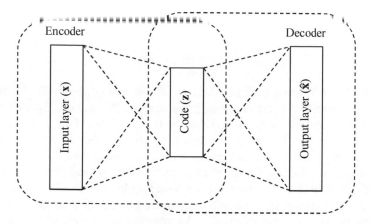

Fig. 8.5 A basic structure of autoencoder

As we can observe Fig. 8.5, we encode the input data $\mathbf{x} \in \mathbb{R}^p$ to compressed data $\mathbf{z} \in \mathbb{R}^q$ where $\mathbf{z}$ have lower dimensionality than $\mathbf{x}$. It is possible to reconstruct $\mathbf{x}$ from $\mathbf{z}$ faithfully. The compressed data $\mathbf{z}$ is referred to as a code. The code is decoded to $\hat{\mathbf{x}}$ that approximate the original data $\mathbf{x}$. The encoding and decoding process is expressed as follows:

$$\text{Encoding: } \mathbf{z} = \mathbf{W}_1 \mathbf{x} + \mathbf{b}_1, \tag{8.1}$$

$$\text{Decoding: } \hat{\mathbf{x}} = \mathbf{W}_2 \mathbf{z} + \mathbf{b}_2 \tag{8.2}$$

where $\mathbf{W}_1$ and $\mathbf{W}_2$ are weight matrices and $\mathbf{b}_1$ and $\mathbf{b}_2$ are bias vectors. The encoding outcome $\mathbf{z}$ can be regarded as a compressed representation of the input vector $\mathbf{x}$. Basically, weights and biases are randomly initialized and iteratively updated during training. The autoencoder is trained to minimize squared errors using backpropagation of the error. In order to have approximation $\hat{\mathbf{x}}$, the objective function (or loss function) can be defined as follows:

$$
\begin{aligned}
\text{Objective function: } J(\mathbf{x}, \hat{\mathbf{x}}) &= \|\hat{\mathbf{x}} - \mathbf{x}\|^2 \\
&= \|\mathbf{W}_2 \mathbf{z} + \mathbf{b}_2 - \mathbf{x}\|^2 \\
&= \|\mathbf{W}_2(\mathbf{W}_1 \mathbf{x} + \mathbf{b}_1) + \mathbf{b}_2 - \mathbf{x}\|^2.
\end{aligned}
\tag{8.3}
$$

We can minimize the objective function using stochastic gradient descent. In this approximation, the activation function of the hidden layer is linear. Thus, we call this a linear autoencoder. If the data is nonlinear, we can define activation function $\sigma()$ such as sigmoid function or rectified linear unit. The encoding and decoding process is modified as follows:

$$\mathbf{h} = \sigma(\mathbf{W_1x} + \mathbf{b_1}), \tag{8.4}$$

$$\hat{\mathbf{x}} = \sigma'(\mathbf{W_2h} + \mathbf{b_2}) \tag{8.5}$$

where $\sigma()$ and $\sigma'()$ are activation functions. The decoding parameters σ', $\mathbf{W_2}$ and $\mathbf{b_2}$ may be not related to the encoding parameters σ, $\mathbf{W_1}$ and $\mathbf{b_1}$. Figure 8.5 illustrates a simple layer encoder and decoder, but it is possible to have a multiple layer encoder and decoder. The multiple layered encoder and decoder provide us with many advantages such as decreasing the number of training data.

C. E. Shannon published the landmark paper "A Mathematical Theory of Communication" [8] in 1948. At that time many researchers in the area of communications wanted to know how to measure information and how much information can send via a channel. He measured information by adopting the concept of entropy which was used in thermodynamics. Entropy of information theory means a level of the uncertainty of a random variable. In addition, he constructed the communication architecture as shown in Fig. 8.6. This communication architecture was innovative because it enables a communication system designer to treat each component of the end-to-end communication system separately.

Based on Shannon's communication architecture, a simple system model with a discrete channel is considered as shown in Fig. 8.7 [9].

In this communication model, the source produces a discrete message m_i (where $i = 1, \ldots, M$) as a random variable. We assume that the probability the message m_i appears is a priori probability $P(m_i)$. The transmitter generates a signal s_i mapping from the message m_i. The signal s_i is transmitted via the discrete channel. The

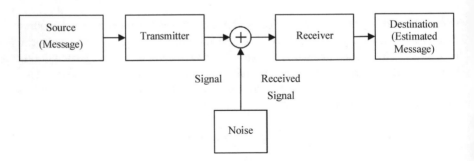

Fig. 8.6 Shannon's communication architecture

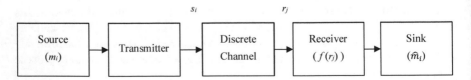

Fig. 8.7 Communication system model with a discrete channel

received signal r_j is expressed as a conditional probability $P(r_j|s_i)$ that means the probability that the received signal r_j is arrived at the receiver when the transmitter sends the signal s_i. We estimate the original messages $\hat{m}_i$ using the decision rule $f()$. The decision rule can be mathematically expressed as follows: $f(r_j) = \hat{m}_i$ where the function is mapping from the received signal to the most probable value. The probability of error is defined as follows:

$$P_\varepsilon = P(\hat{m}_i \neq m_i). \tag{8.6}$$

The receiver is constructed to minimize the probability of error P_ε. If the receiver can minimize the probability of error, we call this an optimum receiver.

Those two concepts can be combined. Conventional wireless communication is divided into multiple signal processing blocks. They are individually optimized. It is not easy to achieve global optimization of the transmitter and receiver by designing individual block separately. The autoencoder can help us to optimize the end-to-end physical layer block chain. The optimization using the autoencoder enables us to have a better performance than the traditional physical layer [1, 6]. The autoencoder adds redundancy and minimizes the errors for a wireless channel while adopting learning mechanism in physical layer blocks. Figure 8.8 illustrates the end-to-end physical layer architecture using an autoencoder. We consider the simplest communication system composed of a transmitter, an AWGN noise, and a receiver. In a transmitter, a message $s \in \mathbb{M} = \{1, 2, \ldots, M\}$ is generated. A transmitted symbol $\mathbf{x} \in \mathbb{C}^n$ with a power constrain $\|\mathbf{x}\|^2 \leq n$ and an amplitude constraint $|x_i| \leq 1$ for all i is produced from a message. The message length is $k = \log_2 M$ [bits]. This message is transmitted to a receiver via n channel uses. Thus, the coding rate R is k/n [bits/channel use]. As the channel is described in Fig. 8.7, the channel can be represented by conditional probability density function $p(\mathbf{y}|\mathbf{x})$ where $\mathbf{y} \in \mathbb{C}^n$ is a noisy received symbol. The receiver finds the estimate $\hat{s}$ of the message. The block error probability is defined as follows:

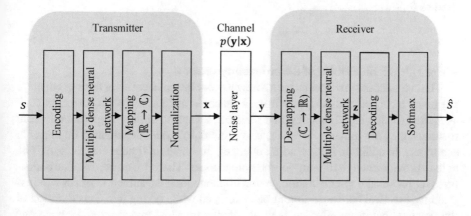

Fig. 8.8 End-to-end physical layer using an autoencoder

$$P_e = \frac{1}{2^k} \sum_s P(\hat{s} \neq s|s). \tag{8.7}$$

The original purpose of the autoencoder is to reduce the dimensionality of the inputs. However, the autoencoder of wireless communication application finds a robust representation $\mathbf{x}$ in terms of channel impairments. The receiver should be able to recover the original message from the representation with the low probability of error. Thus, the autoencoder may not reduce dimensionality but add redundancy. As we can observe Fig. 8.5, the message passes through each neural network and a robust representation of the message is searched. The transmitter of the autoencoder is composed of encoding, neural network layer, mapping and normalization as shown in Fig. 8.8. In the encoding block, the message s is transformed to M dimensional vector whose elements are all zeros except the sth element. That is to say. It is a one-hot vector $1_s \in \mathbb{R}^M$. The encoded inputs are fed to a feedforward neural network and then the n-dimensional complex valued vectors are generated. The complex value is concatenated by real and imaginary parts mapped from the encoded inputs. In the normalization block, the power constraint on the representation $\mathbf{x}$ is satisfied. The channel is expressed as an AWGN channel or a noise layer with fixed or random noise power. In addition, multiple path effect, frequency offset, phase offset, or timing offset can be included in the channel layer. The receiver of the autoencoder consists of demapping, neural network layer, decoding and softmax as shown in Fig. 8.8. In the demapping block, the complex values are transformed to the real values. They are fed to a feedforward neural network. The output of the neural networks is a probability vector $\mathbf{z} \in (0, 1)^M$ assigning a probability over the possible messages. The estimate $\hat{s}$ is decoded by finding the index of the most probable element of $\mathbf{z}$. In the last layer, softmax activation function is performed. The autoencoder can be trained by stochastic gradient descent (SGD) on the set of all possible messages. During the training, the estimate values should be equal to the original values. In the end-to-end physical layer redesign using the autoencoder, the communication problem is regarded as the classification problem. The cross-entropy loss function is used as follows:

$$J_l = -\log z_s \tag{8.8}$$

where z_s is the sth element of the probability vector $\mathbf{z}$.

The autoencoder provides us with a new view of the physical layer design, but there are many technical challenges. Wireless channels include many impairments such as fading, shadowing, frequency, and phase offset and so on. Using gradient descent methods, they are trained. However, it is difficult to model the channel impairment accurately. In addition, in fast fading channel, the coherent time is short. It won't be able to provide enough training for the autoencoder. There are no hardware architectures for practical implementation. When implementing the autoencoder for wireless systems, practical hardware should be developed while taking into consideration of the practical hardware effect as well as regulation. Most importantly, we have one big assumption that the training data distribution is same as the test data distribution.

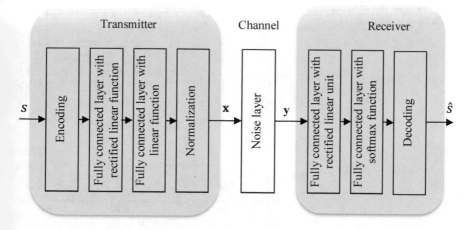

Fig. 8.9 Fully connected autoencoder for end-to-end physical layer

In order to satisfy, the training should be performed in many different environments such as different fading, different SNR, different power allocations, and so on. This is a big challenge. A proper solution should be developed to implement it to an actual physical layer system.

Example 8.1 End-to-End Physical Layer Using the Autoencoder (I) Consider a fully connected autoencoder for end-to-end physical layer as shown in Fig. 8.9.

In the encoding of the transmitter, one-hot vector of length M is generated. In the decoding of the receiver, the classification is performed by finding the most probable value of the length M vector. The autoencoder is trained with the following parameters:

- Number of channel uses (n): 4
- Message length (k): 4
- Normalization: Energy
- Noise layer: AWGN
- $E_b N_0 = 2$ and 4 dB
- Initial learning rate: 0.01
- Maximum epochs: 15
- Minibatch size: 320.

Train the autoencoder and compare the loss and accuracy in terms of different energy (2 and 4 dB).

Solution

The Matlab 5G and deep learning toolbox [10] is exploited. Using computer simulation, the autoencoder is trained at a low bit energy ($E_b N_0 = 2$ dB) and a high bit energy ($E_b N_0 = 4$ dB). We can obtain loss and accuracy in terms of different energy as shown in Fig. 8.10.

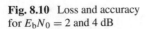

Fig. 8.10 Loss and accuracy
for $E_bN_0 = 2$ and 4 dB

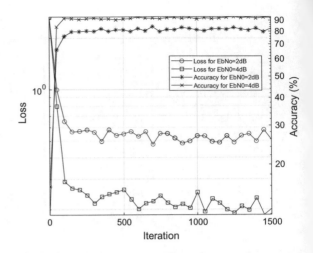

As we can observe Fig. 8.10, the accuracy of a low bit energy and a high bit energy is converging at 80% and 90%, respectively. Their loss is decreasing accordingly. At the low energy, the training can be converged even if some error occurs. In order to maintain good level of accuracy, the training should be performed at the high energy level.

∎

Example 8.2 End-to-End Physical Layer Using the Autoencoder (II) Consider a fully connected autoencoder for end-to-end physical layer as shown in Fig. 8.9. The autoencoder is trained with the following parameters:

- Number of channel uses (n): 2
- Message length (k): 2
- Normalization: Energy
- Noise layer: AWGN
- $E_bN_0 = 2$ and 6 dB
- Initial learning rate: 0.01
- Maximum epochs: 15
- Minibatch size: 320.

Train the autoencoder and compare signal constellation in terms of different energy (2 and 6 dB).

Solution

The Matlab 5G and deep learning toolbox [10] are exploited. Using computer simulation, the autoencoder is trained at $E_bN_0 = 2$ and 6 dB. We can obtain loss and accuracy in terms of different energy as shown in Fig. 8.11.

As we can observe Fig. 8.11, the signal constellation is QPSK constellation. This is because we have $M = 2^k = 2^2 = 4$. The constellation positions of the learned signal constellation are different from a regular QPSK signal constellation.

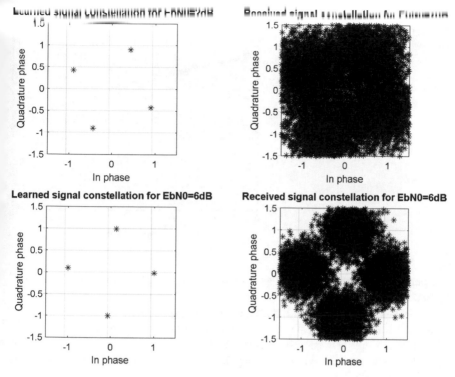

Fig. 8.11 Signal constellation comparison for $E_bN_0 = 2$ and 6 dB

The transmitted symbols **x** are composed of n complex symbols representing one message, and they are correlated over time. In addition, as we can imagine, the received signal constellation positions at a higher energy stick together.

∎

Example 8.3 End-to-End Physical Layer Using the Autoencoder (III) Consider a fully connected autoencoder for end-to-end physical layer as shown in Fig. 8.9. The autoencoder is trained with the basic configuration of Example 8.2. We generate random integers in the range [0 $M - 1$] where $M = 2^k$ and then encode these into complex values. The autoencoder (2,2) with $k = 2$ and $n = 2$ and the autoencoder (7,4) with $k = 4$ and $n = 7$ are considered. They are fed into the trained neural network. The normalized symbol is transmitted to the receiver via AWGN channel. In the receiver, the decoding process is performed.

Compare block error rates (BLER) of autoencoder (2,2), autoencoder (7,4), and theoretical BLER of QPSK modulation and (7,4) Hamming code.

Solution

The Matlab 5G and deep learning toolbox [10] are exploited. We perform Monte Carlo simulation and obtain the BLER of autoencoder (2,2) and (7,4). The block sizes of autoencoder (2,2) and (7,4) are 2 bits and 4 bits, respectively. They are

Fig. 8.12 Comparison of
BLER performance for
autoencoder (2,2) and (7,4)

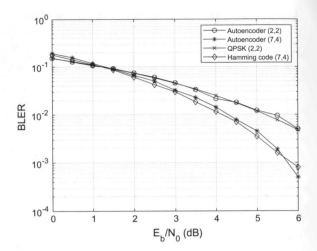

compared with theoretical BLER as shown in Fig. 8.12. Hamming code is decoded
by maximum likelihood method.

As we can observe Fig. 8.12, both autoencoders (2,2) and (7,4) performances are
almost same as the theoretical values of QPSK and (7,4) Hamming code, respectively.
From this example, we can know that autoencoder can obtain coding grain. Fully
trained autoencoders for end-to-end physical layer can be used for channel coding
as well as modulation.   ∎

Summary 8.2. End-to-End Physical Layer Re-design with Autoencoder

1. An autoencoder is an unsupervised neural network to reconstruct unla-
    belled data by training the neural network to reduce their dimension and
    then learn how efficiently we compress them. It was used for image
    processing in order to have the compressed representation. The autoen-
    coder of wireless communication application finds a robust representation
    of the transmitted symbols in terms of channel impairments.
2. The autoencoder and the conventional communication architecture can
    be combined. The autoencoder can help us to optimize the end-to-end
    physical layer block chain.
3. The autoencoder adds redundancy and minimizes the errors for a wireless
    channel while adopting learning mechanism in physical layer blocks.
4. When adopting autoencoder for wireless systems, there are many technical
    challenges such as accurate channel layer design, short coherent time,
    hardware implementation architecture and so on. Most importantly, we
    have one big assumption that the training data distribution is same as the

test data distribution. In order to satisfy, the training should be performed in many different environments such as different fading, different SNR, different power allocations and so on. This is a big challenge. A proper solution should be developed to implement it to an actual physical layer system.

8.3 Wireless Channel Models

Classical wireless channels are regarded as the unknown black box. They are unpredictable noises. However, in modern wireless communications, they are categorized as Gaussian noise, jitter, path loss, fading, shadowing, phase and frequency noise, interferences, and so on. These different channel impairments damage the transmitted signals differently. Acquiring accurate channel state information (CSI) becomes an essential part of the modern wireless communication systems. Wireless system designers make a big effort on channel modelling. However, the wireless channel models are inherently approximation of the real wireless channel. The parameters of the wireless channel models should be estimated. The estimation errors are inevitable. The physical layer is designed to overcome those channel impairments. Each component of the physical layer mitigates the different types of channel impairments and recovers the transmitted signals. For example, one transmitted signal arrives at a receiver via multiple wireless channel paths. Individual wireless channel is experiencing different levels of channel effects. Diversity techniques use those multiple received signals from different paths and mitigate multipath fading. They take an important role in physical layer to improve the reliability. Small-scale fading is categorized as flat fading ($B_s < B_c$), frequency selective fading ($B_s > B_c$), fast fading ($T_s > T_c$), and slow fading ($T_s < T_c$) where B_s, B_c, T_s, and T_c denote the signal bandwidth, coherent bandwidth, signal time interval, and coherent time, respectively. They can be modelled by random and deterministic components. The stochastic channel model with both components is typically built by measurement data over different environments. However, this model can't be used for predicting the time-varying channels in a real world. The deterministic components can be obtained by the knowledge of the radio propagations under the given environments, but it is difficult to forecast the random components. 6G systems will resort to AI techniques to obtain the channel knowledge including random components and predict time-varying channels. They enable a wireless system to learn channel models and adapt to new channel conditions. In order to implement AI algorithms, we need to collect a large amount of channel measurement data, train AI algorithms models, and predict channel knowledge such as channel impulse response and so on. The channel knowledge will be used in many different physical and data link layer tasks including MIMO, beamforming, resource allocation, power control, and so on. Based on channel model characterizing real channel environments, physical layer requirements are defined.

The channel models are developed in complex channel scenarios with many types of noise, nonlinearities, and imperfections. The physical layer design depends on the channel conditions to overcome the channel impairments.

When adapting AI algorithms to channel modelling, one of big advantages are not to develop the model and focus on the relationship between channel input and output. Channel state information such as received power, root mean squares (RMS) delay, angle spread are predicted using feed forward neural networks and radial basis function neural network [11]. ANN can be used to remove the noise from channel impulse response model and principal component analysis (PCA) enable us to exploit the features of channel model [12]. Multiple path components are clustered by unsupervised learning algorithms such as K-mean clustering and fuzzy C-means algorithms [13]. Convolutional neural networks can be applied to identify different wireless channels [14]. The input parameters of CNN are multipath components. After training from measurement data, we can obtain the classification output of the different wireless channels. In [15], autonomous supervised learning was proposed to model wireless channels and protect personal privacy. Pre-training approach enables us to learn deterministic properties of wireless channels from automatically generated huge amount of labelled data. The trained channel model is capable of tuning the model. Therefore, it can save training time. Using the pre-training approach, we can obtain the channel knowledge in the given environment and predict it for the time-varying channels. The performance of the pre-training approach depends on the good quality of the training data and neural networks. As we reviewed in Part I, recurrent neural network (RNN) deals with time sequence problems because its connections form directed cycles. RNN will be suitable for pre-training channel modelling. However, the pre-training channel model couldn't be generalized. The generalization of more realistic channel scenarios would be one big research challenge. The pre-trained channel can extract the features of the channels and update the weights of the model by maximizing the probabilities of the correct labelled data. It is possible to train any task-specific channel models including beamforming, user paring, and so on.

Summary 8.3. Wireless Channel Models

1. There are different channel impairments such as Gaussian noise, jitter, path loss, fading, shadowing, phase and frequency noise, interferences and so on. They damage the transmitted signals differently.
2. Wireless channel models are inherently approximation of the real wireless channel. The estimation errors are inevitable.
3. AI algorithms will be helpful for obtaining the channel knowledge including random components and predicting time-varying channels.
4. AI algorithms for wireless channels focus on the relationship between channel input and output.

8.4 Signal Detection and Modulation

The key function of physical layer is to determine how information is sent from a transmitter to a receiver over a physical channel. In this process, the signal detection plays an important role to identify what information was sent from the transmitter. In order to detect a transmitted signal in a receiver, signal detection is essential. Decision theories are widely used in not only wireless communication systems but also other applications. The decision theory of wireless communication systems is typically developed to minimize the probability of errors. The matched filter can be used for the coherent detection of modulation schemes [9]. We consider a coherent M-ary phase shift keying (MPSK) modulation. The transmitted signal $s_i(t)$ is

$$s_i(t) = \sqrt{\frac{2E_s}{T}} \cos\left(2\pi f_0 t - \frac{2\pi i}{M}\right) \qquad (8.9)$$

where $0 \leq t \leq T$, $i = 1, 2, \ldots, M$, f_0 is carrier frequency, and E_s is the signal energy for symbol duration T. We assume the two-dimensional signal space. The orthonormal basis functions $\phi_1(t)$ and $\phi_2(t)$ are

$$\phi_1(t) = \sqrt{\frac{2}{T}} \cos(2\pi f_0 t), \quad \phi_2(t) = \sqrt{\frac{2}{T}} \sin(2\pi f_0 t). \qquad (8.10)$$

Therefore, the transmitted signal $s_i(t)$ is

$$s_i(t) = s_{i1}\phi_1(t) + s_{i2}\phi_2(t), \qquad (8.11)$$

$$s_i(t) = \sqrt{E_s} \cos\left(\frac{2\pi i}{M}\right)\phi_1(t) + \sqrt{E_s} \sin\left(\frac{2\pi i}{M}\right)\phi_2(t). \qquad (8.12)$$

When $M = 4$, we have quadrature phase shift keying (QPSK) modulation. The transmitted signal $s_i(t)$ has 4 different variants by combining two orthonormal basis functions $\phi_1(t)$ and $\phi_2(t)$. Thus, the transmitted signal $s_i(t)$ is

$$s_i(t) = \sqrt{E_s} \cos\left(\frac{\pi i}{2}\right)\phi_1(t) + \sqrt{E_s} \sin\left(\frac{\pi i}{2}\right)\phi_2(t). \qquad (8.13)$$

The transmitted signal $s_i(t)$ can be illustrated in the two-dimensional signal space as shown in Fig. 8.13.

As we can observe Fig. 8.13, the two-dimensional signal space is divided into 4 areas. If the received signal $r(t)$ falls in area 1, we decide the transmitter sent $s_1(t)$. If it falls in area 2, we decide the transmitter sent $s_2(t)$. We decide $s_3(t)$ and $s_4(t)$ in the same way. The signal estimator needs M correlators for the demodulation of MPSK. The received signal $r(t)$ of MPSK system is

Fig. 8.13 Signal space and decision areas for QPSK modulation

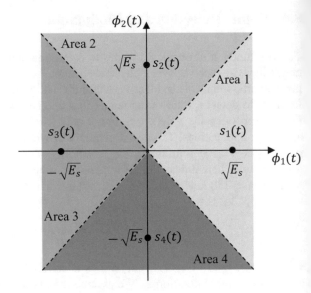

$$r(t) = \sqrt{\frac{2E_s}{T}} \left(\cos\left(\frac{2\pi i}{M}\right) \cos(2\pi f_0 t) + \sin\left(\frac{2\pi i}{M}\right) \sin(2\pi f_0 t) \right) + n(t) \quad (8.14)$$

where $0 \leq t \leq T$, $i = 1, 2, \ldots, M$ and $n(t)$ is white Gaussian noise. The demodulator of MPSK system can be designed by the matched filter as shown in Fig. 8.14.

In Fig. 8.14, the output of one integrator I is the in-phase part of the received signal and the output of the other integrator Q is the quadrature-phase part of the received signal. The output of arctan function block is compared with a prior value θ_i. For QPSK modulation, it has 4 values such as 0, $\pi/2$, π and $3\pi/2$ in radian. The smallest the phase difference is chosen as the estimated signal $\hat{s}_i(t)$. When dealing

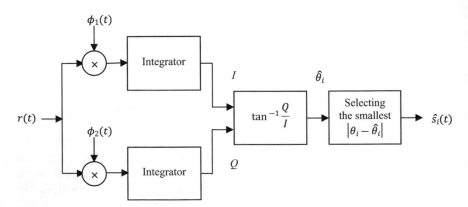

Fig. 8.14 Demodulator of MPSK modulation

Fig. 8.15 Example of signal
space and decision areas for
QPSK modulation after
training the neural network

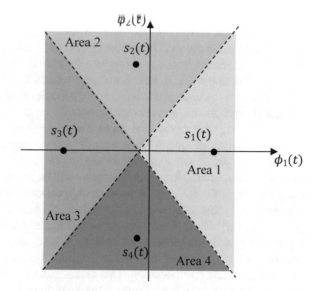

Fig. 8.15 Example of signal space and decision areas for QPSK modulation after training the neural network

with non-coherent detection system, we should have different approach. Since the receiver does not have any knowledge about a reference (such as the carrier phase and the carrier frequency) of the transmitted signal, the non-coherent detection cannot use the matched filter. The differential phase shift keying (DPSK) is basically classified as non-coherent detection scheme. When detecting the differentially encoded signal, we do not need to estimate a phase of carrier and use the phase difference between the present signal phase and the previous one.

If we use an AI technique for signal detection and train the neural network, we may have different signal detection area as shown in Fig. 8.15. This figure has asymmetric shapes and is different from optimal detection areas using detection theory as shown in Fig. 8.13. The optimal detection areas for AWGN channels are determined by the Euclidean distance between the received signal points and the closest signal points. On the other hands, the detection areas of the neural network are determined by training dataset. If we have an optimal detection algorithm in a receiver, the trained detection areas by the neural network cannot outperform it. The error probabilities of both detection areas would be almost same. However, in practice, some of the received signals can appear at the long-tailed Gaussian distribution occasionally. The trained neural network can handle these signal points well. In addition, the trained neural network can rediscover how the noise distribution is changing and redefine the detection areas accordingly. It might give us more accurate outputs at some channel environments. The main contributions to signal detection by AI techniques would be as follows: (1) There are many optimal solutions in communication theory. However, most of them have unrealistic assumptions or high complexity. Thus, it is not easy to implement them in practical systems. AI algorithms are trained by practical data set and may provide us with practical solution with reliable complexity. (2) Some signal detection algorithms are optimal as stand-alone. However, signal detection is one block in physical layer. In terms of end-to-end physical layer design, they may be inadequate. AI algorithms learn from the system outputs and may be better performance in the physical layer block chain. (3) Basically, a physical layer system has errors. The errors are unavoidable. The physical layer component including signal detection should provide some level of fault tolerance. AI algorithms will be able to give robustness to the unpredicted errors or behaviours because they may be fed into the neural layer and the strange behaviours may be trained.

Modulation and demodulation problems can be regarded as clustering or classification problems. Defining decision boundary as we can observe Fig. 8.13 is directly related to the performance of modulation and demodulation. Clustering techniques including unsupervised/supervised learning will be able to improve the performance. For example, in the NN classification model, we consider a Euclidean distance and compute the decision boundaries. The decision boundary between any two data points is a straight line. It shows us how input space is divided into classes. Each line segment is equidistant between two data points of classes. It is clear that the decision boundary of 1-NN is part of Voronoi diagram. Voronoi diagram is typically used in modulation and demodulation. This classification model can be applied in more general signal constellation of various modulation and demodulation. In [16], signal classification technique using deep learning was proposed. Automatic modulation classification technique is trained to classify 11 modulation schemes using LSTM at a high SNR [16]. At a low SNR, CNN is used to classify. The process is composed of pre-training and find tuning. The system is composed of LSTM and CNN and achieve good performance in various SNR regions. In [17, 18], CNN is applied to classify different modulations and have capability of distinguishing signal constellations.

Modulation classification (or modulation recognition) is a function to recognize modulation schemes from the received signals. It enables us to facilitate a better commination among different types of communications and also monitor other communications. This technique is based on decision-theoretic approach and pattern recognition approach [19]. The common process consists of pre-processing, feature extraction and classification. This is in line with AI algorithm process. In [19], modulation classifier using the neural network is proposed to discriminate the modulated signals with distortion from 13 types of digital modulation and 3 types of analogue modulation.

Example 8.4 Classifying Modulation Scheme Using Convolutional Neural Network Consider 4 different modulation schemes: BPSK, QPSK, 16QAM, and 64QAM. We generate 1024 modulation symbols at one frame and train convolutional neural network (CNN) recognizing 4 different modulation schemes. The symbols are distorted by multiple channel impairments as follows:

- Carrier frequency: 900 MHz
- Sample rate: 200 kHz

- Rician multipath channel Delays = [0 0 0.00000 0.00017], Average gain = [0 −2 −10], K Factor (Ratio between the power in the direct path and the power in the other) = 4, Maximum Doppler shift: 4 Hz
- AWGN.

The CNN comprises 28 layers: convolutional, ReLU, pooling, normalization, dropout, softmax, input and output layers. When CNN is trained with 100 frames with multiple modulation schemes, test 7 frames and compare the classification accuracy with a low SNR (4 dB) and a high SNR (30 dB).

Solution

The Matlab 5G and deep learning toolbox [10] are exploited. The trained CNN receives 7 frames with 1024 QPSK samples and predict the modulation types of the samples without any prior knowledge. The CNN includes a learning capability to extract the features and classify modulation schemes or number of carriers as shown in Fig. 8.16a. What we want to classify is modulation schemes. Thus, we train CNN at different SNRs in terms of modulation classification. Figure 8.16b illustrates the waveform of 4 different modulation schemes. As we can observe them, they have different curves of real and image part signals. The CNN learns their patterns and has capability to recognize them.

The received signal waveform can be used for the training data of CNN. When transmitting QPSK modulation, we observe the classification results at a low and

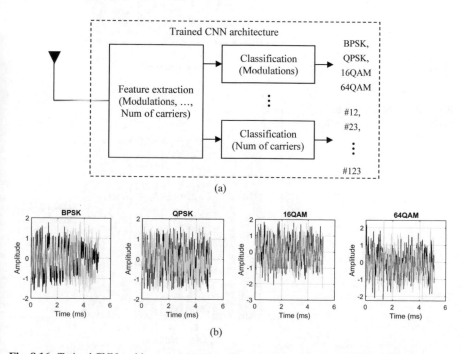

(a)

(b)

Fig. 8.16 Trained CNN architecture (**a**) and received signal waveform at SNR = 30 dB (**b**)

high SNR. Using the trained CNN, we obtain the classification results as shown in Fig. 8.17 at a low SNR and Fig. 8.18 at a high SNR.

As we can expect, the variance of the received signal constellation at a low SNR is high. It is difficult to detect the signal correctly. Thus, the classification error is high as shown in Fig. 8.17. However, at a high SNR, the variance of the received signal constellation is low and the location of the constellation point is clear. Thus, the classification error is low, and the transmitted QPSK modulation is well classified in the receiver as we can observe Fig. 8.18.

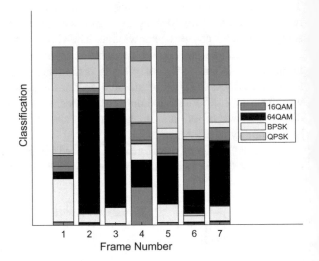

Fig. 8.17 Classification of modulation schemes at SNR = 5 dB

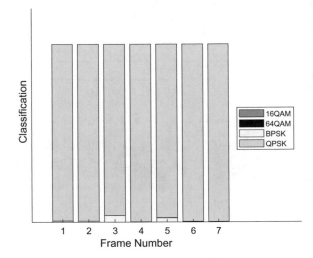

Fig. 8.18 Classification of modulation schemes at SNR = 30 dB

Summary 8.4. Signal Detection and Modulation

1. Decision theories are widely used in not only wireless communication systems but also other applications. The decision theory of wireless communication systems is typically developed to minimize the probability of errors.
2. The main contributions of AI algorithms to signal detection would be as follows: (1) AI algorithms are trained by practical data set and may provide us with practical solution with reliable complexity. (2) AI algorithms learns from the system outputs and may be better performance in the physical layer block chain. (3) AI algorithms will be able to give robustness to the unpredicted errors or behaviours because they may be fed into the neural layer and the strange behaviours may be trained.
3. Modulation and demodulation problems can be regarded as clustering or classification problems. Defining decision boundary is directly related to the performance of modulation and demodulation.

8.5 Channel Estimation

Channel estimation is getting more important in wireless communications because it is used in many physical layer components. The accuracy of channel estimation is directly related to efficient and reliable transmission. If we can have large amounts of pilots signals, the accuracy increases but the overall throughput decreases because they are redundant in terms of a user message transfer. Thus, it is important to reduce the number of pilot signals and obtain good performance. There are two types of channel estimation methods: pilot-based and blind channel estimation. Pilot-based channel estimation is based on dedicated signals. In the receiver side, we know which signals are transmitted. Thus, we compare the received signals and the known transmitted signals and obtain channel information from the comparison. If we have a long pilot symbol, we can estimate a wireless channel accurately. However, a long pilot symbol requires a high redundancy and brings a low spectral efficiency. Thus, finding good trade-off between them is one important issue in physical layer systems design. On the other hand, blind channel estimation do not use any dedicated signal and do not have any redundancy. It is based on statistical information extracting from the received signals. Typically, pilot-based channel estimation is good performance but the redundancy brings lower spectral efficiency. Blind channel estimation does not degrade spectral efficiency but accuracy is not high. Thus, it is important to find good trade-off point between redundancy and accuracy as well. AI algorithms will be able to help us find good trade-off. In order to adopt AI algorithm to channel estimation and learn channel characteristics of different channel environments, one important research challenge is to overcome time-varying condition and the limited training data size.

In [20], support vector machines (SVM) are adopted to perform channel estimation in time-varying and fast-fading channels. Deep learning can be used to learn channel structure in mmWAVE massive MIMO systems [21]. In [22], channel estimation using off-grid sparse Bayesian learning is used to exploit spatial sparse structure and identify the angles and gains of channel paths in mmWAVE massive MIMO uplink. Channel estimation is based on regression. Regression is one of AI algorithm applications. They are well matched. The conventional channel estimation using orthogonal pilots has the limited overhead resources. Practically, channel state information (CSI) is correlated over time, and this is known as channel aging [23]. Channel aging is the variation of the channel caused by the user mobility. AI algorithms enable us to adopt relevant regression techniques to forecast CSI. It is a time series learning problem which is discussed a lot in a AI research field. A recurrent neutral network (RNN) is exploring the hidden pattern within CSI variations.

Example 8.5 Channel Estimation Using Convolutional Neural Network Consider 5G resource grid with one numerology and antenna port. One resource element is the smallest unit of the resource grid as one subcarrier in one OFDM symbol. A 5G NR resource block (RB) is defined as N_{sc}^{RB} consecutive 12 subcarriers in frequency domain. N_{RB}^{μ} is the number of resource block with different configurations. Figure 8.19 illustrates one example of 5G NR resource block and resource element.

5G NR physical layer includes multiple physical signals: Synchronization signals such as primary synchronization signal (PSS) and secondary synchronization signal (SSS) and reference signals such as demodulation reference signal (DM-RS), phase tracking reference signal (PT-RS), channel state information reference signal (CSI-RS), and sounding reference signal (SRS). Among reference signals, DM-RS is used for channel estimation in downlink and uplink. A base station allocates radio resource to a mobile station and configures link adaptation. Some known reference symbols are inserted in the resource block in the transmitter. The receiver collects their channel responses, and the rest of the reference symbols are interpolated using the known reference symbols.

Select one configuration of DM-RS, train CNN, and then compare the channel estimation performance of the linear regression and CNN in terms of different SNRs.

Solution

The Matlab 5G and deep learning toolbox [10] are exploited. We consider single antenna OFDM system and train CNN with multiple SNRs. We generate DM-RS symbols in the resource block with 612 subcarriers and 14 OFDM symbols as shown in Fig. 8.20. The DM-RS symbols are included in PDSCH resource grid and transmitted via the channel. The channel is defined as AWGN, tapped delay lines, and single-input/single-output (SISO) [24, 25]. In the receiver, perfect synchronization is assumed. Using the received channel response by the known reference symbols, channel estimation is performed using linear regression and CNN.

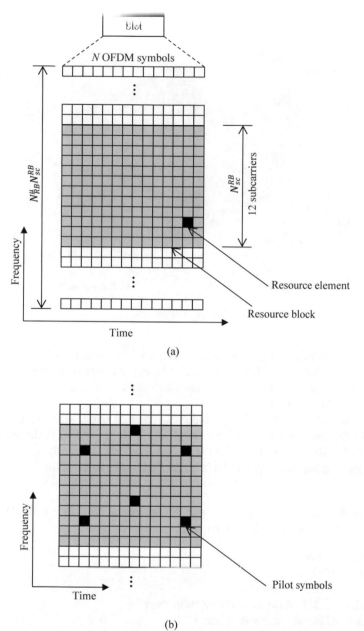

Fig. 8.19 Example of 5G NR resource grid and resource blocks (**a**) and pilot symbols (**b**)

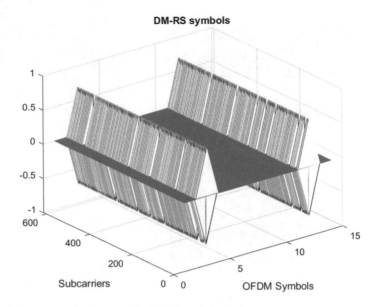

Fig. 8.20 RM-RS symbol mapping in 5G NR resource grid

The CNN is trained with 256 training data set consisting of 612 subcarriers and 14 OFDM symbols. The CNN comprises 11 layers: convolutional, ReLU, input and output layers. The CNN is trained on various channel configurations: different delay profiles and SNR at 5 dB and 10 dB.

As we can observe Fig. 8.21a, b, the mean squared error (MSE) values of linear regression at 5 dB and 10 dB are 0.56566 and 0.17904, respectively. The MSE values of CNN at 5 dB and 10 dB are 0.037486 and 0.019558, respectively. The channel estimation performance using CNN is better than linear regression.   ∎

Example 8.6 Channel Estimation and Equalization Consider a single antenna OFDM system with the following parameters

- Single cell model
- Duplex: FDD
- Number of subcarriers: 180
- Number of OFDM symbols in one subframe: 14
- Number of transmit antenna ports: 1
- Cyclic prefix: Normal
- Modulation: QPSK
- Channel: Extended Vehicular A Model (EVA5) fading channel [24, 25].

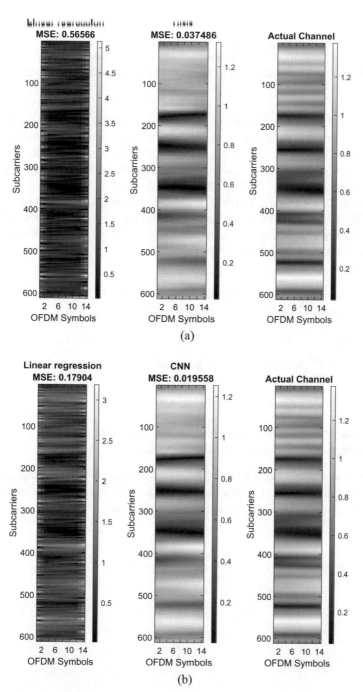

Fig. 8.21 Comparison of linear regression, CNN, and actual channel at SNR 5 dB (**a**) and 10 **dB** (**b**)

In the transmitter, 10 subframes are randomly generated in one frame. The channel estimation is performed by CNN using similar configuration of Example 8.5. After channel estimation, MMSE equalization is performed. Compare the received resource grids and the equalized resource grids at SNR 5, 10, and 20 dB.

Solution

The Matlab 5G and deep learning toolbox [10] are exploited. The CNN model is trained with similar configuration of Example 8.5, and then MMSE equalization is performed at SNR 5, 10, and 20 dB. Figure 8.22 illustrates the signal constellation of the receive symbols, the received resource grids, and the equalized resource grid by MMSE equalization at SNR 5, 10, and 20 dB.

The error vector magnitude (EVM, also called relative constellation error), is used to quantify the performance of the transceiver. It simply measures how far the signal constellation points are from the ideal locations. The RMS EVM between them are calculated at SNR 5, 10, and 20 dB as follows:

At 5 dB,

- Percentage RMS EVM before equalization: 135.718%
- Percentage RMS EVM after equalization: 62.029%.

At 10 dB,

- Percentage RMS EVM before equalization: 127.700%
- Percentage RMS EVM after equalization: 43.185%.

At 20 dB,

- Percentage RMS EVM before equalization: 124.270%
- Percentage RMS EVM after equalization: 18.217%.

As we can observe Fig. 8.22 and the RMS EVM calculations, the signal constellation points are well grouped at a high SNR. The equalization results are better at a high SNR as well. The distorted signal is well recovered.

∎

Fig. 0.22 Comparison of
Fig. 8.22 Comparison of the signal constellation of the receive symbols, the received resource grids, and the equalized resource grid by MMSE equalization at SNR 5 dB (**a**), 10 dB (**b**), and 20 dB (**c**)

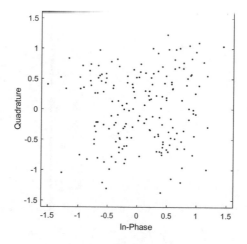

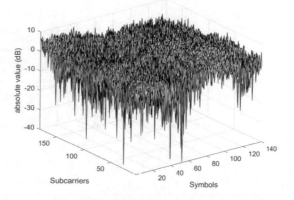

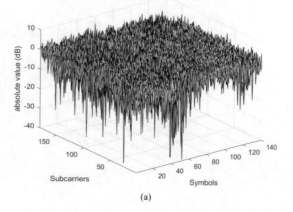

(a)

Fig. 8.22 (continued)

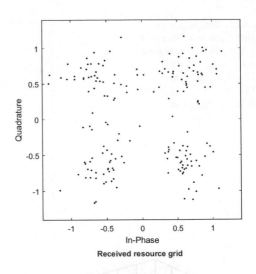

Received resource grid

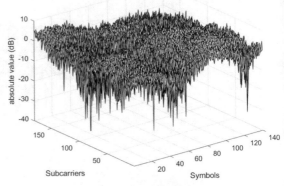

Equalized resource grid

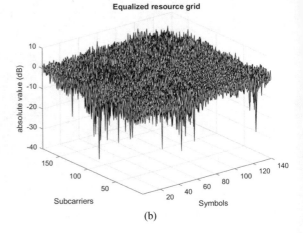

(b)

Fig. 8.22 (continued)

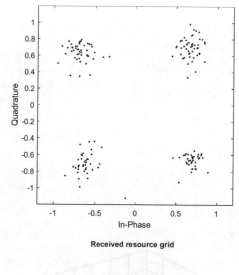

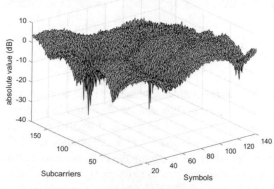

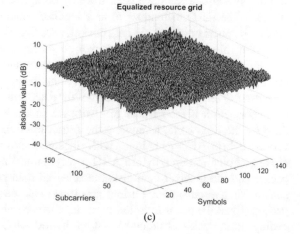

(c)

Summary 8.5. Channel Estimation

1. The accuracy of channel estimation is directly related to efficient and reliable transmission.
2. Pilot based channel estimation exploits dedicated signals. We compare the received signals and the known transmitted signals and obtain channel information from the comparison. On the other hands, blind channel estimation do not use any dedicated signal and do not have any redundancy. Blind channel estimation do not degrade spectral efficiency but accuracy is not high.
3. Channel estimation is based on regression. Regression is one of AI algorithm applications. AI algorithms enable us to adopt relevant regression techniques to forecast CSI.
4. AI algorithms for channel estimation will be able to find good trade-off point between redundancy and accuracy.

8.6   Error Control Coding

Error control coding schemes are an essential part of modern wireless communication systems. They can be classified into forward error correction (FEC), automatic repeat request (ARQ) and hybrid ARQ. A FEC encoder sends codewords composed of information and redundancy to a FEC decoder via wireless channels. The FEC decoder recovers the original information from the corrupted codewords. Turbo codes and low-density parity check (LDPC) codes are mostly used for 4G and 5G systems. They have been the key disciplines in modern communications systems. The ARQ techniques have a feedback channel and error detection techniques. When errors occur at a data packet, a receiver detects errors and then requests that a transmitter resends the data packet. As the name said, hybrid ARQ is combination of FEC and ARQ. In conventional ARQ, redundancy is added to the information using error detection codes such as cyclic redundancy check (CRC) codes. However, hybrid ARQ techniques use FEC codes for detecting an error. The receiver tries to make correction if the number of errors are small. However, if the error can't be fixed by the FEC code, it requests the retransmission to the transmitter. There are various hybrid ARQ techniques with different combinations and mechanisms. In chase combination-type hybrid ARQ, every retransmission includes the same data packet and the receiver combines the received data packet with the previous data packet. Each retransmission brings an effect to increase energy of the received data packet. Since the retransmission is same, chase combining is regards as repetition coding. In incremental redundancy-type hybrid ARQ, redundancy is changed and retransmitted with different puncturing configuration. The different redundancy is encoded from one information. Each retransmission includes different codewords

Fig. 8.23 Example of a
Tanner graph

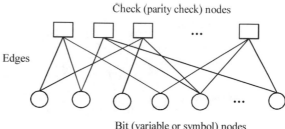

Check (parity check) nodes

Edges

Bit (variable or symbol) nodes

from the previous transmission. Thus, the receiver obtains additional codewords at each retransmission. Typically, the hybrid ARQ brings a better performance than conventional ARQ at poor channel conditions.

LDPC codes are linear block codes with a sparse parity check matrix $\mathbf{H}$ [26–28]. The sparseness of $\mathbf{H}$ means that $\mathbf{H}$ contains relatively few 1's among many 0's. The sparseness enables LDPC codes to increase the minimum distance. Typically, the minimum distance of LDPC codes increases linearly according to the codeword length. LDPC codes are same as conventional linear block codes except the sparseness. The difference between LDPC codes and conventional linear block codes is decoding mechanism. The conventional linear block codes is generally decoded using the maximum likelihood decoding method. The decoder receives n bits codeword and decides the most likely k bits message among the 2^k possible messages. Thus, the codeword length and the decoding complexity of conventional linear block codes are short and low, respectively. On the other hands, LDPC codes are iteratively decoded using a graphical representation (Tanner graph) of $\mathbf{H}$. The Tanner graph is composed of bit (or variable, symbol) nodes and check (or parity check) nodes. The bit nodes and the check nodes mean codeword bits and parity equations, respectively. The edge denotes connection between bit nodes and check nodes if and only if the bit is involved in the corresponding parity check equation. Thus, the number of edges in a Tanner graph means the number of 1's in the parity check matrix $\mathbf{H}$. Figure 8.23 illustrates an example of a Tanner graph. The squares represent check nodes (or parity check equations), and the circles represent bit nodes as shown in Fig. 8.23.

The regular LDPC code by Gallager [27] is denoted as (n, b_c, b_r) where n is a codeword length, b_c is the number of parity check equations, and b_r is the number of coded bits. The decoding of LDPC codes is performed iteratively between bit nodes and check nodes in the Tanner graph. A decoding scheme of LDPC codes is known as a message passing algorithm that is passing messages forward and backward between the bit nodes and check nodes. There are two types of message passing algorithms: the bit flipping (BF) algorithm based on a hard decision and the belief propagation (BP) algorithm based on a soft decision. The BP algorithm calculates the maximum a posteriori (MAP) probability. It calculates the probability $P(c_i|E)$, which means we find a codeword c_i on the event E (all parity check equations are satisfied.) In the BP algorithm, the message denotes the belief level (probability) of the received codewords. Each bit node passes a message to each check node

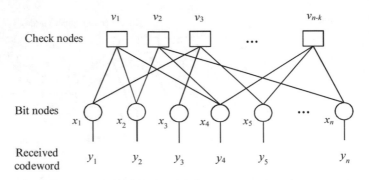

Fig. 8.24 Example of Tanner graph for LDPC decoding

connected to the bit node. Each check node passes a message to each bit node connected to the check node. In the final stage, a posteriori probability of each codeword bit is calculated. We should calculate many multiplication and division operations for the belief propagation algorithm. Thus, the complexity is high. In order to reduce the complexity, it is possible to implement it using log likelihood ratios (LLR). Multiplication and division are replaced by addition and subtractions, respectively. We call this a sum-product decoding algorithm. In order to explain the belief propagation algorithm, the Tanner graph is modified. Figure 8.24 illustrates an example of the Tanner graph for LDPC decoding.

In Fig. 8.24, v_j, x_i, and y_i represent check nodes, bit nodes, and received codeword bits, respectively. We express the received codeword as follows:

$$y_i = x_i + n_i \tag{8.15}$$

where n_i is Gaussian noise with zero mean and standard deviation σ. Two messages $q_{ij}(x)$ and $r_{ji}(x)$ are defined where $q_{ij}(x)$ and $r_{ji}(x)$ represent the message from the bit node x_i to the check node v_j and the message from the check node v_j to the bit node x_i, respectively. They can be expressed in the Tanner graph as shown in Fig. 8.25.

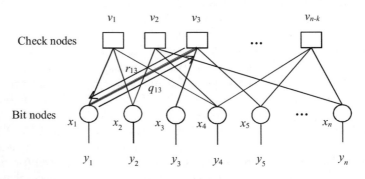

Fig. 8.25 Two messages $q_{ij}(x)$ and $r_{ji}(x)$ in the Tanner graph

The message $q_{ij}(x)$ means the probability $P(x_i = x|y_i)$ or the probability $x_i = x$ satisfying all check node equations except v_j. The message $r_{ji}(x)$ means the probability the parity check node (parity check equation) v_j is satisfied when all bit nodes have x except x_i. Belief propagation algorithm for LDPC codes can be summarized as follows:

- Step 1: Initialization of $q_{ij}{}^{\text{initial}}(x)$ is

$$q_{ij}{}^{\text{initial}}(x) = P(x_i = x|y_i) = \frac{1}{1 + e^{-\frac{2xy_i}{\sigma^2}}}. \tag{8.16}$$

- Step 2: The message $r_{ji}(x)$ from check nodes to bit nodes is

$$r_{ji}(+1) = \frac{1}{2} + \frac{1}{2} \prod_{i' \in V_{j/i}} (1 - 2q_{i'j}(-1)), \tag{8.17}$$

$$r_{ji}(-1) = 1 - r_{ji}(+1) \tag{8.18}$$

where $V_{j/i}$ denotes a bit node set connected to the check node v_j except x_i.
- Step 3: The message $q_{ij}(x)$ from bit nodes to check nodes is

$$q_{ij}(+1) = \alpha_{ij}(1 - p_i) \prod_{j' \in C_{i/j}} r_{j'i}(+1) \tag{8.19}$$

$$q_{ij}(-1) = \alpha_{ij} p_i \prod_{j' \in C_{i/j}} r_{j'i}(-1) \tag{8.20}$$

where $C_{i/j}$ denotes a check node set connected to the bit node x_i except v_j.
- Step 4: The a posteriori probability (APP) ratio for each codeword bit is

$$Q_i(+1) = \alpha_i(1 - p_i) \prod_{j \in C_i} r_{ji}(+1) \tag{8.21}$$

$$Q_i(-1) = \alpha_i p_i \prod_{j \in C_i} r_{ji}(-1). \tag{8.22}$$

- Step 5: The hard decision output is

$$\hat{x}_t = \begin{cases} +1, & \text{if } Q_i(+1) \geq 0.5 \\ -1, & \text{if } Q_i(+1) < 0.5 \end{cases}. \tag{8.23}$$

The iteration process continues until the estimated codewords satisfy the syndrome condition or the maximum number of iteration is finished.

There are many efforts to adapt AI algorithms for error control coding. For example, belief propagation decoder of LDPC codes can be viewed as folded neural networks. It can apply for other linear block codes such as Bose–Chaud-huri–Hocquenghem (BCH) codes. Relying on information theory, we construct FEC codes and calculate key properties of FEC codes such as minimum distance and extrinsic information transfer (EXIT) chart. Their coding gains can be derived math-ematically. Theoretical background of information theory is solid. Thus, informa-tion theory may provide us with solid foundation for AI algorithms and establish the theoretical foundation of AI algorithms. It may offer performance analysis and new methods such as the mechanism to obtain the optimal learning rate, find the relationship between the number of the training data and the optimality of the model, or the trade-off between computational complexity and performance accuracy. We can find common ground between FEC algorithms and AI algorithms and straight-forward applications of combining AI concepts and FEC algorithms. Firstly, the FEC algorithms are processed while focusing on a bit level. The bit level represen-tation can be simply treated as the input or output of AI algorithms. Secondly, it is relatively easy to generate training data set and labelled outputs due to the known variations of the codewords. In addition, the noised or interfered codewords data set can be easily generated. Since the codeword generation can be performed randomly, we can avoid overfitting problems. Comparing with convention FEC algorithms, the decoding complexity and latency of AI algorithms might be much lower if we have the trained AI algorithms. Generally speaking, in the AI application point of view, the FEC decoder can be regarded as a classifier. A Multi-layer perceptron (MLP) can be used for a channel decoder [29]. The received codewords are fed into the trained MLP-based decoder where they are classified in one of candidate codewords. If the MLP decoder is trained well, the performance is close to the maximum a posteriori (MAP) performance [29]. However, this approach is based on the well-trained model and the well interpolation to the time series of codewords. It doesn't work well if the codeword length is short or the wireless channel is fast fading. In addition, the channel coding scheme in cellular networks should support different configuration such as multiple code rate, different SNR, and different data packet length. When one MLP-based decoder is trained in one configuration, the trained MLP-based decoder doesn't provide us with good performance in other configura-tions. Thus, it is very difficult to have the well trained MLP decoder to satisfy various configuration. A neural network may improve the performance of belief propaga-tion (BP) algorithm for LDPC decoding [30]. Basically, neural network has similar structure to the unfolding Tanner graphs of the BP algorithm. The nodes and connec-tions of the neural networks can be regarded as the edge nodes and message passing process of the Tanner graph, respectively. If we can weigh the reliability properly, the approximation error can be reduced and the small cycle effect can be mitigated [30]. In the BP algorithm using the neural network, the basic structure is similar to the conventional BP algorithm. The neural network is trained by the noised code-words. After training the codewords with the multiple SNR ranges, the performance of the trained BP algorithm using the neural network is close to the performance of the maximum likelihood algorithm. The RNN can be used for sequential decoding for convolutional codes or turbo codes because the decoding process of sequential codes can be represented as a hidden Markov model (HMM) and Trellis representation of HMM is widely used for decoding structure of sequential decoding. Convolutional codes are widely used in wireless communications such as GSM and deep space communications. Viterbi algorithm as a decoder of convolutional codes is based on dynamic programming as one of AI algorithms. It solves the problem of the HMM efficiently. Using the similarity between RNN structure and HMM structure, a two-layer gated recurrent unit is adopted to solve the problem of HMM [31]. The decoding scheme using RNN can't improve the decoding performances of both conventional decoders for convolutional codes and turbo codes. This is because the original decoding structure of both sequential codes already contains both iteration and adaptation concept and is close the optimal structure. In [32], the neural network is adopted to estimate the log likelihood ratios (LLR). The LLR is widely used for FEC algorithms such as turbo codes and LDPC codes. The calculations of exact LLR for wireless channel are complicated due to the limited channel knowledge. Consider simple communication system with simple modulator and demodulator. The encoded symbol $\mathbf{c} = [c_1, c_2, \ldots, c_M]^T$ is mapped into the modulated symbol $\mathbf{s} = [s_1, s_2, \ldots, s_N]^T$ that are selected from an arbitrary finite set of constellation points such as BPSK, QPSK and so on. The modulated symbol $\mathbf{s}$ is transmitted via wireless channels. In the receiver, the modulated symbol $\mathbf{s}$ is estimated as $\hat{\mathbf{s}} = \mathbf{s} + \mathbf{n}$ where $\mathbf{n}$ is the N-dimensional Gaussian noise vector. When we demap the observed symbols $\hat{\mathbf{s}}$ into estimates of M bit LLR, the LLR is defined for the ith coded bit as follows:

$$z_i = \log \frac{P\left(c_i = 0|\hat{\mathbf{s}}\right)}{P\left(c_i = 1|\hat{\mathbf{s}}\right)}, \quad i = 1, 2, \ldots, M \tag{8.24}$$

where $P\left(c_i = 0 \text{ or } 1|\hat{\mathbf{s}}\right)$ is a priori conditional probability that $c_i = 0$ or 1 given that the observed symbols is $\hat{\mathbf{s}}$. c_i is the encoded bit of the N bit symbol. The exact computation of the LLR expression is

$$z_i = \log \frac{\sum_{s \in C_i^0} \exp\left(-\frac{\|\hat{\mathbf{s}} - \mathbf{s}\|_2^2}{\sigma^2}\right)}{\sum_{s \in C_i^1} \exp\left(-\frac{\|\hat{\mathbf{s}} - \mathbf{s}\|_2^2}{\sigma^2}\right)}, \quad i = 1, 2, \ldots, M \tag{8.25}$$

where $C_i^{0 \text{ or } 1}$ is the subset of constellation points and σ^2 is noise variance. Exponential and logarithmic function are computationally expensive. Thus, max-log approximation is widely used as follows:

$$z_i \approx \frac{1}{\sigma^2}\left(\min_{s \in C_i^1}\|\hat{\mathbf{s}} - \mathbf{s}\|_2^2 - \min_{s \in C_i^0}\|\hat{\mathbf{s}} - \mathbf{s}\|_2^2\right), \quad i = 1, 2, \ldots, M. \tag{8.26}$$

The LLR using the neural network with small number of hidden layers [32] estimates the exact LLR for a given SNR. Figure 8.26 illustrates the neural network

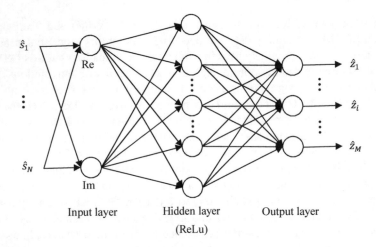

Fig. 8.26 LLR architecture using the neural network

architecture for LLR estimation.

As we can observe Fig. 8.26, the observed symbols $\hat{s}$ with the real and the imaginary parts is fed into the input layer. Two input nodes are connected to one hidden layer with a rectified linear unit (ReLU). The outputs of the hidden layer are connected to the linear output layer estimating the ith bit LLR. Using gradient descent method, the neural network is trained and the mean squared error (MSE) can be used as the loss function

$$J(\mathbf{\theta}) = \frac{1}{B} \sum_{b=1}^{B} \|\hat{\mathbf{z}}^b - \mathbf{z}^b\|_2^2 \qquad (8.27)$$

where $\mathbf{\theta}$ and B are trainable parameters and a batch of received symbols, respectively. Using conventional learning mechanism, the trainable parameters are updated as follows:

$$\mathbf{\theta} \leftarrow \mathbf{\theta} - \alpha \nabla_{\mathbf{\theta}} J(\mathbf{\theta}) \qquad (8.28)$$

where $\alpha > 0$ is the learning rate.

Example 8.7 LLR Calculation Comparison for Channel Coding Consider CP-OFDM system with 16 QAM modulation under AWGN channel. Compare the LLR estimation calculations of log-MAP, max-log-MAP, and LLR using the neural network.

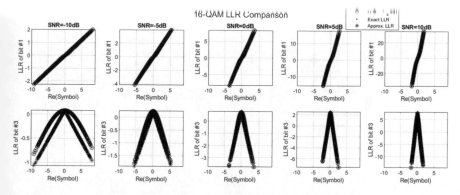

Fig. 8.27 Comparison of the LLR estimation calculations of exact LLR (log-MAP), approx. LLR (max-log-MAP), and LLR using the neural network

Solution

The Matlab 5G and deep learning toolbox [10] are exploited. The neural network for LLR estimation is composed of one input layer (with 2 nodes), one hidden layer (with 8 nodes), and output layer (with 4 nodes) as shown in Fig. 8.25. The neural network is trained with 10,000 received symbols for SNR $= -10, -5, 0, 5,$ and 10. The MSE is used for the loss function. Figure 8.27 illustrates the comparison of LLR estimation for log-MAP, max-log MAP, and LLR using neural network.

16 QAM modulation contains 4 bits. The first bit and the third bits of LLR are placed in the real part. They are plotted in Fig. 8.27 at SNR $= -10, -5, 0, 5,$ and 10 dB. As we can observe Fig. 8.27, the LLR estimation results of log-MAP, max-log-MAP and the neural network usage are similar at a high SNR. However, the LLR estimation using the neural network is almost same as the log-MAP estimation, but max-log-MAP estimation is not accurate.

∎

Example 8.8 LDPC Decoding Using Neural Network Consider 2/3 LDPC code and 16 Amplitude and phase-shift keying (APSK) modulation under AWGN channel. LDPC codes are based on DVB-S.2 standard [33]. The channel coding scheme of DVB-S.2 is composed of LDPC code as the inner code and BCH code as the outer code. Compare the PER performances of log-MAP, max-log-MAP, and LLR using the neural network.

Solution

The Matlab 5G and deep learning toolbox [10] are exploited. In order to decode LPDC codes with code rate 2/3, 16 APSK demodulation generates soft input for LDPC decoder and LLR should be calculated. The neural network LLR calculation is trained with 10,000 received symbols for SNR range from 7 to 9. The number of bit per packet is 1504. Belief propagation (BP) decoding of LDPC codes is used. Figure 8.28 illustrates BER performance of DVB-S.2 channel coding.

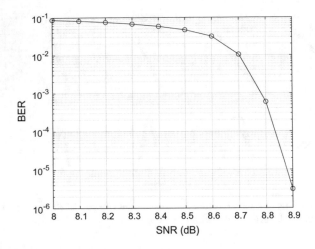

Fig. 8.28 BER performance of DVB-S.2 channel coding

Figure 8.29 illustrates the packet error rate performance comparison of exact LLR (log-MAP), approx. LLR (max-log-MAP), and LLR using the neural network.

As we can observe Fig. 8.29, the LLR estimation using the neural network is similar to the estimation of log-MAP. The trained LLR estimation using the neural network doesn't require high computational complexity such as logarithm and exponential.   ∎

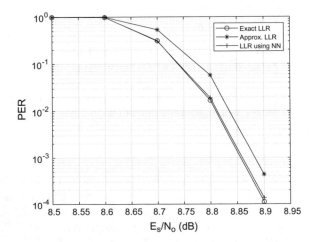

Fig. 8.29 PER comparison of exact LLR (log-MAP), approx. LLR (max-log-MAP), and LLR using the neural network

Summary 8.6. Error Control Coding

1. Error control coding can be classified into forward error correction (FEC), automatic repeat request (ARQ) and hybrid ARQ.
2. LDPC codes are linear block codes with a sparse parity check matrix **H**. The sparseness enables LDPC codes to increase the minimum distance. LDPC codes are same as conventional linear block codes except the sparseness. LDPC codes are iteratively decoded using a graphical representation (Tanner graph) of **H**. The decoding of LDPC codes is performed iteratively between bit nodes and check nodes in the Tanner graph. A decoding scheme of LDPC codes is known as a message passing algorithm that is passing messages forward and backward between the bit nodes and check nodes.
3. We can find common ground between FEC algorithms and AI algorithms and straightforward applications of combining AI concepts and FEC algorithms. Generally speaking, in the AI application point of view, the FEC decoder can be regarded as a classifier.
4. The RNN can be used for sequential decoding for convolutional codes or turbo codes because the decoding process of sequential codes can be represented as a hidden Markov model (HMM) and Trellis representation of HMM is widely used for decoding structure of sequential decoding.
5. A neural network has similar structure to the unfolding Tanner graphs of the BP algorithm. The nodes and connections of the neural networks can be regarded as the edge nodes and message passing process of the Tanner graph, respectively.

8.7 MIMO[1]

The multiple-input multiple-output (MIMO) techniques are widely employed in cellular systems due to significant performance improvement in terms of diversity gain, array gain, and multiplexing gain. These gains are related to different types of system performances [26]. The diversity gain improves link reliability and transmission coverage by mitigating different types of multipath fading. The array gain improves transmission coverage and QoS. The multiplexing gain increases spectral efficiency by transmitting independent signals via different antennas. There is a trade-off between these usages. The goal of spatial diversity is to improve the reliability using one data stream. Space time coding scheme is developed for this purpose. On the other hands, the goal of spatial multiplexing is to improve the transmission rate while maintaining certain level of reliability using multiple data streams.

---

[1] Refer to the book [9], the Sect. 8.7 is written and new contents are added.

Vertical bell laboratories layered space time (V-BLAST) and MIMO precoding are used for this purpose. Typically, space time coding provides us with good performance at a low SNR and spatial multiplexing shows us good performance at a high SNR. Cellular systems takes advantage of various types of MIMO techniques such as massive MIMO, multi-user MIMO, and beamforming [26]. The basic idea of MIMO techniques is based on space correlation properties of the wireless channels and obtains multiple uncorrelated signal replicas. Space diversity of MIMO systems can be achieved by multiple transmit and receive antennas. The multiple transmit antennas allow us to achieve transmitter space diversity and obtain uncorrelated fading signals. The transmit diversity can be defined as the number of independent channels exploited by the transmitter. The multiple receive antennas enable us to achieve receiver space diversity and obtain independent fading signals at the receiver. The receiver diversity can be defined as the number of independent channels exploited by the receiver. Thus, the performance of MIMO techniques depends on channel correlation.

Space time coding is a joint design of channel coding, modulation, and diversity scheme. The space time block codes (STBCs) are constructed from an orthogonal matrix over antennas and time and simply decoded by maximum likelihood decoding. They can achieve full diversity but show us a lack of coding gain. On the other hands, the space time trellis codes (STTCs) transmits multiple and diverse data sequences over antennas and time and reconstruct the actual data sequence at the receiver. They can achieve both diversity gain and coding gain. However, the decoding complexity is higher because the decoding process is based on a joint maximum likelihood sequence estimation. In order to design STBCs [26], we consider the point-to-point MIMO channel with N_t transmit antennas and N_r receive antennas as shown in Fig. 8.30 [26].

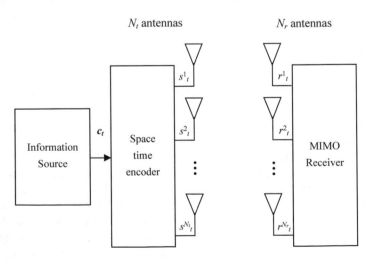

Fig. 8.30 MIMO system for space time coding

The information source block generates m symbols as follows:

$$\mathbf{c}_t = \left(c_t^1, c_t^2, \ldots, c_t^m\right). \tag{8.29}$$

In the space time encoder, m symbols $\mathbf{c}_t$ are mapped into N_t modulation symbols from a signal set of $M = 2^m$ and the transmit vector is represented as follows:

$$\mathbf{s}_t = \left(s_t^1, s_t^2, \ldots, s_t^{N_t}\right)^{\mathrm{T}}. \tag{8.30}$$

The memoryless MIMO channel is assumed. The MIMO system operates over a slowly varying flat-fading MIMO channel. The transmit vector has L frame length at each antenna. The $N_t \times L$ space time codeword matrix is defined as follows:

$$\mathbf{S} = [\mathbf{s}_1, \mathbf{s}_2, \ldots, \mathbf{s}_L] = \text{Space} \downarrow \begin{bmatrix} s_1^1 & s_2^1 & \cdots & s_L^1 \\ s_1^2 & s_2^2 & & s_L^2 \\ \vdots & & \ddots & \vdots \\ s_1^{N_t} & s_2^{N_t} & \cdots & s_L^{N_t} \end{bmatrix} \quad \overset{\text{Time} \rightarrow}{} \tag{8.31}$$

and the MIMO channel matrix is represented by $\mathbf{H}_t \in \mathbb{C}^{N_r \times N_t}$ matrix as follows:

$$\mathbf{H}_t = \begin{bmatrix} h_{11}^t & h_{12}^t & \cdots & h_{1N_t}^t \\ h_{21}^t & h_{22}^t & & h_{2N_t}^t \\ \vdots & & \ddots & \vdots \\ h_{N_r 1}^t & h_{N_r 2}^t & \cdots & h_{N_r N_t}^t \end{bmatrix}. \tag{8.32}$$

Each column of $\mathbf{S}$ is transmitted at a given channel. Since m symbols in L channels are transmitted, the STBC rate is $R = m/L$ symbols/s. If we have a constellation of order M, the bit rate is $R_b = \frac{m}{L} \log M$. Linear STBCs are considered, and the mapping between symbols and transmitted matrices is linear as follows:

$$\mathbf{S} = \sum_{k=1}^{m} \left(\mathbf{A}_k c_k + \mathbf{B}_k c_k^*\right) \tag{8.33}$$

where c_k^* is the complex conjugate of c_k and the $N_t \times L$ matrices $\mathbf{A}_k$ and $\mathbf{B}_k$ are fixed. The mapping rule is designed to optimize diversity gain and multiplexing gain and determine the performance of space time coding. For example, if Alamouti code ($L = 2$, $m = 2$) is defined, both matrices are as follows:

$$\mathbf{A}_1 = \begin{bmatrix} 1 & 0 \\ 0 & 0 \end{bmatrix}, \mathbf{A}_2 = \begin{bmatrix} 0 & 0 \\ 1 & 0 \end{bmatrix}, \mathbf{B}_1 = \begin{bmatrix} 0 & 0 \\ 0 & 1 \end{bmatrix}, \mathbf{B}_2 = \begin{bmatrix} 0 & -1 \\ 0 & 0 \end{bmatrix}. \tag{8.34}$$

Another representation is possible. We divide m complex symbols into their real parts Re() and imaginary parts Im() as follows:

$$r_k = \begin{cases} \text{Re}(c_k), & k = 1, \ldots, m \\ \text{Im}(c_k), & k = m+1, \ldots, 2m \end{cases} \tag{8.35}$$

where space time codewords can be represented as follows:

$$\mathbf{S} = \sum_{k=1}^{m} \left(\mathbf{A}_k c_k + \mathbf{B}_k c_k^* \right) = \sum_{k=1}^{2m} \mathbf{C}_k r_k \tag{8.36}$$

where $\mathbf{C}_k$ is expressed as follows:

$$\mathbf{C}_k = \begin{cases} \mathbf{A}_k + \mathbf{B}_k, & k = 1, \ldots, m \\ j(\mathbf{A}_{k-m} - \mathbf{B}_{k-m}), & k = m+1, \ldots, 2m \end{cases}. \tag{8.37}$$

For example, if Alamouti code ($L = 2$, $m = 2$) is defined, the matrix $\mathbf{C}_k$ is as follows:

$$\mathbf{C}_1 = \begin{bmatrix} 1 & 0 \\ 0 & 1 \end{bmatrix}, \mathbf{C}_2 = \begin{bmatrix} 0 & -1 \\ 1 & 0 \end{bmatrix}, \mathbf{C}_3 = \begin{bmatrix} j & 0 \\ 0 & -j \end{bmatrix}, \mathbf{C}_4 = \begin{bmatrix} 0 & j \\ j & 0 \end{bmatrix}. \tag{8.38}$$

Orthogonal space time block codes (OSTBCs) are one special class of linear STBCs. Attractive features of OSTBCs are summarized as (1) full diversity gain, (2) simple receiver structure by maximum likelihood decoding, (3) maximum SNR or minimum MSE, (4) easy to cancel ISI due to orthogonality. The OSTBCs are optimal in terms of diversity gain and receiver complexity but sub-optimal in terms of rate. The space time codeword matrix of OSTBCs has the following property:

$$\mathbf{S}^H \mathbf{S} = \|\mathbf{c}_t\|^2 \mathbf{I} \tag{8.39}$$

where

$$\|\mathbf{c}_t\|^2 = \sum_{k=1}^{m} |c_t^k|^2 \tag{8.40}$$

and the rows of the codewords are orthogonal with norm $\|\mathbf{c}_t\|^2$. In addition, the matrix $\mathbf{C}_k$ of the complex OSTBCs must satisfy

$$\mathbf{C}_k^H \mathbf{C}_l = \begin{cases} \mathbf{I}, & k = l \\ -\mathbf{C}_l^H \mathbf{C}_k, & k \neq l \end{cases} \quad k, l = 1, \ldots, 2m. \tag{8.41}$$

The maximum likelihood decoding scheme and perfect CSI are assumed at the receiver. Thus, the receiver has the following decision metric:

$$\sum_{t=1}^{L}\sum_{j=1}^{N_r}\left|r_t^j - \sum_{i=1}^{N_t}h_{ji}^t s_t^i\right|^2.\qquad(8.42)$$

The maximum likelihood decoder finds the codewords to minimize (8.42).

The STTC encoder is similar to the TCM encoder. Figure 8.31 illustrates a STTC encoder [26].

As we can observe Fig. 8.31, the input sequence c_t is a block of information (or coded bits) at time t and is denoted by $(c_t^1, c_t^2, \ldots, c_t^m)$. The kth input sequence c_t^k goes through the kth shift register and is multiplied by the STTC encoder coefficient

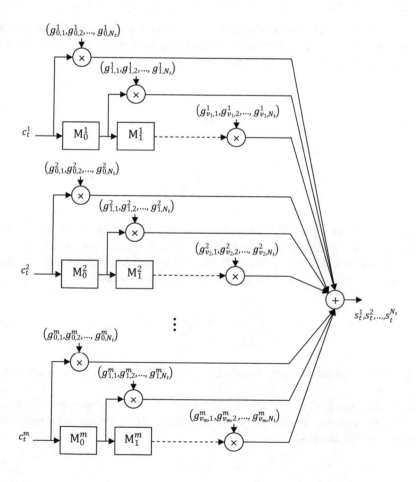

Fig. 8.31 Space time trellis encoder

set $\mathbf{g}^k$. It is defined as follows:

$$\mathbf{g}^k = \left[\left(g^k_{0,1}, g^k_{0,2}, \ldots, g^k_{0,N_t}\right), \left(g^k_{1,1}, g^k_{1,2}, \ldots, g^k_{1,N_t}\right), \ldots, \left(g^k_{v_m,1}, g^k_{v_m,2}, \ldots, g^k_{v_m,N_t}\right)\right].$$
(8.43)

Each $g^k_{l,i}$ represents an element of M-ary signal constellation set and v_m represents the memory order of the kth shift register. If QPSK modulation is considered, it has one of signal constellation set $\{0, 1, 2, 3\}$. The STTC encoder maps them into an M-ary modulated symbol and is denoted by $(s^1_t, s^2_t, \ldots, s^{N_t}_t)$. The output of the STTC encoder is calculated as follows:

$$s^i_t = \sum_{k=1}^{m} \sum_{l=0}^{v_k} g^k_{l,i} c^k_{t-j} \bmod 2^m$$
(8.44)

where $i = 1, 2, \ldots, N_t$. The outputs of the multipliers are summed modulo 2^m. The modulated symbols s^i_t are transmitted in parallel through N_t transmit antennas. The total memory order of the STTC encoder is

$$v = \sum_{k=1}^{m} v_k$$
(8.45)

and v_k is defined as follows:

$$v_k = \left\lfloor \frac{v + k - 1}{m} \right\rfloor.$$
(8.46)

The trellis state of the STTC encoder is 2^v. We assume that r^j_t is the received signal at the received antenna j at time t, the receiver obtains perfect CSI, and the branch metric is calculated as the squared Euclidean distance between the actual received signal and the hypothesized received signals as follows:

$$\sum_{j=1}^{N_r} \left| r^j_t - \sum_{i=1}^{N_t} h^t_{ji} s^i_t \right|^2.$$
(8.47)

The STTC decoder uses the Viterbi algorithm to select the path with the lowest path metric.

Consider a spatial multiplexing system with N_t transmit antennas and N_r receive antennas and assume $N_r \geq N_t$. The received signal $\mathbf{r}$ ($N_r \times 1$ column vector) can be expressed as follows [26]:

$$\mathbf{r} = \mathbf{Hs} + \mathbf{n}$$
(8.48)

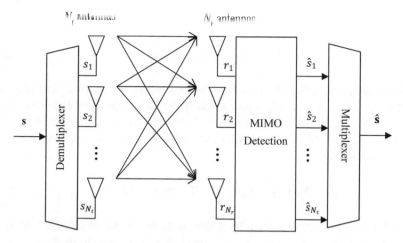

Fig. 8.32 MIMO system for spatial multiplexing

where $\mathbf{H}$, $\mathbf{s}$ and $\mathbf{n}$ represent $N_r \times N_t$ MIMO channel matrix, $N_t \times 1$ transmit signal and $N_r \times 1$ white Gaussian noise vector, respectively. The element of $\mathbf{r}$, r_j, is expressed as a superposition of all elements of $\mathbf{s}$. Figure 8.32 illustrates the MIMO system for spatial multiplexing.

The maximum likelihood detection is to find the most likely $\hat{\mathbf{s}}$ as follows:

$$\hat{\mathbf{s}}_{ML} = \arg \min_{s \in S} \|\mathbf{r} - \mathbf{Hs}\|^2 \tag{8.49}$$

where $\|\cdot\|$ denotes the norm of the matrix. The most likely $\hat{\mathbf{s}}$ is an element of set $\mathbf{S}$. Thus, it is simple solution to search all possible elements of set S and select one to satisfy (8.49). This is known as non-deterministic polynomial (NP) hard problem. The complexity increases exponentially according to the number of transmit antennas and the modulation order. Linear detection techniques such as matched filter (MF), zero forcing (ZF), and minimum mean squared error (MMSE) use a MIMO channel inversion. The transmitted symbol estimation is calculated by the MIMO channel version, multiplication, and quantization. The MF detection technique is one of the simplest detection techniques as follows:

$$\hat{\mathbf{s}}_{MF} = \text{Qtz}(\mathbf{H}^H \mathbf{r}) \tag{8.50}$$

where $\text{Qtz}()$ represents quantization. The estimated symbols are obtained by multiplying the received symbols by Hermitian operation of the MIMO channel matrix. The ZF detection technique uses the pseudo-inverse of the MIMO channel matrix. When the MIMO channel matrix is square ($N_t = N_r$) and invertible, the estimated symbols by the ZF detection is expressed as follows:

$$\hat{\mathbf{s}}_{ZF} = \text{Qtz}(\mathbf{H}^{-1} \mathbf{r}). \tag{8.51}$$

When the MIMO channel matrix is not square ($N_t \neq N_r$), it is expressed as follows:

$$\hat{\mathbf{s}}_{ZF} = \text{Qtz}\left((\mathbf{H}^H\mathbf{H})^{-1}\mathbf{H}^H\mathbf{r}\right). \tag{8.52}$$

As we can observe (8.51) and (8.52), the ZF detection technique forces amplitude of interferers to be zero with ignoring a noise effect. Thus, the MMSE detection technique considering a noise effect provides a better performance than the ZF detection technique. The MMSE detection technique minimizes the mean squared error. It is expressed as follows:

$$\hat{\mathbf{s}}_{MMSE} = \text{Qtz}\left((\mathbf{H}^H\mathbf{H} + N_0\mathbf{I})^{-1}\mathbf{H}^H\mathbf{r}\right). \tag{8.53}$$

where I and N_0 represent an identity matrix and a noise. As we can observe (8.53), an accurate estimation of the noise is required. If the noise term N_0 is equal to zero, the MMSE detection technique is same as the ZF detection technique. In linear detections, an accurate estimation of the MIMO channel matrix is an essential part and these detections are useful at a high SNR. The successive interference cancellation (SIC) detection technique is located between ML detection and linear detection. It provides a better performance than linear detections. The SIC detection uses nulling and cancellation to extract the transmitted symbols from the received symbols. When one layer (a partial symbol sequence from one transmit antenna) is detected, an estimation of the transmitted layer is performed by subtracting from the detected layers in the previous time. The nulling and cancellation are carried out until all layers are detected. The SIC detection shows us good performance at a low SNR. One disadvantage is that the SIC suffers from error propagation. When a wrong decision is made in any layer, the wrong decision affects to the other layers. Thus, the ordering technique is used to minimize the effect of error propagation. In the nulling and cancellation with the ordering technique, the first symbol with a high SNR is transmitted as the most reliable symbol. Then, the symbols with a lower SNR are transmitted. Both the linear detection and the SIC detection cannot achieve the ML detection performance even if they have a lower complexity than the ML detection.

Massive MIMO (also known as large-scale antenna system, full-dimensional MIMO, or very large MIMO) was introduced in 2010 [34]. Massive MIMO is an essential part of 5G systems especially in sub-6 GHz bands. The network configuration of massive MIMO systems is regarded as a base station equipped large antenna arrays simultaneously serving many mobile devices. Comparing with 4G MIMO techniques, 5G massive MIMO techniques provides us with higher spectral efficiency by spatial multiplexing of many mobile devices and energy efficiency by a reduction of radiated power. The key technical characteristics of massive MIMO can be summarized as follows [26]: (1) TDD operation is required due to the reciprocity, (2) channel hardening removes fast fading effect. The link quality varies slowly. It is helpful for resource allocation problems, (3) closed loop link budget in proportional to the number of antennas in a base station. The link budget improvement is helpful for overall QoS improvement in a cell, (4) relatively low complexity of precoding and decoding algorithms, (5) full digital processing is helpful for both spectral efficiency and energy efficiency. In addition, it reduces accuracy and resolution of RF design. If we can have unlimited antennas, unlimited capacity can be achieved under some conditions [34, 35]: (1) spatially correlated channels, (2) MMSE channel estimation, and (3) optimal linear combining. Despite of those advantages, there are practical limitations of massive MIMO. For example, more antennas bring us a better performance, but the performance limit depends on coherence time and bandwidth. Especially, under a high mobility environment, it requires smaller coherence time and more pilot signal. In addition, it has limited orthogonal pilots due to pilot contamination. The narrow beam generated by massive MIMO is very sensitive to mobile devices mobility or antenna array sway.

The MIMO systems are implemented in cellular systems in two manners: single-user MIMO and multi-user MIMO. As the name said, the signal user MIMO system includes single transmitter and single receiver. Multiple antennas with sufficient spacing are equipped with one base station or one user equipment. The multi-user MIMO system includes single or multiple transmitters and single or multiple receivers with single or multiple antennas. The multiple antennas as virtual MIMO can be distributed across different locations. In order to obtain the gains in MIMO systems, there are multiple research challenges. For example, if the MIMO channel has a low rank channel, it is difficult to obtain enough multiplexing gain. Antenna spacing and channel estimation overhead are key research challenges. In a user equipment, it is not possible to implement with large number of antennas. In addition, massive MIMO requires large size of pilots. It causes a big overhead of channel estimation. The multi-user MIMO can overcome the problems of signal user MIMO. The multi-user MIMO enables a base station to have large number of antennas and multiple mobile devices to have single antenna. In 6G systems, massive MIMO systems are considered as well because of many advantages. There are many approaches to adopt AI techniques to MIMO systems and solve those problems. The main design approaches to adopt AI algorithms to MIMO systems can be summarized as shown in Table 8.1.

As we discussed in Sects. 8.4 and 8.5, AI algorithms will be helpful for acquiring accurate CSI. An accurate CSI will bring a big impact to improve the multiple gains of MIMO systems. The huge data property of massive MIMO systems is good environment to adopt AI algorithms. AI algorithms will be helpful for analysing huge data generated by massive MIMO systems. Massive antenna array has a large size of matrix, and signal detection and channel estimation of massive antenna array require high computation power. Especially, pilot contamination of massive MIMO systems is directly related to the performance. It comes from the interference between neighbouring cells and degrades accuracy of channel state information. In [41], using sparse Bayesian learning method, channel state information of massive MIMO systems are obtained. It provides us with a better performance in the pilot contamination environment. In MIMO detection, the iterative methods provide us with good performance but they may require a prior knowledge such as channel distribution. It may not work well at fast fading channel or complex channel environments. We

Table 8.1 Design approaches of AI-enabled MIMO systems

MIMO techniques	Design approaches
Beam selection	CNN for classification of beam channel information among multiple classes representing beam numbers [36]
Radio modulation recognition	CNN using two different dataset for highly accurate Automatic modulation recognition (AMR) [37]
Precoding and CSI acquisition	Using trained neural network and both instantaneous and statistical CSI, precoding vectors is designed to maximize the sum rate subject to a total transmit power constraint. The complexity of MIMO precoding is substantially reduced compared with the existing iterative algorithm [38]
Antenna selection	CNN for MIMO antenna selection and hybrid beamformer design [39]
MIMO detection	MIMO detection problem is regarded as clustering problem of unsupervised learning [40]

can combine AI algorithm and MIMO detection algorithm and improve the performance. For example, the MIMO detection algorithm can reduce the number of the variables for training and the trained AI algorithm can reduce the convergence time. This approach can be helpful for especially massive MIMO systems that have many parameters under time-varying channels.

As we discussed the MIMO detection including ML, MF, ZF, and MMSE for spatial multiplexing, the ML detector is the optimal detector in terms of minimum joint error probability for all symbols detection. It can be implemented by the sphere decoder or others. In practice, the complexity of MIMO detector is important to implement. The search algorithms like sphere decoder have a high computational complexity. Thus, the sub-optimal detection algorithms such as MF, ZF, and MMSE are widely implemented in cellular systems. The basic concept of detection theory is in line with classification of AI algorithms. Both of them make a decision or find a pattern from noisy data set. The main difference between them are design approach. The detection theory is based on a probabilistic model of the cellular system. The AI algorithms have data-driven approach. In the MIMO detection, a system model is given and it can generate synthetic data set. The detection theory finds the best estimate from noisy data set, whereas AI algorithms find the best rule to be applied. In terms of implementation, the computational complexity increases as we have new data set. However, the heavy computational part can be performed by off-line when implementing AI algorithms. Once the best rule is found, the implementation cost is affordable. In [40], a MIMO receiver uses the natural property that the received signals form clusters representing the transmitted symbols. Using machine learning, MIMO transceivers are redesigned. The MIMO detection problem is regarded as clustering problem of unsupervised learning. The basic approach is to apply the expectation maximization algorithm to cluster the data set described by a Gaussian mixture model. There are two main challenges to implement this approach. Firstly, the clustering results indicate the received signal set belonging to the same transmitted signals, and it would't explain correspondence between the following results and the transmitted signals. Secondly, each cluster is estimated with its own pattern. The number of clusters should be much smaller than the number of received signals for clustering. In order to overcome two challenges, label-assisted transmission and modulation constrained Gaussian mixture model are adopted. The label-assisted transmission enables us to have the correspondence between the clusters and the transmitted symbols, but it requires the overhead to label. The modulation constrained Gaussian mixture model allows us to minimize the number of parameters for accurate clustering.

The massive MIMO will be one of key techniques in 6G systems. The 6G base station equipped with hundreds of antennas in a centralized or distributed manner will reduce multi-user interference and improve the system throughput. In order to implement the massive MIMO systems at base stations, it is important to obtain accurate channel state information (CSI). In LTE MIMO systems, the downlink CSI is used and vector quantization or codebook are adopted to reduce the overhead of the feedback channel. However, if the number of antennas increases, the feedback channel has a big burden to MIMO systems. Thus, this approach can't be used for massive MIMO. Reducing the overhead of the feedback channel is one of key research challenges when implementing massive MIMO systems. In 5G systems, compressive sensing methods are considered. The basic concept is based on transformation from correlated CSI to uncorrelated sparse vectors. The compressive sensing method enables us to obtain an accurate estimate of a sparse vector. However, this approach has multiple problems such as sparse channel assumption, random projection, and slow signal reconstruction time. The deep learning approaches will be helpful for reconstructing natural images from compressive sensing [42]. The CSI encoder using deep learning was adopted in a closed-loop MIMO system [43]. Using the deep learning concept, CSI can be recovered with good reconstruction quality [43].

Summary 8.7. MIMO

1. The Multiple Input Multiple Output (MIMO) techniques are widely employed in cellular systems due to significant diversity gain, array gain and multiplexing gain.

2. The Space time block codes (STBCs) are constructed from an orthogonal matrix over antennas and time and simply decoded by maximum likelihood decoding. They can achieve full diversity but show us a lack of coding gain.

3. The space time trellis codes (STTCs) transmits multiple and diverse data sequences over antennas and time and then reconstruct the actual data sequence at the receiver. They can achieve both diversity gain and coding gain.

4. The maximum likelihood (ML) detection for spatial multiplexing is

$$\hat{s}_{ML} = \arg \min_{s \in S} \|\mathbf{r} - \mathbf{Hs}\|^2$$

5. The matched filer (MF) detection for spatial multiplexing is

$$\hat{s}_{MF} = \mathrm{Qtz}(\mathbf{H}^H \mathbf{r})$$

6. The zero forcing (ZF) detection for spatial multiplexing is

$$\hat{s}_{ZF} = \mathrm{Qtz}\left((\mathbf{H}^H \mathbf{H})^{-1} \mathbf{H}^H \mathbf{r}\right).$$

7. The minimum mean squared error (MMSE) detection for spatial multiplexing is

$$\hat{s}_{MMSE} = \mathrm{Qtz}\left((\mathbf{H}^H \mathbf{H} + N_0 \mathbf{I})^{-1} \mathbf{H}^H \mathbf{r}\right).$$

8. The SIC detection uses nulling and cancellation to extract the transmitted symbols from the received symbols.
9. The huge data property of massive MIMO systems is good environment to adopt AI algorithms. AI algorithms will be helpful for analysing huge data generated by massive MIMO systems and so on.
10. AI algorithms will be able to contribute to 6G MIMO system design including Beam selection, Radio modulation recognition, Precoding and CSI acquisition, Antenna selection, MIMO detection and so on.

8.8 Problems

8.1 Describe key research challenges to adopt AI algorithms to physical layer.
8.2 Compare the model-based approach of communication theory and the data-driven approach of AI algorithm.
8.3 The belief propagation of LDPC decoding is based on iteration concept. Describe which 5G physical layer components are adopting two key concepts adaptation and iteration.
8.4 Explain the pros-and-cons of the block-by-block physical layer design concept.
8.5 Compare the required computational powers of an user equipment and a base station.
8.6 Explain the HW-SW co-design pros-and-cons when implementing AI algorithms at physical layer.

8.7   Explain how and where maximum likelihood (ML), maximum à poste-
        riori (MAP), minimum mean squared error (MMSE), are zero-forcing (ZF)
        algorithms are used in physical layer.

8.8   There are many physical layer components such as digital pre-distortion,
        channel coding, OFDM modulation, channel estimation, synchronization,
        and so on. Discuss which AI design approach is suitable for each physical
        layer component.

8.9   Compare the autoencoder architecture and the conventional communication
        architecture in terms of performance and complexity.

8.10   Describe key research challenges when adopting autoencoder to wireless
        systems.

8.11   There are different channel impairments such as Gaussian noise, jitter, path
        loss, fading, shadowing, phase and frequency noise, interferences, and so on.
        They damage the transmitted signals differently. Explain how they affect to
        the transmitted signals and how to compensate.

8.12   There are many diversity techniques in terms of time, frequency, and space.
        Describe what physical layer components exist for these diversities.

8.13   Explain the difference of small-scale fading: flat fading (Bs < Bc), frequency
        selective fading ($B_s > B_c$), fast fading ($T_s > T_c$), and Slow fading ($T_s < T_c$)
        where B_s, B_c, T_s, and T_c denote the signal bandwidth, coherent bandwidth,
        signal time interval, and coherent time, respectively.

8.14   AI algorithms for wireless channels focus on the relationship between channel
        input and output. The generalization of more realistic channel scenarios would
        be one big research challenge. Discuss how to train AI algorithms efficiently
        and generalize wireless channels.

8.15   Explain where decision theory is used in real day-to-day life.

8.16   Explain the difference between coherent detection and non-coherent detec-
        tion.

8.17   AI algorithms will be able to give robustness to the unpredicted errors or
        behaviours because they may be fed into the neural layer and the strange
        behaviours may be trained. Describe how deep learning handles unpredicted
        errors.

8.18   Defining decision boundary is directly related to the performance of modula-
        tion and demodulation. Describe the difference between model-based modu-
        lation schemes and data-driven AI algorithms and explain their pros-and-cons.

8.19   Compare pilot-based channel estimation and blind channel estimation

8.20   Channel estimation is based on regression. Regression is one of AI algorithm
        applications. AI algorithms enable us to adopt relevant regression techniques
        to forecast CSI. Select one of regression techniques in AI algorithms and
        adopt to channel estimation in wireless communications. Compare the perfor-
        mances of conventional wireless channel estimation and AI-based channel
        estimation.

8.21   There is common ground between FEC algorithms and AI algorithms. The trained neural network will be able to reduce the convergence time of iterative decoding. Assume specific conditions like SNR level and compare the training time and the reduced convergence time of deep learning.

8.22   The Multiple Input Multiple Output (MIMO) techniques are widely employed in cellular systems due to significant diversity gain, array gain and multiplexing gain. Explain how they affect to wireless systems.

8.23   Compare the pros-and-cons of ML, MF, ZF, and MMSE detections for MIMO spatial multiplexing.

8.24   AI algorithms will be able to contribute to 6G MIMO system design including beam selection, radio modulation recognition, precoding and CSI acquisition, antenna selection, MIMO detection, and so on. Find and discuss examples of AI-enabled MIMO systems in these topics.

References

1. T.J. O'Shea, J. Hoydis, An introduction to deep learning for the physical layer. IEEE Trans. Cognitive Commun. Netw. **3**(4), 563–575 (2017)
2. H. Ye, G.Y. Li, B.H. Juang, Power of deep learning for channel estimation and signal detection in OFDM systems. IEEE Wirel. Commun. Lett. **7**(1), 114–117 (2018)
3. H. Huang, Y. Song, J. Yang, G. Gui, F. Adachi, Deep-learning-based millimeter-wave massive MIMO for hybrid precoding. IEEE Trans. Veh. Technol. **68**(3), 3027–3032 (2019)
4. F. Aoudia, J. Hoydis, End-to-end learning of communications systems without a channel model. arXiv:1804.02276 (2018)
5. T.D. Vo-Huu, G. Noubir, Fingerprinting Wi-Fi devices using software defined radios, in *Proceedings of 9th ACM Conf. Security & Privacy in Wireless and Mobile Networks* (2016), pp. 3–14
6. S. Dörner, S. Cammerer, J. Hoydis, S.T. Brink, Deep learning based communication over the air. IEEE J. Sel. Topics Signal Process. **12**(1), 132–143 (2018). https://doi.org/10.1109/JSTSP. 2017.2784180
7. Y. Bengio, P. Lamblin, D. Popovici, H. Larochelle, Greedy layer-wise training of deep networks, in *Advances in Neural Information Processing Systems* (2007)
8. C.E. Shannon, A mathematical theory of communication. Bell Syst. Tech. J. **27**, 379–423 & 623–656 (1948)
9. H. Kim, *Wireless Communications Systems Design* (Wiley, 2015). ISBN 978-1-118-61015-2
10. https://www.mathworks.com/
11. J. Huang, C.-X. Wang, L. Bai, J. Sun, Y. Yang, J. Li, O. Tirkkonen, M. Zhou, A big data enabled channel model for 5G wireless communication systems. IEEE Trans. Big Data (2019) (in press)
12. X. Ma, J. Zhang, Y. Zhang, Z. Ma, Data scheme-based wireless channel modeling method: Motivation, principle and performance. J. Commun. Inf. Netw. **2**(3), 41–51 (2017)
13. R. He, B. Ai, A.F. Molisch, G.L. Sẗuber, Q. Li, Z. Zhong, J. Yu, Clustering enabled wireless channel modeling using big data algorithms. IEEE Commun. Mag. **56**(5), 177–183 (2018)
14. H. Li, Y. Li, S. Zhou, J. Wang, Wireless channel feature extraction via GMM and CNN in the tomographic channel model. J. Commun. Inf. Netw. **2**(1), 41–51 (2017)
15. Y. Huangfu, J. Wang, C. Xu, R. Li, Y. Ge, X. Wang, H. Zhang, J. Wang, Realistic channel models pre-training. https://arxiv.org/abs/1907.09117

16. J. Karunman, W. Meert, D. (functions, V. Lenders, S. Pollin, Deep learning models for wireless signal classification with distributed lowcost spectrum sensors. IEEE Trans. Cognitive Commun. Netw. **4**(3), 433–445 (2018)

17. S. Peng, H. Jiang, H. Wang, H. Alwageed, Y. Yao, Modulation classification using convolutional neural network based deep learning model, in *Proceedings of 26th Wireless and Optical Communication Conference (WOCC 2017)*, Newark, NJ, USA, Apr. 2017, pp. 1–5

18. T.J. O'Shea, T. Roy, T.C. Clancy, Over-the-air deep learning based radio signal classification. IEEE J. Sel. Topics Signal Process. **12**(1), 168–179 (2018)

19. A.K. Nandi, E.E. Azzouz, Algorithms for automatic modulation recognition of communication signals. IEEE Trans. Commun. **46**(4), 431–436 (1998)

20. A. Charrada, A. Samet, Joint interpolation for LTE downlink channel estimation in very high-mobility environments with support vector machine regression. IET Commun. **10**(17), 2435–2444 (2016)

21. H. He, C.-K. Wen, S. Jin, G.Y. Li, Deep learning-based channel estimation for beamspace mmWave massive MIMO systems. IEEE Wirel. Commun. Lett. **7**(5), 852–855 (2018)

22. H. Tang, J. Wang, L. He, Off-grid sparse Bayesian learning based channel estimation for mmWave massive MIMO uplink. IEEE Wireless Commun. Lett. **8**(1), 45–48 (2019)

23. N. Palleit, T. Weber, Time prediction of non flat fading channels, in *Proceedings of IEEE ICASSP*, May 2011, pp. 2752–2755

24. 3GPP TR 38.901, Study on channel model for frequencies from 0.5 to 100 GHz in *The 3rd Generation Partnership Project; Technical Specification Group Radio Access Network*

25. 3GPP TS 36.104, Evolved Universal Terrestrial Radio Access (E-UTRA); Base Station (BS) radio transmission and reception, in *The 3rd Generation Partnership Project; Technical Specification Group Radio Access Network*

26. H. Kim, *Design and Optimization for 5G Wireless Communications* (Wiley, 2020). ISBN 9781119494553

27. R.G. Gallager, *Low-Density Parity-Check Codes* (MIT Press, Cambridge, MA, 1963)

28. D.J.C. MacKay, R.M. Neal, Near Shannon limit performance of low density parity check codes. Electron. Lett. **32**, 1645–1646 (1996)

29. T, Gruber, S. Cammerer, J. Hoydis, On deep learning-based channel decoding, in *The 51st Annual Conference on Information Sciences and Systems* (2017), pp. 1–6. https://doi.org/10.1109/CISS.2017.7926071

30. E. Nachmani, Y. Be'ery, D. Burshtein, Learning to decode linear codes using deep learning, in *54th Annual Allerton Conference on Communication, Control, and Computing* (2016), pp. 341–346. https://doi.org/10.1109/ALLERTON.2016.7852251

31. H. Kim, Y.H. Jiang, R. Rana, Communication algorithms via deep learning. https://arxiv.org/abs/1805.09317 (2018)

32. O. Shental, J. Hoydis, Machine LLRning: Learning to softly demodulate, in *IEEE Globecom Workshops 2019*, HI, USA (2019), pp. 1–7

33. ETSI Standard EN 302 307 V1.1.1: Digital Video Broadcasting (DVB), *Second Generation Framing Structure, Channel Coding and Modulation Systems For Broadcasting, Interactive Services, News Gathering and other Broadband Satellite Applications (DVB-S2)* (European Telecommunications Standards Institute, Valbonne, France, 2005–03)

34. T.L. Marzetta, Noncooperative cellular wireless with unlimited numbers of base station antennas. IEEE Trans. Wirel. Commun. **9**(11), 3590–3600 (2010)

35. E. Björnson, J. Hoydis, L. Sanguinetti, *Massive MIMO Networks: Spectral, Energy, and Hardware Efficiency* (Now Publishers, 2018). https://doi.org/10.1561/2000000093

36. H. P. Tauqir, A. Habib, Deep learning based beam allocation in switched-beam multiuser massive MIMO systems, in *2019 Second International Conference on Latest trends in Electrical Engineering and Computing Technologies (INTELLECT)* (2019) pp. 1–5

37. Y. Wang, S. Member, M. Liu, Data-driven deep learning for automatic modulation. IEEE Trans. Veh. Technol. **68**(4), 4074–4077 (2019)

38. J. Shi, W. Wang, X. Yi, X. Gao, G.Y. Li, Robust precoding in massive MIMO: a deep learning approach. arXiv:2005.13134 (2020)

39. A.M. Elbir, K.V. Mishra, Joint antenna selection and hybrid beamformer design using unquantized and quantized deep learning networks. arXiv:1905.03107 (pre-printed, 2019)
40. Y. Huang, P.P. Liang, Q. Zhang, Liang, A machine learning approach to MIMO communications, in *2018 IEEE International Conference on Communications (ICC)* (2018), pp. 1–6. https://doi.org/10.1109/ICC.2018.8422211
41. C.K. Wen, S. Jin, K.K. Wong, J.C. Chen, P. Ting, Channel estimation for massive MIMO using Gaussian-mixture Bayesian learning. IEEE Trans. Wirel. Commun. **14**(3), 1356–1368 (2015)
42. A. Mousavi, G. Dasarathy, R.G. Baraniuk, DeepCodec: adaptive sensing and recovery via deep convolutional neural networks. http://arxiv.org/abs/1707.03386 (2017)
43. T.J. O'Shea et al., Deep learning based MIMO communications. https://arxiv.org/abs/1707.07980 (2017)

Chapter 9

AI-Enabled Data Link Layer

Wireless communications and networks technologies are changing rapidly. 6G systems will be dynamic, heterogeneous, large scale, and complex networks and require higher performance of multiple applications and services. In order to support new requirements, general purpose protocols may be replaced by application tailored protocols. The current protocol design is based on empirical control rules or network configurations with specific thresholds that can be adjusted by network operators. The network configuration is decided in a trial-and-error way depending on the network environments and scenarios. In order to satisfy higher network system requirements and provide a mobile user with better services, we should overcome varying wireless channels and dynamically changing network characteristics. Thus, more intelligent network management and operation are required. The data link layer of wireless systems is designed to take the data, convert to packets of the data, and send them over the underlying hardware for reliable transmission. The main purpose of the data link layer in cellular systems is to maximize the utilization of the radio resources. The key functions of the data link layer are data streams multiplexing, data frame detection, medium access, radio resource allocation, and error control for reliable connectivity. In the context of wireless systems, the learning capability of AI algorithms covers context awareness and intelligence of wireless systems. The data-driven approach has gained popularity. It will be able to facilitate more intelligent behaviours of wireless systems. This approach can be applied to data link layers including resource allocations, scheduling, data aggregation, localization, and so on. In this chapter, we discuss how a radio resource allocation is performed, how AI algorithms will improve the performance of data link layer, and what research challenges we have. In particular, we focus on the topic of the resource allocation.

H. Kim, *Artificial Intelligence for 6G*,
https://doi.org/10.1007/978-3-030-95041-5_9

9.1   Design Approaches of AI-Enabled Data Link Layer

In 5G new radio (NR), the 5G layer 1 (L1) is physical layer. The 5G layer 2 (L2) is data link layer consisting of Medium Access Control (MAC), Radio Link Control (RLC) and Packet Data Convergence Protocol (PDCP). The main role of the MAC layer is mapping information between logical channels and transport channels. Logical channels represent the type of information such as control channels and traffic channels. Transport channels indicate how to multiplex the logical data and transport the information via the radio interface. The other key functions of the MAC layer are summarized as follows: scheduling information report in uplink, hybrid automatic repeat request (HARQ) in both downlink and uplink, and logical channel prioritisation in uplink. The RLC has three different modes of operation: Transparent Mode (TM), Un-acknowledge Mode (UM) and Acknowledge Mode (AM). The RLC serves different logical channels in terms of their requirements and the main functions of the RLC depend on the transmission modes. The functions of RLC layer can be summarized as follows: transfer of upper layer in one of three transmission modes, concatenation, segmentation and reassembly of RLC Service Data Units (SDUs) in UM and AM modes, HARQ in AM data transfer mode, reordering of RLC in UM and AM modes, duplication detection in UM and AM modes, resegmentation of RLC in AM mode, RLC SDU discard, RLC reestablishment and so on. The PDCP layer provides the following services to upper or lower layers: transfer of user and control plane data, header compression and decompression, ciphering and deciphering, integrity protection and verification of control plane data, reordering, and so on. The main functions of 5G layer 2 are similar to 4G layer 2 functions. The data link layer of 6G systems may be similar to 5G data link layer. Based on the structure of 5G data link layer, 6G data link layer may employ new functions such as terahertz communication support, distributed communication support, cell-free access, and so on. Figure 9.1 illustrates 5G NR L1 and L2 protocol stack.

The function of the data link layer can be regarded as mapping from a perceived information about the network conditions to the choice of network configurations. This is similar to reinforcement learning or deep learning approaches. In this point of view, reinforcement learning or deep learning structure can be constructed in data link layer. In wireless systems, AI algorithms will be able to learn from data and then extract patterns or predict the future data. The learning ability can be used for improving the performance or creating a new application in wireless systems. The patterns or predictions by AI algorithms can be exploited to adjust the parameters of the wireless networks and operate the wireless networks efficiently. Using the collected data from mobile devices, AI algorithms can generate a new information that can be exploited in many different applications such as localization, analysis of mobile data usage and so on. In particular, 5G and 6G systems generate many different types of big data in their networks. When monitoring the network as well as communicating between mobile devices and networks, 6G systems can collect data about throughput, latency, packet loss, jitter, spectrum usages, and so on. AI algorithms can analyse big data, optimize wireless network configurations

Fig. 9.1   5G link protocol
structure

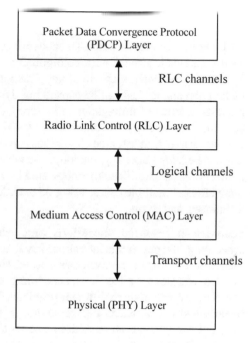

and improve the performances and services for end users. Depending on data type, input data, network types, applications, and so on, AI contributions to 6G systems will be different. The key functions of data link layer are to make connectivity between mobile devices and networks and share the radio resource efficiently. This is one of key functions in wireless systems. Due to the time-varying wireless channels and network environments, it is very challenging to satisfy all requirements and conditions. The MAC layer takes a responsibility to manage them. There are many types of multiple access schemes such as Time Division Multiple Access (TDMA), Carrier Sense Multiple Access/Collision Avoidance (CSMA/CA), Code Division Multiple Access (CDMA), and Orthogonal Frequency Division Multiple Access (OFDMA). Due to the high system capacity requirement, 5G systems adopted OFDMA. 6G systems may use Orthogonal Multiple Access (OMA) like OFDMA or Non-Orthogonal Multiple Access (NOMA). The NOMA is one of promising multiple access techniques. It was considered as one of 5G multiple access schemes due to high spectral efficiency. In order to accommodate multiple users on the radio resources, the NOMA uses both different power levels and different codes for multiplexing. In the NOMA transmitter, superposition coding is adopted to accommodate multiple users. In the NOMA receiver, successive interference cancellation (SIC) receiver can separate the multiple users. If the NOMA technique is proved in practical systems and achieves a higher technology readiness level (TRL) such as TRL 8 or 9, it will be the key function of 6G systems to improve system capacity. Another important feature of data link layer for 6G systems is network automation that enables the cellular network to become more intelligent, gain more cost efficiency, and craft a tailor-made service for mobile users. The network automation needs several key functions such as automatic provisioning, configuration, and orchestration of network services. The resource allocation as one of key functions in data link layer is a part of automatic provisioning and can be carried out dynamically for different radio resource requests in terms of throughput and latency, different number of users and heterogeneous cells. The key AI contributions to 6G data link layer can be summarized as follows: data-driven resource allocations and scheduling, AI-enabled adaptive MAC layer, AI-enabled data aggregation, AI-enabled system failure detection, cognitive radio and MAC identification empowered by AI, AI-enabled interference detection, more efficient multiple access using AI techniques, handover optimization using AI techniques, and so on.

Data-driven resource allocations and scheduling: The resource allocations, scheduling, and power allocations are key functions of MAC layer. They are directly related to the system performance and efficiency. Due to a high complexity, the problem of resource allocations, scheduling, and power allocations can be decomposed by individual three different problems. Each individual solution is combined linearly and performed in a base station. Sometime, a joint solution considering different wireless system conditions and requirements is developed. However, it is still not close to an optimal solution because we can't consider all conditions and requirements of wireless systems. The joint solution can be optimal at only a specific condition and requirement. In order to solve complex optimization problems and find the optimal or sub-optimal solutions, exhaustive and greedy approaches are widely used. AI techniques will be helpful for reducing the complexity of resource allocations and obtaining the global optimal points in a data-driven way. They enable a resource allocation function to have the learning ability, adapt, and optimize for various channel conditions. In [1], CNN is used to optimize the resource allocation in cellular networks. The CNN is trained by observing the radio resource usages of a particular network slice. The trained CNN predicts the required capacity to accommodate the future radio resources associated to network slices. The network can achieve the optimal resource allocation.

AI-enabled adaptive MAC layer: Some adaptive MAC layer was discussed in IEEE802.15 standard. The adaptive MAC layer can be classified into three types such as traffic load estimation, adaptive wake-up interval, and adaptive time slot allocation. In [2], adaptive MAC layer using decision trees is proposed. A reconfigurable MAC layer can be changed at run time. The trained MAC layer chooses the most suitable MAC protocol in terms of network conditions and applications. The selection of MAC protocol is a classification problem. The classifier is trained by received signal strength indicator (RSSI) statistics and traffic patterns through interpacket interval (IPI) and application requirements.

AI-enabled data aggregation: Data aggregation is a technique to collect and combine the useful data in wireless networks. It is a key procedure to reduce energy consumption and utilize network resources. In particular, it is very useful when multiple sensor nodes detect the same data in IoT networks. This problem is known as data overlapping problem. In [3], a data aggregation using clustering technique is proposed to reduce energy and computational complexity in wireless sensor networks. The clustering techniques enable the nodes to form clusters sensing similar values within a threshold. Thus, only one sensor data in a cluster is transmitted to a gateway. We can reduce the number of transmission and improve energy efficiency significantly.

AI-enabled system failure detection: The problem of anomaly detection including abnormal system behaviours can be solved by AI algorithms. When we have enough amount of labelled or unlabelled data, supervised or unsupervised algorithm can be used to solve classification or clustering problems that can be transformed to anomaly detection problems. There are many examples such as transmission outage, fault detection, and intrusion detection in wireless systems. A Denial-of-Service (DoS) attack makes networks shut down by flooding with traffics or sending some information triggering a crash. A legitimate mobile devices or IoT sensor nodes can't transmit their data packets to a base station or a gateway. Wireless systems can be the target of DoS attacks. A secure MAC layer should be designed to protect the network from DoS attacks. In [4], the neural network functions of MAC layer monitor variations of MAC layer parameters such as collision rate, average waiting time of MAC bugger, and arrival rate of Request-To-Send (RTS) packets. If the traffic variations exceed a certain threshold, the MAC layer considers it as abnormal behaviour. The trained neural network detects the DoS attacks and the networks are switched off temporarily. In [5], the neural network is trained in a way of online learning on sensor nodes in wireless sensor networks. An extreme learning machine algorithm is used to classify system failure, intrusion, and abnormal behaviours.

Cognitive radio and MAC identification empowered by AI: The efficiency of radio spectrum usage is one of key metrics in wireless systems. Many stakeholders want more efficient spectrum usages. Cognitive radio is a concept to use the radio spectrum more efficiently by intelligently detecting empty spectrum in channels and using vacant spectrum while avoiding any interferences. It provides us with an opportunistic spectrum use by unlicensed users (or secondary users) where they should cause any harmful interferences to licensed users (or primary users). Figure 9.2 illustrates key cognitive radio functions in physical layer and MAC layer.

In physical layer of cognitive radio, spectrum sensing and data transmission are performed. MAC layer of cognitive radio includes spectrum sensing, sharing, decision and mobility. AI algorithms will take a key role of spectrum decision. However, there are many disadvantages. For example, there are no accurate detection of vacant spectrum without any interferences to licensed users. Basically, mobile operators with licensed spectrum expect zero interferences from unlicensed users. This technique requires multi-band antenna to search. The QoS can't be guaranteed. The wireless channel may be jammed or eavesdropped. The security can't be guaranteed. Thus, this technique was studied in academy but doesn't be adopted in commercial cellular systems. In order to implement cognitive radio, the key questions can be summarized

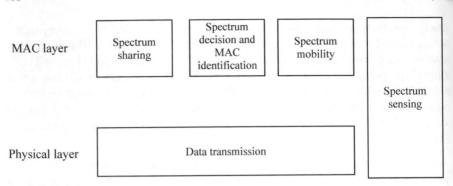

Fig. 9.2 Key cognitive radio functions in physical and MAC layer

as follows: How can we detect channel idle or busy accurately? Which channels can be aggregated in multi-channel environments? How long do we use the idle channel and manage co-channel interference? How to maintain the performance of the unlicensed users? AI algorithms will be able to help cognitive radio systems find the empty spectrum slot accurately and transmit their packets accordingly. In addition, AI techniques allow us to identify different types of MAC protocols. This benefit will improve the network throughput. For example, in [6], the temporal features of three different WiFi networks (802.11 b, g and n) MAC layer are analysed and KNN classifier distinguishes them. In [7, 8], CNN is used to detect types of MAC protocols. They convert the data into the form of a spectrogram and combine the spectrogram and CNN in order to identify MAC protocols such as TDMA, CSMA/CA, and ALOHA. Another key contribution of AI algorithms to cognitive radio is to predict the empty spectrums. It will be helpful for using the spectrum more efficiently. There are many approaches to predict the spectrum usages. One approach is to measure channel usages for a certain time and predict the future empty spectrum slots using RNN. Another approach is to observe the quality of channels in terms of idle or busy durations and then find the most suitable channels with good quality. In [9, 10], the neural networks are used to predict the spectrum. The neural network learns from history of spectrum usages, predicts the spectrum usage of other neighbouring networks, and optimizes the spectrum usage. In [11], the neural network predicts the future activities of primary users from the past channel occupancy data and improves the throughputs of secondary users while alleviating collision between them. In [12], a channel selection problem is regarded as a binary classification problem. Unsupervised learning algorithms such as K-means clustering do not require any channel information from the primary users and identify the channel states. In [13], reinforcement learning is applied to license assisted access (LAA) configuration. In the base station with LAA, the reinforcement learning helps us to select an unlicensed carrier and find suitable transmission duration.

AI-enabled interference detection: In 6G heterogeneous networks, interference problems will be severe and degrade the network throughput. Thus, interference detection and management will be one of 6G key functions. In MAC layer, we should be able to identify interfered channels and packets and quantify the interferences. Thus, we can find appropriate solutions to manage the interferences. For example, in [14], MAC layer observes energy variations of the received packet and monitors the link quality indicator. A decision tree classifier finds candidates and detects interferences.

More efficient multiple access using AI techniques: AI techniques will be helpful for having more efficient transmission by learning from the system transmission history. For example, the CSMA/CA is one of multiple access schemes based on carrier sensing. The devices attempt to avoid collision. They can transmit their packet when the channel is empty. The collision avoidance technique is used to improve the performance of the transmission. The basic mechanism is as follows: Before data packet transmission, devices observe the channel to determine whether or not other devices are transmitting their data packet. We call this process carrier sensing. If the channel is busy, the devices wait for a certain period of time and then transmit their data packet. An exponential backoff is a method to determine how long devices should backoff before they retransmit. Typically, they attempt to request exponentially while increasing the waiting time until a maximum backoff time. When the collision is detected, the retransmission process is performed. If retransmission happens frequently, the energy consumption and transmission delay increase. Thus, it is important to avoid the collision. The CSMA/CA is widely used in wired communications but it is adopted in wireless LAN systems. The channel characteristics of wireless systems are significantly different from wired systems. In particular, transmission power, receiver sensitivity, channel impairments, and so on. There is a hidden node problem because a device couldn't detect the transmission of other devices well. AI techniques will be able to reduce the retransmission by predicting a proper time slot from a history of collisions.

Handover optimization using AI techniques: A handover is a key function of cellular networks that should maintain the average throughput of the users while minimizing any service interruptions. In 6G systems, it will be more important to manage the QoS in heterogeneous networks. The 6G heterogeneous networks will be mixed by mmWAVE femtocell, macrocell, IoT networks, ultra-dense networks, and so on. This network environment such as ultra-dense heterogeneous networks will lead to more frequent handovers that can be defined as channel and resource block change of users or cell association while maintaining a session. Satisfying these new requirements will be more challenging tasks for a network designer to develop a 6G handover mechanism. The dense networks will be a key type of urban cellular networks where the density of base stations increases to support a high throughput service to users. The small cell concept enables us to facilitate more base station deployments. Higher mobility of mobile devices needs to perform more handover. Thus, more frequent handovers are required in 6G systems. Typically, there is an inverse proportional relationship between the average throughput of mobile users and the number of handovers. The frequent handovers cause the QoS degradation in the network. Service interruptions may happen during the handovers. Thus, the main research challenges in the topic of handover are to minimize the number of handover among cells, reduce the service interruptions, or reduce the cost of the handover management. AI techniques will be helpful for predicting the handover of users and enhance the network performance. They will play an important role in a handover optimization and a base station selection by reducing the number of handover, computational complexity and delays.

Summary 9.1. AI-Enabled Data Layer Design

1. In 5G NR, The 5G layer 2 is about data link layer consisting of Medium Access Control (MAC), Radio Link Control (RLC) and Packet Data Convergence Protocol (PDCP).
2. Based on the structure of 5G data link layer, 6G data link layer may employ new functions such as terahertz communication support, distributed communication support, cell-free access, and so on.
3. The function of the data link layer can be regarded as mapping from a perceived information about the network conditions to the choice of network configurations. This is similar to reinforcement learning or deep learning approaches.
4. The learning ability of AI techniques can be used for improving the performance or creating a new application in wireless systems. The patterns or predictions by AI algorithms can be exploited to adjust the parameters of the wireless networks and operate the wireless networks efficiently.
5. The key AI contributions to 6G data link layer can be summarized as follows: Data-driven resource allocations and scheduling, AI-enabled adaptive MAC layer, AI-enabled data aggregation, AI-enabled system failure detection, Cognitive radio and MAC identification empowered by AI, AI-enabled interference detection, More efficient multiple access using AI techniques, and Handover optimization using AI techniques.

9.2 Radio Resource Allocation and Scheduling

Two fundamental aspects of cellular networks are the limited radio resources and the dynamic nature of wireless environments. This is because of the limited bandwidth, many users who want to use the bandwidth, varying channel conditions, and a user equipment mobility. In order to address these issues, radio resource allocations and scheduling to multiple users should be discussed. They are designed using channel knowledge at the transmitter. The channel knowledge can be obtained in various ways such as duplex system characteristics, feedback channel, and pilot signal measurement. The channel knowledge is useful for resource allocation and scheduling as well as power control and adaptive modulation and coding. The radio resource allocation among mobile users influences on wireless system performances and flexible services. A base station should include a transmission allocation task while adopting channel conditions and system requirements. Cellular networks have a limited radio resource that should be shared by multiple mobile users where the limited radio resources can be a transmit power, bandwidth, transmission time, and spatial antenna. Depending on how efficiently the radio resources are used, key system performances such as energy efficiency, end user throughput, user QoS guarantee level, network capacity, and system latency are determined. Thus, the main purpose of the radio resource allocation task is to allocate the limited radio resources to mobile users according to the requirements of the individual users and achieve the efficient operation of the cellular networks. The optimization problems of the radio resource allocation can be formulated in terms of different cost functions such as system throughput, energy efficiency, connection density, and so on. Classical resource allocation schemes include maximum throughput, proportional fair, exponential rule, and so on. They work well at simple voice services in cellular systems. However, they couldn't satisfy various service requirements in cellular network supporting multiple user data services. In 6G systems, various service requirements should be considered in terms of delay, user priorities, user mobility, pack loss rate, energy efficiency, service class, and so on. The resource allocation and scheduling functions should take into account them all. In 4G LTE, service class-based resource allocation and scheduling is used. As the first step, admission controller admits or rejects active mobile users depending on available radio resources where active mobile users are defined as a user to be allocated and scheduled. Multiple services (voice, video, game, constant bit rate (CBR) and so on) are ready for resource allocation and scheduling to multiple users. The next step is to allocate and distribute radio resources to different service classes. As the final step, the packet scheduler considers multiple metrics about service class specifications and sorts the data packet as the radio resources. Figure 9.3 illustrates the steps of resource allocation and scheduling in 4G LTE.

The mechanism of the resource allocation and scheduling relies on multiplexing schemes in physical layer. Comparing with CDMA and TDMA schemes, the OFDMA scheme provides us with more flexibility for the resource allocation. The radio resources in OFDMA systems are divided into multiple time–frequency resource blocks for transmission. The time–frequency resource blocks are composed of a large number of closely spaced sub-channels. Key advantages include robustness against multipath fading and intersymbol interferences, simple receiver equalization, good combination with MIMO, and so on. In particular, MIMO-OFDM technique is a powerful combination and widely used in wireless systems. The MIMO-OFDM technique achieves high spectral efficiency and even makes us optimal use of the wireless channel at specific conditions. In this section, we discuss radio resource allocation problem formulation, performance measures, utility functions, fairness, and AI technique contributions to radio resource allocation problem.

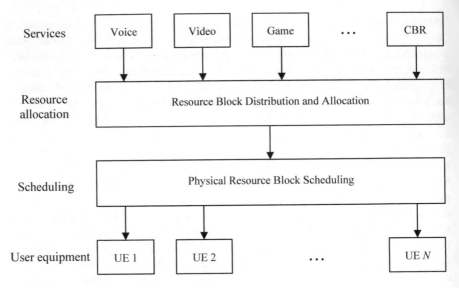

Fig. 9.3 Resource allocation and scheduling steps in 4G LTE

9.2.1   Resource Allocation Problems in Wireless Networks and Convex Optimization

A number of formulations for radio resource allocation have been studied in terms of multiple system parameters. One approach is to minimize the average transmission powers in a cellular network. A mobile device relies on a battery life. Due to the limited power supply, the usage of the mobile device is limited. Thus, minimizing the transmit power results in more usage and power saving of the mobile device. The objective of the resource allocation problem can be the minimization of the total transmission power under the target data rate for each user [15]. The target data rates of each user can be determined by a fair queuing algorithm or a dynamic queuing algorithm in terms of user's channel conditions. Another approach is to maximize sum rate under the given total transmission power [16, 17]. This is one of common approaches. In [18], both minimum and maximum date rates for each user are targeted while taking into account multiple system constraints. In [19], some users who are sensitive to delay but have a fixed target data rate while other best effort users have no data rate constraints are considered. Thus, radio resource allocation and scheduling is to maximize the sum rate of the best effort users while satisfying the rate targets of the delay sensitive users. In [20], a radio resource allocation problem for a single cell uplink is formulated in the frameworks of Nash Bargaining and an iterative algorithm is used as a solver. In real world, wireless channel includes multiple channel impairments such as interferences. If we consider the interference as a channel impairment, resource allocation problems become more complex. In

~~mobile networks~~, power control can be used for interference management. In downlink, a mobile device can receive both transmitted signals and interferences from
adjacent cells. In uplink, a base station can experience adjacent channel interferences from mobile devices in neighbouring cells and co-channel interferences from
mobile devices in the same cells. The signal-to-interference noise ratio (SINR) is
used to deal with the interferences. In [21], radio resource allocation and interference problems in multi-cell downlink are considered. They manage the interference
among multiple cells by radio resource coordination. As we briefly reviewed in the
above, there are many different types of radio resource allocation and scheduling
problems formulation. Many resource allocation problems targeting sum-rate maximization are known as NP-hard problems. The complexity is very high. It is not
easy to find an optimal solution. In order to solve the resource allocation problems,
systematic methods including branch-and-bound methods are used. However, the
computational complexity is high, and it takes long time to converge. Thus, they are
far away from practical use. Common approaches are to reformulate the resource
allocation problems as a convex optimization problem and then solve them using
the known solver like interior point methods. Thus, one of effective approaches to
deal with resource allocation problem is convex optimization. Figure 9.4 illustrates
examples of convex function and concave function.

A standard form of the convex optimization problem is expressed as the following
standard form [22, 23]:

$$\min_{\mathbf{x}} f_0(\mathbf{x}) \tag{9.1}$$

subject to

$$f_i(\mathbf{x}) \leq 0, i = 1, 2, \ldots, m$$
$$h_j(\mathbf{x}) = 0, j = 1, 2, \ldots, p \tag{9.2}$$

where the objective function f_0 and the constraint functions $f_1, \ldots, f_m$ are convex
satisfying

$$f_i(\alpha\mathbf{x} + \beta\mathbf{y}) \leq \alpha f_i(\mathbf{x}) + \beta f_i(\mathbf{y}) \tag{9.3}$$

for all $\mathbf{x}, \mathbf{y} \in \mathbb{R}^n$ and all $\alpha, \beta \in [0, 1]$ with $\alpha + \beta = 1$. The equality constraint
function $h_j(\mathbf{x}) = \mathbf{a}_j^T\mathbf{x} - b_j$ where $\mathbf{a}_j$ is a column vector and b_j is a real number
must be affine. Convex equality constraints are linear. Roughly speaking, a function
is convex when satisfying the condition: If $z = \alpha x + \beta y$ varies over the line $[x, y]$,
the point $(z, f(z))$ is located below the line connecting two points $(x, f(x))$ and
$(y, f(y))$. Generally, a hierarchical structure of convex optimization problems is
as follows [23]: Convex optimization problems (CP) $\supset$ Semidefinite programming
(SDP) $\supset$ Second-order cone programming (SOCP) $\supset$ Convex quadratic programming (CQP) $\supset$ Linear programming (LP). There are no generalized analytical solutions for convex optimization problems. Finding a generalized solution is still active

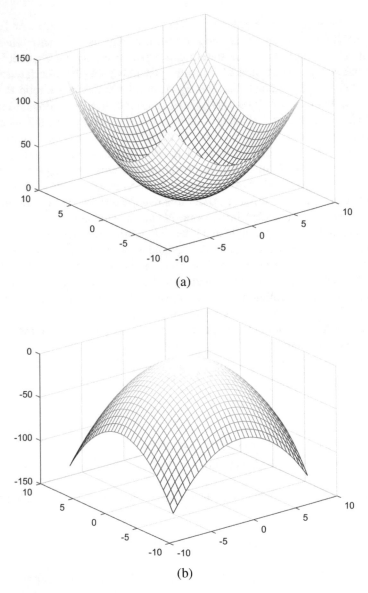

Fig. 9.4 Example of convex function (**a**) and concave function (**b**)

research area. Conventional optimization problems in cellular networks are difficult to find a solver. However, convex optimization problems can be reliably tackled by transforming the optimization problems into convex expressions. A global optimal solution to the convex optimization problem can be found by interior point method, ellipsoid method, sub-gradient method and so on. Among them, the interior point method (IPM) is a very useful tool for many convex optimization problems. It is widely applied in practical systems. However, it requires unavoidable computation of the objective function and constraint functions. If we have a large size of the problem, computation is very high and it is not efficient. Gradient descent methods may be better with medium accuracy in large size convex optimization problems. The IPM relies on Newton's method. The IPM starts searching a solution from the interior of the feasible region and then updating until converging to an optimal point. Thus, the interior of the feasible region should not be empty and also most of iterations should be in the interior of the feasible region. The basic approach is to approximate the constraints of the optimization problems using Newton's method. Firstly, the optimization problem is converted in a standard from. Secondly, the optimal solution in the feasible region is iteratively searched by Newton's method. There are two algorithms of the interior point methods: barrier method and primal–dual method [23]. The primal–dual interior point method is more efficient because it is more accurate with similar iteration and enables us to solve larger optimization problems. The barrier method solves a sequence of linearly constrained (or unconstrained) minimization problems. The detail of both algorithms is described in [23]. In this section, we summarize their procedures. We compute an optimal solution $\mathbf{x}^*(t)$ for increasing values of $t > 0$ (a penalty parameter about the accuracy of the approximation), until $m/t \leq \epsilon$ where ϵ is tolerance. It guarantees to have an ϵ-suboptimal solution. We start at $t = t^{(0)} > 0$ and solve the problem using Newton's method at centring step to produce $\mathbf{x}^{(0)} = \mathbf{x}^*(t)$. Then, we increase t for a barrier parameter $\mu > 1$. We can summarize the steps of the barrier method as follows [23]:

Initial conditions: Strictly feasible $\mathbf{x}$, $t = t^{(0)} > 0$, $\mu > 1$ and $\epsilon > 0$.

Repeat

- *Step 1*: Centring step. Compute $\mathbf{x}^*(t)$ solving $\min_{\mathbf{x}} t f_0(\mathbf{x}) + \phi(\mathbf{x})$ subject to $\mathbf{A}\mathbf{x} = \mathbf{b}$ where ϕ is the logarithmic barrier function with $\mathbf{dom}\phi = \{\mathbf{x} \in \mathbb{R}^n | f_i(\mathbf{x}) < 0, i = 1, \ldots, m\}$, constraint functions are convex and twice continuously differentiable, and the matrix $\mathbf{A} \in \mathbb{R}^{p \times n}$ has rank $\mathbf{A} = p$.
- *Step 2*: Update. Produce $\mathbf{x}^{(k)} = \mathbf{x}^*(t)$ at $t = t^{(k)}$ for $k = 1, 2, \ldots$
- *Step 3*: Stop condition. Stop if $m/t \leq \epsilon$.
- *Step 4*: Increase penalty. $t^{(k+1)} = \mu t$.

Until it converges: $m/t \leq \epsilon$.

The step 1 is regarded as outer iterations and computing $\mathbf{x}^*(t)$ using Newton's method is regarded as inner iterations. It terminates with $f_0(\mathbf{x}^*(t)) - p^* \leq \epsilon$. There are two parameters we should carefully choose: μ and $t^{(0)}$. The choice of μ means a trade-off between number of inner iterations and outer iterations. If μ is small, more outer iterations is required. If μ is big, Newton's method requires many iterations to converge. Typical values of μ are 10–20. The choice of $t^{(0)}$ means a trade-off between number of inner iterations within the first outer iteration and number of outer iterations. If $t^{(0)}$ is small, many outer iterations are required. If $t^{(0)}$ is large, the first outer iteration needs more iterations to compute $\mathbf{x}^{(0)}$. Primal–dual interior point methods are similar to the barrier methods. In the primal–dual method, both the primal variable $\mathbf{x}$ and the associated dual variables $\boldsymbol{\lambda}, \boldsymbol{v}$ are updated. We can summarize the steps of the primal–dual interior point method as follows [23]:

Initial conditions: Start with a feasible point $\mathbf{x}$ with $f(\mathbf{x}) < 0, \lambda > 0, \mu > 1,$ tolerance $\epsilon_f > 0, \epsilon > 0$.

Repeat

- ***Step 1***: Determine t. Set $t = \mu m / \hat{\eta}$.
- ***Step 2***: Compute the primal–dual search direction $\Delta \mathbf{y}_{pd}$

- ***Step 3***: Line search on $\lambda, f(\mathbf{x})$ and $\|r_t\|$ where $r_t(\mathbf{x}, \lambda, \boldsymbol{v}) = \begin{bmatrix} r_d \\ r_c \\ r_p \end{bmatrix}$ where r_d, r_c

  and r_p are called dual residual, centrality residual and primal residual, respectively.

  o ***Step 3a***: Start with the step size $s = 0.99 s^{max}$ where $s^{max} = \min\{1, \min\{-\lambda_i/\Delta\lambda_i | \Delta\lambda_i < 0\}\}, s > 0$.
  o ***Step 3b***: Continue $s = \beta s$ until $f(\mathbf{x} + s\Delta\mathbf{x}) < 0, \|r_t(\mathbf{y} + s\Delta\mathbf{y}_{pd})\| > (1 - \alpha s)\|r_t(\mathbf{y})\|$ where α and β are typically in the range [0.01 0.1] and [0.3 0.8], respectively.

- ***Step 4***: Update. $\mathbf{y} = \mathbf{y} + s\Delta\mathbf{y}_{pd}$

Until $\|r_p\| \le \epsilon_f, \|r_d\| \le \epsilon_f$ and $\hat{\eta} \le \epsilon$.

Example 9.1 Convex optimization

Consider the following optimization problem

$$\min_{x,y}\left(xe^{-(x^2+y^2)} + \left(1.5x^2 + y^2\right)/10\right)$$

subject to

$$xy + (x+2)^2 + (y-2)^2 \le 2.$$

Find the optimal point with the start point $(x, y) = (-3, 2)$.

Solution

In this optimization problem, the objective function $f(x, y) = \left(xe^{-(x^2+y^2)} + \left(1.5x^2 + y^2\right)/10\right)$ and the constraint function $g(x, y) = xy + (x+2)^2 + (y-2)^2 - 2$ are convex. We can solve this optimization problem using interior point methods. The initial value is given as $(-3, 2)$. Using computer simulation, we can obtain the Table 9.1 after 10 iterations.

First-order optimality represents a measure of how close the point is to be optimal. As we can observe Table 9.1, the objective function is saturated at 6th iteration and the first-order optimality measure is minimum at 10th iteration. The constraint function is satisfied within the value of the constraint tolerance. Thus, the optimization process is complete, we obtained the optimal point (−0.0020, 0.7702) and the optimal value −0.0362. Figure 9.5 illustrates the convex optimization including starting point, optimal point, objective function, and constraint function.

Summary 9.2. Radio Resource Allocation and Convex Optimization

1. A standard form of the convex optimization problem is expressed as the following standard form

$$\min_{\mathbf{x}} f_0(\mathbf{x})$$

subject to

$$f_i(\mathbf{x}) \leq 0, \quad i = 1, 2, \ldots, m$$

$$h_j(\mathbf{x}) = 0, \quad j = 1, 2, \ldots, p$$

where the objective function f_0 and the constraint functions $f_1, \ldots, f_m$ are convex satisfying $f_i(\alpha \mathbf{x} + \beta \mathbf{y}) \leq \alpha f_i(\mathbf{x}) + \beta f_i(\mathbf{y})$ for all $\mathbf{x}, \mathbf{y} \in \mathbb{R}^n$ and all $\alpha, \beta \in [0, 1]$ with $\alpha + \beta = 1$.

2. Generally, a hierarchical structure of convex optimization problems is as follows: Convex optimization problems (CP) $\supset$ Semidefinite programming (SDP) $\supset$ Second order cone programming (SOCP) $\supset$ Convex quadratic programming (CQP) $\supset$ Linear programming (LP).

Table 9.1 Objective functions and optimality after 10 iterations

Iteration	Objective function	First-order optimality
0	1.75E + 00	9.00E−01
1	9.43E−01	5.94E−01
2	7.87E−02	3.70E−01
3	1.74E−01	1.21E−01
4	−2.41E−02	9.56E−02
5	−1.64E−02	1.99E−02
6	−3.61E−02	4.79E−03
7	−3.60E−02	3.88E−04
8	−3.60E−02	2.00E−04
9	−3.62E−02	4.00E−05
10	−3.62E−02	4.01E−07

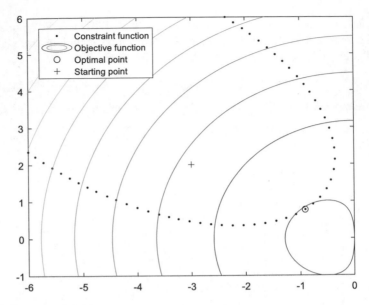

Fig. 9.5 Convex optimization

3. A global optimal solution to the convex optimization problem can be found by interior point method, ellipsoid method, sub-gradient method and so on.
4. The basic approach of interior point methods is to approximate the constraints of the optimization problems using Newton's method. Firstly, the optimization problem is converted in a standard from. Secondly, the optimal solution in the feasible region is iteratively searched by Newton's method.
5. There are two algorithms of the interior point methods: barrier method and primal–dual method. The barrier method solves a sequence of linearly constrained (or unconstrained) minimization problems. The primal–dual interior point method is more efficient because it is more accurate with similar iteration and enables us to solve larger optimization problems.

9.2.2   Resource Allocation Models and Performance Measure

Optimal resource allocations in MIMO-OFDM systems are investigated in [24]. The resource allocation problems and solutions are discussed in MIMO-OFDM systems. They assume linear transmission and reception, perfect synchronization, flat-fading channels, perfect channel state information, centralized optimization, and ideal transceiver hardware. Some of assumptions are relaxed further at some model.

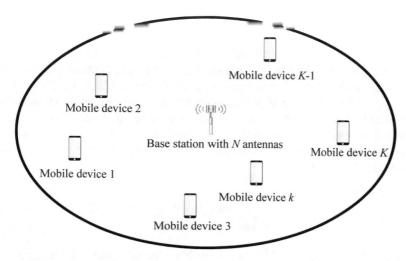

Fig. 9.6 Single cell downlink model

They allocate powers and radio resources to mobile users and select beamforming to maximize system utilization. Consider a single cell downlink model where a base station has N antennas and K mobile devices (or mobile users) with single antenna belong to the single cell as shown in Fig. 9.6.

This system model is a multi-user MISO communication. We assume a flat-fading channel with the channel response $\mathbf{h}_k \in \mathbb{C}^N$ where $\mathbb{C}^N$ is the set of complex valued N vectors. The channel response is known at a base station and also constant during the transmission. The received signal at mobile devices $y_k \in \mathbb{C}$ can be expressed as follows:

$$y_k = \mathbf{h}_k^H \mathbf{x} + n_k \tag{9.4}$$

where $\mathbf{x} \in \mathbb{C}^N$ and $n_k \sim CN(0, \sigma^2)$ are the transmitted signal and the additive noise with complex Gaussian distribution, respectively. σ^2 is the noise variance (noise power). The transmitted signal contains data signal for each mobile device as follows:

$$\mathbf{x} = \sum_{k=1}^{K} \mathbf{s}_k \tag{9.5}$$

where $\mathbf{s}_k \in \mathbb{C}^N$ is each mobile device signal. They can be modelled as signal correlation matrices with zero mean $\mathbf{S}_k \in \mathbb{C}^{N \times N}$

$$\mathbf{S}_k = E[\mathbf{s}_k \mathbf{s}_k^H]. \tag{9.6}$$

This represents a linear multi-stream beamforming. We call it signal correlation matrices. Each matrix represents a transmission strategy and determines what transmission is received at multiple mobile devices. Finding a transmission strategy $\mathbf{S}_k$ and power constraints is radio resource allocation. The rank of the matrix $\mathbf{S}_k$ denoted Rank $(\mathbf{S}_k)$ means the number of streams. Trace of the matrix $\mathbf{S}_k$ denoted $\mathrm{tr}(\mathbf{S}_k)$ is the average transmit power allocated to each mobile device. The eigenvalues and eigenvectors of $\mathbf{S}_k$ mean the spatial distribution of the transmit power. The total transmission power at a base station is basically limited because of the transmit power regulations and interference management. The power constraints are defined as follows:

$$\sum_{k=1}^{K} \mathrm{tr}(\mathbf{Q}_{lk}\mathbf{S}_k) \le q_l \tag{9.7}$$

where $l = 1, \ldots, L$, $\mathbf{Q}_{lk} \in \mathbb{C}^{N \times N}$ are Hermitian semi-definite weighting matrices, and the limit $q_l \ge 0$ for $\forall l, k$. One condition $\sum_{l=1}^{L} \mathbf{Q}_{lk} > \mathbf{0}_N$ for $\forall k$ represents constraints in all spatial directions. If $L = 1$ and $\mathbf{Q}_{1k} = \mathbf{I}_N$ (where $\mathbf{I}_N$ is $N \times N$ identity matrix.) for $\forall k$, q_1 means maximal total power. If $L = N$ and $\mathbf{Q}_{lk}$ has only nonzero element at the lth diagonal $\mathbf{Q}_{1k} = \mathrm{diag}(1, 0, \ldots, 0), \ldots, \mathbf{Q}_{Nk} = \mathrm{diag}(0, 0, \ldots, 0, 1)$, this constraint means maximal power at lth antenna. The weighting matrices $\mathbf{Q}_{lk}$ can be same for all mobile devices or different for specific mobile devices. Under the condition of no disturbance to neighbouring system, the corresponding constraints can be defined as soft-shaping [24]. The shape of transmission is only affected if the power level without the constraint would have exceeded the threshold q_l [24]. If the interuser interference caused to mobile devices should not exceed q_l, $\mathbf{Q}_{li} = \mathbf{h}_k\mathbf{h}_k^{H}$ for $\forall i \ne k$ can be set [24]. This is related to zero-forcing transmission (or zero interuser interference). The linear power constraints in the Eq. (9.7) can be rewritten as per-user constraints

$$\mathrm{tr}(\mathbf{Q}_{lk}\mathbf{S}_k) \le q_{lk} \tag{9.8}$$

where $l = 1, \ldots, L$, $k = 1, \ldots, K$, and $q_{lk} \ge 0$ for $\forall l, k$. In order to satisfy (9.7), the per-user power limits should satisfy the following conditions:

$$\sum_{k=1}^{K} q_{lk} \le q_l \tag{9.9}$$

where $l = 1, \ldots, L$. The linear sum power constraints will be an useful tool for designing optimal transmit power allocations. Finding q_{lk} is related to optimize power allocation for each user. In the single cell downlink model, frequency reuse is assumed and the interferences between cells are negligible. A base station determines radio resource allocation for each user in the cell while not considering any cooperation or interference among neighbouring cells. However, many new features such as multi-cells, multiple antennas, and cooperative processing would be incorporated in 6G heterogeneous networks. The multi-cell downlink model can use the same frequency bands simultaneously while managing the interferences. The multi-cell model will be efficient to maximize spectral efficiency. Coordinated multi-point (CoMP) is operated in this model. In [25, 26], co-processing among base stations is investigated and multiple networks operation is regarded as one large multi-user MISO with distributed transmit antennas. The mobile users are served by a joint transmission from multiple base stations. Optimal spectral efficiency can be obtained under the given ideal conditions. However, this approach requires global channel knowledge sharing among base stations, coherent joint transmission, accurate synchronization, joint interference management, and centralized resource allocation. These conditions are unrealistic. In the multi-cell downlink model, the channel responses from base stations to mobile devices can be defined as follows:

$$\mathbf{h}_k = \left[\mathbf{h}_{1k}^T, \ldots, \mathbf{h}_{\mathcal{K}k}^T\right]^T \tag{9.10}$$

where $\mathbf{h}_{jk} \in \mathbb{C}^{N_j}$ is channel response from the jth base station with N_j antennas and the multi-cell model has $\mathcal{K}$ base stations. The total number of antenna is $N = \sum_{j=1}^{\mathcal{K}} N_j$. In this scenario, each mobile device is jointly served by a subset of base stations and interferences among adjacent base stations are managed. This scenario can be modelled by dynamic cooperation clusters [27] with the following properties: (1) a base station has channel estimates to mobile devices in outer circle $\mathcal{C}_j$ and interference from mobile device $i \neq \mathcal{C}_j$ is negligible. (2) In inner circle $\mathcal{D}_j$, mobile devices are served with data. Namely, the inner circle $\mathcal{D}_j$ describes data from base stations and the outer circle $\mathcal{C}_j$ describes coordination from base stations. The user association with base stations will be changed dynamically in terms of resource allocation and network operation. It is important to choose inner and out circles for optimal resource allocation. Figure 9.7 illustrates multi-cell downlink model based on

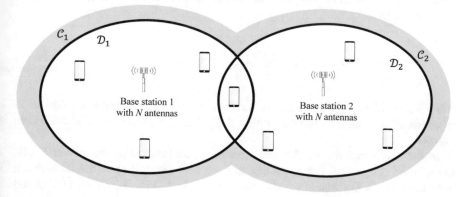

Fig. 9.7 Multi-cell downlink model based on dynamic cooperation clusters

dynamic cooperation clusters. The coordination work does not deal with the outside circle.

In this multi-cell scenario based on dynamic cooperation clusters, certain channel elements of $\mathbf{h}_k$ are related to data transmission. They are selected by the diagonal matrices $\mathbf{D}_k \in \mathbb{C}^{N \times N}$ and $\mathbf{C}_k \in \mathbb{C}^{N \times N}$ as follows:

$$\mathbf{D}_k = \begin{pmatrix} \mathbf{D}_{1k} & \cdots & 0 \\ \vdots & \ddots & \vdots \\ 0 & \cdots & \mathbf{D}_{Kk} \end{pmatrix} \tag{9.11}$$

where

$$\mathbf{D}_{jk} = \begin{cases} \mathbf{I}_{N_j}, & \text{when } k \in \mathcal{D}_j \\ \mathbf{0}_{N_j}, & \text{when } k \notin \mathcal{D}_j \end{cases} \tag{9.12}$$

and

$$\mathbf{C}_k = \begin{pmatrix} \mathbf{C}_{1k} & \cdots & 0 \\ \vdots & \ddots & \vdots \\ 0 & \cdots & \mathbf{C}_{Kk} \end{pmatrix} \tag{9.13}$$

where

$$\mathbf{C}_{jk} = \begin{cases} \mathbf{I}_{N_j}, & \text{when } k \in \mathcal{C}_j \\ \mathbf{0}_{N_j}, & \text{when } k \notin \mathcal{C}_j \end{cases} \tag{9.14}$$

where $\mathbf{I}_{N_j}$ and $\mathbf{0}_{N_j}$ are the $N_j \times N_j$ identity matrix and the $N_j \times N_j$ matrix with only zero elements, respectively. $\mathbf{h}_k^H \mathbf{D}_k$ and $\mathbf{h}_k^H \mathbf{C}_k$ represents the channel carrying data to the kth mobile device and the channel carrying non-negligible interference, respectively. The received signal at the kth mobile device in this multi-cell scenario is expressed as follows:

$$y_k = \mathbf{h}_k^H \mathbf{C}_k \sum_{i=1}^{K} \mathbf{D}_i \mathbf{s}_i + n_k \tag{9.15}$$

where $n_k \sim CN(0, \sigma_k^2)$ is a noise term representing both Gaussian noise and weak interference from all base stations with $k \notin \mathcal{C}_j$. The Eq. (9.15) can be illustrated in Fig. 9.8.

In the single cell model, the transmission is constrained by the power limits. In the multi-cell model, $\mathbf{D}_i \mathbf{s}_i$ is related to the actual transmitted signal. Thus, each weighting matrix $\mathbf{Q}_{lk}$ should satisfy the condition that $\mathbf{Q}_{lk} - \mathbf{D}_k^H \mathbf{Q}_{lk} \mathbf{D}_k$ is diagonal for all l and k [24]. This condition means that the power should be allocated to the

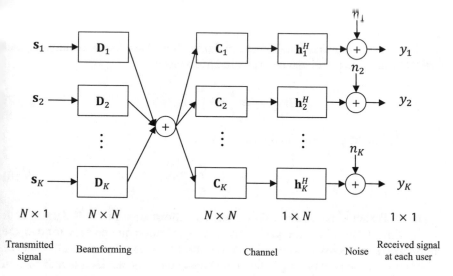

Fig. 9.8 Multi-cell downlink model with K single antenna mobile devices and N antennas

selected subspaces for the specific purpose. Each power constraint only affects the signals from one of base stations.

As we discussed in the previous section, there are multiple metrics and approaches to optimize the radio resource allocation. The common performance measurements in multi-cell resource allocation scenario can be summarized as user performance metrics (such as mean square error (MSE), bit error rate (BER), information rate, and so on) and system utility functions (such as sum performance, proportional fairness, harmonic mean, max–min fairness and so on). The user performance metrics depend on SINR functions regarding MSE, BER, information rate, and so on. Each mobile user can have quality measurement about the user performance function for the kth user $g_k : \mathbb{R}_+ \to \mathbb{R}_+$ of SINR [24]. This function depends on the current service such as throughput and delay and means how much the mobile users are satisfied. The priority can be given by the user requirements. The function $g_k()$ should be selected to measure the performance. The MSE is defined as the average of the squares of the errors. To put it simply, this is the difference between transmitted signals and received signals as follows:

$$\text{MSE}_k = E\left[\left\|\hat{\mathbf{s}}_k - \mathbf{s}_k\right\|_2^2\right] \tag{9.16}$$

where $\mathbf{s}_k$ and $\hat{\mathbf{s}}_k$ are the transmitted signal and the received signal (or an estimate of $\mathbf{s}_k$), respectively. $E[\mathbf{X}]$ is the expectation of the matrix $\mathbf{X}$. When we have Rank$(\mathbf{S}_k) \leq M$ where M is data streams, $\text{MSE}_k = M - \frac{\text{SINR}_k}{1-\text{SINR}_k}$ [24]. We should minimize the metric. In other words, we should maximize $g_k(\text{SINR}_k) = \frac{\text{SINR}_k}{1-\text{SINR}_k}$. The BER is more widely used in physical layer. The bit errors are defined as the received bits over a channel that have been distorted by noises, interferences, distortion, and so on. The BER is defined by the number of bit errors divided by the number of transmitted bits during the transmission. It can be calculated by Monte Carlo methods. In AWGN channel, the BER can be expressed as a function of E_b/N_0. If we have QPSK modulation and AWGN channel, the BER can be expressed as follows:

$$\text{BER}_{\text{QPSK}} = \frac{1}{2}\text{erfc}\left(\sqrt{E_b/N_0}\right) \tag{9.17}$$

where erfc() is the complementary error function

$$\text{erfc}(z) = \frac{2}{\sqrt{\pi}} \int_z^\infty e^{-t^2} dt. \tag{9.18}$$

The BER should be minimized as well. The information rate R is defined as $R = rH$, where r and H are the rate of message generation and entropy (or average information), respectively. The information rate means average number of bits of information per second or simply bits per channel use. The achievable information rate is $g_k(\text{SINR}_k) = \log_2(1 + \text{SINR}_k)$ and means the number of bits that can be carried to the kth user per channel use with arbitrarily low decoding error probability [28]. The signal correlation matrices are related to the user performance metrics and will affect all users that are characterized by channel gain regions. When we have a signal with correlation matrix $\mathbf{S}_k$, the received signal power at the ith user is $x_{ki}(\mathbf{S}_k) = \mathbf{h}_i^H \mathbf{C}_i \mathbf{D}_k \mathbf{S}_k \mathbf{D}_k^H \mathbf{C}_i^H \mathbf{h}_i$. The channel gain region of the signal is expressed as $\Omega_k = \left\{(x_{k1}(\mathbf{S}_k), \ldots, x_{kK}(\mathbf{S}_k)) | \mathbf{S}_k \succeq \mathbf{0}_N, \text{tr}(\mathbf{Q}_{lk}\mathbf{S}_k) \leq q_{lk} \text{ for } \forall l\right\}$. As we can observe Ω_k, the channel gain region relies on the signal correlation matrices $\mathbf{S}_k$ and user power constraint q_{lk}. The channel gain region is compact and convex [24]. The upper boundaries in different directions mean maximization of the received signal power at different users. Choosing signal correlation matrices means resource allocation. Any choice of the function $g_k()$ can be used to find an optimal resource allocation. In the mobile devices point of view, each user has own objective to be optimized. We have K different objectives that might be conflict each other. Thus, multi-objective optimization problems are solved by finding their trade-off points in terms of performance, cost, and so on. Based on the standard form of convex optimization, multi-objective resource allocation problem can be formulated as follows:

$$\max_{\mathbf{S}_1 \succeq \mathbf{0}_N, \ldots, \mathbf{S}_K \succeq \mathbf{0}_N} (g_1(\text{SINR}_1), \ldots, g_K(\text{SINR}_K)) \tag{9.19}$$

subject to

$$\sum_{k=1}^{K} \text{tr}(\mathbf{Q}_{lk}\mathbf{S}_k) \leq q_l \text{ for } \forall l. \tag{9.20}$$

In (9.19), $\mathbf{S}_k \succeq \mathbf{0}_N$ represents that $\mathbf{S}_k$ is positive semi-definite. (9.19) and (9.20) represent to find a transmission strategy maximizing the performance $g_k(\text{SINR}_k)$

In all multiple devices under the given power constraints. Each user typically has different requirements, environments and power constraints. Thus, it is difficult to find a single transmission strategy to satisfy all users. In order to investigate the conflict among objectives, we need to consider all feasible operating points in (9.19) and (9.20) that are called the performance region [29]. The achievable performance region $\mathcal{R} \subseteq \mathbb{R}_+^K$ is expressed as follows [24]:

$$\mathcal{R} = \{(g_1(\text{SINR}_1), \ldots, g_K(\text{SINR}_K)) | (\mathbf{S}_1, \ldots, \mathbf{S}_K) \in \mathbb{S}\} \tag{9.21}$$

where $\mathbb{R}_+^K$ is the set of non-negative members of $\mathbb{R}^K$ and $\mathbb{S}$ is the set of feasible transmission strategies as follows:

$$\mathbb{S} = \left\{ (\mathbf{S}_1, \ldots, \mathbf{S}_K) | \mathbf{S}_k \succeq \mathbf{0}_N, \sum_{k=1}^{K} \text{tr}(\mathbf{Q}_{lk}\mathbf{S}_k) \leq q_l \quad \text{for } \forall l \right\}. \tag{9.22}$$

The achievable performance region means the guaranteed performance that the user can be served simultaneously. The K dimensional performance region is non-empty and its shape strongly relays on the channel vectors, power constraints and dynamic cooperation clusters [24]. Generally, $\mathcal{R}$ might not be convex set but we can prove that $\mathcal{R}$ is compact and normal [30]. $\mathcal{R}$ can be convex when the users are weakly coupled and concave when they are strongly coupled. In practice, the performance regions are mixed. In addition, the performance region can have an upper bound by a box denoted $[\mathbf{a}, \mathbf{b}]$ for some $\mathbf{a}, \mathbf{b} \in \mathbb{R}^K$. The box is the set of all $\mathbf{g} \in \mathbb{R}^K$ such that $\mathbf{a} \leq \mathbf{g} \leq \mathbf{b}$ (componentwise operation, namely, $a_k \leq g_k \leq b_k$ for all vector indices k.). The performance region $\mathcal{R}$ is a subset of the box $[\mathbf{0}, \mathbf{u}]$. It can be denoted as $\mathcal{R} \subseteq [\mathbf{0}, \mathbf{u}]$ where $\mathbf{u} = [u_1, \ldots, u_K]^T$ is the utopia point. The element u_k is the optimum of the single user optimization problem as follows [24]:

$$\max_{\mathbf{S}_k \succeq \mathbf{0}_N} g_k \left(\frac{\mathbf{h}_k^H \mathbf{D}_k \mathbf{S}_k \mathbf{D}_k^H \mathbf{h}_k}{\sigma_k^2} \right) \tag{9.23}$$

subject to

$$\text{tr}(\mathbf{Q}_{lk}\mathbf{S}_k) \leq q_l \quad \text{for } \forall l. \tag{9.24}$$

The utopia point is the unique solution of (9.19) and (9.20) when multi-objective optimization problems are decomposed and all users can reach maximal performance. Generally, it represents an unattainable upper bound. Figure 9.9 illustrates an example of two users performance region and an unattainable upper bound (utopia point).

As we can observe Fig. 9.9, the utopia point is located outside the performance region. We can find only subjective solutions. We can only achieve a set of tentative vector solutions to (9.19) and (9.20) that are mutually unordered [24]. All operating points in $\mathcal{R}$ are tentative solutions. They are not dominated by any other feasible points that care called Pareto optimal [24], which means that the performance cannot

Fig. 9.9 Example of
performance region with two
users

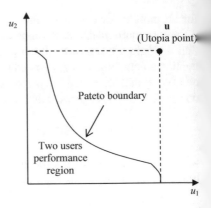

be improved for any user without degrading at least one other users. In general, Pareto
optimality means that we can achieve optimal performance in a way that one user
performance cannot be improved without degrading another user performance. In
this performance region $\mathcal{R}$, each solution means a trade-off among the user performances. There are two properties of the optimal multi-objective resource allocations:
sufficiency of single-stream beamforming and conditions for full power usage [24].
The potential solutions in (9.19) and (9.20) can be found by $\mathbf{S}_k = \mathbf{v}_k \mathbf{v}_k^H$ where
$\mathbf{v}_k \in \mathbb{C}^{N \times 1}$ for $\forall k$ is beamforming vectors. (9.19) and (9.20) can be reformulated as
follows:

$$\max_{\mathbf{v}_1,\dots,\mathbf{v}_K} (g_1(\text{SINR}_1), \dots, g_K(\text{SINR}_K)) \tag{9.25}$$

subject to

$$\text{SINR}_k = \frac{\left|\mathbf{h}_k^H \mathbf{C}_k \mathbf{D}_i \mathbf{v}_i\right|^2}{\sigma_k^2 + \sum_{i \neq k} \left|\mathbf{h}_k^H \mathbf{C}_k \mathbf{D}_i \mathbf{v}_i\right|^2} \quad \text{for } \forall k, \tag{9.26}$$

$$\sum_{k=1}^{K} \mathbf{v}_k^H \mathbf{Q}_{lk} \mathbf{v}_k \leq q_l \quad \text{for } \forall l. \tag{9.27}$$

The reformulated problem is easier to find optimal solution and implement
than (9.19) and (9.20) by successive interference cancellation if $\text{rank}(\mathbf{S}_k) > 1$. In
the general power constraints of wireless networks, we should consider a balance
between channel gain increase and the interference limitation. The power usage
condition on linear independence can be relaxed to the existence of at least one user
in $\mathcal{D}_j$ with a channel linearly independent to all other users in $\mathcal{C}_j$ that are scheduled
[24]. A wireless system designer should select its own system utility function to find
subjectively optimal solutions. Consider a system utility function $f : \mathcal{R} \rightarrow \mathbb{R}$ that
can take any point in the performance region as an input and generates a scalar output.
A big value of output means high preference of the wireless system. A system utility

function is defined as $f_{(k)}(\text{SINR}_1)$, ..., $g_K(\text{SINR}_K))$. The following system utility functions can be used [24]

$$\text{Weighted sum utility}: f(\mathbf{g}) = \sum_k w_k g_k, \tag{9.28}$$

$$\text{Weighted proportional fairness}: f(\mathbf{g}) = \prod_k g_k^{w_k}, \tag{9.29}$$

$$\text{Weighted harmonic mean}: f(\mathbf{g}) = \left(\sum_k \frac{w_k}{g_k}\right)^{-1}, \tag{9.30}$$

and

$$\text{Weighted max} - \text{min fairness}: f(\mathbf{g}) = \min_k \frac{g_k}{w_k} \tag{9.31}$$

where $w_k \geq 0$ is a weighting factor satisfying $\sum_{k=1}^{K} w_k = 1$. The weighting factor can compensate for channel conditions of multiple users, delay constraints management, and so on. The system utility functions from weighted sum utility, weighted proportional fairness, weighted harmonic mean, to weighted max–min fairness gradually abandon aggregate utility and satisfy more fairness. In addition, there are many other system utility functions such as α-proportional fair [31], weighted utility for best effort users [32] and so on. The system utility function allows us to convert the multi-objective optimization problem (9.25), (9.26) and (9.27) to the single-objective optimization problem as follows:

$$\max_{\mathbf{v}_1,\dots,\mathbf{v}_K} f(g_1(\text{SINR}_1), \dots, g_K(\text{SINR}_K)) \tag{9.32}$$

subject to

$$\text{SINR}_k = \frac{\left|\mathbf{h}_k^H \mathbf{C}_k \mathbf{D}_i \mathbf{v}_i\right|^2}{\sigma_k^2 + \sum_{i \neq k} \left|\mathbf{h}_k^H \mathbf{C}_k \mathbf{D}_i \mathbf{v}_i\right|^2} \quad \text{for } \forall k, \tag{9.33}$$

$$\sum_{k=1}^{K} \mathbf{v}_k^H \mathbf{Q}_{lk} \mathbf{v}_k \leq q_l \quad \text{for } \forall l. \tag{9.34}$$

As we can observe (9.32), (9.33) and (9.34), the utility function resolves the interest of user performance conflict in the multi-objective resource allocation optimization problem. The user performance functions $g_k()$ are continuous and strictly monotonically increasing and the system utility function $f()$ is Lipschitz continuous and monotonically increasing [24] where Lipschitz continuity represents uniform continuity for functions. When the problem (9.32), (9.33) and (9.34) is convex, we can solve it efficiently. Based on a prior knowledge about the performance region

$\mathcal{R}$, we should select the utility function very carefully before finding optimal points. Selecting the utility function and formulating the problem are the starting point of radio resource allocation optimization. Depending on wireless system designers' subjective views, the resource allocation problems can be formulated and the solvers can be found accordingly. Key approaches can be summarized as follows [33]: (1) The resource allocation optimization problem (9.32), (9.33), and (9.34) can be solved using the weighted utility functions (9.28), (9.29), (9.30), and (9.31). (2) When we have a certain system utility function and a reference point is given in advance, the optimal solution can be found by minimizing the error between the estimate and the reference. This approach is useful if we have a prior knowledge about the performance region and the preference on the solution. (3) We create sample points on the Pareto boundary that are found by solving the resource allocation optimization problem (9.32), (9.33), and (9.34). We select one sample point among them. (4) The final approach is a mixed form of the second and the third approaches. Based on a prior knowledge or experience, we create new sample points on the Pareto boundary by iteration and then select one sample point among them. This interactive approach has a benefit in terms of convergence in general.

Example 9.2 Radio Resource Allocation Problem Formulation and Classification.

Formulate a resource allocation problem of wireless networks using (9.32), (9.33), and (9.34) and classify the resource allocation problems.

Solution
The radio resource allocation problems can have two different approaches: user view and network view. There is a fundamental difference between the two points of view. In mobile networks, users want to have a high performance such as high bandwidth, low delay, and so on. Network operators want to serve mobile users with good QoS while minimizing the cost. The single-objective optimization problem is useful to satisfy both requirements by maximizing the utility functions subject to some constraints such as power limitation and so on. Based on both the standard form of the convex optimization problem (9.1) and (9.2) and the single-objective optimization problem (9.32), (9.33), and (9.34), the radio resource allocation problem can be formulated as follows:

$$\min_{\mathbf{g}} -f(\mathbf{g})$$

subject to

$$\mathbf{g} \in \mathcal{R}$$

where $\mathbf{g} = (g_1(\mathrm{SINR}_1), \ldots, g_K(\mathrm{SINR}_K))^T$ is the user performance metrics as the optimization variables, $-f()$ is the system utility function as cost function, and $\mathcal{R}$

A the achievable performance region an the feasible region. As we can bee the above questions, the radio resource allocation problem is to find the user performance metrics in the achievable performance region in order to minimize the cost function. The radio resource allocation problem can be solved by two approaches: (1) Traditional optimization theory and (2) AI techniques such as deep learning based approach. Depending on the type of problems and conditions, we can find suitable approaches. In [22, 30], the optimization problems are classified as follows:

- Linear problem: If the cost function and the constraint functions are linear, the feasible set is a convex polytope in $\mathbb{R}^n$.
- Convex problem: If the cost function and constraint functions are convex functions, the feasible set is a convex set in $\mathbb{R}^n$.
- Quasi-convex problem: If the cost function and constraint functions are quasi-convex functions, the feasible set is a convex set in $\mathbb{R}^n$.
- Monotonic problem: If the cost function and constraint functions are monotonic functions, the feasible set is a mutually normal set.

Generally, a hierarchical structure of the problems is as follows: Monotonic problems $\supset$ quasi-convex problems $\supset$ convex problems $\supset$ linear problems. The convex problem needs a polynomial time to solve. This is practically solvable. The monotonic problem needs an exponential time to solve. Thus, we need to reformulate it using approximation theory. As we discussed in Sect. 9.2.1, it is not easy to formulate a real world problem to one of above types problem. Thus, we should transform the real world problem to one of the problems and find an optimal point using the given solver or should create a new solver for the dedicated problem. ∎

The single-objective optimization problem (9.32), (9.33), and (9.34) has convex constraint functions. The problem classification as linear, convex, quasi-convex, or monotonic problem relies on the cost function $-f(g_1(\text{SINR}_1), \ldots, g_K(\text{SINR}_K))$. The system utility function $f()$ depends on SINRs. The SINR constraints are non-convex functions of the beamforming vectors $\mathbf{v}_k$ due to the multiplication between the SINR value at the kth mobile device and the interuser inference caused to the kth mobile device. We introduce the auxiliary optimization variables $\gamma_k = \text{SINR}_k$ and rewrite (9.32), (9.33), and (9.34) as follows:

$$\min_{\mathbf{v}_k, \gamma_k \forall k} -f(g_1(\gamma_1), \ldots, g_K(\gamma_K)) \tag{9.35}$$

subject to

$$\gamma_k \left(\sigma_k^2 + \sum_{i \neq k} \left| \mathbf{h}_k^H \mathbf{C}_k \mathbf{D}_i \mathbf{v}_i \right|^2 \right) \leq \left| \mathbf{h}_k^H \mathbf{C}_k \mathbf{D}_i \mathbf{v}_i \right|^2 \text{ for } \forall k, \tag{9.36}$$

$$\sum_{k=1}^{K} \mathbf{v}_k^H \mathbf{Q}_{lk} \mathbf{v}_k \leq q_l \text{ for } \forall l. \tag{9.37}$$

(9.36) denotes the auxiliary SINR constraints $\gamma_k \leq \text{SINR}_k$. The optimal solution gives equality in the auxiliary SINR constraints. Now, the cost function $-f(g_1(\gamma_1), \ldots, g_K(\gamma_K))$ is convex in terms of the auxiliary optimization variables γ_k. In order to resolve the non-convexity of SINR constraints, there are three approaches [25]: (1) fix the interuser interference caused to each user, (2) fix the SINR value at each user and (3) turn the multiplication into addition by change of variables. By these approaches, we can have the radio resource allocation problem as convex problem. The convex optimization problem can be solved by interior point methods as we discussed in Sect. 9.2.1.

Summary 9.3. Resource Allocation Models and Performance Measures

1. In radio resource allocation, the user performance metrics depend on SINR functions regarding MSE, BER, information rate and so on. System utility functions can be sum performance, proportional fairness, harmonic mean, max–min fairness and so on.

2. A multi-objective resource allocation problem can be formulated as follows:

$$\max_{\mathbf{S}_1 \succeq \mathbf{0}_N, \ldots, \mathbf{S}_K \succeq \mathbf{0}_N} (g_1(\text{SINR}_1), \ldots, g_K(\text{SINR}_K))$$

subject to

$$\sum_{k=1}^{K} \text{tr}(\mathbf{Q}_{lk}\mathbf{S}_k) \leq q_l \quad \text{for } \forall l.$$

where $\mathbf{S}_k \succeq \mathbf{0}_N$ represents that $\mathbf{S}_k$ is positive semi-definite. The function $g_k()$ should be selected to measure the performance. The problem represents to find a transmission strategy maximizing the performance $g_k(\text{SINR}_k)$ for all mobile devices under the given power constraints.

3. Pareto optimality means that we can achieve optimal performance in a way that that one user performance cannot be improved without degrading another user performance.

4. The system utility functions from weighted sum utility, weighted proportional fairness, weighted harmonic mean, to weighted max–min fairness gradually abandon aggregate utility and satisfy more fairness.

5. The single-objective optimization problem is useful to satisfy both requirements by maximizing the utility functions subject to some constraints such as power limitation and so on. The radio resource allocation problem can be formulated as follows:

$$\min_{\mathbf{g}} - f(\mathbf{g})$$

subject to

$$\mathbf{g} \in \mathcal{R}$$

where $\mathbf{g} = (g_1(\mathrm{SINR}_1), \ldots, g_K(\mathrm{SINR}_K))^T$ is the user performance metrics as the optimization variables, $-f()$ is the system utility function as cost function, and $\mathcal{R}$ is the achievable performance region as the feasible region.

9.2.3 Utility Functions and Fairness of Resource Allocation

Fairness is an important metric in the radio resource allocation. If wireless networks do not provide mobile users with the radio and network resources fairly, the mobile users will spell their complaints out and may request the service termination to mobile operators. However, fairness may be different for mobile users in different situations. In wireless networks, fairness means the distribution of network resources among applications and users while avoiding no throughput at specific flow and considering a trade-off between efficiency and user throughput. The efficiency of the radio resources can be measured by the aggregate throughput of the flows in the networks. The aggregate throughput can be calculated by the sum of the allocated bandwidth over all network flows. The key question in the resource allocation is how to allocate bandwidths to network flows fairly and efficiently. Now, we consider how resources are allocated to the different links to satisfy a given criterion. Figure 9.10 illustrates one example of resource allocation with different link capacities.

As we can observe Fig. 9.10, there are three nodes (Node 1, 2, and 3), two different links with different capacities (Link 1 with $C_1 = 1$ and Link 2 with $C_2 = 2$), and

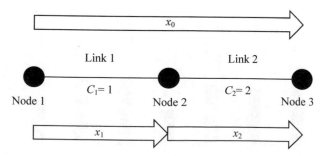

Fig. 9.10 Example of resource allocation with different links and capacities

three different user flows (x_0, x_1 and x_2). We assume equal radio resource sharing among users. In the link 1, user flow x_0 and user flow x_1 should share the capacity evenly. Thus, capacity of x_0 is 0.5 and capacity of x_1 is 0.5. The capacity of x_0 should be maintained in the link 2. The capacity of x_0 is still 0.5 in the link 2. Thus, we can allocate the capacity 1.5 to user flow x_2. When the resource allocation is regarded as a vector $\mathbf{x} = [x_0, \ldots, x_n]$, the resource allocation in Fig. 9.10 can be expressed as $\mathbf{x} = [x_0, x_1, x_2] = [0.5\ 0.5\ 1.5]$. The total amount of user flows in the networks is $\sum_i x_i = 2.5$. If we consider another resource allocation $\mathbf{x} = [x_0, x_1, x_2] = [0.1\ 0.9\ 1.9]$, the total amount of user flows in the networks is $\sum_i x_i = 2.9$. Although the fairness among users is not achieved, the aggregate throughput is 0.4 greater than the equal sharing resource allocation. We can observe the trade-off between fairness and aggregate throughput. We need to consider a balance between them. The fairness is one of important metrics in wireless networks. There are many kinds of fairness in the resource allocations such as max–min fairness, maximum throughput scheduling, proportional fairness, minimum potential delay fairness, and so on. In the example, the resource allocation $\mathbf{x} = [x_0, x_1, x_2] = [0.5\ 0.5\ 1.5]$ is based on max–min fairness. The max–min fairness is to maximize the minimum of users. The basic approach of the max–min fairness is that small users obtain all they want and large users evenly split the rest. It is like pouring a water into different levels of vessel for equalization strategy. Figure 9.11 illustrates an example of max–min fairness. As we can observe the figure, the data rate at a certain level C is roofed. In terms of fairness, the data rates of c_1, c_3, c_5, and c_6 are equal. The lower data rates c_2, c_4, and c_7 can be increased.

The basic procedure is as follows: (1) the resource is allocated to users in order of increasing demand. (2) Users basically do not receive the resource they requested. (3) Users with unsatisfied demands split the remaining resource. The advantages of max–min fairness are summarized as follows: (1) simple algorithm, (2) small players can get a big share, and (3) all players get treated equally. On the other hands, the disadvantages are summarized as follows: (1) not suitable for complex systems, (2) big players can get a small share, and (3) not continuous function. The maximum throughput scheduling is to maximize the total throughput of the wireless network.

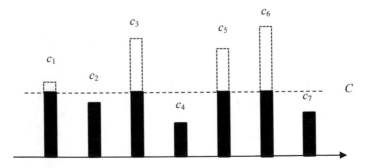

Fig. 9.11 Example of max–min fairness

In the example, the maximum throughput allocation is $[x_0, x_1, x_2] = [0\ 1\ 2]$. The total amount of user flows in the networks is $\sum_i x_i = 3$. The total throughput is the maximum but the user x_1 is not served by the network. Proportional fairness is a compromise between the max–min fairness and the maximum throughput scheduling. It tries to maximize the total throughput like the maximum throughput scheduling while all users are served with a minimum level of throughput. The basic approach is to allocate a certain data rate to each user data flow or give it priority that is inversely proportional to the anticipated resource consumption. In the example, when the minimum throughput is 0.25, the proportional fairness allocation is $[x_0, x_1, x_2] = [0.25\ 0.75\ 1.75]$. The total amount of user flows in the networks is $\sum_i x_i = 2.75$. The total amount of user flows is bigger than the max–min fairness and also the minimum requirement of user throughput is satisfied.

Let $U_r(x_r)$ be the utility function of rate x_r to the rth user. The utility function $U_r(x_r)$ is assumed to be continuous differentiable concave function. Assume that the objective of resource allocation is to maximize the aggregate utility function of all users subjective to the capacity constraints function of the links and also all link capacities is fixed. The resource allocation problem can be formulated as follows:

$$\max \sum_{x_r \in S} U_r(x_r) \tag{9.38}$$

subject to

$$C_1 : \sum_{r, l \in r} x_r \le c_l, \quad l \in L, \tag{9.39}$$

$$C_2 : x_r \ge 0, \quad r \in S. \tag{9.40}$$

In this problem, S and L are the set of users and the set of all links in the wireless network, respectively. The rth user can be identified with its rate x_r that can be expressed as its flow. The first constraint C_1 represents that the aggregate data rate of all flows through any link $l \in L$ could not exceed the capacity of the link c_l. The second constraint C_2 represents that the rate x_r is non-negative. The approach of the resource allocation is to maximize the concave objective function (This is same as minimizing the convex function) subject to two linear constraint functions. This is a convex optimization problem.

In wireless networks, congestion problems happen when the receiver is bottleneck. The bottleneck in the networks is natural because some applications are slow or fast, some processors are slow or fast, or each mobile user want different throughputs. There are two congestion controls: end-to-end congestion control and network assisted congestion control. In the end-to-end congestion control, mobile devices use a combination of flow and congestion control for correct amount of data. There are no explicit feedbacks from IP layer networks. Thus, we call it implicit congestion control. On the other hands, the network assisted congestion control has explicit feedbacks from routers or other network components. The feedbacks indicate direction, congestion, and so on. Thus, we call it explicit congestion control. Senders learn of congestion at queue from the feedback. The resource allocation plays an important role in the congestion control. The political policies for resource sharing are targeting to a high average throughput and a better utilization of the resources than equal sharing policies. Thus, the max–min fairness is a well-known concept in political science. The usage in communications and networks was proposed by Bertsekas and Gallager [34]. The max–min fairness is Pareto efficient. It tries to provide maximum to the user who has the least amount of resources. The formal definition of max–min fairness is as follows: a rate vector $\mathbf{x}$ is max–min fairness if for any set of rates z_r satisfying the following capacity constraints is true: if $z_s > x_s$ for some $s \in S$ then there is $a, p \in S$ such that $x_p \leq x_s$ and $z_p \leq x_p$. A link l becomes a bottleneck link for a user r if the link is fully used and r has the largest rate among all users using the link l. we can express it mathematically as follows: $\sum_{s \in S_l} x_s = c_l$ and $x_s \leq x_r$ for $\forall s$ such that $x_s \in S_l$ where S_l represents all flows via the link l. Proportional fairness in communication systems was proposed by Kelly, Maulloo and Tan [35]. The proportional fair resource allocation is widely used in wireless networks. Briefly speaking, a network is regarded as proportional fairness if and only if small change in resource allocation has a negative effect on the average throughput of a flow in the network. Consider a network with a set of resources J. C_j is the finite capacity of the resource $j \in J$. A route $r \in R$ is a non-empty subset of J. R is the set of possible routes. The resource allocation problem is formulated as follows:

$$\max \sum_{r \in R} U_r(x_r) \tag{9.41}$$

subject to

$$\mathbf{Ax} \leq \mathbf{C}, \quad \mathbf{x} \geq 0 \tag{9.42}$$

where x_r is the transmission rate of a route r, $\mathbf{C}$ is a vector about the capacities of all links, and the matrix $\mathbf{A}$ has the element A_{jr} with the value 1 when link j lies on the route r or the value 0 otherwise. $U_r(x_r)$ is the utility function of node r transmitting at rate x_r. The utility function means the degree of user satisfaction. Assume that all links have a fixed capacity and the utility function are concave and then solved the problem using an iterative distributed algorithm. This problem is to maximize aggregate utility subject to capacity constraints. The utility function for the weighted proportional fairness is given as follows:

$$U_r(x_r) = w_r \log x_r \tag{9.43}$$

where w_r is the weight. The weighted proportional fairness resource allocation problem is formulated as follows:

$$\max \sum_{r \in R} w_r \log x_r, \tag{9.44}$$

subject to

$$\mathbf{Ax} \le \mathbf{C}, \quad \mathbf{x} \ge 0 \tag{9.45}$$

This problem means that the network maximizes a logarithmic utility function with the weight w_r selected by users. When the resource allocation is performed using the proportional fairness, x_r^* is the resource allocation vector in terms of proportional fairness. For any other resource allocation x_r, we have the inequality as follows:

$$\sum_{r \in R} \frac{x_r - x_r^*}{x_r^*} \le 0. \tag{9.46}$$

(9.46) indicates that if we deviate from the rate x_r^* to some other feasible resource allocation x_r, the aggregate of the proportional changes in the user rates is zero or negative. When considering the weighted proportional fairness, the inequality can be written as follows:

$$\sum_{r \in R} w_r \frac{x_r - x_r^*}{x_r^*} \le 0. \tag{9.47}$$

A set of the rate x_r^* solves the network resource allocation problem (9.44) and (9.45) if and only if the rates x_r^* are the weighted proportional fair. For the example in Fig. 9.10, we formulate the proportional fair resource allocation problem as follows:

$$\max(\log x_0 + \log x_1 + \log x_2) \tag{9.48}$$

subject to

$$C_1 : x_0 + x_1 \le 1, \tag{9.49}$$

$$C_2 : x_0 + x_2 \le 2, \tag{9.50}$$

$$C_3 : x_0, x_1, x_2 \ge 0. \tag{9.51}$$

This problem is convex. As we discussed in Sect. 9.2.1, the convex optimization problem can be solved by interior point methods. We solve this optimization problem using the interior point method based on Lagrange multipliers. The variables of Lagrange multipliers are regarded as the inequality constraints. In convex optimization, the complementary slackness conditions should be satisfied by the optimal solution. The variables of Lagrange multipliers can be expressed as λ_1 and λ_2 corresponding to the constraints C_1 and C_2, respectively. They are positive. Thus,

Lagrangian function of the problem can be expressed as follows:

$$L(\mathbf{x}, \boldsymbol{\lambda}) = \log x_0 + \log x_1 + \log x_2 + \lambda_1(1 - x_0 - x_1) + \lambda_2(2 - x_0 - x_2) \quad (9.52)$$

where $\mathbf{x}$ and $\boldsymbol{\lambda}$ are the resource allocation vector and Lagrange multipliers, respectively. We should find stationary point of the Lagrangian function. By setting $\frac{\partial L}{\partial x_r} = 0$ for $\forall r$, we obtain as follows:

$$\frac{\partial L}{\partial x_0} = \frac{1}{x_0} - \lambda_1 - \lambda_2 = 0, \quad (9.53)$$

$$\frac{\partial L}{\partial x_1} = \frac{1}{x_1} - \lambda_1 = 0, \quad (9.54)$$

and

$$\frac{\partial L}{\partial x_2} = \frac{1}{x_2} - \lambda_2 = 0. \quad (9.55)$$

From (9.53), (9.54), and (9.55), we have the resource allocation vector as Lagrange multipliers as follows:

$$x_0 = \frac{1}{\lambda_1 + \lambda_2}, x_1 = \frac{1}{\lambda_1} \text{ and } x_2 = \frac{1}{\lambda_2}. \quad (9.56)$$

As we can observe (9.56), the optimal point depends on the Lagrange multipliers. We can regard the Lagrange multipliers as the link congestion measures. They are positive or zero when the link is fully occupied or not used, respectively. The flow rate does not rely on the congestion of links that are not on its path but the congestion of the links in its route. The flow rates in the proportional fairness are inversely proportional to the aggregate of the congestion of their routes. From (9.56) and two constraints $x_0 + x_1 = 1$ and $x_0 + x_2 = 2$, we obtain

$$\frac{1}{\lambda_1 + \lambda_2} + \frac{1}{\lambda_1} = 1 \text{ and } \frac{1}{\lambda_1 + \lambda_2} + \frac{1}{\lambda_2} = 2 \quad (9.57)$$

and

$$\lambda_1 = \sqrt{3} \text{ and } \lambda_2 = \frac{\sqrt{3}}{\sqrt{3} + 1}. \quad (9.58)$$

From (9.58) and (9.56), we obtain the optimal resource allocation under the proportional fair as follows:

$$x_0 = \frac{\sqrt{3}-1}{2\sqrt{3}+3} = 0.4226, \ x_1 = \frac{1}{\sqrt{3}} = 0.5774 \ \text{ and } \ x_2 = \frac{\sqrt{3}+1}{\sqrt{3}} = 1.5774.$$

$$(9.59)$$

Another fairness form is the minimum potential delay fairness [36–38]. We consider that a user r is to send a file size w_r. The utility function can be formulated as follows:

$$U_r(x_r) = -\frac{w_r}{x_r} \tag{9.60}$$

where w_r is the weight associated with the rate x_r. We is trying to minimize the time taken to complete a transfer. Namely, higher the allocated rate means smaller the transfer time. The term $1/x_r$ indicates the delay in associated with the file transfer complete because the delay can be defined as follows: the file size divided by the rate allocated to the user r. Thus, we call this the minimum potential delay fairness. The objective is to minimize the total time to complete all file transfers. When the weight is 1, the optimization problem in the minimum potential delay fairness can be regarded as maximizing the utility function $U_r(x_r) = -\frac{1}{x_r}$. For the example in Fig. 9.10, we formulate the minimum potential delay fairness resource allocation problem as follows:

$$\max \sum_r \left(-\frac{1}{x_r} \right) \tag{9.61}$$

subject to

$$C_1 : x_0 + x_1 \le 1, \tag{9.62}$$

$$C_2 : x_0 + x_2 \le 2, \tag{9.63}$$

$$C_3 : x_0, x_1, x_2 \ge 0. \tag{9.64}$$

Like the proportional fair, Lagrangian function of the problem can be expressed as follows:

$$L(\mathbf{x}, \boldsymbol{\lambda}) = -\frac{1}{x_0} - \frac{1}{x_1} - \frac{1}{x_2} + \lambda_1(1 - x_0 - x_1) + \lambda_2(2 - x_0 - x_2). \tag{9.65}$$

By setting $\frac{\partial L}{\partial x_r} = 0$ for $\forall r$, we obtain as follows:

$$\frac{\partial L}{\partial x_0} = \frac{1}{x_0^2} - \lambda_1 - \lambda_2 = 0, \tag{9.66}$$

$$\frac{\partial L}{\partial x_1} = \frac{1}{x_1^2} - \lambda_1 = 0, \tag{9.67}$$

and

$$\frac{\partial L}{\partial x_2} = \frac{1}{x_2^2} - \lambda_2 = 0. \tag{9.68}$$

We have the resource allocation vector as Lagrange multipliers as follows:

$$x_0 = \sqrt{\frac{1}{\lambda_1 + \lambda_2}}, x_1 = \sqrt{\frac{1}{\lambda_1}} \text{ and } x_2 = \sqrt{\frac{1}{\lambda_2}}. \tag{9.69}$$

Similar to the proportional fair, we obtain the optimal resource allocation under the proportional fair as follows:

$$x_0 = 0.4864, x_1 = 0.5136 \text{ and } x_2 = 1.5136. \tag{9.70}$$

Now, we consider general utility functions with different fairness criteria by choosing the parameter α that takes values between 0 and ∞. The utility function with the parameter α is expressed as follows [39]:

$$U_r(x_r) = \begin{cases} w_r \frac{x_r^{1-\alpha}}{1-\alpha}, & \alpha \geq 0, \alpha \neq 1 \\ w_r \log x_r, & \alpha = 1 \end{cases} \tag{9.71}$$

where α is a free parameter. The α fairness is one of the most common fairness metrics. Depending on different α value, we have different fairness. We call this α-fair allocations. Figure 9.12 illustrates utility function with different α values when the weight is 1.

We formulate α fairness resource allocation problem as follows:

$$\max \sum_r \left(w_r \frac{x_r^{1-\alpha}}{1-\alpha} \right) \tag{9.72}$$

subject to

$$C_1 : \sum_r \left(A_{jr} x_r \right) \leq C_j, \quad j \in J \tag{9.73}$$

$$C_2 : x_r \geq 0, \quad r \in R. \tag{9.74}$$

The utility is maximized at $\mathbf{x}^*$ that lies in a feasible set C_1 and C_2 of the form $\mathbf{Ax} \leq \mathbf{c}$ where $\mathbf{c}$ is a capacity vector for p of resources, $\mathbf{A}$ is a binary user (p, n) resource constraint matrix for n users, and $\mathbf{w}$ is a positive weight vector. Depending

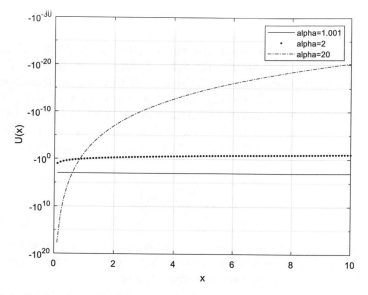

Fig. 9.12 Utility functions with different α values

on a particular value of α, we have a particular fairness concept as follows: maximum overall throughput when $\alpha = 0$, proportional fairness when $\alpha = 1$, minimum potential delay when $\alpha = 2$, and max–min fairness when $\alpha \rightarrow \infty$. Table 9.2 summarizes the different fairness depending on a particular value of α.

Figure 9.13 illustrates particular fairness concepts as specific choice of $\boldsymbol{\alpha}$.

Table 9.2 Specific choice for α

α	Different fairness	$\max_{x_r} \sum U_r(x_r)$
0	Maximum overall throughput	$\max_{x_r} \sum w_r x_r$
1	Proportional fairness	$\max_{x_r} \sum w_r \log x_r$
2	Minimum potential delay	$\max_{x_r} \sum \frac{w_r}{x_r}$
∞	Max–min fairness	$\max_{x_r} \min x_r$

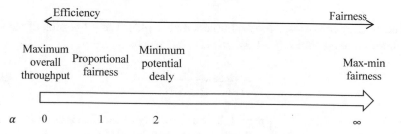

Fig. 9.13 Different fairness with different α values

The fairness could be measured in multiple ways. We define the fairness measure as follows: $f(\mathbf{X}) : \mathbb{R}_n^+ \rightarrow \mathbb{R}^+$ where $\mathbf{X}$ and n are the resource allocation and the number of connections, respectively. The basic requirements of the fairness measure should be satisfied as follows [40]: (1) $f(\mathbf{X})$ should be continuous on $\mathbf{X} \in \mathbb{R}_n^+$, (2) $f(\mathbf{X})$ should be independent of n, (3) The range of $f(\mathbf{X})$ should be easily mapped on to [0,1], (4) $f(\mathbf{X})$ should be easily extendable to multi-resource case, (5) $f(\mathbf{X})$ should be easy to implement, and (6) $f(\mathbf{X})$ should be sensitive enough to the variation of $\mathbf{X}$. The fairness index (or Jain's index) [41] is widely used for evaluating the fairness of resource allocation schemes in the network. The fairness index is defined as follows:

$$f(\mathbf{X}) = \frac{\left(\sum_{i=1}^{n} x_i\right)^2}{n \sum_{i=1}^{n} x_i^2} \tag{9.75}$$

where $0 \leq f(\mathbf{X}) \leq 1$ and the value 1 of $f(\mathbf{X})$ represents the maximal fairness. x_i is the normalized throughput of the ith flow. A larger value of the fairness index represents fairer resource allocation in terms of the network.

Example 9.3 Fairness index Consider the following network conditions:

- Measure throughputs: $(t_1, t_2, t_3) = (5\,\text{Mbps}, 3\,\text{Mbps}, 5\,\text{Mbps})$
- Fair throughputs by a specific fairness resource allocation: $(o_1, o_2, o_3) = (5\,\text{Mbps}, 1\,\text{Mbps}, 1\,\text{Mbps})$.

Compute the fairness index of the network.

Solution
The normalized throughputs are calculated as follows:

$$(x_1, x_2, x_3) = \left(\frac{t_1}{o_1}, \frac{t_2}{o_2}, \frac{t_3}{o_3}\right) = \left(\frac{5}{5}, \frac{3}{1}, \frac{5}{1}\right) = (1, 3, 5).$$

From (9.75), we calculate the fairness index as follows:

$$f(\mathbf{X}) = \frac{(1 + 3 + 5)^2}{3(1^2 + 3^2 + 5^2)} = \frac{9^2}{3(1 + 9 + 25)} = 0.81. \blacksquare$$

The entropy can be used for the fairness metric [42]. The proportion of resource are allocated to n connections $\mathbf{P} = (p_1, \ldots, p_n)$ where

$$p_i = \frac{x_i}{\sum_{i=1}^{n} x_i} \tag{9.76}$$

where $0 \leq p_i \leq 1$ and $\sum_{i=1}^{n} p_i = 1$. The uncertainty of the distribution $\mathbf{P}$ is called the entropy of the distribution $\mathbf{P}$ and is measured by $H(\mathbf{P}) = H(p_1, \ldots, p_n)$ as follows: [40].

$$H(\mathbf{P}) = \sum_{i=1}^{n}\left(p_i \log_2 p_i^{-1}\right). \tag{9.77}$$

When $H(\mathbf{P})$ is used for a fairness metric, the absolute resource values of $\mathbf{X}$ in $f(\mathbf{X})$ are replaced by the resource proportions $\mathbf{P}$ in $H(\mathbf{P})$.

Example 9.4 Max–min Fairness Resource Allocation Consider the network flow as shown in Fig. 9.14.

As we can observe Fig. 9.14, four flows (x_0, x_1, x_2, x_3) should be shared the link 1 with 10 Mbps throughput. When the demands of each flow are $x_0 = 2$ Mbps, $x_1 = 2.5$ Mbps, $x_2 = 3$ Mbps, $x_3 = 4$ Mbps, allocate the throughput to each flow using max–min fairness.

Solution
The procedure of max–min fairness resource allocation can be summarized as follows: (1) All throughputs of each flow starts at 0. (2) We increase the throughput equally until some flow is limited by a bottleneck. (3) The remaining throughputs are allocated to the flow that didn't reach the demand. (4) We stop allocation until there is no resource. In this example, the first step is to increase the throughput equally to four flows. Thus, we have the first allocation as follows:

$$\text{The bottleneck throughput} = 10\,\text{Mbps.}$$
$$x_8 = 2\,\text{Mbps}, x_1 = 2\,\text{Mbps}, x_2 = 2\,\text{Mbps}, x_3 = 2\,\text{Mbps.}$$
$$\text{The remaining throughput} = 2\,\text{Mbps.}$$

In the first iteration, the demand of the flow x_0 is satisfied. The remaining throughput is 2Mbps, and they are equally allocated to the others. 2 Mbps/3 = 0.6667 Mbps can be allocated to the others. Thus, the second allocation is as follows:

$$x_8 = 2\,\text{Mbps}, x_1 = 2.5\,\text{Mbps}, x_3 = 2.6667\,\text{Mbps}, x_3 = 2.6667\,\text{Mbps.}$$
$$\text{The remaining throughput} = 0.1667\,\text{Mbps.}$$

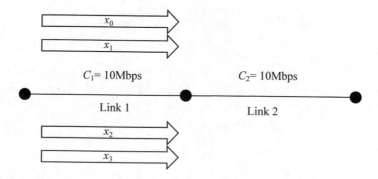

Fig. 9.14 Example of max–min resource allocation

In the second iteration, the demand of the flow x_1 is satisfied. The remaining throughput is 0.1667 Mbps, and they are equally allocated to the others. 0.1667 Mbps/2 = 0.08335 Mbps can be allocated to the others. Thus, the third allocation is as follows:

$$x_8 = 2\,\text{Mbps}, x_1 = 2.5\,\text{Mbps}, x_2 = 2.75\,\text{Mbps}, x_3 = 2.75\,\text{Mbps}.$$

There is no remaining throughput. This is the final resource allocation. As we can observe the final resource allocation, the demands of the flows x_0 and x_1 are satisfied. Although the demands of the flows x_2 and x_3 are not satisfied, we give the highest possible throughput to them, and they are maximized fairly.   ∎

Summary 9.4. Utility Functions and Fairness of Resource Allocation

1. In wireless networks, fairness means the distribution of network resources among applications and users while avoiding no throughput at specific flow and considering a trade-off between efficiency and user throughput.
2. There are many kinds of fairness in the resource allocations such as max–min fairness, maximum throughput scheduling, proportional fairness, minimum potential delay fairness, and so on.
3. The max–min fairness is Pareto efficient. The max–min fairness is to maximize the minimum of users. The basic approach of the max–min fairness is that small users obtain all they want and large users evenly split the rest. It is like pouring a water into different levels of vessel for equalization strategy.
4. The maximum throughput scheduling is to maximize the total throughput of the wireless network.
5. Proportional fairness is a compromise between the max–min fairness and the maximum throughput scheduling. It tries to maximize the total throughput like the maximum throughput scheduling while all users are served with a minimum level of throughput. The basic approach is to allocate a certain data rate to each user data flow or give it priority that is inversely proportional to the anticipated resource consumption.
6. Minimum potential delay fairness is trying to minimize the time taken to complete a transfer. Namely, Higher the allocated rate means smaller the transfer time.
7. We consider general utility functions with different fairness criteria by choosing the parameter α that takes values between 0 and ∞. The utility function with the parameter α is expressed as follows:

$$U_r(x_r) = \begin{cases} w_r \frac{x_r^{1-\alpha}}{1-\alpha}, & \alpha \geq 0, \alpha \neq 1 \\ w_r \log x_r, & \alpha = 1 \end{cases}$$

where α is a free parameter. The α fairness is one of the most common fairness metrics. Depending on different α value, we have different fairness. We call this α-fair allocations.

9.2.4 Resource Allocation Using AI Techniques

As we discussed in the previous sections, the purpose of the resource allocation is to maximize the utilization of the limited radio and network resources under the given demands and requirements. Many approaches of the traditional resource allocations assume static networks and depend on the fixed network models. The ideal assumptions enable us to deal with a tractable network problem. However, in real worlds, the network status is changing dynamically. The resource allocation methods that are developed under the assumption do not work well in real world or cause a large performance degradation. In order to overcome them, AI techniques will be helpful for resource allocation and scheduling in data link layer. The reinforcement learning and the deep learning are based on the data-driven approaches. The reinforcement learning could learn a resource allocation policy using the cost from the environment. Based on the trained policy, it is possible to make a quick decision for dynamic networks. The complexity of the resource allocation can be reduced by deep learning. The deep learning could learn the popular contents and use them for cache resource. The problem of the resource allocation can be solved by supervised learning and unsupervised learning methods. The conventional optimization solvers have a high complexity. The data-driven approach of AI algorithms enables us to train a model on a large amount of data, learn to approximate the relationship between input and output, and find an optimal resource allocation. The resource allocation and scheduling problems can be formulated as a convex problem, a mixed integer nonlinear problem, or others. The neural network will be helpful for solving the problem, optimizing the resource allocation procedure, and achieving a better efficiency.

The traditional optimization techniques of resource allocations have a high complexity. Optimization techniques with high complexity can be replaced by deep learning. For example, a neural network can be adopted to approximate a weighted minimum mean-squared error (WMMSE) algorithm for power control [43]. The key idea is to deal with the input and the output of a resource allocation as an unknown nonlinear mapping and approximate it using a DNN. The WMMSE is used to provide the desired output for each sample. The training data set for the neural network can be generated by running the WMMSE in different channel environments. The performance of the DNN is close to the WMMSE while maintaining much lower complexity. Due to the approximation by the DNN, the resource allocation can be performed in real time. In [44], a power control using CNN is proposed for approximating WMMSE. When training the CNN, spectral efficiency or energy efficiency are used for the cost function. The performance of CNN is similar or even higher than the WMMSE while maintaining faster computational speed. In [45], the neural networks are used for power allocation. The training data is generated by solving a sum-rate maximization problem under different channel information [45]. It takes the channel information and location indicators as the inputs representing whether or not a cell edge user is and generates the output representing the resource allocation. In [46], distributed reinforcement learning is adopted to manage intertier interference in heterogeneous networks. When femtocells and macrocells are sharing the same frequency bands, interference management is more important. In particular, it is very challenging to meet the QoS constraints when operating both femtocells and macrocells. In the system model of distributed reinforcement learning, the femtocells act as agents and optimize their capacity while satisfying the system requirements. Due to dynamic channel environments, the resource allocation is different, and the power control policy by the femtocells may violate the constraints of the networks. Thus, the macrocells inform the femtocells of the resource block scheduling plan facilitating the power control knowledge in different environments. Thus, the femtocells avoid interference to users in macrocells. The distributed reinforcement learning showed us a better average capacity than the traditional power controls.

Consider a simple resource allocation problem in an interference channel. The wireless network contains K single-antenna transceivers pairs where $h_{kk} \in \mathbb{C}$ and $h_{kj} \in \mathbb{C}$ represent the direct channel response between the kth transmitter and the kth receiver and the interference channel response between the jth transmitter and the kth receiver, respectively. The channels are stationary in the resource allocation. The transmitted symbol of the kth transmitter is assumed as a Gaussian random variable with zero mean and variance. The transmission power of the kth transmitter is denoted as p_k. The transmitted symbols are independent each other. The noise power at the kth receiver is denoted as σ_k^2. The signal-to-interference noise ratio (SINR) at the kth receiver is defined as follows:

$$\text{SINR}_k = \frac{|h_{kk}|^2 p_k}{\sum_{j \neq k} |h_{kj}|^2 p_j + \sigma_k^2}. \tag{9.78}$$

What we want to do is to allocate a proper power to the transmitters in order to maximize the weighted system throughput. The power allocation problem can be formulated as follows:

$$\max_{p_1, \dots, p_K} \sum_{k=1}^{K} \alpha_k \log(1 + \text{SINR}_k) \tag{9.79}$$

subject to

$$0 \leq p_k \leq P_m, \quad k = 1, \dots, K \tag{9.80}$$

where α_k and P_m are the weights and the maximum power of each transmitter, respec ively. This is a popular weighted sum-rate maximization problem. The power alloca- ion problem (9.79) and (9.80) is non-convex and known as NP-hard [47]. The power illocation using the WMMSE algorithm is one of popular algorithms. It is possible :o transform the weighted sum-rate maximization problem to a higher dimensional space. Thus, we can easily solve the problem using MMSE-SINR equality [48] as follows:

$$SINR_k = \frac{1}{MMSE_k} - 1 = \frac{1}{\left[\left(I_p + \frac{1}{m}H^H H\right)^{-1}\right]_{kk}} \tag{9.81}$$

where I_p, H, m, and H^H are a $p \times p$ identity matrix, channel matrix, the number of receive antenna, and the Hermitian transpose of H, respectively. The problem (9.79) and (9.80) can be transformed to the weighted MSE minimization (WMMSE) problem [43, 49] as follows:

$$\max_{[w_k, u_k, v_k]_{k=1}^{K}} \sum_{k=1}^{K} \alpha_k (w_k e_k - \log w_k) \tag{9.82}$$

subject to

$$0 \le v_k \le \sqrt{P_k}, \quad k = 1, \ldots, K \tag{9.83}$$

where w_k is a positive weight variable. v_k and u_k are the transmit and receive beam- formers. They are the optimization variables as real numbers. e_k is the mean square estimation error [43, 49] as follows:

$$e_k = (u_k |h_{kk}| v_k - 1)^2 + \sum_{j \ne k} (u_k |h_{kj}| v_j)^2 + \sigma_k^2 u_k^2. \tag{9.84}$$

It has been shown in [43, 49] that WMMSE can reach a stationary solution of the problem. The WMMSE algorithm solves the problem (9.82) and (9.83) by optimizing one set of variables while fixing the rest. We can summarize the WMMSE algorithm for the scalar interference channel as follows:

Initialize v_k^0 such that $0 \le v_k^0 \le \sqrt{P_k}$, $k = 1, \ldots, K$

$$\text{Compute} \quad u_k^0 = \frac{|h_{kk}| v_k^0}{\sum_{j=1}^{K} |h_{kj}|^2 \left(v_j^0\right)^2 + \sigma_k^2}, \quad k = 1, \ldots, K$$

$$\text{Compute} \ w_k^0 = \frac{1}{1 - u_k^0 |h_{kk}| v_k^0}, \quad k = 1, \ldots, K$$

$$t = 0$$

Repeat

$$t = t + 1$$

Update v_k

$$v_k^t = \left[\frac{\alpha_k w_k^{t-1} u_k^{t-1} |h_{kk}|}{\sum_{j=1}^{K} \alpha_j w_j^{t-1} \left(u_j^{t-1}\right)^2 |h_{kj}|^2} \right]_0^{\sqrt{P_m}}, \quad k = 1, \ldots, K$$

Update u_k

$$u_k^t = \frac{|h_{kk}| v_k^t}{\sum_{j=1}^{K} |h_{kj}|^2 \left(v_j^t\right)^2 + \sigma_k^2}, \quad k = 1, \ldots, K$$

Update w_k

$$w_k^t = \frac{1}{1 - u_k^t |h_{kk}| v_k^t}, \quad k = 1, \ldots, K$$

Until Converge to a stationary point of the problem (9.82) and (9.83).

Output $p_k = (v_k)^2, \quad k = 1, \ldots, K$.

As we can observe the above pseudo code of the WMMSE algorithm, it is convex and it is easier to handle. We optimize the variables (u, v, w) while holding others fixed and then find the solution about the power allocation. Now, the neural network is adopted to approximate the WMMSE algorithm. The well-known universal approximation for multiplayers feedforward networks [50] are used for this problem. A continuous mapping function is defined as follows:

$$y = f(x), \quad x \in \mathbb{R}^m, y \in \mathbb{R}^n \tag{9.85}$$

when $y_i = f_i(x)$ is a continuous function for all i and y_i is the ith coordinate of y. The universal approximation for iterative algorithm like WMMSE can be expressed by the neural network. A finite step iterative algorithm can be expressed as follows:

$$x^{t+1} = f^t(x^t, z), \quad t = 0, 1, \ldots, T \tag{9.86}$$

where $f^t()$ is a continuous mapping function at the tth iteration. $z \in Z$ and $x^t \in X$ are the elements of the parameter space and the feasible region of the problem, respectively. Assume Z and X are certain compact sets. The mapping function from problem parameter z and initialization x^0 to final output x^T

$$x^T = f^{''}(f^{''-1}(\dots f^{-1}(f^{''}(x'', z), z), z)\dots, z), z) - f^{-1}(x^0, z) \qquad (9.87)$$

can be approximated by the neural network with N sigmoid activation functions at one hidden layer $NN_N(x^0, z)$. For any given error $\epsilon > 0$, there is a positive constant N large enough [43] such that

$$\sup_{(x^0, z) \in X \times Z} \| NN_N(x^0, z) - F^T(x^0, z) \| \leq \epsilon. \qquad (9.88)$$

Namely, (9.88) refers to the largest value $\| NN_N(x^0, z) - F^T(x^0, z) \|$ could get to as (x^0, z) varies. The value is less than the error ϵ. It is possible to learn the mapping $z \to x^t$ for a fixed initialization x^0. Each iteration represents a continuous mapping and the optimal variables lie in a compact set. If we assume that the channel realization h_{ij} lies in a compact set, the WMMSE algorithm can be approximated by the neural network. Consider an input channel vector $h_{ij} \in \mathbb{R}^{K^2}$, a variable v_i at tth iteration by WMMSE $v(h)_i^t$, the minimum and maximum channel strength $H_{\min}, H_{\max} > 0$, a minimum power $P_{\min}$ of each transmitter, and a given positive number $V_{\min} > 0$. WMMSE is randomly initialized with

$$\left(v_k^0\right)^2 \leq P_m \text{ and } \sum_{i=1}^{K} v(h)_i^0 \geq V_{\min}. \qquad (9.89)$$

It is performed for T iterations. The admissible channel set $\mathcal{H}$ lies in a compact set. The channel realization are lower bounded by $H_{\min}$ and upper bounded by $H_{\max}$. The WMMSE sequences are satisfying $\sum_{i=1}^{K} v(h)_i^t \geq V_{\min}$. Admissible channel realizations are defined as follows [43]:

$$\mathcal{H} = \left\{ h | H_{\min} \leq |h_{jk}| \leq H_{\max}, \forall j, k, \sum_{i=1}^{K} v(h)_i^t \geq V_{\min}, \forall t \right\}. \qquad (9.90)$$

The input of the neural network is $h_{ij} \in \mathbb{R}^{K^2}$ and $v^0 \in \mathbb{R}_+^K$ and the output of the neural network is $NN(h, v^0) \in \mathbb{R}_+^K$. The number of the neural network layers [43] is

$$O\left(T^2 \log\left(\max\left(K, P_m, H_{\max}, \frac{1}{\sigma}, \frac{1}{H_{\min}}, \frac{1}{P_{\min}} \right) \right) \right) + T \log \frac{1}{\epsilon}. \qquad (9.91)$$

The number of ReLUs and binary units [43] is

$$O\left(T^2 K^2 \log\left(\max\left(K, P_m, H_{\max}, \frac{1}{\sigma}, \frac{1}{H_{\min}}, \frac{1}{P_{\min}} \right) \right) \right) + T K^2 \log \frac{1}{\epsilon}. \qquad (9.92)$$

We have the following relationship [43]:

$$\max_{h \in \mathcal{H}} \max_i \left| \left(v(h)_i^t \right)^2 - NN\left(h, v^0\right)_i \right| \le \epsilon. \tag{9.93}$$

(9.93) holds true if the WMMSE has a fixed initialization with $\left(v_k^0 \right)^2 = P_m$ and h is an input of the neural network. The resource allocation techniques other than WMMSE can be approximated in similar way if they can be expressed by basic operations such as binary search, threshold operations, and so on. Now, we construct the neural network for WMMSE approximation with one input layer, multiple hidden layers, and one output layer as shown in Fig. 9.15.

As we can observe Fig. 9.15, the input layer is the magnitude of the channel vector h_{ij}, the hidden layers have ReLU as activation function, and the output layer is the power allocation p_k with the activation function $y = \min(\max(x,0), P_m)$ to incorporate the power constraint. The training data of the channel realization $h_{kj}^{(i)}$ is generated where the index (i) the training sample. Firstly, P_m and σ_k for $\forall k$ are fixed. Secondly, for each tuple $\left(P_m, \sigma_k, h_{kj}^{(i)} \right)$, the corresponding power vector $p_k^{(i)}$ is generated by WMMSE with $v_k^0 = \sqrt{P_m}$, $\forall k$ as initialization. $\left(\left| h_{kj}^{(i)} \right|, p_k^{(i)} \right)$ is the ith training sample. We repeat this process until obtaining the entire training data $\mathcal{T}$ and the validation data set $\mathcal{V}$. In the training state, $\left(\left| h_{kj}^{(i)} \right|, p_k^{(i)} \right)_{i \in \mathcal{T}}$ is used to find the weights of the neural network. The mean-squared error (MSE) between $p_k^{(i)}$ and the output of the neural network is used for the cost function. Mini-batch stochastic gradient descent is used to optimize the weight [43]. We obtain the trained neural network. In the test stage, the channel vectors with the same distribution as the training data are generated [43]. For each channel realization, the optimized power is obtained by passing them through the trained network. Finally, the sum rate of

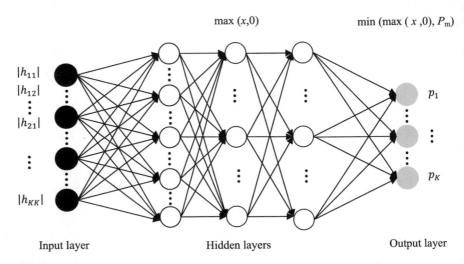

Fig. 9.15 Neural network for WMMSE approximation

he power allocation by the neural network is computed and then they are computed with the WMMSE.

Example 9.5 Resource allocation using the neural network Consider single cell model with channel realizations generated by a standard normal distribution. We will compare the resource allocation schemes: WMMSE, Deep learning, random power allocation ($p_k \sim$ uniform$(0, P_m)$) and maximum power allocation ($p_k = P_m$). The neural network has one input layer, three hidden layers and one output layer. Each hidden layer has 200 neuron nodes. As we can see Fig. 9.15, the input of the neural network is channel realization and the output is the power allocations. Evaluate the sum-rate performances of different resource allocation schemes when we have different users ($K = 10, 20$ and 50) and different training data size (5000 and 20,000).

Solution
Using open source simulation tool [43], the performances of the different resource allocation schemes are evaluated. First of all, the training data sets (5000 and 20,000) are generated. The neural network is trained. We can obtain the following Figs. 9.16, 9.17, 9.18 and 9.19.

As we can observe figures, the sum-rate performance of deep learning (DNN) is close to the one of WMMSE when the number of user is small. The MSEs are evaluated on the validation set. Large batch size enables us to converge slowly. However, if the number of users in the single cell increases, the performance is degraded. This is because of the mismatch between training data and testing data. We assumed the same distribution between training data and test data. In real world, the number of users are not constant, network configuration is varying, and the distribution is changing. It is not possible to train a different neural network for each network configuration and varying wireless channels. This is one of important research challenges in order to adopt AI algorithms to wireless networks. ∎

Fig. 9.16 Sum-rate
performance comparison
when training data = 5000
and $K = 10$ (**a**), 20 (**b**) and
50 (**c**)

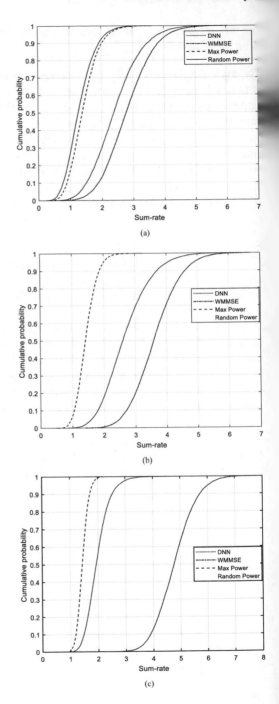

Fig. 9.17 MSE comparison
when training data = 5000
and $K = 10$ (**a**), 20 (**b**) and
50 (**c**)

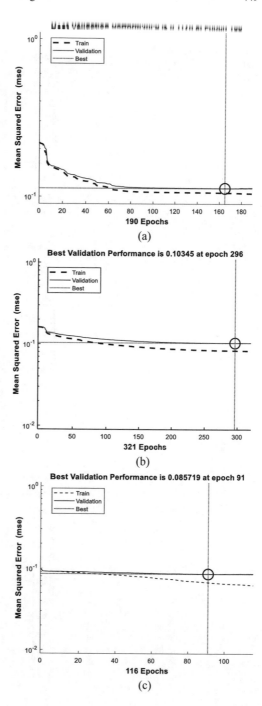

(a)

(b)

(c)

Fig. 9.18 Sum-rate
performance comparison
when training data = 20,000
and $K = 10$ (**a**), 20 (**b**), and
50 (**c**)

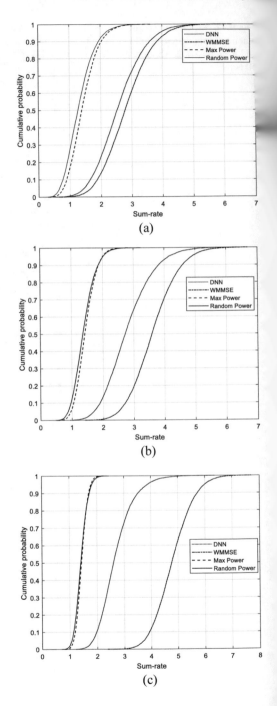

Fig. 4.19 MSE comparison when training data = 20,000 and $K = 10$ (**a**), 20 (**b**) and 50 (**c**)

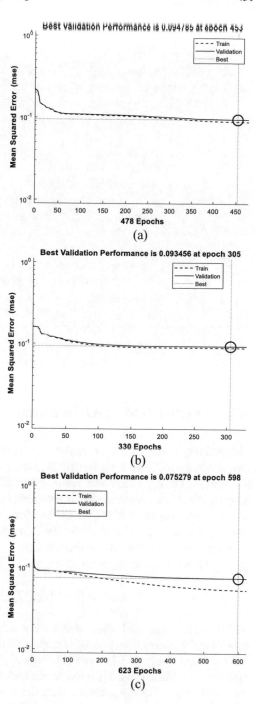

(a)

(b)

(c)

Summary 9.5. Resource Allocation Using AI Techniques

1. The data-driven approach of AI algorithms enables us to train a model on a large amount of data, learn to approximate the relationship between input and output, and find an optimal resource allocation.
2. In resource allocation, optimization techniques with high complexity can be replaced by deep learning. Due to the approximation by the DNN, the resource allocation can be performed in real time.
3. The resource allocation and scheduling problems can be formulated as a convex problem, a mixed integer nonlinear problem, or others. The neural network will be helpful for solving the problem, optimizing the resource allocation procedure, and achieving a better efficiency.
4. There are many attempts to apply AI techniques to resource allocations and scheduling in wireless networks. We assumed the same distribution between training data and test data. In real world, the number of users are not constant, the distribution is changing, and wireless system conditions and configurations are varying. It is not possible to train a different neural network for each network configuration and varying channels. This is one of important research challenges in order to adopt AI algorithms to wireless networks.

9.3 Handover Using AI Techniques

The mobility management of wireless networks is one of key functions. The main purpose of the mobility management is to trace where users exist and deliver services to the users. During a handover event, interruption or delay in cellular networks can happen. We should minimize the interruption or delay. In particular, it will be critical when we provide users with URLLC services. In the URLLC services, the mobility interruption time (MIT) defined by the 3GPP standard should be minimized where MIT represents the time that a mobile user cannot exchange user plan data packet during a handover. During a handover, the MIT T_{MIT} is defined as follows [51]:

$$T_{\mathrm{MIT}} = (1 - P_{\mathrm{HOF}})T_{\mathrm{HIT}} + P_{\mathrm{HOF}}T_{\mathrm{HOF}} \tag{9.94}$$

where T_{HIT}, T_{HOF}, and P_{HOF} denote the interruption time during a successful handover, the interruption time of a handover failure, and the probability of a handover failure, respectively. In LTE network, typical T_{HIT} is about 50 ms and T_{HOF} is about 100 ms ~ 2 s [52]. Thus, T_{HOF} has much higher impact on the T_{MIT}. It implies that we should minimize P_{HOF}. Figure 9.20 illustrates the general handover process and prediction based handover process in a cellular network.

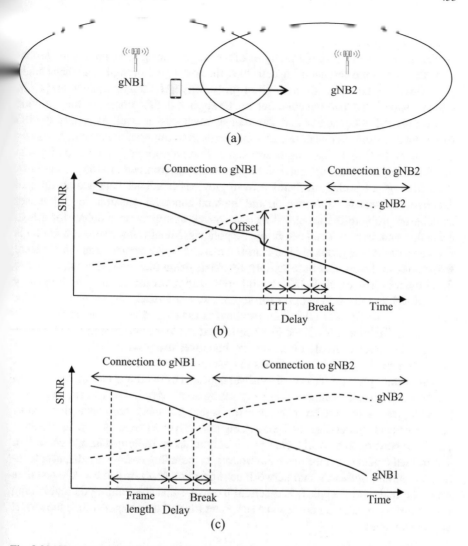

Fig. 9.20 Handover from gNB1 to gNB2 (**a**), the current handover process (**b**) and the prediction based handover process (**c**) in a cellular network

In the current handover as shown in Fig. 9.20b, when there is the signal strength offset (approximately 1–3 dB) between two base stations (gNBs), the handover process starts and causes a delay such as the time to trigger (TTT, approximately 200–300 ms), and so on. The task of the mobility management adjusts these parameters to optimize the handover process. They do not depend on the mobile device speed, channel condition, signal propagation, trajectory of user movement, and so on. In order to have more customized handover, 5G systems consider more mobility parameters. Using AI techniques, the prediction based handover as shown in Fig. 9.20c enables us to improve a handover performance by learning from a mobility environment and optimizing the process of the handover. AI-enabled handover mechanism enables us to collect the signal strengths of the neighbouring cells, predict the probability that one of the neighbouring cells have the best signal strength, and then decide to handover to the cell with the highest probability. AI techniques such as LSTM [53] and supervised learning classifier can be applied in this process. It may provide us with enough adaptability and low management cost as well as satisfy the 6G requirements. Reinforcement learning can be used to improve the current handover mechanism. In [54], Q learning is adopted to find an optimal handover policy that maximize the future throughput under a pedestrian environment. In [55], a handover is decided by a learning mechanism from previous handover data and an internal data between vehicles' information and the final handover decision. In [56], a user association problem is formulated as non-convex optimization problem for a load balancing handover. A deep deterministic policy gradient reinforcement learning is adopted to solve the problem. The learning method is to associate users in different trajectories to the optimal base stations by maximizing their sum rate and reducing the number of handovers. In [57], a joint optimization for minimizing the frequency of handover and maximizing user throughput was proposed. The system model for the handover and power allocation problem is developed as a cooperative multi-agent task. The model is trained in a centralized manner after decentralized polices are obtained. As we briefly discussed the handover using AI techniques, there are many attempts to apply AI techniques to handover managements. They all provides us with meaningful results as well as novel approaches but they faces many research challenges such as collection of proper training data, privacy and security, real-time process, generalized handover mechanism, signal overhead, load balancing and so on. In particular, generalization is one of big research challenges. As we discussed in Part I, the performance of AI techniques highly relies on training data and condition. The trained model under the given environments including channel model, user location, the number of users, and network configuration can't be same as the test data set in the real world. Thus, it is essential to ensure that the training data obtaining from a history of user handover must represent the test data to generalize the model of AI techniques.

Summary 9.6. Handover Using AI Techniques

1. The main purpose of the mobility management is to trace where users exist and deliver services to the users.
2. During a handover, the mobility interruption time (MIT) T_{MIT} is defined as follows:

$$T_{MIT} = (1 - P_{HOF})T_{HIT} + P_{HOF}T_{HOF}$$

where T_{HIT}, T_{HOF}, and P_{HOF} denote the interruption time during a successful handover, the interruption time of a handover failure, and the probability of a handover failure, respectively.

3. AI-enabled handover mechanism enables us to collect the signal strengths of the neighbouring cells, predict the probability that one of the neighbouring cells have the best signal strength, and then decide to handover to the cell with the highest probability.

9.4 Problems

9.1   Describe the main functions of 5G layer 2.

9.2   Based on 5G protocol, 6G data link layer may employ new functions such as terahertz communication support, distributed communication support, cell-free access, and so on. Discuss the key requirements and changes of data link layer in order to include them.

9.3   Using the collected data from mobile devices, AI algorithms can generate a new information that can be exploited in many different applications such as localization, analysis of mobile data usage and so on. Discuss the key data link layer functions or services that can be improved by AI techniques. In addition, describe the key research challenges to implement them.

9.4   The channel knowledge can be obtained in various ways such as duplex system characteristics, feedback channel, and pilot signal measurement. Discuss how they can be obtained and which functions use the channel knowledge in 5G systems.

9.5   The optimization problems of the radio resource allocation can be formulated in terms of different cost functions such as system throughput, energy efficiency, latency, connection density, and so on. They all can't be optimized at the same time. Discuss the relationship among throughput, energy efficiency, latency, and connection density when designing the radio resource allocation function.

9.6   The mechanism of the resource allocation and scheduling relies on multiplexing schemes in physical layer. Discuss different radio resource allocation schemes of TDMA, CDMA and OFDMA.

9.7   There are no generalized solutions for convex optimization problems. Discuss why it is difficult to have generalized solutions.

9.8   A global optimal solution to the convex optimization problem can be found by interior point method, ellipsoid method, sub-gradient method and so on. Describe the pseudo codes of the solvers.

9.9   Describe the pros and cons of barrier method and primal–dual method.

9.10 Consider the following optimization problem

$$\min_{x,y}\left(xe^{-(x^2+y^2)} + \left(3x^2 + 2y^2\right)\right)$$

subject to

$$xy + (x+2)^2 + (y-2)^2 \le 2.$$

Find the optimal point with the start point $(x, y) = (-4, 4)$.

9.11 Describe the significance of the concepts: Utopia point, pateto boundary and utility functions.

9.12 The system utility function allows us to convert the multi-objective optimization problem to the single-objective optimization problem. Describe the pros and cons of the utility functions: weighted sum utility, weighted proportional fairness, weighted harmonic mean, and weighted max–min fairness in resource allocation of wireless systems.

9.13 Compare the fairness and aggregate throughput of max–min fairness, maximum throughput scheduling, proportional fairness, and minimum potential delay fairness in 5G NR systems.

9.14 In α-fair allocations, we have different fairness depending on different α value. Describe the effect of α values in terms of different fairness.

9.15 Consider the following network conditions:

- Measure throughputs: $\begin{aligned}(t_1, t_2, t_3, t_4, t_5)\\ = (5\,\text{Mbps}, 3\,\text{Mbps}, 5\,\text{Mbps}, 1\,\text{Mbps}, 2\,\text{Mbps})\end{aligned}$
- Fair throughputs by a specific fairness resource allocation: $(o_1, o_2, o_3, o_4, o_5) = (5\,\text{Mbps}, 1\,\text{Mbps}, 1\,\text{Mbps}, 1\,\text{Mbps}, 2\,\text{Mbps})$.

Compute the fairness index of the network.

9.16 There are many attempts to apply AI techniques to resource allocations in wireless networks. We assumed unrealistic channel condition and network configurations. Discuss the gap between theoretical works and practical implementation in resource allocation of wireless networks.

9.17 The prediction based handover using AI techniques enables us to improve a handover performance by learning from a mobility environment and optimizing the process of the handover. In order to implement, it is important to collect training data set representing test data set. Assume different offset of distribution mismatch between them and then compare the effect of the mismatch when performing handover in multiple cells.

References

1. D. Bega, M. Gramaglia, M. Fiore, A. Banchs, X. Costa-Perez, Deepcog: cognitive network management in sliced 5g networks with deep learning, in *Proceedings of the IEEE INFOCOM*, Paris, France, 29 April–2 May 2019
2. M. Sha, R. Dor, F. Hackmann, C. Lu, T. Kim, R. Park, Self-adapting mac layer for wireless sensor networks, in *Proceedings of the 2013 IEEE 34th Real-Time Systems Symposium*, Vancouver, BC, Canada, 3–6 December 2013, pp. 192–201
3. S. Yoon, C. Shahabi, The clustered aggregation (cag) technique leveraging spatial and temporal correlations in wireless sensor networks. ACM Trans. Sens. Net. **3**(3) (2007)
4. R.V. Kulkarni, G.K. Venayagamoorthy, Neural network based secure media access control protocol for wireless sensor networks, in *Proceedings of the Neural Networks*, Atlanta, GA, USA, 14–19 June 2009, pp. 1680–1687
5. B. Huang, Q. Zhu, C. Siew, Extreme learning machine: A new learning scheme of feedforward neural networks, in *Proceedings of the 2004 IEEE International Joint Conference on Neural Networks*, Budapest, Hungary, vol 2, 25–29 July 2004, pp. 985–990
6. S. Rajab, W. Balid, M. Kalaa, H. Refai, Energy detection and machine learning for the identification of wireless mac technologies, in *Proceedings of the 2015 International Wireless Communications and Mobile Computing Conference (IWCMC)*, Dubrovnik, Croatia, 24–28 August 2015, pp. 1440–1446
7. Y. Zhou, S. Peng, Y. Yao, Mac protocol identification using convolutional neural networks, in *Proceedings of the 2020 29th Wireless and Optical Communications Conference (WOCC)*, Newark, NJ, USA. 1–2 May 2020, pp. 1–4
8. X. Zhang, W. Shen, J. Xu, Z. Liu, G. Ding, A mac protocol identification approach based on convolutional neural network, in *Proceedings of the 2020 International Conference on Wireless Communications and Signal Processing (WCSP)*, Nanjing, China, 21–23 October 2020, pp. 534–539
9. R. Mennes, M. Camelo, M. Claeys, S. Latre, A neural-network-based mf-tdma mac scheduler for collaborative wireless networks, in *Proceedings of the 2018 IEEE Wireless Communications and Networking Conference (WCNC)*, Sydney, Australia, 18–21 May 2018, pp. 1–6
10. R. Mennes, M. Claeys, F. Figueiredo, I. Jabandžic, I. Moerman, S. Latré, Deep learning-based spectrum prediction collision avoidance for hybrid wireless environments. IEEE Access **7**, 45818–45830 (2019)
11. Y. Zhang, J. Hou, V. Towhidlou, and M. Shikh-Bahaei, A neural network prediction based adaptive mode selection scheme in full-duplex cognitive networks. IEEE Trans. Cog. Comm. Net. **5**, 540–553 (2019)
12. K. M. Thilina, K. W. Choi, N. Saquib, E. Hossain, Machine learning techniques for cooperative spectrum sensing in CRNs, *IEEE JSAC* (2013)
13. A. Galanopoulos, F. Foukalas, T.A. Tsiftsis, Efficient coexistence of LTE with WiFi in the licensed and unlicensed spectrum aggregation. IEEE TCCN **2**(2), 129–140 (2016)
14. A. Hithnawi, H. Shafagh, S. Duquennoy, Tiim: technology-independent interference mitigation for low-power wireless networks, in *Proceedings of the Proceedings of the 14th International Conference on Information Processing in Sensor Networks*, Seattle, WA, USA, 14–16 April 2015, pp. 1–12
15. Y. Zhang, K. Letaief, Adaptive resource allocation and scheduling for multiuser packet-based OFDM networks, *IEEE ICC*, vol. 5 (2004)
16. J. Jang, K. Lee, Transmit power adaptation for multiuser OFDM systems. IEEE J. Sel. Areas Commun. **21**(2), 171–178 (2003)
17. H. Yin, H. Liu, An efficient multiuser loading algorithm for OFDM-based broadband wireless systems, *IEEE Globecom* (2000)
18. W. Jiao, L. Cai, M. Tao, Competitive scheduling for OFDMA systems with guaranteed transmission rate, *Elsevier Computer Communications*, special issue on Adaptive Multicarrier Communications and Networks, 29 August, 2008

19. M. Tao, Y. C. Liang, F. Zhang, Resource allocation for delay differentiated traffic in multiuser OFDM systems. IEEE Trans. Wirel. Commun. **7**(6), 2190–2201 (2008)
20. Z. Han, Z. Ji, K. Liu, Fair multiuser channel allocation for OFDMA networks using Nash bargaining solutions and coalitions. IEEE Trans. Comm. **53**(8), 1366–1376 (2005)
21. N. Damji, T. Le-Ngoc, Dynamic resource allocation for delay-tolerant services in downlink OFDM wireless cellular systems, *IEEE ICC*, vol. 5 (2005), pp. 3095–3099
22. S. Boyd, L. Vandenberghe, *Convex Optimization* (Cambridge University Press, 2004). ISBN 978-0-521-83378-3
23. H. Kim, *Design and Optimization for 5G Wireless Communications* (Wiley, 2020). ISBN:9781119494553
24. E. Björnson, E. Jorswieck, Optimal resource allocation in coordinated multi-cell systems. Foundations Trends Commun. Inf. Theory **9**(2–3), 113–381 (2013)
25. M. Karakayali, G. Foschini, R. Valenzuela, Network coordination for spectrally efficient communications in cellular systems. IEEE Wirel. Comm. Mag. **13**(4), 56–61 (2006)
26. H. Zhang, H. Dai, Cochannel interference mitigation and cooperative processing in downlink multicell multiuser MIMO networks. EURASIP J. Wirel. Commun. Netw. **2**, 222–235 (2004)
27. E. Björnson, N. Jaldén, M. Bengtsson, B. Ottersten, Optimality properties, distributed strategies, and measurement-based evaluation of coordinated multicell OFDMA transmission. IEEE Trans. Signal Process. (2011)
28. T. Cover, J. Thomas, *Elements of Information Theory* (Wiley, 1991)
29. J. Branke, K. Deb, K. Miettinen, R.S. (Eds.), *Multiobjective Optimization: Interactive and Evolutionary Approaches* (Springer, 2008)
30. H. Tuy, Monotonic optimization: problems and solution approaches. SIAM J. Optim. **11**(2), 464–494 (2000)
31. J. Mo, J. Walrand, Fair end-to-end window-based congestion control. IEEE/ACM Trans. Network. **8**(5), 556–567 (2000)
32. Z. Jiang, Y. Ge, Y. Li, Max-utility wireless resource management for best-effort traffic. IEEE Trans. Wirel. Commun. **4**(1), 100–111 (2005)
33. J. Branke, K. Deb, K. Miettinen, R.S. (Eds.) *Multiobjective Optimization: Interactive and Evolutionary Approaches* (Springer, 2008)
34. D. Bertsekas, R. Gallager, *Data Networks* (Prentice Hall, 1987)
35. F.P. Kelly, A. Maulloo, D. Tan, Rate control for communication networks: shadow prices, proportional fairness and stability. J. Oper. Res. Soc. **49**, 237–252 (1998)
36. E.M. Rogers, *Diffusion of Innovations*, 4th edn. (Free Press, New York, 1962)
37. R. Srikant, *The Mathematics of Internet Congestion Control* (Birkhauser, 2004)
38. L. Massoulié, J. Roberts, Bandwidth sharing: objectives and algorithms. IEEE INFOCOM **3**, 1395–1403 (1999)
39. J. Mo, J. Walrand, Fair end-to-end window-based congestion control. IEEE/ACM Trans. Network. (ToN) **8**(5), 556–567 (2000)
40. H. SHI, R. V. Prasad, E. Onur, I.G.M.M. Niemegeers, Fairness in wireless networks: issues, measures and challenges. IEEE Commun. Surv. Tutor. **16**(1), 5–24 First Quarter 2014. https://doi.org/10.1109/SURV.2013.050113.00015
41. R. Jain, D. Chiu, W. Hawe, A quantitative measure of fairness and discrimination for resource allocation in shared systems, digital equipment corporation, *Technical Report DEC-TR-301*, Tech. Rep. (1984)
42. A. Renyi, On measures of entropy and information, in *Proceedings of the 4th Berkeley Symposium on Mathematics, Statistics and Probability* (1960), pp. 547–561
43. H. Sun, X. Chen, Q. Shi, M. Hong, X. Fu, N.D. Sidiropoulos, Learning to optimize: training deep neural networks for interference management. IEEE. Trans. Signal Process. **66**(20), 5438–5453 (2018)
44. W. Lee, M. Kim, D. Cho, Deep power control: transmit power control scheme based on convolutional neural network. IEEE Comm. Lett. **22**(6), 1276–1279 (2018)
45. K.I. Ahmed, H. Tabassum, E. Hossain, Deep learning for radio resource allocation in multi-cell networks (2018) arXiv:1808.00667v1

46. A. Galindo-Serrano, L. Giupponi, G. Auer, Distributed learning in multiuser OFDMA femtocell networks, *IEEE VTC 2011*, Yokohama, Japan, , May 2011, pp. 1–6
47. Z.-.Q. Luo, S. Zhang, Dynamic spectrum management: Complexity and duality. IEEE J. Sel. Top. Signal Process. **2**(1), 57–73 (2008)
48. Ping Li, D. Paul, R. Narasimhan, J. Cioffi, On the distribution of SINR for the MMSE MIMO receiver and performance analysis. IEEE Trans. Inf. Theory **52**(1), 271–286 (2006). https://doi.org/10.1109/TIT.2005.860466
49. Q. Shi, M. Razaviyayn, Z.-Q. Luo, C. He, An iteratively weighted MMSE approach to distributed sum-utility maximization for a MIMO interfering broadcast channel. IEEE Trans. Signal Process. **59**(9), 4331–4340 (2011)
50. K. Hornik, M. Stinchcombe, H. White, Multilayer feedforward networks are universal approximators. Neural Netw. **2**(5), 359–366 (1989)
51. G. T. 38.913, Study on scenarios and requirements for next generation access technologies, Tech. Rep. v14.3.0. Release 14 (2017)
52. G. T. 36.300, Evolved universal terrestrial radio access (e-utra) and evolved universal terrestrial radio access network (e-utran); overall description; stage 2 (release 14) (2017)
53. C. Wang, L. Ma, R. Li, T. S. Durrani, H. Zhang, Exploring trajectory prediction through machine learning methods. IEEE Access **7**, 101 441–101 452 (2019)
54. Y. Koda, K. Yamamoto, T. Nishio, M. Morikura, Reinforcement learning based predictive handover for pedestrian-aware mmwave networks, in *IEEE Conference on Computer Communications Workshops* (INFOCOM WKSHPS) (2018), pp. 692–697
55. L. Yan, H. Ding, L. Zhang, J. Liu, X. Fang, Y. Fang, M. Xiao, X. Huang, Machine learning-based handovers for sub-6 ghz and mm wave integrated vehicular networks. IEEE Trans. Wirel. Commun. **18**(10), 4873–4885 (2019)
56. S. Khosravi, H.S. Ghadikolaei, M. Petrova, Learning-based load balancing handover in mobile millimeter wave networks (2020). arXiv:2011.01420. [Online]. Available: http://arxiv.org/abs/2011.01420
57. D. Guo, L. Tang, X. Zhang, Y.-C. Liang, 'Joint optimization of handover control and power allocation based on multi-agent deep reinforcement learning. IEEE Trans. Veh. Technol. **69**(11), 13124–13138 (2020)

Chapter 10
AI-Enabled Network Layer

In order to explore websites and communicate with mobile devices, we should inter-connect among different networks. The interconnection tasks among networks take place at the network layer. The main function of the traditional network layer is to establish the connection between different networks by forwarding network data packets to network routers discovering the best path across the networks. It uses typically Internet Protocol (IP) addresses to transfer them to the destination and controls the operations of the networks. The routers as one of network components operate in network layer to forward network data packets between different networks. The key tasks in the network layer can be summarized as follows: finding the path for the network data packets, checking whether or not network components in other networks is up and running, and then transmitting and receiving the network data packets from other networks. In order to perform these tasks, the network layer is equipped with different protocol stacks about establishing connection, testing, routing, encryption, and so on. As we discussed in Part I, finding an optimal path, classifying data, and predicting data are key tasks of AI techniques. They are in line with the network layer tasks. We expect that AI techniques will improve the performance of network layers. In this chapter, we discuss how AI techniques contribute network layer tasks, review cellular systems and networking, and investigate one selected topic about the network traffic prediction in the classical and AI approach.

10.1 Design Approaches of AI-Enabled Network Layer

The Open Systems Interconnection (OSI) model is a conceptual model characterizing the communication and network functions for interoperability of different communication and network systems. It provides us with good guideline to understand telecommunication and network systems, but it is not perfectly matched with the protocol stacks of cellular networks. The 5G protocol stacks consist of a user plane and a control plane. The control plane takes care of signalling or controlling data. In the user plane, the user data are exchanged. This separation enables mobile operators to scale their functions independently and deploy networks flexibly. The protocol stacks of 5G systems inherit the basic structure of 4G systems. The 6G protocol stacks may be similar to the 5G protocol stacks. In 5G systems, signalling mechanism is simplified to support an URLLC application. User-centric protocols enable 5G system to support various applications. In 5G system, layer 1 is physical layer. Layer 2 includes MAC, RLC, PDCP, and SDAP. Layer 3 includes radio resource control (RRC) layer and non-access stratum (NAS) layer. Figure 10.1 illustrates 5G protocol layers for user plane and control plane.

Layer 3 is not user plane protocol but control plane protocol. In RRC layer, 5G networks configure radio resources to establish connections between mobile devices and base stations. The main role of RRC layer is to manage the connection status (connected, idle, or inactive) of next-generation NodeB (gNB) and user equipment (UE) by adjusting the configuration values in terms of the connection status. The functions of RRC layer can be summarized as follows [1]: broadcasting of access stratum (AS) and non-access stratum (NAS) system information, paging by NG-RAN or 5GC, establishment, maintenance and release of RRC connection between UE and NG-RAN, Security and mobility functions (key management, handover and context transfer, cell addition and release, etc.), UE measurement and control reporting, QoS management functions, and so on. The main role of NAS layer is to manage signalling between UE and core network (CN). The key network components of the CN are access and mobility management function (AMF) and session management function (SMF). In NAS layer, 5G network manages the mobility by signalling to AMR. It establishes communication sessions and maintains continuous communications with the UE by signalling SMF. In 5G networks, the eNB coexists with gNB in a certain period and is gradually replaced by gNB. The RAN nodes connect to the 5G core (5GC) network including AMF, SMF, and user plane function. Figure 10.2 illustrates 5G NR RAN and GC.

As we can observe Fig. 10.2, the next-generation RAN (NG-RAN) consists of gNBs and ng-eNBs (upgraded version of the 4G LTE base station). 5GC provides us with NFV, SDN, cloud service, network slicing, and so on. They are flexible and scalable and also are interconnected with each other via Xn interface and connects to AMF and UPF via NG interface. The key functions of 5GC are to provide Internet connectivity, ensure the QoS in this connection, and track the user mobility and usage for billing. Network virtualization is one of key concepts in 5G systems. 6G will accelerate the adaptation of virtualization concept. In this virtualization approach, the network components such as servers, routers, and data centres can be virtualized as software components. They can be connected among geographically unrelated network components. It enables the network to be scaled flexibly and improve the network performance and efficiency. The routers of 6G systems should be flexible and cost effective. Thus, virtual routing was one of key research challenges a long time ago. A virtual router was developed as software function that can replicate the hardware-based Internet Protocol routing. It is a part of network function virtualization (NFV) and has more benefits such as higher interoperability, flexibility, and cost efficiency. Since the virtual router is implemented as software,

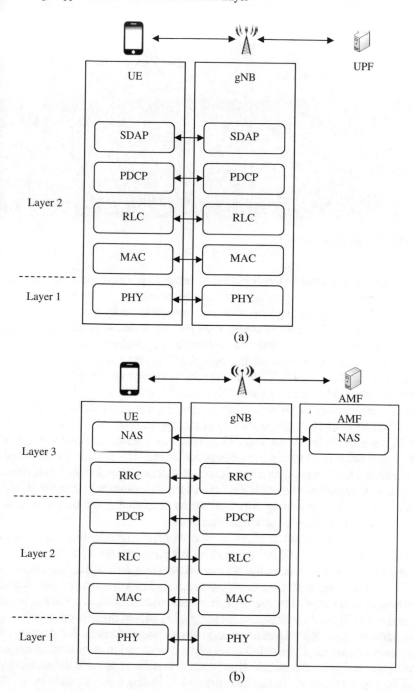

Fig. 10.1 5G protocol stacks for user plane (**a**) and control plane (**b**)

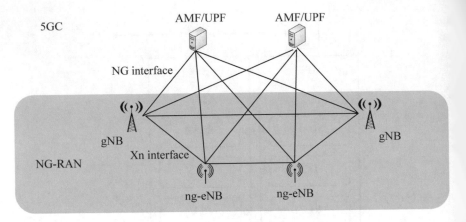

Fig. 10.2 5G NR RAN and CN

the routing functions can be freely placed in networks or data centres and also allow us to configure dynamically in terms of network or user requirements. It is possible to deploy distributed routing functions and manage them by a centralized control plan. 5G networks require to support multiple and different needs from multiple vertical industries such as healthcare, automotive, agriculture, factory automation, energy, and so on. 5G introduced the concept of "service-based architecture" based on network virtualization [1]. The network slicing enables mobile operators to deploy different networks having customized services of applications (eMBB, mMTC, and URLLC) over a common network infrastructure. Each network slicing is isolated and also tailored to satisfy the requirements of the applications. Network slices can be regarded as a set of virtual routers (or virtual machines) under the same management. The virtualization is a key concept of network slicing. The virtualized 5G networks are connected to IP-based data network such as Internet. The IP-based data network still plays an important role in 5G network and is the common bearer for 5G applications. The IP packet routing is performed by determining the best path passing through a series of hardware or virtualized routers.

As we discussed in Part I, the basic approach of AI techniques is to exploit hidden patterns in training data and use the patterns for grouping, anticipating, or decision making in the test data. Definitely, 5G and 6G networks deal with a huge amount of data. The computing powers and storage capabilities in 6G networks enable us to train the voluminous data and create new applications and services for 6G systems. In terms of the network layer, AI techniques can analyse big data of the network traffics, optimize cellular network configurations, and improve the performance of the networks. The current approach of the cellular networks design and management is based on time series analysis and heuristic algorithms. It is difficult to scale the network and meet the requirements. As the cellular network is getting complex and new services are adopted, a new network design and management is essential. AI-enabled network design and management will be able to achieve a better network design and management by finding the network traffic pattern, predicting the demand, and detecting a network fault. The process of AI-enabled networks design and management will be similar to conventional process of AI techniques implementation: Formulate the network problem, collect data, train the AI model, and apply the trained model to the problem. In the problem formulation phase, the network problems are classified as finding the patterns as clustering or classifying and predicting the future events. The network problems to be solved by AI techniques are about finding the type of the network traffic data, extracting the features of the network data, and predicting the future volume of the traffics or detecting the network faults. In the phase of the data collection, we can use historical network data as offline or real-time network data as online. Practically, we can collect the data set using open-source tools (such as tshark, tcpdump), commercial tools, or network component setup (such as flow monitoring). The types of the network data are data flow, packet captures, log data, telemetry, network component configurations, network topology data, and so on. Using these data, the features can be extracted. For example, the features of flow can be duration of a flow, the number of packets at a flow, data volume at a flow, and so on. The features of packets can be statistical information such as mean, variance, RMS, and so on. The other features are throughput, session time, TPC window size, and so on. The feature extraction can be performed by people with domain expertise. In the training phase, the training and test data sets are generally split into an 80:20 or 70:30 ratio. AI techniques can be implemented using AI platform such as Keras, Tensorflow, or PyTorch, or network designers can develop their own techniques for a specific problem. AI techniques should be trained while maintaining spatial and temporal diversity among the data sets. The trained model should be evaluated in terms of accuracy, complexity, reliability, and stability. In particular, accuracy is an important metric and accuracy should be evaluated as the difference between actual values and prediction. If we have a well-trained model, the network problem can be solved efficiently. The key AI contributions to 6G network layer can be summarized as follows: AI contribution to network traffic prediction, network traffic classification by AI algorithms, optimal routing path by AI algorithms, queueing management by AI algorithms, AI contributions to network fault management, and QoS management by AI algorithms.

AI contribution to network traffic prediction: The objective of the network traffic prediction is to predict characteristics of the subsequent network traffics from the previous network traffic data. This technique plays an important role in the network operation and management in 6G heterogeneous networks. The results of the network traffic prediction can be used for network monitoring, resource management, fault detection, and others. The traditional approach of the time series forecasting is based on a regression model to find correlation between the past traffic data and future traffic data. The autoregressive integrated moving average (ARIMA) model is widely used for the time series forecasting. It is generalization of the autoregressive moving average by adding the concept of integration. For time series stationary, the integration typically uses the subtraction between the observation at the current time step and the observation at the previous time step. Thus, ARIMA has a temporal patterned structure and provides us with a simple but powerful tool for analysing and forecasting time series data. However, as the complexity of the network and the volume of the traf fics is getting higher and also the network operators require to reduce the overhead o the packet, AI techniques become an important method for the time series forecasting of the network traffic. The network forecasting problem can be regarded as a time series analysis problem. The network traffics are varying in terms of user require- ments, system configurations, applications, and others. In [2], the neural network is employed for improving accuracy of autoregressive methods. Different types of the neural networks are adopted for the time series analysis problems [3–5]. In [3], the bandwidth prediction on a given path using the neural networks is proposed for grid environment. In [4], a neural network ensemble is adopted for real-time forecasting and compared with autoregressive models. The delay and computation complexity is compared with them. They showed us that the neural network ensemble is more suitable for real-time forecasting. In [5], a hybrid training algorithm is proposed by combining artificial bee colony (ABC) algorithm and particle swarm optimization (PSO) as an evolutionary search algorithm. They are implemented with a neural network. The combination of ABC and PSO provides us with a faster training time. In [6], interdata centre network traffic prediction is investigated. They focus on the prediction of incoming and outgoing dominated traffic volume by elephant flows contrary to a time series forecasting, where the elephant flows are an extremely large continuous flows. The elephant flows can occupy unbalanced share of the bandwidth over a certain period. Typically, they occupy at least 10% of the total bandwidth and 10% longest active flows. Thus, it should be detected and managed for a traffic balance. The neural network with simple gradient descent is used to capture the traffic features in time and frequency. In [7], Gaussian processes regression, online Bayesian moment matching, and neural network are used for traffic flow size predic- tion and elephant flow detection. The problem was formulated as an online machine learning to adjust the traffic flow changes. Those attempts provide us with meaningful accuracy of network traffic prediction. The neural network is useful for forecasting traffic volumes from past data, but complexity reduction is still challenging. In addi- tion, the time series forecasting relies on the past observation. If we can't measure enough past data at a high-speed flow, the accuracy of the network traffic prediction can't be guaranteed. Thus, achieving good balance among accuracy, computational complexity, and latency is still an important topic. In Sect. 10.3, we discuss classical network traffic prediction and RNN-based network traffic prediction.

Network traffic classification by AI algorithms: The network operators typically classify the type of network traffics into sensitive traffics, best effort traffics and undesired traffics. The sensitive traffic is a time-sensitive traffic including VoIP, teleconference, online gaming, and so on. The network operators can prioritize the sensitive traffic and guarantee the QoS. The best effort traffic is not sensitive and not detrimental traffic including email service and so on. The undesired traffic is detrimental traffic including spam email, malicious attacks, and so on. The network operators should block this type of traffic. Sometimes, they often distinguish latency critical traffics, application oriented traffics, and others. Depending on the packet classification, a predefined network policy can be applied to the traffics and provide

users with proper services. Classifying data is one key application of AI techniques
as we discussed in Part I. The network traffic classification is in line with the work
AI algorithm can contribute. The network traffic classification using AI techniques
has been widely studied. The network traffic classification is the foundation for identifying the traffic types of applications and enabling traffic shaping and policing on
the network. The mobile and network operators need to classify the network traffics
in order to perform performance monitoring, resource provisioning, QoS management, capacity management, and so on. In particular, prioritizing the network traffics
will be one important task in 6G systems. If the network operators can identify
the network traffics and classify them into latency critical applications or normal
applications, it is possible to provide users with the prioritized services and achieve
more efficient resource management. The traditional method of the network traffic
classification can be summarized as port numbers based, packet payload based, host
behaviour based, and flow statistics based [8]. The traditional approach like using
port numbers may not be efficient in 6G systems due to the usage of dynamic port
negotiation, tunnelling, and so on. AI-based traffic classification techniques will
be more useful for complex network systems. The payload-based network traffic
classification uses the application payload information excluding the packet header
information of the network traffics. The traffic classification using packet payload
should search the application patterns at the packet payload and require high computational complexity and large storage. It is not an efficient way if the packets have
dynamic behaviour or are encrypted. In [9], AI clustering technique is used to perform
payload-based traffic classification. The classifier is trained using a list of partially
correlated protocols that are modelled by distribution of sessions. Agglomerative
hierarchical clustering is performed to cluster the protocols and distinguish them.
The host behaviour-based network traffic classification is based on the inherent host
behaviour. Depending on applications generated by different communication signa-
tures, it investigates the network traffics between hosts and is useful when there
are unregistered port numbers and encrypted payload. For example, a peer-to-peer
host and a webserver have different behaviours. In [10], a table of (IP address, port
numbers) pairs for each flow is used for classification. This approach is helpful for
identifying the unclassified flows. In [11], a SVM classifier uses the probability
mass function of the number of peers and is trained to find a peer-to-peer application patterns. However, the accuracy of the host behaviour-based network traffic
classification relies on the monitoring system location [12]. Flow statistic-based
traffic classification is based on complete flows observations. A complete flow is a
unidirectional exchange of consecutive packets on the network between a port at an
IP address and another port at a different IP address using a particular application
protocol [13]. A sub-flow is a subset of a complete flow and can be collected over
a time window in an ongoing session [13]. Thus, a complete flow contains information about session setup, data transfer, and session termination. Flow statistic-based
traffic classification is based on flow features such as packet length, flow duration,
number of packets, and so on and exploits these characteristics of the traffic generated
by different applications. In 5G systems, network function virtualization (NFV) is
adopted and makes the network flexible and scalable. It brings us to another research

challenge. Since the type of traffics can be changed in the NFV-based networks, the accuracy of classifiers may vary significantly. In [14], a NFV-based traffic-driven learning framework is proposed for the network traffic classification. The framework contains a controller, a set of machine learning classifiers, and feature collectors as virtual network functions. The framework is designed to make a balance between accuracy and speed. Since extracting different features varies from one another, we should take two important steps: feature collection identification on data plane or control plane and centralized view of network resources. The controller should maintain the offline trained machine learning model and select the most suitable classifier and flow features. The controller selects effective machine learning classifiers and adjusts cost-efficient flow features for a flow protocol such as TCP, UDP, and others. In [15], machine learning-based traffic classification procedure is proposed as shown in Fig. 10.3.

In the first step, traffic samples are collected, statistical features are extracted, and then samples are labelled with its corresponding class. In the second step, the statistical features are scaled. In the third step, machine learning model is trained and the test data is evaluated. In the final step, the performances such as accuracy and others are measured. There are other traffic classification produces using machine learning. Their approaches are similar to Fig. 10.3. In 6G systems, NFV and SDN will be widely used and provide us with more flexibility and efficiency of the network. In the dynamic and encrypted network environment, AI techniques will significantly contribute to improve the performance of the network traffic classification.

Optimal routing path by AI algorithms: Finding an optimal routing path and transmitting the data packet via the optimal path is a fundamental task in the network. The efficient routing requires suitable policies that can adapt to the dynamically

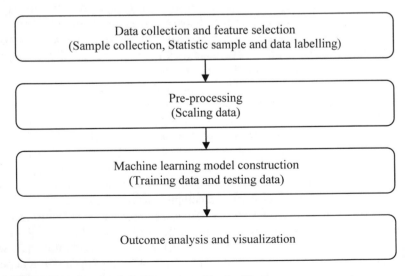

Fig. 10.3 Machine learning-based four-step traffic classification

changing networks In terms of traffic loading, traffic pattern, network topologies, and so on. They relies on the network policy and operational objectives in terms of cost, capacity, latency, QoS provision, and so on. The routing methods play an important role in exchanging the packets to avoid congestions and delays. AI techniques such as reinforcement learning are well matched with a pathfinding problem. As we discussed reinforcement learning in Chap. 5, in reinforcement learning, we face a game-like situation to make a sequence of decision. The agent learns to achieve a goal by taking suitable actions to maximize rewards in a complex situation. Taking suitable actions is regarded as finding the best possible path. More specifically, in the reinforcement learning, we can define system model consisting of a set of states S, a set of action per state $A(s_t)$, and the corresponding reward r_t. In this system model, a state s_t can be the status at time step t of all nodes and links when S is related to the network or can be the status of the node holding the packet at time step t when S is related to the packet to be routed. $A(s_t)$ can be all possible neighbouring nodes that the packet can be routed to the destination node. When taking an action or selecting a path, r_t can be defined in terms of multiple metrics such as available bandwidth, queuing delay, packet loss, energy consumption, and so on. In [16], Q learning algorithm is applied to a packet routing problem. They called this adaptive routing algorithm Q-routing. This distributed reinforcement learning algorithm improved packet routing in the network. In the Q-routing, each node decides its routing and stores Q values estimating the quality of the alternative routes. The values are updated at each time when transmitting a packet from one node to its neighbouring nodes. Consider that a routing decision is made at each node x using a table of values $Q_x(y, d)$ where each value for a neighbouring node y and destination node d is an estimate of how long it takes for a packet to reach d through y including any time the packet has to spend in node x's queue and transmission time over the link between x and y. We should decide the routing path as the value $Q_x(y, d)$ is minimum. After the packet is received, x gets back y's estimate for the remaining routing delay as follows:

$$Q_y(\hat{z}, d) = \min_{z \in N(y)} Q_y(z, d) \tag{10.1}$$

where $N(y)$ and z are a set of neighbouring of the node y and neighbours of the node y, respectively. If a packet spends q_x units of time in x's queue, we have the following update rule:

$$\Delta Q_x(y, d) = \alpha \left(Q_y(\hat{z}, d) + q_x - Q_x(y, d) \right) \tag{10.2}$$

where α is the learning rate, the term $Q_y(\hat{z}, d) + q_x$ is a new estimate, and $Q_x(y, d)$ is an old estimate. During forward exploration, the sending node x updates its $Q_x(y, d)$ value pertaining to the remaining path of the packet via the node y [17]. During backward exploration, the receiving node y updates its $Q_y(x, s)$ value pertaining the traversed path of the packet via the node x [17]. Using this update rule, we can obtain an optimal routing policy after convergence of the algorithm. The Q-routing outper forms when a network topology is changing dynamically under heavy loads. Based on Q-routing, there are further research challenges: (1) improving the performance by reducing convergence speed, (2) reducing the complexity, or (3) satisfying further network requirements and achieving global optimal points. In the current mechanism of the reinforcement learning for routing problem, we can apply in distributed way or centralized way. In distributed way, each routing node as an agent makes local deci-sion from the environment independently or collaboratively. It enables us to have a real-time traffic management. In centralized way, a controller operates the reinforce-ment learning and allows us to avoid an elephant flow or congestion. However, it may cause delayed routing. In addition, depending on the network requirements, the reinforcement learning evaluates different routing policies against a network utility function as a reward. The utility functions can be load balancing, QoS, throughput, latency, and so on. Under the given network constraints, we should find good trade-off between the benefits and sacrifices.

Queueing management by AI algorithms: In general, queue management is the process to minimize the user's waiting experience. In the network, a queue means collection of packets waiting to be transmitted by network components. The network operators want to smooth out bursty arrived packets and increase the network utiliza-tion. In the network routers and switches point of view, the conventional queue management is a trail drop that is used by network schedulers to decide when the packets is dropped. Based on first-in-first-out (FIFO), each queue has a maximum length to accept input packets. If the buffer is full, the subsequent input packets are dropped until it is available again. However, this approach causes many prob-lems such as inefficient network utilization, a long queueing delay if a buffer is full continuously, unfairness, and so on. Active queue management (AQM) is the policy of dropping packets at a buffer before the buffer is full. Thus, the main goal is to control average queuing delay, reduce network congestion and maintain high network utilization. In addition, queuing management targets to improve fairness of network resource, reduce unnecessary packet dropping, and accommodate tran-sient congestion. Random early detection (RED) is one of well-known active queue managements by randomly dropping or marking packet with increasing probability before the buffer is full. Random early detection is also known as random early discard or drop. It brings us the advantages such as lower average queuing delay, desynchro-nizing coexisting flows, and others. The main goal is to maintain average queue size by setting maximum and minimum queue size in times of normal congestion. In practice, the algorithm is very sensitive to set the parameters, and it is very difficult to find the suitable parameters. Many AQM methods have been developed to improve these problems. However, they depend on the fixed parameters that are not sensitive to varying network conditions. AI techniques can be applied for improving AQM methods. They are capable of managing the queue and parameters for the network traffics. In [18], predictive AQM is developed using predicting future traffic volume. Their approach depends on the normalized least mean square in order to compute the linear minimum mean square error. The LMMSE-based controller achieved good capacity to enhance the stability of the queue length. Using the predictions, it adjusts the packet dropping probability and provides us with high link utilization while maintaining low packet loss. In [19], AQM method based on a proportional–integral–derivative (PID) controller is developed. The PID controller uses an adaptive neuron to turn the parameters. It can adjust the queue length at a given target and show us a better performance than conventional AQM methods. To sum up, the conventional queue management schemes couldn't provide us with enough performance under the time-varying and nonlinear network environments. Thus, AI contributions to queue management give us a new opportunity for better queue length stabilization. In particular, the AI capability to predict the future value will be helpful for adjust the packet drop probability.

AI contributions to network fault management: In network management, fault management is the component of the network functions concerned with detection, isolation, and collection of malfunctions. The network operators should have enough knowledge of the entire network, observe the problems of the network, analyse the situation, resolve the problems, and then log the errors. Since the fault management increases the network reliability and availability, it is getting more important in order to manage complex networks and also faces more challenging problems due to network virtualization, dynamic, and heterogeneity. The basic process of fault management can be summarized as follows: (1) fault detection to determine whether or not network faults occur, (2) isolation of the root causes of the faults, (3) notification to administrators, and (4) collection of the problem. In addition to this, fault prediction can be added. This step can prevent future faults by predicting them and reacting rapidly to minimize the performance degradation. AI techniques will play an important role for fault management especially fault prediction. In [20], fault prediction using machine learning approaches is developed by continuously learning the difference between normal and abnormal network behaviours. The continuous learning enables the fault prediction and diagnostic measure to deal with continuously changing network without explicit control. In [21], fault detection for cellular systems is developed to find faults at different base station, sector, carrier, and channel. Fault detection covers fault classification using network symptoms, performance degradation, and others. They use a statistical hypothesis to test framework combining parametric and nonparametric test statistics and then model the behaviours. In parametric statistical tests, significant deviations from the expected are observed and a fault is detected. In nonparametric statistical tests, hypothesis test combining empirical data and statistical correlations is performed when the expected distribution is unknown. In [22], fault detection in a wireless sensor network is developed using recurrent neural network (RNN). The nodes of the hidden layers in the RNN model sensor nodes in the wireless sensor network. The weights are based on confidence factors of the received signal strength indicators. The outputs are approximation of the operation of the wireless sensor network. Fault detection is performed by discrepancies between approximation and real values of the wireless sensor network. The RNN can detect faults successfully for small-size wireless sensor network. In [23], decision tree is used to diagnose faults in large Internet sites. The decision trees are trained on the request traces from time periods in that user failures exist. Path through the decision tree are ranked in terms of their degree of correlation with failure and nodes are merged in terms of the observed partial order of components. In order to reduce convergence time, an early stopping criteria is used and also the learning algorithm follows the most suspicious path in the decision tree. When applying AI techniques to fault management, one important research challenge is lack of fault data in the network. As we discussed in Part I, the performance of AI techniques relies on the training data set. If the fault data set is not enough, the accuracy of AI techniques will be low.

QoS management by AI algorithms: The Quality of Service (QoS) management is directly related to user experience. The key metrics about QoS are throughput, roundtrip time, jitter, packet loss, and error rate. The measure of the network QoS is performed using the data from network management system. The quality of experience (QoE) is a metric of the overall user satisfaction. The QoE is different from the QoS. The QoE focuses on the actual user experience. The mean opinion score (MOS) is widely used for assessing the quality of media based on the user's view. In networks, the MOS is a ranking metric about the quality of voice or video. Typically, users can give the media a score of 1 (bad) to 5 (excellent). The scores are averaged. Depending on these metrics, the QoS level can be determined as success, degradation, or failure of the service. The QoS management functions should oversee the QoS parameters to deliver high service quality. They should be guaranteed at some level to run application under the limited network resources. We should determine boundaries and priorities for different data flows and assign the various types of the network data flows with different priority levels. For example, if we have time critical data transmission, the network operators need to identify applications and put them in higher priority on the network flows. One conventional way to guarantee the QoS of the network is to mark data packets for service types, create separate queues for the service, assign their priorities to them, and then reserve the network resource for critical applications or others. In [24], machine learning classifiers such as Naive Bayes, SVM, KNN, decision tree, random forest, and neural network are used to model QoS and QoE correlation. A video traffic flows including different delay, jitter, and packet loss are generated from a network emulator. The MOS are calculated. Cross-validation is performed to estimate the machine learning classifiers. The decision tree and random forest provide us with a slightly better performance than others in terms of a mean absolute error. In [25], deep learning is adopted for real-time quality assessments of video streaming. For general condition networks and extremely lossy networks, their assessment method is evaluated and the network condition impact is studied. In the study, delay, jitter, and throughput are treated as independent conditions, but these impairments are correlated. One important research approach in this area is to predict QoS or QoE. For multiple media types, different assessment tools using AI techniques have been developed. In addition, since there is a lack of quantitative measure on network QoS and QoE, many research groups are investigating to apply AI technique to measure QoS or QoE accurately and create a new metric or an efficient methodology for them.

Table 10.1   AI contributions to network problems and applications

Network problems and applications	Objectives	Data set	Traditional approach	AI techniques
Network traffic prediction	Estimate the traffic volume and control congestion	Flow or packet data with statistical information such as number of packets	ARIMA	Supervised learning, Deep learning,
Network traffic classification	Identify applications and services	Flow or packet data with labelled application classes	Port or payload-based method	Supervised learning, Deep learning
Optimal routing path	Find an optimal routing path	Log data	Graph theory	Reinforcement learning
Queue management	Control congestion	Flow or packet data with statistical information	AQM	Supervised learning, Deep learning
Network fault management	Security, network troubleshooting	Flow data, packet headers, telemetry data, log data	Threshold or rule-based method, signature-based method, Manual detection	Unsupervised learning, Supervised learning
QoS management	Guarantee the service quality	Flow data, packet data, log data	Manual management, Threshold or rule-based method	Supervised learning, Deep learning

Table 10.1 summarizes the AI contributions to some network problems and applications.

Summary 10.1. AI-Enabled Network Layer Design

1.   The main function of the traditional network layer is to establish the connection between different networks by forwarding network packets to network routers discovering the best path across the networks.

2.   The 5G protocol stacks consist of a user plane and a control plane. The control plane takes care of signalling or controlling data. In the user plane, the user data are exchanged. This separation enables mobile operators to scale their functions independently and deploy networks flexibly. The 6G protocol stacks in the network layer may be similar to the 5G protocol stacks.

3.   The process of AI-enabled networks implementation will be similar to conventional process of AI techniques implementation: Formulate the

network problem, collect data, train the AI model, and apply the trained model to the problem.

4. The key AI contributions to 6G network layer can be summarized as follows: AI contribution to network traffic prediction, network traffic classification by AI algorithms, optimal routing path by AI algorithms, queueing management by AI algorithms, AI contributions to network fault management, and QoS management by AI algorithms.

10.2   Cellular Systems and Networking

10.2.1   Evolution of Cellular Networks

A cellular network (also called a mobile network) is a cluster of land areas that are known as a cell. The cells as geographic areas include base stations with transceivers and are connected to voice networks such as public switched telephone network (PSTN) or data networks such as Internet. In the 1G and 2G of cellular networks, the connectivity through voice calls was the main goal and the networks are connected to a circuit switched network. However, the tremendous growth of Internet makes the cellular network support the data service. From 3G networks, the data transmission becomes the main goal. From 4G networks, the circuited switched networks are replaced by the IP-based packet switched networks. The main difference is that circuit switched network is connection oriented and the packet switched network is connectionless. Since packet switching is more affordable and efficient than circuit switching, packet switched networks become the data transmission network. In packet switched networks, the data is broken into packets for more efficient transmission and the packets should find their own data routes to the destination. Figure 10.4 illustrates comparison of circuit switched network and packet switched network.

Fig. 10.5 illustrates conceptual diagram of cellular network connecting to Internet and telephone network. This is the fundamental network architecture of cellular systems. The 1G cellular system provides mobile users with voice service only and has the simple network architecture.

As we can observe Fig. 10.5, geographic area is divided into cells with base stations. The mobile switching centre (MSC) manages base stations and serves as a gateway to the backbone networks such as PSTN, Integrated Services Digital Network (ISTN) and Internet. In general, the base station located at the centre of each cell is equipped with antenna, controller, and transceivers. Each base station defines a single cell. The controller reserves frequency bands, controls paging and handover among base stations, and manages the call process between the mobile devices and other network components. A number of mobile devices are connected to the base station and can move inside the cell or handover to the neighbouring cells. There are two types of channels: control channels (or control plane) and traffic channels

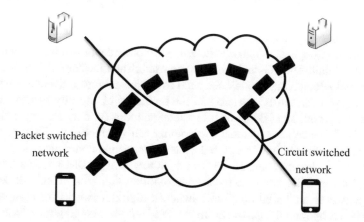

Fig. 10.4 Example of circuit switched network and packet switched network

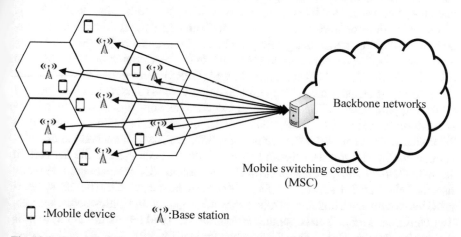

Fig. 10.5 Conceptual diagram of cellular network connecting to Internet and telephone network

(or user plane). Using control channels, connection between the mobile devices and the network is established and setting up the connection and maintaining call is managed. Traffic channel delivers the user data. The MSC manages calls between mobile devices and establishes the connection between a fixed users to the PSTN and a mobile user to the cellular networks. In addition, it assigns channels to cells, manage handovers, and monitor connections, traffics, and so on. The MSC contains a visitor location register (VLR) and a home location register (HLR). The HLR is the main database where the user information is stored. The VLR stores the information of the visitors from the HLR when they roam in a different place. The HLR manages authentication using an authentication centre (AuC) with encryption keys. When a call is initiated, the HLR checks a genuine mobile user with valid international mobile equipment identity (IMEI) and then the call is established between users.

As evolving cellular networks from 1G to 5G, new features and additional capabil
ities are continuously included in the network and the network structure are changed
In 1973, Martin Cooper and John F. Mitchell of Motorola demonstrated the first public
mobile phone call. The 1G analogue system established the foundation of cellular
networks and adopted key techniques such as frequency reuse, licenced spectrum
and so on. However, the 1G system had the limitation of capacity because FDMA
system is inefficient. The FDMA of 1G systems supports for only 1 user per channel
In 1991, the 2G of cellular systems was commercially launched in Finland. The 2G
systems can be divided into the global system for mobile communications (GSM)
using TDMA technology and IS-95 (or cdmaOne) using CDMA technology. The
main purpose of 2G systems is to provide mobile users with voice call services.
The 2G networks are based on circuit switched digital networks for compatibility
between PSTN and cellular networks. In 2G GSM radio access networks, base station
subsystem (BSS) is composed of base station controller (BSC) and base transceiver
subsystem (BTS). The BSC manages radio resource, handover, power management,
and signalling for BTSs. In 2G core network (CN), network subsystem (NSS) is
composed of the mobile switching centre (MSC) and database for mobile user information. In addition, it includes gateway mobile switching centre (GMSC) and interconnection of GSM network and PSTN/ISDN. In the next version of 2G GSM (also
called 2.5G), the General Packet Radio Service (GPRS) is developed to support data
connectivity. It includes the Serving GPRS Support Node (SGSN) and the Gateway
GPRS Support Node (GGSN) for packet switched network. Wideband Code Division Multiple Access (WCDMA) has a better utilization of radio resources than
2G systems. Based on WCDMA technology, NTT DoCoMO launched the first pre-
commercial 3G network in 1998 and deployed the first commercial 3G network in
Japan in October 2001. Based on cdma200 technology, SK Telecom commercially
launched the first 3G network in South Korea in January 2002. The 3G systems
provides us with much higher data rate, better voice quality, and multi-media services.
The Universal Mobile Telecommunications Service (UMTS) as the 3G system is
defined by the 3rd Generation Partnership Project (3GPP). The 3G radio access
network is called the Universal Terrestrial Radio Access Network (UTRAN). Based
on the 2G radio access network, new functionalities (such as rate adaptation, closed-
loop power control, congestion control, and so on) included at the eNodeB (This
is same as a base station.) and the radio network controller (RNC). In 3G CN, it is
extended from 2G network in order to provide mobile users with a high data rate
service and supports both circuit switched networks and packet switched networks.
In 2009, 4G systems were developed on a new network architecture. Voice services
and data services are no longer separated. All IP core networks of 4G systems support
both voice service and high-speed data services including multimedia service, mobile
TV, video conference, and so on. The Evolved-UTRAN (E-UTRAN) as 4G LTE radio
access networks and the Evolved Packet Core (EPC) as 4G LTE core networks are
defined by the 3GPP. In order to improve the system capacity and the radio resource
management, Orthogonal Frequency Division Multiplexing Access (OFDMA) technique is used for a radio interface. In 4G networks, single-core network can handle
both voice and data traffic and manage heterogeneous traffic with different QoS

features, In 4G radio access networks, the E-UTRAN includes only a set of eNodeB entities and splits control plane and user plane. In 4G core networks, the EPC includes the Serving Gateway (SGW) and the Packet Data Network Gateway (PGW). The main functions of the SGW are to route and forward user data packets and also act as a local mobility anchor when we need a handover. The main functions of the PGW are to allocate IP addresses to the user equipment (UE) during default bearer setup and also support policy enforcement features, packet filtering, and evolved charging. The Mobility Management Entity (MME) in the control plane takes in charge of security procedure, signalling procedure, and location management using Home Subscriber Server (HSS). The 4G network architecture becomes easier management and scalability. The main goal of the cellular networks from 1G to 4G is to offer a stable voice service and a fast and reliable data service to mobile users. However, 5G has expanded this scope to provide mobile users with wider range of services. The 5G systems can be classified into three main communication systems: (1) enhanced mobile broadband communication (eMBB), (2) ultra-reliable and low latency communication (URLL), and (3) massive machine-type communication (mMTC). The 5G networks architecture is more intelligent and includes flexible and virtual concepts. As we briefly discussed in Chap. 7, the virtualization concept makes the network more flexible and scalable. It allows us to create new services and applications. In particular, open RAN enables us to have virtualized network components and deploy multiple vendors equipment with off-the-shelf hardware due to easier interoperability. The mobile operators can react rapidly as user requirements are changed and capacity grows. In addition, network slicing can be realized by enabling multiple virtual networks to run simultaneously in a shared network infrastructure. The mobile operators can more efficiently manage 5G applications with different requirements and demands by creating end-to-end virtual networks. The 5G core network is based on a cloud-aligned service-based architecture (SBA) and control plane and user plane separation (CUPS) to support 5G functions and interactions such as authentication, session management, security, and aggregation of traffics. As 5G systems includes new features such as millimetre wave, massive MIMO, network slicing, and so on, the 5G core network is different from the 4G EPC. In 4G EPC, SGW and PGW are decomposed to SGW/PGW-C and SGW/PGW-U to provide mobile users with an efficient scaling of services independently. The decomposition continues in 5G networks. The user plane function (UPF) manages user traffic processing and the control plane function (CPF) and session management function SMF deals with all other signal processing. Figure 10.6 illustrates evolution of cellular networks.

As we reviewed 5G networks and evolution of cellular networks in this section and the previous section, the cellular network technology evolves to meet the increasing demands and requirements. 6G networks will face new challenges and opportunities. First of all, the network services will minimize uncertainty and provide mobile users with high reliability. Secondly, in order to build various ecosystems with vertical industries and support customized services, openness and customization should be adopted in 6G networks. Thirdly, in order to meet high requirements of 6G systems, AI techniques should be adopted to 6G networks. 6G networks should facilitate the network intelligence and automation. Fourthly, in order to achieve 100% coverage,

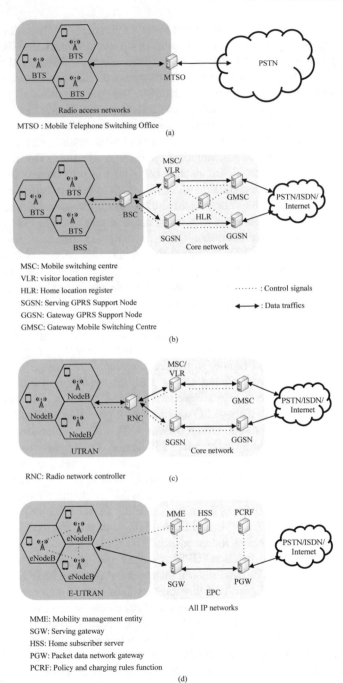

MTSO : Mobile Telephone Switching Office

(a)

MSC: Mobile switching centre
VLR: visitor location register
HLR: Home location register   ······· : Control signals
SGSN: Serving GPRS Support Node   ◄──► : Data traffics
GGSN: Gateway GPRS Support Node
GMSC: Gateway Mobile Switching Centre

(b)

RNC: Radio network controller   (c)

MME: Mobility management entity
SGW: Serving gateway
HSS: Home subscriber server
PGW: Packet data network gateway
PCRF: Policy and charging rules function

(d)

Fig. 10.6   Evolution of cellular networks: 1G (**a**), 2G and 2.5G (**b**), 3G (**c**), and 4G (**d**)

network integration with satellites is expected. It will be helpful for 6G network services at a high mountain, arctic area, oceans, and so on. Fifthly, more efficient spectrum usage is expected. The radio spectrum is very valuable and expensive resource. Using dynamic spectrum, AI, and relevant techniques, we should manage 5G spectrum more efficiently, flexibly, and intelligently. Sixthly, network security is one of key techniques in many 6G applications and services. In particular, cellular networks will support autonomous vehicles. If a high level of security couldn't be guaranteed, we can't achieve the level 5 of autonomous vehicles. In 6G networks, new security techniques such as post quantum cryptography and quantum key distribution may be adopted to ensure the network security. Lastly, the Sustainable Development Goals (SDGs) were adopted by all United Nations Member States in 2015. SDG 13 is about climate action "Take urgent action to combat climate change and its impacts". Cellular network infrastructure spends high energy and carbon as increasing network throughputs. 6G networks should be deployed to achieve the goal of a low-carbon and energy-saving.

Summary 10.2. Evolution of Cellular Networks

1. A cellular network is a cluster of land areas that are known as a cell.
2. From 4G networks, the circuited switched networks are replaced by the IP based packet switched networks. The main difference is that circuit switched network is connection oriented and the packet switched network is connectionless.
3. The main goal of the cellular networks from 1G to 4G is to offer a stable voice service and a fast and reliable data service to mobile users. However, 5G has expanded this scope to provide mobile users with wider range of services: (1) Enhanced Mobile broadband communication (eMBB), (2) Ultra-reliable and low latency communication (URLL), and (3) Massive machine type communication (mMTC).
4. In order to meet the increasing demands and requirements. 6G networks will face new challenges and opportunities: high reliability, openness and customization, AI techniques adaptation, better coverage, more efficient spectrum usage, a high level of security, and a higher energy efficiency.

10.2.2 Concept of Cellular Systems

A traditional mobile telephone system was similar to radio broadcasting that has one powerful transceiver at the highest place and serve mobile users in a large area. This approach enables us to involve small numbers of channels. The system capacity is very low. Typically, it supports about 25 channels with a radius of 80 km. On the other hand, a cellular system has many lower-power transceivers to cover a large

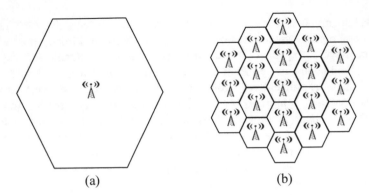

Fig. 10.7 Traditional mobile telephone system (**a**) and cellular system (**b**)

area and increase the system capacity significantly. It was a breakthrough to solve the problem of spectral shortage and system capacity. Figure 10.7 illustrates comparison of traditional mobile telephone system and cellular system.

A cellular system which has hexagonal cells covering a whole area without overlaps was introduced in the paper "The Cellular Concept" by V. H. MacDonald [26]. This paper produced a landmark cellular concept and overcame many wireless communication system problems such as power consumption, coverage, user capacity, spectral efficiency, interference, and so on. The hexagon geometry enables us to have maximum coverage. Hexagonal cell shape is now universally used in cellular networks. Namely, the smallest number of cells can cover a geographic area using hexagon geometry. For example, when we have three different geometry with same radius R as shown in Fig. 10.8, the area of hexagon provides us with the maximum coverage. The distance between the cell centre and each adjacent cell is $\sqrt{3}R$. The coverage distance is $r = \frac{\sqrt{3}}{2}R$. In practice, the cell coverage is known as the footprint. The actual cell coverage is decided from propagation prediction model by simulation or field test measurements. Typically, base stations are deployed at

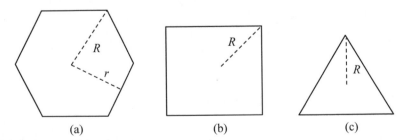

Fig. 10.8 Example of different geometry to cover a geographic area: hexagon (**a**), square (**b**), and triangle (**c**)

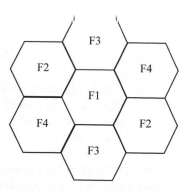

the centre or cell vertices. Although cells are geographically separated, the signals can interfere with each other. The same frequency use in a cellular system should be limited, and a new method should be considered to support a large number of channels.

The frequency reuse is a key concept in the cellular network and enables us to achieve a high capacity. The coverage of the cellular system is divided into hexagonal cells which are assigned different frequencies (F1–F4) as shown in Fig. 10.9. Typically, 10 to 50 frequency bands are assigned to cells depending on the mobile traffic.

As we can observe Fig. 10.9, each cell does not have adjacent neighbouring cells with same frequency. The transmission power of a base station at each cell should be adjusted very carefully. Thus, we can expect that intercell interferences are reduced, cell capacity is increased, and cell coverage is extended. The reuse distance should be enough so that the inferences by a mobile user using same or adjacent frequency in neighbouring cells are low sufficiently. In each cell, it is necessary to have a multiple access scheme that enables users to access a base station. The cells can be grouped. A cell cluster is the group of cells where all available frequencies are totally consumed. It is important to minimize the interference among cells using the same frequencies. Consider a cellular system with the following parameters: D = minimum distance between centres of cells that use the same band of frequencies, R = cell radius, d = distance between centres of adjacent cells, and N = number of cells in a repetitious pattern. In a hexagon cell, N can have the following values:

$$N = i^2 + j^2 + ij \tag{10.3}$$

where i and j are non-negative integers. The values of N can be 1, 3, 4, 7, 9, 12, 13, and so on. The common clustering sizes are 4, 7, and 12. The signal-to-co-channel interference ratio can be expressed as a function of R and D. The spatial separation among co-channel cells is related to coverage distance as follows:

$$D = 2r\sqrt{i^2 + j^2 + ij} = 2r\sqrt{N} = R\sqrt{3N} \tag{10.4}$$

and the co-channel reuse ratio Q is

$$Q = \frac{D}{R} = \sqrt{3N}. \tag{10.5}$$

A large value of Q improves the transmission quality because co-channel interference is small, whereas a small value of Q means a small value of N and larger capacity because the cluster size is small. Thus, a trade-off should be found between them. We have the relationship $D/d = \sqrt{N}$ as well. Figure 10.10 illustrates example of frequency reuse patterns.

In order to increase cell capacity, there are multiple approaches as follows: (1) new channel allocation. When a cellular system is deployed, all channels are not used. Depending on the requirements, additional channels can be added. (2) Dynamic frequency allocation. Depending on the number of users or traffics in cells, additional frequency can be assigned to busy cells. (3) Cell splitting and smaller cell. The distribution of base stations and traffics is not uniform. Depending on the requirements, the cells can be split into smaller cells. As the cell size is getting smaller, more base stations and more frequent handover are required. This will be additional cost. The smaller cells will be more useful for urban areas requiring the high traffics. The transmit power of the split cell must be smaller than the original power. For example, when the split cell size is half of the original cell and the path loss exponent is $n = 4$, the received powers of the split cell and the original cell can be calculated as P_r(orignal cell boundary) $\propto P_t$(orignal cell)R^{-4} and P_r(split cell boundary) $\propto P_t$(split cell)$(R/2)^{-4}$, respectively. Since the interference level shouldn't be changed after splitting and their received powers should be same like P_r(orignal cell boundary) $= P_r$(split cell boundary), we have P_t(split cell) $= P_t$(orignal cell)$/16$. In addition to this, interference coordination and management techniques such as intercell interference coordination (ICIC) and coordinated multipoint transmission (CoMP) enable us to improve the system capacity. In order to use spectrum efficiently and minimize interference, channel assignment techniques are widely used in cellular systems to assign time frequency resources to users. The channel assignment strategies can be classified into fixed channel assignment, dynamic channel assignment, and hybrid channel assignment. The choice of the channel assignment technique affects the cellular system performance. The fixed channel assignment assigns each cell to a predetermined set of channels. The users can only be served by unused channels. If all channels in the cell are occupied, further call is blocked and additional users can't be served. The blocking probability would be high in the strategy. Thus, in order to overcome this, a borrowing strategy can be adopted. In this approach, a cell can borrow channels from neighbouring cells when all channels in the cell is occupied. The mobile switching centre (MSC) manages the procedure so that the borrowing channels do not degrade the neighbouring cells and the interferences do not occur. In a dynamic channel assignment, there are no permanent channel assignments. The base stations that need a channel request channels from the MSC. The MSC takes into account the blocking probability within the cell, the frequency reuse, the reuse distance of channels, and operational

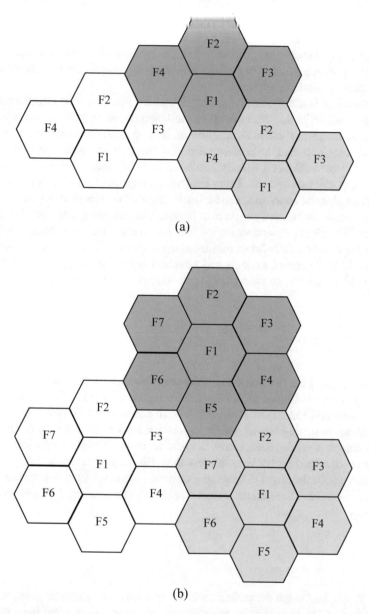

Fig. 10.10 Example of frequency reuse pattern in a cellular system: $N = 4$ (**a**) and $N = 7$ (**b**)

cost and then assigns the channels to the cells. The MSC should monitor the real-time data about channel occupancy, traffics and the radio signal strength indications (RSSI) of all channels. Thus, dynamic channel assignment is more complex than fixed channel assignment, but it can decrease the probability of blocking. The hybrid channel assignment is a mixed form of the fixed channel assignment and dynamic channel assignment. The total number of channels is divided into the fixed set and the dynamic set. When a user needs a channel, the fixed set of channels is assigned. However, if all fixed channels are occupied, the dynamic set of channels can be used. Thus, this approach can achieve a high efficiency by assigning co-channels to cells that are closely located each other.

Interferences is a major impairment of cellular systems. The interferences are caused by the mobile devices in the same cell, neighbouring base stations operating in the same frequency, other radio systems, and so on. The interference results in cross talk, degradation of the performance, blocking call, and so on. The interference is getting worse as the cell density increases in 6G systems and becomes a major bottleneck in cell deployment. There are two main types of interferences in cellular systems: co-channel interference and adjacent channel interference. As we discussed, frequency reuse means that there are cells with the same frequency band in a given coverage. We call this co-channel cells. The interference between them is called co-channel interference. In order to minimize the co-channel interferences, co-channel cells should be separated enough and maintain sufficient isolation between them. The signal-to-interference ratio (SIR) for a desired mobile device can be defined as follows:

$$\text{SIR} = \frac{S}{I} = \frac{S}{\sum_{i=1}^{i_0} I_i} \tag{10.6}$$

where S, I_i, and i_0 are the desired signal power from a base station, the interference power by the base station at the ith interfering co-channel cell, and the number of interfering co-channel cells, respectively. In a hexagonal shaped cellular system model, there are six co-channel interfering cells in the first tier and $i_0 = 6$. Most of the co-channel interferences comes from the first tier and the amount of the interferences from the second and higher tiers are less than 1%. In particular, the interference in small cells will be the major channel impairment. Assume all interfering base stations are equal distance. The average received power P_r at a distance d can be expressed as follows:

$$P_r = P_0 \left(\frac{d}{d_0} \right)^{-n} \tag{10.7}$$

where P_0, d_0, and n are the power received at a close-in reference point in the far field region of the base station, the distance from the transmitting base station to the reference point, and the pass loss exponent (Typically, 2–4 at an urban area), respectively. Figure 10.11 illustrates propagation measurements in a mobile radio channel.

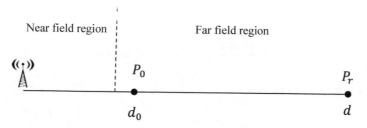

Fig. 10.11 Propagation measurement in a mobile radio channel

When different base stations transmit the same level of power, the SIR can be expressed as follows:

$$\text{SIR} = \frac{S}{I} = \frac{R^{-n}}{\sum_{i=1}^{i_0}(D_i)^{-n}} \tag{10.8}$$

where D_i is the distance between the mobile device and the ith interferer. When considering the first tier of interfering cells and the equal distances between cell centre, $D_i = D$, (10.8) can be rewritten as follows:

$$\text{SIR} = \frac{R^{-n}}{\sum_{i=1}^{i_0}(D)^{-n}} = \frac{(D/R)^n}{i_0} = \frac{\left(\sqrt{3N}\right)^n}{i_0} = \frac{Q^n}{i_0}. \tag{10.9}$$

In a hexagonal-shaped cellular system model, (10.9) can be written as follows:

$$\text{SIR} = \frac{Q^n}{6}. \tag{10.10}$$

Thus, we have

$$Q = (6 \cdot \text{SIR})^{1/n} \tag{10.11}$$

and

$$N = \frac{(6 \cdot \text{SIR})^{2/n}}{3}. \tag{10.12}$$

Example 10.1 Frequency Reuse of Cellular Systems I Consider a cellular system in that $\text{SIR} = 12$ or 18 dB and the pass loss exponent $n = 4$ are required to maintain the voice quality. What should be the frequency reuse factor for the cellular systems?

Solution

From (10.12), we can calculate the frequency reuse factors as follows:
When $\text{SIR} = 12$ and $n = 4$, we have

$$N = \frac{\left(6 \times 10^{12/10}\right)^{2/4}}{3} = 3.2505 \approx 4.$$

When SIR $= 18$ and $n = 4$, we have

$$N = \frac{\left(6 \times 10^{18/10}\right)^{2/4}}{3} = 6.4857 \approx 7.$$

The minimum cluster size N should be at least 7 and 4 in order to achieve SIR $= 18$ dB (AMPS requirement) and 12 dB (GSM requirement), respectively.

∎

Example 10.2 Frequency Reuse of Cellular Systems II Consider a cellular system that can tolerate a SIR $= 15$ dB in the worst case. Find the optimal value of the cluster size N for omnidirectional antennas and $120°$ sectoring when we have a path loss exponent $n = 4$.

Solution

We have the worst case SIR $= 15$ dB as follows:

$$15\,dB = 10\log SIR$$

$$SIR = 31.62.$$

For omnidirectional antennas, $i_0 = 6$. From (10.9) we have

$$31.62 = \frac{\left(\sqrt{3N}\right)^4}{6},$$

$$N = 4.5913 \approx 7.$$

For $120°$ sectoring, $i_0 = 2$. From (10.9) we have

$$31.62 = \frac{\left(\sqrt{3N}\right)^4}{2},$$

$$N = 2.6508 \approx 3.$$

∎

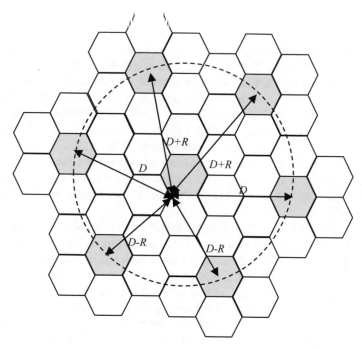

Fig. 10.12 Co-channel cells in the first tier

In worst case, the mobile device is located at the cell boundary. The distance from the two nearest co-channel interferers is *D-R*. The other distances from other interferers are D and $D + R$ as shown in Fig. 10.12.

In Fig. 10.12, the co-channel cells are expressed as shaded colour. When $n = 4$, the SIR can be approximated as follows:

$$\text{SIR} = \frac{R^{-4}}{2(D - R)^{-4} + 2(D + R)^{-4} + 2D^{-4}} \quad (10.13)$$

and (10.13) can be rewritten as follows:

$$\text{SIR} = \frac{1}{2(Q - 1)^{-4} + 2(Q + 1)^{-4} + 2Q^{-4}}. \quad (10.14)$$

Adjacent channel interference is caused by an imperfect receiver filter that allows nearby frequencies to leak into passband and nonlinearity of amplifiers. It results in performance degradation. Adjacent channel interference can be minimized by using modulation techniques with low out-of-band radiation, designing receiver filter carefully, and assigning adjacent channels such that frequency separation between channels is maximized. It is referred to as the near-far problem (or the near-far effect). The near-far problem is the resultant multiple user communications in the cellular system that uses the same transmit power level, the same frequency bands and the same time slots. It happens when a mobile device nearby a base station transmit on a channel nearby one being used by a weak mobile device. The base station may be difficult to distinguish the desired mobile device from the close adjacent channel mobile devices. This is common situation in cellular systems. TDMA and OFDMA are less sensitive to the near-far effect because a mobile user has its own channel assigned by a base station. However, it is a major problem in CDMA systems because CDMA mobile user share the same bandwidth and time slots. It can be solved by dynamic output power adjustment or power control. In cellular systems, a mobile user must transmit the smallest power necessary to maintain a good link quality and avoid an interference. The received power must be sufficiently higher than the noise level. The power control is helpful for avoiding interferences as well as improving the battery life.

Cellular systems rely on a trunking technique to accommodate a large number of mobile users in a limited radio resource. The main purpose of the trunking technique is to provide network access to multiple users simultaneously by sharing the radio and network resources. Typically, users are allocated at a channel per a call. If the call is terminated, the channel is immediately returned to the available channel list. The number of trunks with connections among mobile switching centres represents the number of connection pairs. In cellular systems, it is important to determine the number of trunks required on connections among the mobile switch centres and use them efficiently. The mobile operators use trunking techniques to find the number of channels to be allocated. The fundamentals of trunking theory were developed by Danish mathematician, A. K. Erlang [27]. He studied how a large population can be accommodated in a limited resource and proposed statistical strategies of the arrival and the length of calls. His works contain derivation of the Poisson distribution for the numbers of the calls in a fixed time interval. Based on the trunking theory, the users in cellular systems can be blocked or denied access if all channels are occupied. A queue is used to hold the service for users until channel is available. The grade of service (GOS) should be managed efficiently. The main task of wireless system designer is to estimate the maximum required capacity and allocate channels to meet the GOS. The GOS represents a measure of how a user access a trunked cellular system at the busiest hour during a week, month, or year. It is typically about the likelihood that a call is blocked or longer delayed than a certain time. The traffic can be measured in percentage of occupancy, centrum call seconds (CCS), Erlang, and so on. The Erlang is widely used because wireless system designers need to know the required capacity in a network in order to provision correctly. The Erlang represents traffic intensity carried by a channel that is completely occupied. It is a dimensionless unit about the continuous use of one connection. One Erlang can be defined as one call-hour per hour or one call-minute per minute. Thus, if a channel is occupied for 10 min per an hour, the channel carries 0.1667 (=1/6) Erlangs of traffic. In an ideal case, available channels are equal to the requested users at a given time. However, it is practically not possible to have this situation at all time slots. When we have N simultaneous channels and L users, $L > N$ and $L < N$ mean blocking calls and non-blocking calls, respectively. Thus, key questions in the network traffic management can be ummarized as follows: How much is the blocking probability? How much capacity is required to achieve an upper bound on the blocking probability? How much is the average delay? How much capacity is required to achieve the average delay? A network traffic engineer should consider them and design a network deployment. In order to design a trunked cellular system, it is essential to understand queuing theory. The common terms in queuing theory are defined as follows [28]:

- Set-up Time: The time required to allocate a trunked radio channel to a user.
- Blocked Call: A call that can't be completed at the time of request due to congestion.
- Holding Time: Average duration of a typical call. Denoted by H (in seconds).
- Traffic Intensity: Measure of channel time utilization, that is the average channel occupancy measured in Erlangs. This is dimensionless quantity and may be used to measure the time utilization of single or multiple channels. Denoted by A.
- Load: Traffic intensity across the entire trunked radio system, measured in Erlangs.
- Grade of Service (GOS): A measure of congestion that is specified as the probability of a call being blocked (for Erlang B), or the probability of a call being delayed beyond a certain amount of time (for Erlang C).
- Request Rate: The average number of call requests per unit time. Denoted by λ second^{-1}.

Now, we compute GOS. Each user generates a traffic intensity of A_u as follows:

$$A_u = \lambda H \text{(Erlangs)}. \qquad (10.15)$$

For a cellular system entering U users, the total offered traffic intensity A is

$$A = U A_u \text{(Erlangs)}. \qquad (10.16)$$

If there are C channels in the cellular system and the traffic is equally distributed, the traffic intensity per channel is

$$A_c = U A_u / C \text{(Erlangs)}. \qquad (10.17)$$

There are two types of trunked cellular systems: a blocked calls cleared (BCC) system and a blocked calls delayed (BCD) system. In a BCC system, if there are no available channels, the requesting users are blocked without access and try again later. It is a memoryless system. The BCC system assumes that there is no setup time and the users can get an immediate access to channels if the channel is available. In addition, the BBC system is based on the following assumptions: Traffic (or call) request is described by a Poisson distribution. There are an infinite number of users. Both new and blocked users may request a channel at any time. Service time of a user is exponentially distributed, and there are a finite number of available channels. The BBC system uses M/M/m queuing model. In queuing model A/B/m, m is the

number of servers and A and B can be selected as M (Markov, exponential distribution), D (Deterministic), and G (General, arbitrary distribution). For example, M/M/
queueing system represents that interarrival times and service times are exponentially
distributed, there is only one server, the buffer size is infinite, and the queuing is based
on first-come-first-serve (FCFS). Sometimes, the trunking systems is modelled as
M/M/m/m queuing system. The first M represents a memoryless Poisson process for
traffic requests, the second M represents an exponentially distributed service time
the third m represents the number of available channels, and the last m represents a
hard limit on the number of simultaneous serving users. In order to estimate GOS,
we use the Erlang B formula that determines the probability that a call is blocked.
The Erlang B formula is expressed as follows [28]:

$$\text{GOS} = P_\text{B} = \frac{\frac{A^C}{C!}}{\sum_{k=0}^{C} \frac{A^k}{k!}} \tag{10.18}$$

where C, A, and k are the total number of the trunked channels, the total offered
traffic, and kth channel, respectively. P_B is the call blocking probability. The call
blocking probability is same as the probability that none of C channels are free. This
formula gives us a conservative estimate of GOS. The finite users predict smaller
likelihood of call blocking. In a BCD system, a queue is provided to hold calls that are
blocked. If there are no immediately available channels, the call request is delayed
until there are available channels. This system is modelled as M/M/m/d queueing
systems where the third m represents the maximum number of simultaneous serving
users and d represents the maximum number of calls that are held in the queue. In
this system, GOS is defined as the probability that a call is blocked after the system
waits a certain time in the queue. In order to estimate GOS of the BCD system, we
use the Erlang C formula determining the likelihood that a call access is initially
denied. The Erlang C formula is expressed as follows [28]:

$$\text{GOS} = P_\text{D} = \frac{A^C}{A^C + C!\left(1 - \frac{A}{C}\right) \sum_{k=0}^{C-1} \frac{A^k}{k!}}. \tag{10.19}$$

$P_D = P[\text{delay} > 0]$ represents the likelihood of a call not having immediate access
to a channel. If there are now available channels immediately to access, a call is
delayed. The delayed call should wait more than t seconds. The GOS of a trunked
cellular system where the blocked calls are delayed can be expressed as follows [28]:

$$P[\text{delay} > t] = P[\text{delay} > 0]P[\text{delay} > t|\text{delay} > 0]$$
$$= P[\text{delay} > 0]\exp\left(-\frac{(C - A)t}{H}\right). \tag{10.20}$$

The average delay D for all calls is

$$D = P_D \frac{H}{C - A} \tag{10.21}$$

where the term $\frac{H}{C-A}$ represents the averaged delay for calls that are queued.

Example 10.3 Blocked Calls Cleared System Consider a FDD cellular system with the total bandwidth 24 MHz. It uses each 30 kHz bandwidth for uplink and downlink. Each mobile user generates 0.1 Erlangs. The cellular system has 4 cells and is based on Erlang B formula.

(a) Find the number of channels at each cell.
(b) If each cell has capacity with 90% utilization, find the maximum number of users that can be supported per cell when each base station is equipped with omnidirectional antennas.
(c) Compute the blocking probability when the maximum number of users are available.

Solution

First of all, we can compute the number of channel at each cell as follows:

$$\text{Number of channels at each cell} = \frac{24\,\text{MHz}}{4 \cdot 60\,\text{KHz}} = 100\text{channels/cell}.$$

Secondly, we have 100 channels per cell and each cell has 90% utilization. It carries 90 Erlangs ($= 100 \cdot 90\%$) of traffic. Thus, the number of user is

$$\frac{A}{A_u} = \frac{90}{0.1} = 900.$$

Thirdly, we can have GOS for 100 channels and 90 Erlangs using (10.18) or Erlangs B table [28, 29] as follows:

$$\text{GOS} = 0.03.$$

∎

Example 10.4 Blocked Calls Delayed System Consider a cellular system with a hexagonal shape, 1.387 km radius, 4 cell frequency reuse factor, and 60 channels in total. Each user generates a traffic intensity of $A_u = 0.029$ Erlangs. The average number of call requests per unit time is $\lambda = 1$ call/hour. The Erlang C system has a 5% probability of a delayed call.

(a)   How many users per km^2 are supported?
(b)   Compute the probability that a delayed call should wait for more than 10 s.
(c)   Compute the probability that a call is delayed for more than 10 s.

Solution
   The total number of channels at each cell can be calculated as follows:

$$\frac{60}{4} = 15 \text{ channels.}$$

   The GOS of the Erlang C system is given as 0.05. Using (10.19) or Erlangs C table [28, 29], we have
   The total offered traffic intensity $A = 9$ Erlangs.
   The total number of users is computed as follows:

$$\frac{A}{A_u} = \frac{9}{0.029} = 310 \text{ users.}$$

Thus, the number of users per Km^2 is computed as follows:

$$\frac{310}{\frac{3\sqrt{3}}{2}(1.387)^2} = \frac{310}{5} = 62 \text{ users/km}^2.$$

From (10.15), we have

$$H = \frac{A_u}{\lambda} = \frac{0.029}{1/3600} = 104.4 \text{ s.}$$

From (10.20), the probability that a delayed call should wait for more than 10 s can be computed as follows:

$$P[\text{delay} > 10\,\text{s}|\text{delay} > 0] = \exp\left(-\frac{(15-9)10}{104.4}\right) = 56.3\%.$$

Thus, we can compute the probability that a call is delayed for more than 10 s as follows:

$$P[\text{delay} > 10\,\text{s}] = P[\text{delay} > 0]P[\text{delay} > 10\,\text{s}|\text{ delay} > 0]$$
$$= 0.05 \cdot 0.563 = 2.8\%.$$

∎

   The advantages of cellular systems can be summarized as follows: (1) higher capacity due to frequency reuse, (2) lower transmission power due to small coverage area, (3) lower uncertainty about radio channel due to small coverage area, and (4) more stable network architecture. If one base station is malfunction, it affects the only one cell. On the other hands, the disadvantages of cellular systems are (1) more infrastructure and higher CapEx due to more base station deployment, (2) interferences among cells, (3) handover procedure, and (4) more complex network management.

Summary 10.3. Concept of Cellular Systems

1. The cellular concept was a breakthrough to solve the problem of spectral shortage and system capacity.
2. The frequency reuse is a key concept in the cellular network and enables us to achieve a high capacity.
3. A cell cluster is the group of cells where all available frequencies are totally consumed. The common clustering sizes are 4, 7 and 12.
4. In order to increase cell capacity, there are multiple approaches as follows: (1) New channel allocation. (2) Dynamic frequency allocation. (3) Cell splitting and smaller cell.
5. The interference is getting worse as the cell density increases in 6G systems and becomes a major bottleneck in cell deployment.
6. Cellular systems rely on a trunking technique to accommodate a large number of mobile users in a limited radio resource. It is important to determine the number of trunks required on connections among the mobile switch centres and use them efficiently.
7. There are two types of trunked cellular systems: a blocked calls cleared (BCC) system and a blocked calls delayed (BCD) system.
8. In a BCC system, if there are no available channels, the requesting users are blocked without access and try again later. It is a memoryless system. The BCC system assumes that there is no setup time and the users can get an immediate access to channels if the channel is available.

10.2.3   Cell Planning

The main purpose of cell planning is to minimize the infrastructure costs and operational costs while maintaining good performance of the cellular system. The ultimate goal is to establish an optimal cost effective network. Cell planning was investigated for all generations of cellular systems. The key design parameters of cell planning can be summarized as follows: (1) minimizing the overall network deployment and operation costs, (2) maximizing the overall system capacity, (3) maximizing the overall coverage while satisfying the limit of the interferences, (4) minimizing the energy consumption, and (5) minimizing the control signalling. As the cellular network becomes complex, more control signalling is required. In particular, as the cell size is getting smaller, frequent handovers are required. It causes wasting energy, radio and network resources, and costs. In order to satisfying the design parameters, we should find the optimal location of base stations and the network configuration. The process of cell planning is composed of three steps: initial planning (or dimensioning), comprehensive planning, and network optimization [30]. The process of cell planning is iteratively performed. In the phase of the initial planning, network designers find an initial approximation of the number and placement of base stations to cover service areas and meet the requirements. As an input of the first phase, they collects the required service coverage, capacity and quality and considers population, geographical area, frequency, propagation model, and so on. Based on these data, they compute the initial planning including link budget analysis, traffic analysis, coverage estimation, and capacity estimation. As an output of the first phase, they determine how many base stations are needed, where they deploy, and how their parameters (antenna type, transmit power, height, frequency allocation, and so on) should be configured. In the comprehensive planning, network designers determine the actual location of the base stations within the service areas and obtain further detailed information such as coverage calculation, capacity planning, base station configuration, and so on. The results of the initial planning phase become the input of the comprehensive planning. They create databases including geographical and statistical information and decide base station location and configuration in detail. In the phase of the network optimization, the base stations are deployed and up and running. The output of the comprehensive planning phase becomes the input of the network optimization phase. The performances are measured by a field test. The settings of the base stations are adjusted to achieve a better performance. Typically, the process of cell planning is very complex and the network parameters are a trade-off relationship. For example, maximizing the cell coverages means increasing the transmission power. This is a conflict with minimizing the energy or the interference. It is difficult to find an optimal solution to satisfy the multiple design parameters. One approach is to use a linear combination of different design parameters with different weight factors. Depending on the importance, each weight factor can be different values from 0 to 1. Another approach is to formulate the problem with multiple decision variables and different weights. The weighted multi-objective functions enable us to have more flexibility. If one solution of one objective does not degraded the other objective, it is regarded as an optimal solution. However, it is difficult to solve the multi-objective problem. Typically, a cost-effective network can be found by an iterative process of cell planning. Figure 10.13 illustrates the process of cell planning [30].

The new features such as heterogeneous network, user-centric network deployments, ultra-dense networks, mmWAVE, massive MIMO, multi-connectivity, integration with satellite communications, O-RAN, and others will be incorporated in 6G cellular networks. These features are new challenges to network designers. In particular, O-RAN and cloud-RAN decompose the traditional base stations functions. For example, the base band functions are located in a central node. The low complexity remote radio heads (RRHs) with simple radio transceiver are distributed in a cell. In addition, drones with simple radio transceiver can act like RRHs. Thus,

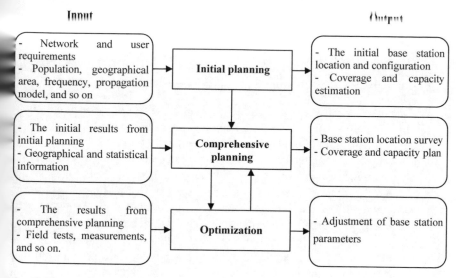

Fig. 10.13 Cell planning process

6G RAN can be flexible, scalable, and adaptive and helpful for reducing deployment costs. However, heavy traffics between the central node and RRHs are required and fronthaul bottleneck could be a new target to be optimized. Another challenge is user-centric approach. One user can be connected by multiple base stations or RRHs. Radio transceivers can be a virtual form. Cell planning should focus on user-centric traffic management. Heterogeneous networks and integration with satellite communications will be another big challenge. The different types of traffics and services makes cell planning more complicated. Seamless coexisting satellite communication and cellular network results in many research problems such as interference, traffic management and latency control. In particular, co-channel interference between satellites communications and cellular networks is one of key challenges because millimetre wave bands are now used in satellite and also this bands have been adopted in cellular small cell networks. The 24–29 GHz bands are allocated in 5G millimetre wave bands. The Ka-band (26.5–40 GHz) is a part of satellite communications. They are overlapping. One of key applications in 6G systems is IoT service. It will require massive data connection and different quality of services. For example, industrial IoT needs massive connectivity as well as ultra-reliable connection. Cell planning should take into account their requirements and coverages. As 6G RAN deals with control traffics and user data traffics separately, cell planning should consider the different characteristics and design the networks. The control traffics should require a high reliable and low throughput. The user data traffics should satisfy the demands about flexible, adaptive, high throughput, low latency, and energy efficient transmission. In order to meet the different requirements, the network configuration should be adjusted. As we briefly reviewed the new challenges of 6G cell planning, many new 6G technologies will be incorporated to meet the high level of 6G requirement and satisfy new demands. Cell planning of 6G networks will be a major challenge.

Summary 10.4. Cell Planning

1. The main purpose of cell planning is to minimize the infrastructure costs and operational costs while maintaining good performance of the cellular system. The ultimate goal is to establish an optimal cost effective network.
2. The key design parameters of cell planning can be summarized as follows: (1) Minimizing the overall network deployment and operation costs, (2) Maximizing the overall system capacity, (3) Maximizing the overall coverage while satisfying the limit of the interferences, (4) Minimizing the energy consumption, and (5) Minimizing the control signalling.
3. The process of cell planning is composed of three steps: initial planning (or dimensioning), comprehensive planning, and network optimization.
4. Many new 6G technologies will be incorporated to meet the high level of 6G requirements and satisfy new demands. Cell planning of 6G networks will be a major challenge.

10.3   Network Traffic Prediction

10.3.1   Classic Network Traffic Prediction

According to Ericsson's mobile data traffic outlook [31], "5G networks will carry more than half of the world's smartphone traffic in 2026. Globally, the average usage per smartphone now exceeds 10 GB, and is forecast to reach 35 GB by the end of 2026. Video traffic currently accounts for 66 percent of all mobile data traffic and is expected to increase to 77 percent in 2026. Populous markets that launch 5G early are likely to lead traffic growth, with large variations between regions." Mobile network traffics continues to grow significantly in the next 5 years. The mobile operators need to collect, sort, and analysis the network data. They should identify the performance of overall networks and individual users and then manage the network operation efficiently. However, continuous network traffic measurements and monitoring are expensive. This process relies on dedicated network components such as packet gateway, serving gateway, radio network controller, and so on. The collected data should be stored, and the huge amount of data should be processed. This is not cost-effective solution. Thus, prediction or forecasting techniques will be useful for network traffic managements. There are three types of forecasting technique: judgmental model, associative model, and time series model. The judgemental model uses subjective inputs and relies on subjective decision of an individual. It is a qualitative method Survey results, expert opinion, panel consensus, and Delphi technique are examples of the judgemental model. The associative model uses explanatory variables to predict and suggest a causal relationship. It analyses the past data to find the relationships between observations and predictions. It is a quantitative method and has numerical and statistical approach. The associative model focuses on finding an association between predictor (or independent) variables and predicted (dependent) variables. Regression techniques are widely used for this model. The time series model uses sequential data of observations. It assumes that the prediction will be similar to the past. There are many time series analysis techniques such as simple average, moving average, exponential smoothing, and so on. The network traffics can be analysed by the time series model. The accurate and real-time network traffic prediction enables them to prevent the network faults, control congestions, and take actions immediately. In addition, it allows us to manage network resources efficiently, improve performances, decrease power consumption, reduce operational costs, and analyses a user mobility. The mobile network traffic prediction is one of key techniques to realize 6G systems and operate 6G mobile network efficiently. In 6G cellular networks, they should establish massive connections and deal with a big data. More robust and faster network connection are essential to provide mobile users with various services. The network traffic is a time series bit transmission that can be modelled by a stochastic or deterministic process. The prediction methods should consider the computational complexity, accuracy, and latency. The classical network traffic predictions can be summarized as follows: Wavelet transform-based prediction and time series-based prediction. Fourier transform decomposes a signal into sinusoidal basis functions of different frequencies. There is no information loss, and we can completely recover the original signal by inverse Fourier transform. However, wavelet transform can decompose a signal into a number of set of mutually orthogonal wavelet basis functions and extract both local spectral and temporal information, where a wavelet is a localized wave-like oscillation in time. The key properties of wavelets are scale and location. The scale (or dilation) represents how a wavelet is stretched or squished. It is related to frequency of wavelets. The location means the position of the wavelet in time or space. The first derivative of Gaussian wavelet can be expressed as follows:

$$y(t) = -(t - b)e^{-\frac{(t-b)^2/2a^2}{\sqrt{2\pi}a^3}} \tag{10.22}$$

where a and b denote the scale and the location of wavelet, respectively. Depending on the value of a, the wavelet can be stretched (large scale and low frequency) or squished (small scale and high frequency). It can capture high- or low-frequency information when the values of a are low or high, respectively. Depending on the value of b, the location of the wavelet is shifted to the left or right. Since the wavelet has nonzero value in a short time, the location is important to analyse a signal. Figure 10.14 illustrates example of wavelet changes in terms of scale and location.

The basic concept of wavelet transform is to calculate how much wavelets exists in a signal for a specific scale and location. We select a wavelet with a specific scale

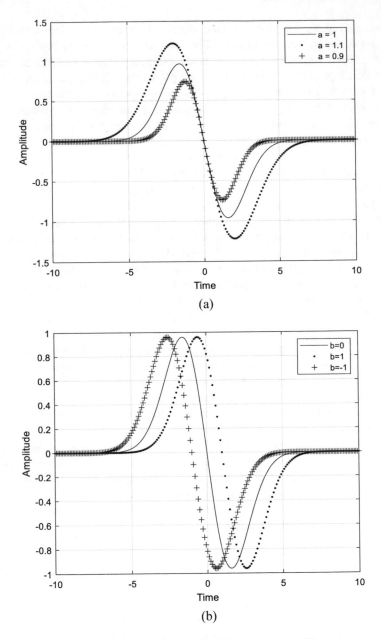

Fig. 10.14 Examples of wavelets: scale change (**a**) and location change (**b**)

and then vary its location across the entire signal. At each time step, we compute the multiplication of the wavelet and the signal and the product of the multiplication provides us with a coefficient for the wavelet scale at the time. We repeat this process as increasing the wavelet scale. In this process, we can use continuous wavelet transform (CWT) or discrete wavelet transform (DWT). The CWT exploits all possible wavelet over an infinite number of scales and locations while the DWT uses a finite set of wavelet over a finite number of scales and locations. Practically, the DWT is more widely used for many applications such as numerical analysis, signal analysis, control analysis, adjustment of audio signals, and so on. The basis functions of wavelet transform are different from sinusoidal basis functions of Fourier transform. Wavelet transform is not a single transform but a set of transform. The most common sets of the wavelet basis functions are Haar and Daubechies set of wavelets. Their important properties can be summarized as (1) wavelet functions are spatially localized, (2) wavelet functions are dilated, translated, and scaled versions of a common mother wavelet, and (3) each set of wavelet functions forms an orthogonal set of basis functions [32]. The DWT signal $x[k]$ can be computed by passing through a series of filters. The signal $x[k]$ is passed through a low-pass filter with an impulse response $g[n]$ and a high-pass filter with an impulse response $h[n]$ as follows:

$$y_l[n] = \sum_{k=-\infty}^{\infty} x[k]g[2n-k], \tag{10.23}$$

$$y_h[n] = \sum_{k=-\infty}^{\infty} x[k]h[2n-k]. \tag{10.24}$$

Their outputs represent the approximation coefficients from the low-pass filter and the detail coefficient from the high-pass filter. Two filters are known as a quadrature mirror filter. The half frequency of the signal was removed according to Nyquist's rule. The filter output is subsampled by 2. Figure 10.15 illustrates the wavelet decomposition.

The decomposition makes the time resolution be half because only half of the filter output represents the signal. However, the outputs have half the frequency of the input and the frequency resolution becomes doubled. The wavelet transform-based

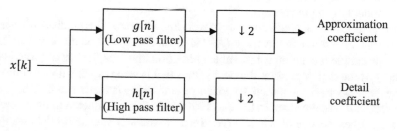

Fig. 10.15 Wavelet decomposition

prediction is suitable for multi-scale prediction as it naturally transforms a signal into multiple resolution [33]. The basic concept of wavelet transform-based prediction is to decompose a signal into a wavelet subspace and then predicts the coefficients in the wavelet space. The predicted values of the signal traffic can be constructed by the predicted coefficients. The wavelet transform-based prediction is based on wavelet decomposition, signal extension, and signal reconstruction [32]. The wavelet decomposition uses a wavelet prototype function and has two steps: a low-pass output (approximation) and a high-pass output (detail). It is performed recursively. Basically, the father wavelet and the mother wavelet approximate the smooth (low frequency) components and the detail (high frequency) components of the signal, respectively. There are many prediction methods using wavelet transforms [34, 35]. One of popular approaches is to decompose a network traffic into wavelet coefficients and scaling coefficients in different time scales and then predict each coefficient independently using normalized least mean square algorithm [34, 35]. The prediction of the network traffic can be reconstructed by the predicted coefficients. In order to analyse time series data, we should understand the characteristic of the time series data. Typically, an observation in a time series can be decomposed into systematic components and non-systematic components. The systematic components have consistency and can be modelled, but the non-systematic components can't be modelled directly. The main systematic components are level, trend and seasonality. The level represents the average value of the time series data. The trend as a long-term direction represents the general tendency of the time series data to increase or decrease in a long period. The overall trend should be upward, downward, or stable. The seasonality is a short-term cycle of the time series data. The seasonal variations are caused by the natural forces or man-made conventions. Examples of the seasonal variations are the crop production, the umbrella sale, the air conditioner sale, and so on. The non-systematic components can be defined as a random variable. They are regarded as a short-term fluctuation or a noise. They are uncontrollable and unpredictable. Examples of the non-systematic components are flood, wars, earthquakes, and so on. The time series data can be expressed as combination of those components. Two different types of the time series models are widely used: additive model and multiplicative model. When those components are independent of each other, additive model can be used as follows: $y(t) = c_1(t) + c_2(t) + \cdots + c_n(t)$. If they are not independent, multiplicative model can be used as follows: $y(t) = c_1(t) \cdot c_2(t) \cdot \ldots \cdot c_n(t)$. Figure 10.16 illustrates examples of time series data as additive model combination of level, trend, random, and seasonality components.

In time series-based prediction, one of well-known time series prediction is autoregressive integrated moving average (ARIMA) model. It will be discussed in detail later. One simple prediction is last value prediction that uses the last observed value for the next interval. It predicts the next values that is same as what it was computed during the previous computation. Exponential smoothing is used to predict the immediate future. It was proposed in the late 1950s [36, 37] and affected to many prediction techniques. As one of window functions, it smooths time series data using the exponential window function. The past observation is equally weighted in a simple moving average but the exponential smoothing gives it to exponentially decreasing

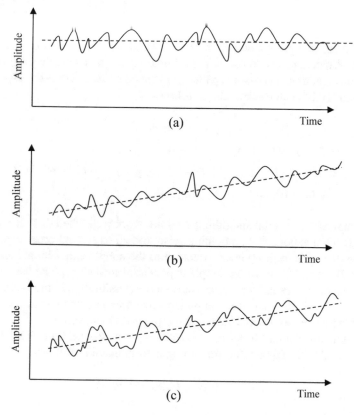

Fig. 10.16 Examples of time series data: Level + random (**a**), Level + trend + random (**b**), Level + trend + seasonality + random (**c**)

weights over time. When we have a sequence of observations, the simple exponential smoothing is formulated as follows:

$$s_t = \alpha x_t + (1 - \alpha)s_{t-1}, \tag{10.25}$$

$$= s_{t-1} + \alpha(x_t - s_{t-1}) \tag{10.26}$$

where α are the smoothing factor and $0 \leq \alpha \leq 1$, x_t as an input is the current observation, and s_t is the output of the exponential smoothing (10.25) represents that we need to balance between the recent observation and the predictive value using the smoothing fact. We can interpret (10.26) as Prediction = Past prediction + (Reliability of the past prediction)(Error of the past prediction). The smoothing factor adjusts the rate about the influence of the observations at the previous time step. If it is close to 1, the most recent past observation is mainly focused. The prediction will have greater responsiveness to the observations and less smoothing.

If it is close to 0, the recent past observation is not so much taken into account when predicting. The prediction will be slower to adjust to prediction errors. As we can observe (10.25), the prediction s_t is based on weighting the current observation x_t and the most recent prediction s_{t-1}. This simple equation is suitable for predicting data with no clear trend or seasonal pattern. As the name said, it uses the exponential window and (10.25) can be rewritten as follows:

$$\begin{aligned}
s_t &= \alpha x_t + (1-\alpha)(\alpha x_{t-1} + (1-\alpha)s_{t-2}) \\
&= \alpha x_t + \alpha(1-\alpha)x_{t-1} + (1-\alpha)^2 s_{t-2} \\
&= \alpha\left(x_t + (1-\alpha)x_{t-1} + (1-\alpha)^2 x_{t-2} + \cdots + (1-\alpha)^{t-1}x_1\right) \\
&\quad + (1-\alpha)^t x_0.
\end{aligned} \tag{10.27}$$

The simplest exponential smoothing does not work well when there is a trend in the data. Double exponential smoothing (also known as second-order exponential smoothing or Holt's method) is an extension to the simple exponential smoothing. It introduces a trend factor to the simple exponential smoothing. The basic concept of double exponential smoothing is to consider the possibility of a time series representing a trend. It includes a level component and a trend component at each period. It assigns exponentially lower weights to the previous observations and add a trend component for estimation where a trend is an average value of the time series increases or decreases. Double exponential smoothing is formulated as follows:

$$s_t = \alpha x_t + (1-\alpha)(s_{t-1} + b_{t-1}), \tag{10.28}$$

$$b_t = \beta(s_t - s_{t-1}) + (1-\beta)b_{t-1} \tag{10.29}$$

where α are the data smoothing factor and $0 \le \alpha \le 1$, β are the trend smoothing factor and $0 \le \beta \le 1$, and b_t is the estimator of the trend at time t. The data smoothing factor controls the rate for the level and the trend smoothing factor control the influence of the change in trend. We need both (10.28) and (10.29) to handle the trend. The trend (10.29) is the expected increase or decrease per a time in the current level. The output of double exponential smoothing is expressed as F_{t+m} an estimate of x_{t+m} at time $m > 0$. The m-step-ahead prediction at time t by double exponential smoothing is

$$F_{t+m} = s_t + mb_t \tag{10.30}$$

and the one-step-ahead prediction at time t is

$$F_{t+1} = s_t + b_t. \tag{10.31}$$

There are several methods to set the initial value b_1. In general, b_0 is set to x_0. We could have the following initial value:

$$b_0 = x_1 - x_0, \tag{10.32}$$

or

$$b_0 = \frac{x_n - x_0}{n}, \text{ for some } n. \tag{10.33}$$

An autoregressive integrated moving average (ARIMA) model is generalization of an autoregressive moving average (ARMA) model. The main purpose of the ARIMA model is to predict future values of the time series data. It is one of popular statistical tools to analyses or predict time series data. The term "ARIMA" is an acronym indicating different meaning: The "AR" part is an autoregressive term representing that the evolving variable is regressed on its lagged values. A model exploits the dependent relationship between observations and some of lagged observations. The "I" part is an integrated term representing a series that needs to be differenced to be made stationary. The differencing process of the raw observation is used to make the time series stationary. The "MA" part is a moving average term representing that the regression error is a linear combination of error terms. A model exploits the dependency between observations and residual errors from a moving average model. These are specified in the model as specific parameters. A standard form of ARIMA is denoted by ARIMA(p,d,q) where p, d, and q are non-negative integers. p (order of the autoregressive part) is the number of lag observations included in the model, d (degree of non-seasonal differences) is the number of times that the raw observations are differenced, and q (order of the moving average part) is the size of the moving average window. There are many ways to build time series prediction models. Stochastic process is one of popular models. In this chapter, we assume time series data can be expressed as a collection of random variables that is referred to as a stochastic process. The observations are referred to as a realization of the stochastic process. Basically, it is not easy to predict time series data because it is naturally non-deterministic. We can't guarantee with certainty what it will happen in the future. However, if the time series data are stationary, it would be a little bit easier to predict them. We can say that the statistical properties of the future time series data are same as the one of the past time series data. A stationary time series has constant statistical parameters (such as mean, variance, autocorrelation, and so on) over a time. Thus, most of prediction methods assume that the time series data are stationary approximately. There are two types of stationary: strictly stationary (or strongly stationary, strict-sense stationary) and weakly stationary (or weak-sense stationarity, wide-sense stationarity). Strictly stationary is defined as follows: A time series $\{X_t\}$ is said to be strictly stationary if the random vectors $(X_{t_1}, X_{t_2}, \ldots, X_{t_n})$ and $(X_{t_1+h}, X_{t_2+h}, \ldots, X_{t_n+h})$ have the same joint distribution for all set of indices $\{t_1, t_2, \ldots, t_n\}$ and for all integer h and $n > 0$. It means that the joint distribution only depends on not the time indices $\{t_1, t_2, \ldots, t_n\}$ but the difference h. In practice, this condition is too strict to exploit. On the other hands, weakly stationary means that the process has the same mean at all time and the covariance between two different point depend on the difference and not on the location of the points. Weakly stationary is defined as follows: A time series $\{X_t\}$ is said to be weakly stationary if $E[X_t^2] < \infty$, $E[X_t] = \mu$, and $\gamma_X(s, t) = \gamma_X(s + h, t + h)$, for all s, t, h are satisfied. Namely the time series should have finite variation, constant mean, and the autocovariance $\gamma_X(s, t)$ only relies on the difference $|s - t|$ and not on the location of s and t. Basically, stationary means weakly stationary. If a process is stationary in the strict sense, we use the term strictly stationary. We can rewrite autocovariance of stationary time series

$$\gamma_X(t + h, t) = E[(X_{t+h} - \mu)(X_t - \mu)] = \text{Cov}(X_{t+h}, X_t)$$
$$= \text{Cov}(X_h, X_0) = \gamma(h, 0) = \gamma(h) \tag{10.34}$$

where h is a time shift and autocorrelation of stationary time series

$$\rho(h) = \frac{\gamma(t + h, t)}{\sqrt{\gamma(t + h, t + h)\gamma(t, t)}} = \frac{\gamma(h)}{\gamma(0)}. \tag{10.35}$$

In general, a time series prediction model can be expressed as follows:

$$y_{t+h} = f(X_t, \theta) + \epsilon_{t+h} \tag{10.36}$$

where y_t is a dependent variable to be predicted, X_t is an independent variable at time t, t is a time, h is the prediction (or forecast) horizon, θ is a parameter of function f, and ϵ_t is an error. In addition, the backshift operator B (or sometimes L for Lag) is defined as follows:

$$BX_t = X_{t-1}. \tag{10.37}$$

The forwardshift operator is expressed as B^{-1} that satisfies $B^{-1}B = 1$. In addition, it can be extended as follows:

$$B^2 X_t = B(BX_t) = B(X_{t-1}) = X_{t-2} \tag{10.38}$$

and we have the generalized form as follows:

$$B^k X_t = X_{t-k}. \tag{10.39}$$

An autoregressive model of order p AR(p) models the current value as a linear combination of the previous p values and a random error as follows:

$$Y_i \quad \phi_i X_{i-i} \mid \phi_j N_j \mid 2 + \cdots \mid \phi_p X_{t-p} + w_t = \sum_{i=1}^{r} \phi_i X_{t-i} + w_t \qquad (10.40)$$

where X_t is a time series variable and stationary, ϕ_i is a model parameter, $w_t \sim wn(0, \sigma_w^2)$ is white noise as normal distribution with zero mean and σ_w^2 variance, and the parameter p represents the length of the direct look back in the series. Using the backshift operator, (10.40) can be concisely rewritten as follows:

$$\phi(B)X_t = w_t \qquad (10.41)$$

where

$$\phi(B) = 1 - \phi_1 B - \phi_2 B^2 - \cdots - \phi_p B^p = 1 - \sum_{j=1}^{p} \phi_j B^j. \qquad (10.42)$$

The simplest autoregression process is AR(0). This is equivalent to a white noise. AR(1) is expressed as follows:

$$X_t = \phi_1 X_{t-1} + w_t. \qquad (10.43)$$

As we can observe (10.43), the output is produced by the previous term in the process and the noise term. If ϕ_1 is close to zero, the output is a noise. If ϕ_1 is close to one, the output is a random walk. A moving average model of order q MA(q) models the current value as a linear combination of the previous q error terms as follows:

$$X_t = w_t + \theta_1 w_{t-1} + \theta_2 w_{t-2} + \cdots + \theta_q w_{t-q} = w_t + \sum_{j=1}^{q} \theta_j w_{t-j} \qquad (10.44)$$

where $w_t \sim wn(0, \sigma_w^2)$ is white noise and θ_i is a model parameter. As we can observe (10.44), unlike the autoregression model, an error term is concerned and it is always stationary because the observation is a weighted moving average over the past prediction errors. The name of this moving average model might be misleading. It should not be confused with the moving average smoothing. Using the backshift operator, (10.44) can be concisely rewritten as follows:

$$X_t = \theta(B)w_t \qquad (10.45)$$

where

$$\theta(B) = 1 + \theta_1 B + \theta_2 B^2 + \cdots + \theta_q B^q = 1 + \sum_{j=1}^{q} \theta_j B^j. \qquad (10.46)$$

Typically, the parameter estimation for the MA model is more difficult than the one for the AR model because the lagged error terms are not observable and the MA process is nonlinear. Autoregressive moving average (ARMA) model is the combination of both the AR model and the MR model. ARMA model considers both the past values and the past error and explains a stationary stochastic process using two polynomials. An ARMA(p,q) is given by

$$X_t = \phi_1 X_{t-1} + \cdots + \phi_p X_{t-p} + w_t + \theta_1 w_{t-1} + \cdots + \theta_q w_{t-q}$$

$$= \sum_{i=1}^{p} \phi_i X_{t-i} + w_t + \sum_{j=1}^{q} \theta_j w_{t-j} \tag{10.47}$$

where $\phi_p \neq 0$, $\theta_q \neq 0$, and $w_t \sim wn(0, \sigma_w^2)$ is white noise. Using the backshift operator, (10.47) can be concisely rewritten as follows:

$$\phi(B)X_t = \theta(B)w_t. \tag{10.48}$$

As we can observe (10.47) and (10.48), it includes both lagged values of X_t and lagged errors. The redundancy of parameters can occur in the model. If $\phi(B) = 0$ and $\theta(B) = 0$ have common factors, it will include redundant parameters and make the model complicated. Thus, we should remove the redundancy and simplify the model. The redundancy can be eliminated by covariance analysis.

Example 10.5 Redundancy of Parameter in ARMA Model Consider the following process

$$X_t - 0.4X_{t-1} - 0.45X_{t-2} = w_t + w_{t-1} + 0.25w_{t-2}.$$

Find the redundancy of the parameter in the process.

Solution
Using the backshift operator, the process can be concisely rewritten as follows:

$$\left(1 - 0.4B - 0.45B^2\right)X_t = \left(1 + B + 0.25B^2\right)w_t.$$

We now check if $\phi(B)$ and $\theta(B)$ are common factors. The polynomial $\phi(B)$ is

$$\phi(B) = 1 - 0.4B - 0.45B^2 = (1 + 0.5B)(1 - 0.9B)$$

and the polynomial $\theta(B)$ is

$$\theta(B) = 1 + B + 0.25B^2 = (1 + 0.5B)^2.$$

Thus, we have common factor $(1 + 0.5B)$ the process can be simplified as follows:

$$(1 + 0.5B)(1 - 0.9B)X_t = (1 + 0.5B)^2 w_t,$$

$$(1 - 0.9B)X_t = (1 + 0.5B)w_t.$$

Thus, we can find the redundancy ϕ_2 and θ_2. It becomes a ARMA(1,1) process. Now, there is no redundant parameters in the simplified model.

∎

(10.47) and (10.48) are not seasonal ARMA. However, the data we have is seasonal. A seasonal ARMA model ARMA$(p,q)_h$ where a lag h is a length of the seasonal period can be expressed as follows:

$$\Phi(B^h)X_t = \Theta(B^h)w_t \tag{10.49}$$

where

$$\Phi(B^h) = 1 - \Phi_1 B^h - \Phi_2 B^{2h} - \cdots - \Phi_p B^{ph} \tag{10.50}$$

and

$$\Theta(B^h) = 1 + \Theta_1 B^h + \Theta_2 B^{2h} + \cdots + \Theta_q B^{qh}. \tag{10.51}$$

They includes seasonal AR operator and seasonal MA operator with the seasonal period of length h. The seasonal ARMA model relies on seasonal lag h and differences to fit the seasonal pattern. If we combine seasonal ARMA with ARMA, the mixed season ARMA can be expressed as follows:

$$\Phi(B^h)\phi(B)X_t = \Theta(B^h)\theta(B)w_t. \tag{10.52}$$

Autoregressive integrated moving average (ARIMA) model is based on a differenced time series. The key property of stationary is not to depend on the time at which the time series is observed. Thus, if we can remove the non-stationary of a time series, it would be easier to handle the process. Differencing is a transformation method applied to a non-stationary time series. It makes the mean stationery (not variance or covariance). The difference between consecutive observations can be computed as follows:

First order differencing : $\nabla X_t = X_t - X_{t-1} = (1 - B)X_t,$ (10.53)

Second order differencing : $\nabla^2 X_t = \nabla(\nabla X_t) = \nabla(X_t - X_{t-1})$
$= (X_t - X_{t-1}) - (X_{t-1} - X_{t-2}) = X_t - 2X_{t-1} + X_{t-2}$

$$= X_t - 2BX_t + B^2 X_t = (1 - 2B + B^2)X_t = (1 - B)^2 X_t. \qquad (10.54$$

Differencing removes the changes in the level of a time series, eliminating trend and seasonality and consequently stabilizing the mean of the time series [38, 39]. In addition, we can difference the observations multiple times. The differences of order d is a generalized form as follows:

$$\nabla^d = (1 - B)^d \text{ for any positve integer } d. \qquad (10.55)$$

The ARIMA model is usually denoted by ARIMA(p,d,q). A process $\{X_t\}$ is said to be ARIMA(p,d,q) if

$$\nabla^d X_t = (1 - B)^d X_t \qquad (10.56)$$

is ARMA(p,q). In general, it can be defined as follows:

$$\phi(B)(1 - B)^d X_t = \theta(B)w_t. \qquad (10.57)$$

There is a general approach of fitting an ARIMA model [39]. Three-stage iterative approach to estimate an ARIMA model is proposed in [40] as follows: Step 1. Model identification and selection: Checking stationary and seasonality, performing differencing, and choosing model specification ARIMA(p,d,q). Step 2. Parameter estimation: Computing coefficients that best fit the selected ARIMA model using maximum likelihood estimation or nonlinear least-squares estimation. Step 3. Statistical model checking: Testing whether the obtained model conforms to the specifications of a stationary univariate process. If it failed, it goes back to step 1. Table 10.2 summarizes special cases of ARIMA models.

Example 10.6 Data Prediction Using ARIMA Consider two time series data as shown in Fig. 10.17.

Table 10.2 Special cases of ARIMA models

ARIMA (p,d,q)	Popular forecasting models
ARIMA(0,0,0)	White noise
ARIMA(0,1,0) with no constant	Random walk
ARIMA(0,1,0) with a constant	Random walk with drift
ARIMA(0,1,1) with no constant	Basic exponential smoothing model
ARIMA(p,0,0)	Autoregression
ARIMA(0,0,q)	Moving average

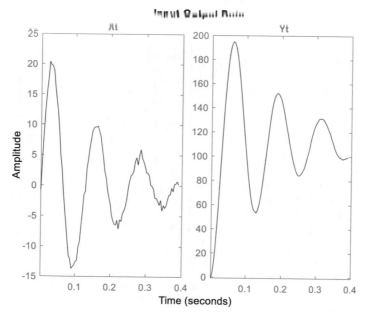

Fig. 10.17 Time series data

One model with a process $\{X_t\}$ includes a noise term but the other model with a process $\{Y_t\}$ has not a noise term. Predict 100- and 200-step-ahead outputs using ARIMA models.

Solution

When we have the following ARIMA models:

$$\left(1 - 1.921B + 0.9244B^2 + 0.06734B^3 - 0.03459B^4\right)X_t = \left(1 - 1.93B + 0.9698B^2\right)w_t,$$

and

$$\left(1 - 2.393B + 1.352B^2 + 0.5339B^3 - 0.4929B^4\right)Y_t = (1 - 0.9737B)w_t,$$

Using a computer simulation, we obtain 85.53% and 96.81% fit to estimate data. Two predictions are illustrated at Figs. 10.18, and 10.19.

∎

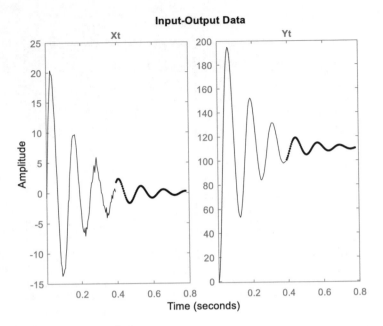

Fig. 10.18   100-step-ahead prediction

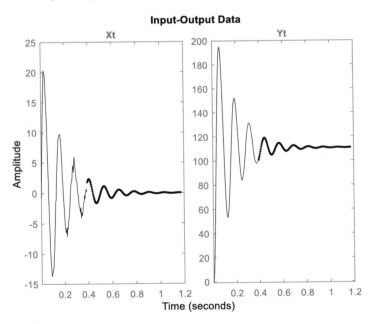

Fig. 10.19   200-step-ahead prediction

10.3.2 AI Enabled Network Traffic Prediction

6G systems will increase the network traffic significantly due to massive connectivity by smartphones, tables, IoT devices, and so on. The network function virtualization (NFV) enables 6G systems to establish flexible and scalable network connections depending on the traffic requests. The management of VNF is one of key challenges to operate 6G core networks efficiently. Adaptive provisioning of VNF could be one of key techniques when traffic volume is increasing and decreasing. In 5G systems, access and mobility management function (AMF) is responsible for handling connection and mobility management tasks. It should scale up or down if the number of resource requests is high or low, respectively. In order to scale up or down the resources of VNF, a fast and accurate decision making is essential. If the decision making is slow and not accurate, users will experience the connection delay or rejection. In order to make a fast and accurate decision, network traffic prediction will be a key technique. Since this technique can forecast the resource requests of users, we can scale up or down properly and avoid the delay or rejection of the requests. Network traffic prediction as a proactive approach plays an important role in the 6G cellular networks. AI techniques are in line with this research topic. In particular, recurrent neural network (RNN) is a power tool for forecasting time series data from the past data. Many research groups adopted the RNN to network traffic prediction. We can apply the RNN to time series data prediction straightforwardly. The main steps for the network traffic prediction can be summarized as follows: collection of training and test data set, data formatting, training the neural network model, and testing the neural network. In the step of data collection, mobile operators should collect the dataset in their own infrastructure. The types of dataset can be flow or packet data with statistical information, log data, and so on. Depending on the purpose of the prediction, specific dataset will be collected. In the step of data formatting, the data set should be formatted for training. This step affects to the accuracy of prediction. Depending of key performance indication, we can format the data set by classifying or labelling. In the steps of training and testing, we should select a proper RNN algorithm and decide configuration of the neural network. In case of LSTM, it allows the data set to be stored while predicting. In 6G systems, RNN will be useful for predicting the required resources of AMFs and managing the traffics efficiently. The NFV orchestrator can include the RNN and find the optimal virtual resource allocation to deploy proactively by forecasting the traffic changes. If we can have a well-trained RNN model, it doesn't need to monitor the performance of VNF frequently and reduce the overhead in control plane.

Example 10.7 Data Prediction Using RNN Consider one small company with the network traffic data for 450 days as shown in Fig. 10.20.
Predict the future 50 days network traffics usage using RNN.

Solution
Using computer simulation, we predict the future 50 days network traffics usage. The simulation configuration can be summarized as follows: RNN algorithm =

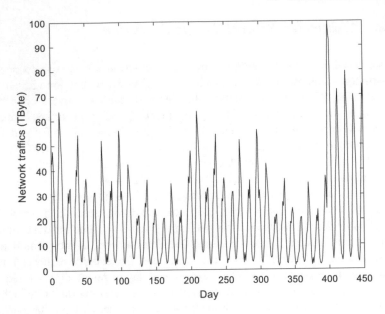

Fig. 10.20 Network traffics usages for 450 days

LSTM, Up to 100 epochs, Up to 100 iterations, Loss function = RMSE, LSTM hidden units = 200, Gradient threshold = 1, Initial learning rate = 0.05, Learning rate drop period = 125, and Learning rate drop factor = 0.2. After performing LSTM, we can obtain future 50 days network traffics usage. Figure 10.21 illustrates comparison of observation and prediction. In the upper of Fig. 10.21, the curve "Observed" means observation (actual data). The curve "Prediction" means the output of LSTM algorithms. The lower of Fig. 10.21 represents the loss between observed data and predicted data.

After updating LSTM network with 100 iterations, we compare of observation and prediction as shown in Fig. 10.22. As we can observe RMSE values (10.3808 and 4.9954) of Figs. 10.22 and 10.23, the LSTM network model update provides us with a better performance.

After having enough training, we can have a better result. Figure 10.23 illustrates the observation for 450 days and the prediction for the future 50 days.

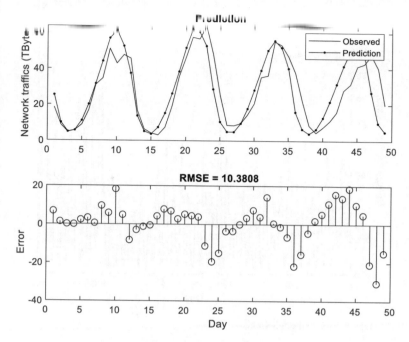

Fig. 10.21 Comparison of observation and prediction for future 50 days

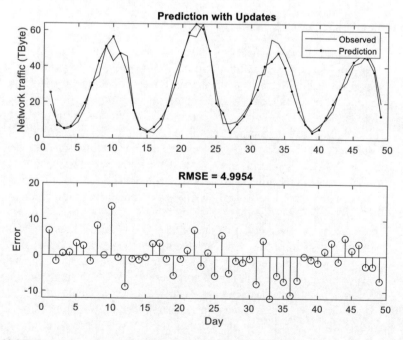

Fig. 10.22 Comparison of observation and prediction for future 50 days after updating

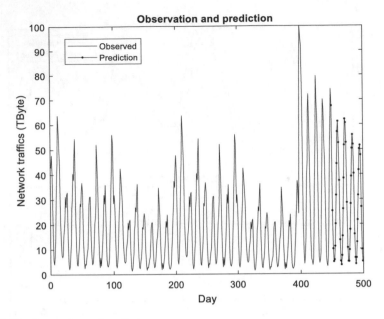

Fig. 10.23 Observation and prediction

Summary 10.5. Network Traffic Prediction

1. The mobile operators need to collect, sort and analysis the network data. They should identify the performance of overall networks and individual users and then manage the network operation efficiently. However, continuous network traffic measurements and monitoring are expensive.

2. Network traffic prediction or forecasting techniques will be useful for network traffic managements. There are three type of forecasting technique: Judgmental model, associative model and time series model.

3. The network traffic is a time series bit transmission that can be modelled by a stochastic or deterministic process. The classical network traffic predictions can be summarized as follows: Wavelet transform based prediction and time series based prediction.

4. In time series based prediction, one of well known time series prediction is autoregressive integrated moving average (ARIMA) model. A standard form of ARIMA is denoted by ARIMA(p,d,q) where p, d, and q are non-negative integers. p (order of the autoregressive part) is the number of lag observations included in the model, d (degree of non-seasonal differences) is the number of times that the raw observations are differenced, and q (order of the moving average part) is the size of the moving average window.

1.   Autoregressive integrated moving average (ARIMA) model is based on a differenced time series. The key property of stationary is not to depend on the time at which the time series is observed.
6.   RNN is a power tool for forecasting time series data from the past data. We can apply the RNN to time series data prediction straightforwardly. The main steps for the network traffic prediction can be summarized as follows: Collection of training and test data set, data formatting, training the neural network model, and testing the neural network.

10.4 Problems

10.1   Describe the 5G network layer tasks and classify them in terms of routing, monitoring, and resource allocation.

10.2   Compare the layer 3 tasks of 4G and 5G systems and discuss the bottleneck to achieve a higher throughput and lower latency.

10.3   Describe the 5G core network tasks and radio access network tasks.

10.4   Network layer problems are about (1) finding the type of the network traffic data, (2) extracting the features of the network data, and (3) predicting the future volume of the traffics or detecting the network faults. Discuss which AI algorithms are suitable for tackling the problem.

10.5   Discuss which AI techniques are suitable for predicting elephant flows.

10.6   Describe why mobile operators need to classify the network traffic data.

10.7   Finding an optimal routing path and transmitting the data packet via the optimal path is a fundamental task in the network. What is the classical approach to find an optimal routing path? Reinforcement learning will be a powerful tool to find the optimal routing path. In 6G virtualized networks, what is key challenges for the routing problem? Compare the pros-and-cons when 6G networks adopts reinforcement learning for the routing problem.

10.8   Discuss how queuing theory can be applied to virtualized networks and what key challenges we face.

10.9   The basic process of fault management can be summarized as follows: (1) fault detection to determine whether or not network faults occur, (2) isolation of the root causes of the faults, (3) notification to administrators, and (4) collection of the problem. In addition to this, fault prediction can be added. Which AI algorithms will be suitable for fault management? Discuss the research challenges when applying to distributed networks and centralized networks.

10.10   Discuss how to find the suitable AI techniques for network layer problems.

10.11   Discuss the pros-and-cons of circuit switched network and packet switched network.

10.12   Review evolution of cellular networks as shown in Fig. 10.6 and discuss the key features of upcoming 6G and beyond networks.

10.13   Discuss 6G techniques for reducing intercell interferences.

10.14   There are multiple approaches to increase cell capacity as follows: (1) New channel allocation. (2) Dynamic frequency allocation. (3) Cell splitting and smaller cell. Select one approach and discuss the pros-and-cons in terms of CAPEX and OPEX.

10.15   Consider a cellular system that can tolerate a SIR $= 10$ dB in the worst case. Find the optimal value of the cluster size N for omnidirectional antennas and $120°$ sectoring when we have a path loss exponent $n = 5$.

10.16   Consider a FDD cellular system with the total bandwidth 100 MHz. It uses each 250 kHz bandwidth for uplink and downlink. Each mobile user generates 0.1 Erlangs. The cellular system has 7 cells and is based on Erlang B formula.

   (a)   Find the number of channels at each cell.
   (b)   If each cell has capacity with 90% utilization, find the maximum number of users that can be supported per cell when each base station is equipped with omnidirectional antennas.
   (c)   Compute the blocking probability when the maximum number of users are available.

10.17   Consider a cellular system with a hexagonal shape, 1.387 km radius, 7 cell frequency reuse factor, and 120 channels in total. Each user generates a traffic intensity of $A\_u = 0.029$ Erlangs. The average number of call requests per unit time is $\lambda = 1$ call/hour. The Erlang C system has a 2% probability of a delayed call.

   (a)   How many users per Km2 are supported?
   (b)   Compute the probability that a delayed call should wait for more than 10 s.
   (c)   Compute the probability that a call is delayed for more than 10 s.

10.18   The new features such as heterogeneous network, user-centric network deployments, ultra-dense networks, mmWAVE, massive MIMO, multi-connectivity, integration with satellite communications, O-RAN, and others will be incorporated in 6G cellular networks. These features are new challenges to network designers. Discuss how the network virtualization affects to cell planning.

10.19   There are three type of forecasting technique: judgmental model, associative model, and time series model. Among them, time series model is widely used. Discuss why it is better than other two models.

10.20   Discuss the difference between ARIMA model and ARMA model.

11.21 Select one special case of ARIMA models in Table 10.2 and prove that the special case is same as the specific model.

10.22 Discuss which AI techniques are suitable for network layer tasks: network traffic classification by AI algorithms, optimal routing path by AI algorithms, queueing management by AI algorithms, AI contributions to network fault management, and QoS management by AI algorithms.

References

1. H. Kim, *Design and Optimization for 5G Wireless Communications* (Wiley, 2020), ISBN:9781119494553
2. E.S. Yu, C.Y.R. Chen, Traffic prediction using neural networks, in *Proceedings of GLOBECOM'93. IEEE Global Telecommunications Conference*, 1993, vol 2, pp. 991–995. https://doi.org/10.1109/GLOCOM.1993.318226
3. A. Eswaradass, X.H. Sun, M. Wu, Network bandwidth predictor (nbp): A system for online network performance forecasting, in *Proceedings of 6th IEEE International Symposium on Cluster Computing and the Grid (CCGRID)* (2006)
4. P. Cortez, M. Rio, M. Rocha, P. Sousa, Internet traffic forecasting using neural networks, in *Proceedings of IEEE International Joint Conference on Neural Networks (IJCNN)*. pp. 2635–2642, 2006
5. Y. Zhu, G. Zhang, J. Qiu, Network traffic prediction based on particle swarm bp neural network. J. netw. **8**(11) (2013), ISSN 1796-2056
6. Y. Li, H. Liu, W. Yang, D. Hu, W. Xu, Inter-data-center network traffic prediction with elephant flows, in *NOMS 2016—2016 IEEE/IFIP Network Operations and Management Symposium*, 2016, pp. 206–213. https://doi.org/10.1109/NOMS.2016.7502814
7. P. Poupart et al., Online flow size prediction for improved network routing, in *2016 IEEE 24th International Conference on Network Protocols (ICNP)*, pp. 1–6, 2016. https://doi.org/10.1109/ICNP.2016.7785324
8. T. Bakhshi, B. Ghita, On Internet traffic classification: a two-phased machine learning approach. J. Comput. Netw. Commun. **2016**, 21p, Article ID 2048302 (2016). https://doi.org/10.1155/2016/2048302
9. J. Ma, K. Levchenko, C. Kreibich, S. Savage, G.M. Voelker, Unexpected means of protocol inference, in *Proceedings of the 6th ACM SIGCOMM Conference on Internet Measurement*, pp. 313–326. (2006). https://doi.org/10.1145/1177080.1177123
10. A.W. Moore, K. Papagiannaki, Toward the accurate identification of network applications, in *Proceedings PAM'05*, pp. 41–54, 2005
11. P. Bermolen, M. Mellia, M. Meo, D. Rossi, S. Valenti, Abacus: Accurate behavioral classification of P2P-tv traffic. Comput. Netw. **55**(6), 1394–1411 (2011)
12. T. Karagiannis, K. Papagiannaki, M. Faloutsos, BLINC: multilevel traffic classification in the dark. ACM SIGCOMM Comput. Commun. Rev. **35**(4), 229–240 (2005). https://doi.org/10.1145/1090191.1080119
13. R. Boutaba, M.A. Salahuddin, N. Limam, S. Ayoubi, N. Shahriar, F. Estrada-Solano, O.M. Caicedo, A comprehensive survey on machine learning for networking: evolution, applications and research opportunities. J Internet Serv. Appl. **9**, 16 (2018). https://doi.org/10.1186/s13174-018-0087-2
14. L. He, C. Xu, Y. Luo, vTC: machine learning based traffic classification as a virtual network function, in *Proceedings of the 2016 ACM International Workshop on Security in Software Defined Networks & Network Function Virtualization*. ACM, pp. 53–56, 2016

15. R.M. AlZoman, M.J.F. Alenazi, A comparative study of traffic classification techniques fo smart city networks. Sensors **21**, 4677 (2021). https://doi.org/10.3390/s21144677
16. J.A. Boyan, M.L. Littman, Packet routing in dynamically changing networks: a reinforcemen learning approach. Adv. Neural inf. Process. Syst. pp. 671–678 (1994)
17. S. Kumar, R. Miikkulainen, Dual reinforcement Q-routing: an on-line adaptive routing algo-rithm, *in Proceedings Artificial Neural Networks in Engineering Conference*, pp. 231–238 1997
18. Y. Gao, G. He, J.C. Hou, On exploiting traffic predictability in active queue manage-ment, in*Proceedings of Twenty-First Annual Joint Conference of the IEEE Computer ana Communications Societies*, vol. 3. pp. 1630–1639, 2002
19. J. Sun, S. Chan, K. Ko, G. Chen, M. Zukerman, Neuron pid: a robust aqm scheme, in *Proceed-ings of the Australian Telecommunication Networks and Applications Conference (ATNAC) 2006*, pp. 259–262, 2006
20. R. A. Maxion, Anomaly detection for diagnosis, in *Digest of Papers. Fault-Tolerant Computing: 20th International Symposium*, pp. 20–27, 1990. https://doi.org/10.1109/FTCS.1990.89362
21. S. Rao, Operational fault detection in cellular wireless base-stations. IEEE Trans. Netw. Serv. Manage. **3**(2), 1–11 (2006). https://doi.org/10.1109/TNSM.2006.4798311
22. A.I. Moustapha, R.R. Selmic, Wireless sensor network modeling using modified recurrent neural networks: application to fault detection, in *2007 IEEE International Conference on Networking, Sensing and Control*, pp. 313–318, 2007. https://doi.org/10.1109/ICNSC.2007. 372797
23. M. Chen, A.X. Zheng, J. Lloyd, M.I. Jordan, E. Brewer, Failure diagnosis using decision trees, in *International Conference on Autonomic Computing, 2004*, pp. 36–43, 2004. https://doi.org/ 10.1109/ICAC.2004.1301345
24. M.S. Mushtaq, B. Augustin, A. Mellouk, Empirical study based on machine learning approach to assess the QoS/QoE correlation, in *2012 17th European Conference on Networks and Optical Communications*, pp. 1–7, 2012. https://doi.org/10.1109/NOC.2012.6249939
25. M.T. Vega, D.C. Mocanu, A. Liotta, Unsupervised deep learning for real-time assessment of video streaming services. Multim. Tools Appl. **76**, 22303–22327 (2017). https://doi.org/10. 1007/s11042-017-4831-6
26. V.H. MacDonald, The cellular concept. Bell Syst. Tech. J. **58**(1), 15–42 (1979)
27. J.F.C. Kingman, The first Erlang century and the next. Queueing Syst. **63**, 3 (2009). https://doi. org/10.1007/s11134-009-9147-4
28. T. S. Rappaport, *Wireless Communications: Principles and Practice*, 2nd Edition, Prentice Hall, 2002, ISBN: 0-13-042232-0.
29. https://www.erlang.com/calculator/
30. H. Kim, *Wireless Communications Systems Design* (Wiley, 2015), ISBN: 978-1-118-61015-2
31. https://www.ericsson.com/en/reports-and-papers/mobility-report/dataforecasts/mobile-tra ffic-forecast
32. A. Graps, An introduction to wavelets. IEEE Comput. Sci. Eng. **2**(2), 50–61 (1995)
33. H. Zhao, N. Ansari, Wavelet transform-based network traffic prediction: a fast on-line approach. J. Comput. Inf. Technol. **20**(1) (2012)
34. X. Wang, X. Shan, A wavelet-based method to predict Internet traffic, in *International Confer-ence on Communications, Circuits and Systems and West Sino Expositions*, pp. 690–694, 2002. https://doi.org/10.1109/ICCCAS.2002.1180710
35. S. ATTALLAH, "Thewavelet transform-domain LMS algorithm: A more practical approach", *IEEE Transaction on Circuits and Systems–Analog and Digital Signal Processing*, vol. 47, No. 3 March, 2000.
36. R.G. Brown, *Statistical Forecasting for Inventory Control* (McGraw/Hill, 1959)
37. C.E. Holt, *Forecasting Seasonals and Trends by Exponentially Weighted Averages* (O.N.R. Memorandum No. 52) (Carnegie Institute of Technology, Pittsburgh USA, 1957)
38. S. Wang, C. Li, A. Lim, Why Are the ARIMA and SARIMA not Sufficient (2021). https:// arxiv.org/abs/1904.07632v3

39. R.J. Hyndman, G. Athanasopoulos, *Forecasting: Principles and Practice*, 2nd edn (OTexts. Melbourne, Australia, 2018), OTexts.com/fpp2
40. G.E.P. Box, G.M. Jenkins, G.C. Reinsel, G.M. Ljung, *Time Series Analysis: Forecasting and Control*, 5th edn. (Wiley, 2015). ISBN: 978-1-118-67502-1

Index

Printed in the United States
by Baker & Taylor Publisher Services